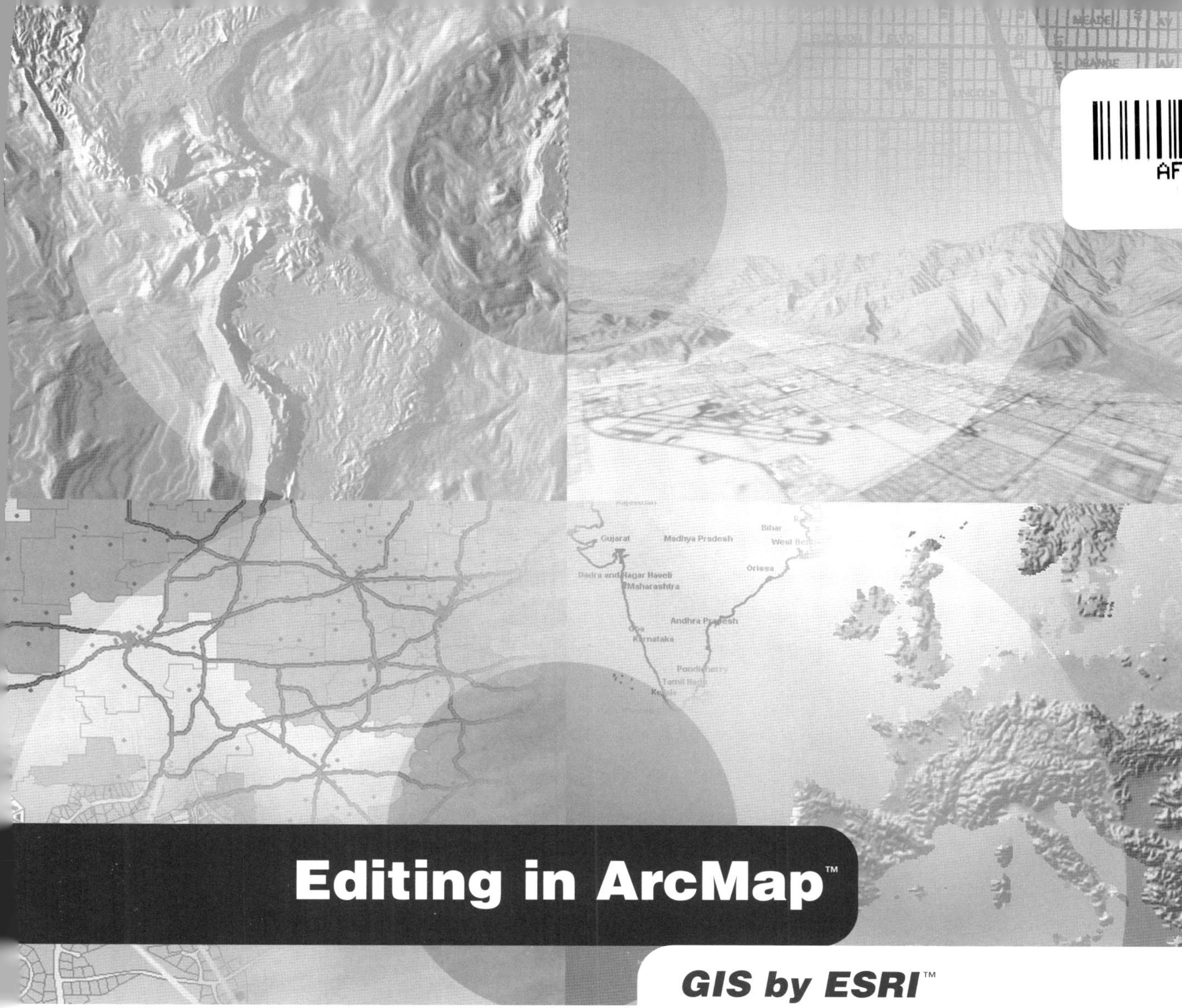

Editing in ArcMap™

GIS by ESRI™

DATA CREDITS

Graphical Editing Map: Wilson, North Carolina
Universal Data Editor Map, Editing in data view and layout view map: Greeley, Colorado
Context menus and shortcut keys map: P.F.R.A., Regina, Saskatchewan, Canada
Quick-Start Tutorial Data: Wilson, North Carolina; Greeley, Colorado

CONTRIBUTING WRITERS

Bob Booth, Jeff Shaner, Scott Crosier, Phil Sanchez, Andy MacDonald

Contents

3 Creating new features 45

Introduction

1

IN THIS CHAPTER

- **Rich suite of graphical editing tools**
- **ArcGIS data editor**
- **Editing in data view and layout view**
- **Tools for editing and managing topologies**
- **Tools for editing and managing networks in a geodatabase**
- **Context menus and shortcut keys for increased productivity**
- **Tools for rubber sheeting, adjusting, and edgematching feature data**
- **Multiuser editing with version management and conflict detection**
- **Remote editing of data checked out from your versioned geodatabase**
- **Editing in projected space**
- **Tips on learning how to edit in ArcMap**

In addition to mapmaking and map-based analysis, ESRI® ArcGIS™ ArcMap™ is the application for creating and editing geographic data and tabular data. With ArcMap, you can edit features in shapefiles and geodatabases with one common user interface. ArcMap contains sophisticated, computer-aided design (CAD)-based editing tools that help you construct features quickly and easily while maintaining the spatial integrity of your geographic information system (GIS) database. ArcEditor™ and ArcInfo™ seats of ArcMap can be used to edit geometric networks and geodatabase topologies as well as simple features. ArcView® seats of ArcMap can be used to edit simple features in shapefiles and geodatabases. ArcView seats also let you create a temporary Map Topology that can be used to edit simple features that share geometry.

Whether you use ArcView or ArcInfo, you use the same editing tools in ArcMap to work on your geographic data. In cases where your organization has multiple users simultaneously editing on a shared geodatabase, ArcMap, in concert with ArcSDE®, provides the tools necessary to manage long editing transactions as well as to manage versions and resolve potential conflicts. ArcEditor and ArcInfo seats of ArcMap can check out features from a master geodatabase to a checkout geodatabase for disconnected editing.

Whether you use ArcView, ArcEditor, or ArcInfo, the goal of this book is to help you learn and use the editing capabilities in ArcMap for any level of geographic database maintenance. The next few pages highlight some of the features you will find invaluable while editing in ArcMap.

Rich suite of graphical editing tools

ArcMap helps you create and edit geographic features quickly and easily by including many of the graphic editing functions popular with the latest CAD editing packages.

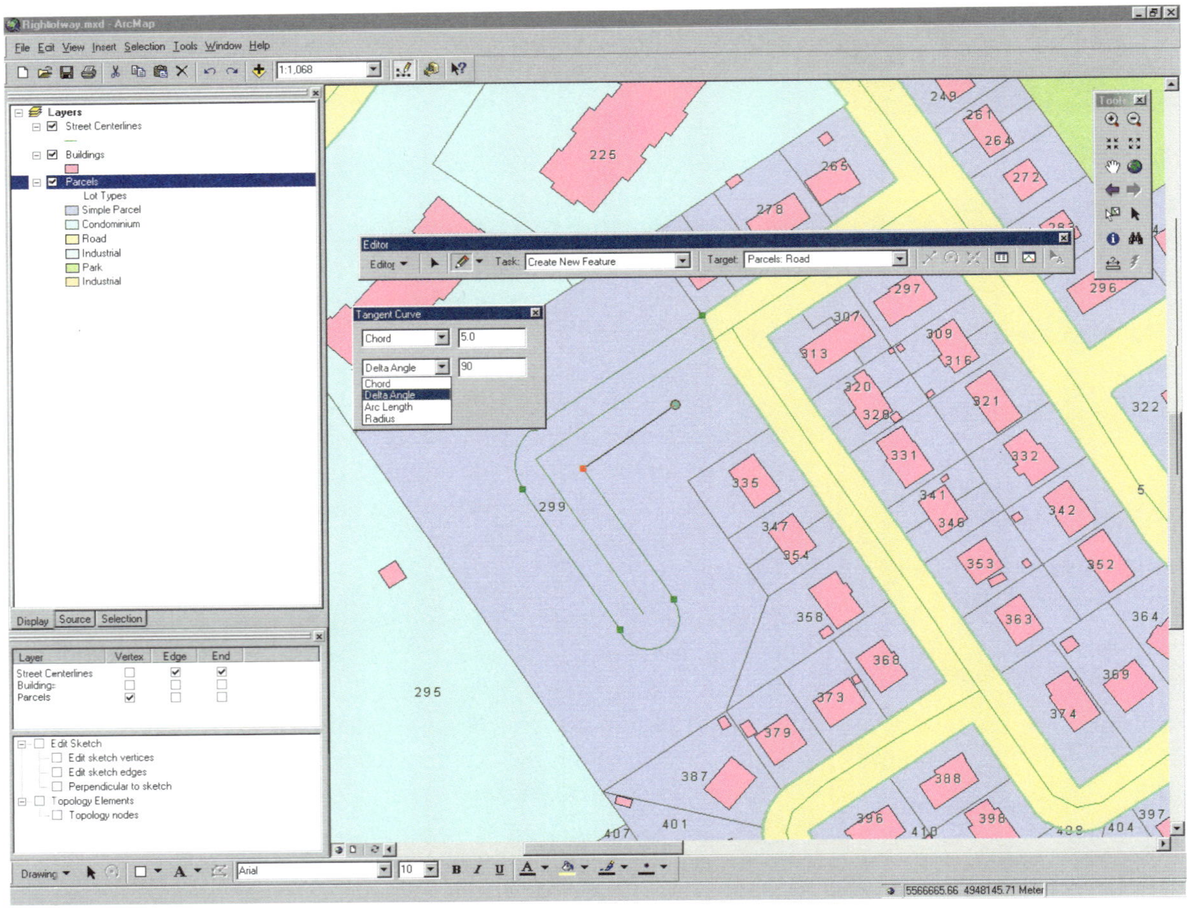

Sketch construction tools in ArcMap will allow you to quickly and accurately edit street rights-of-way.

ArcGIS data editor

ArcMap lets you edit shapefiles and geodatabases. Also, you can edit an entire folder of data at once. ArcEditor and ArcInfo seats can take advantage of a geodatabase's coded value and range domains and validation to make editing attributes quicker and maintain high data quality.

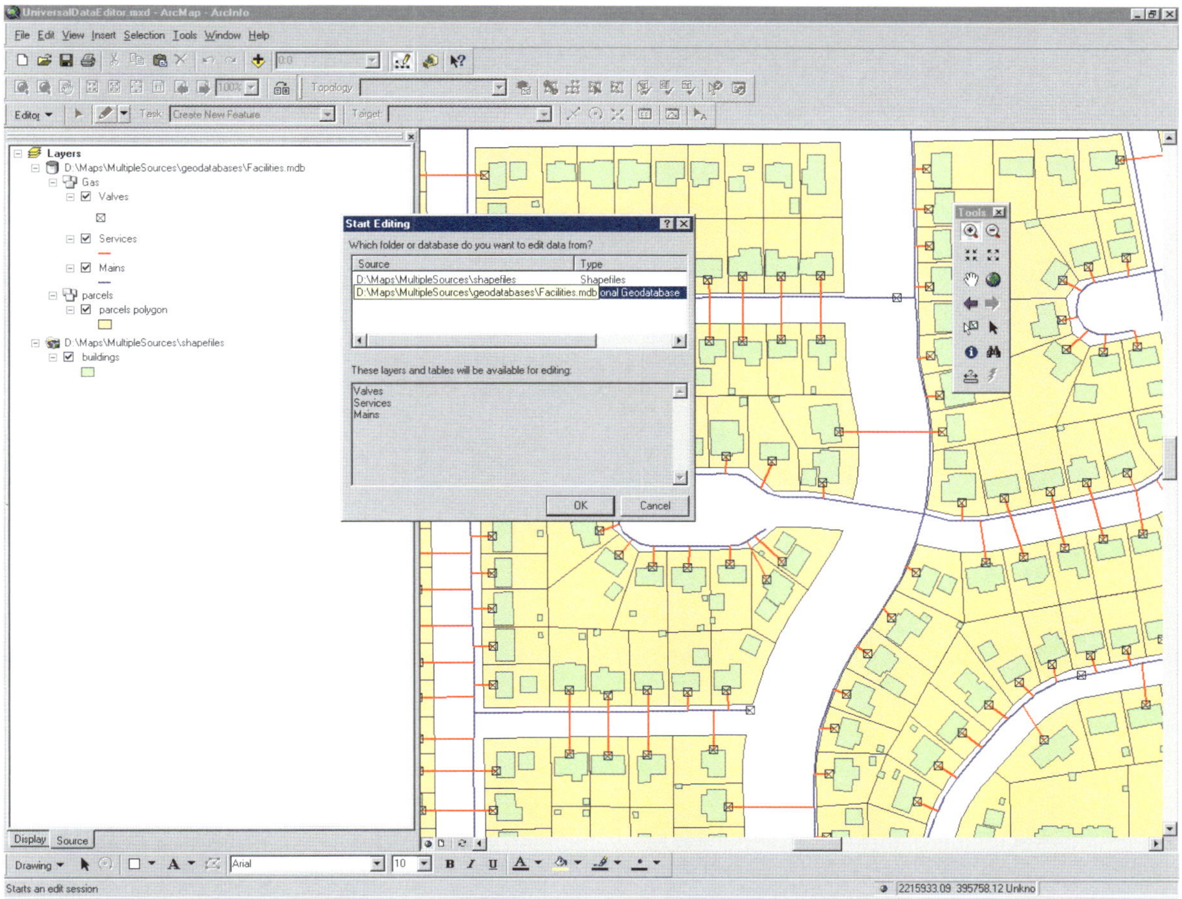

Pick the geodatabase or folder of data that you want to edit when you start editing in ArcMap.

Editing in data view and layout view

ArcMap provides two different ways to view a map: data view and layout view. Each view lets you look at and interact with the map in a different way. Data view hides all of the map elements on the layout such as titles, North arrows, and scale bars. In layout view, you'll see a virtual page upon which you can place and arrange map elements. You can edit your geographic data in either data view or layout view.

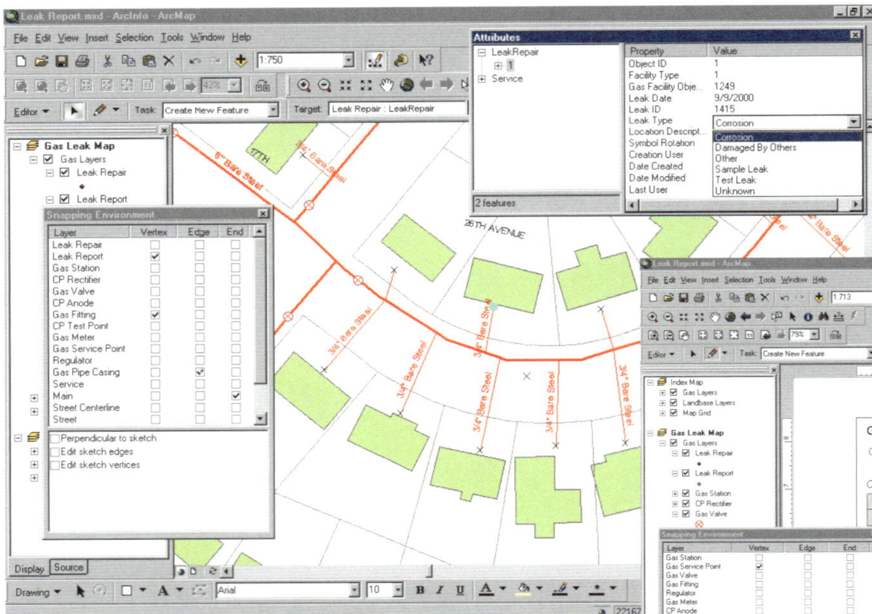

Maximize the display of geographic information when editing using data view.

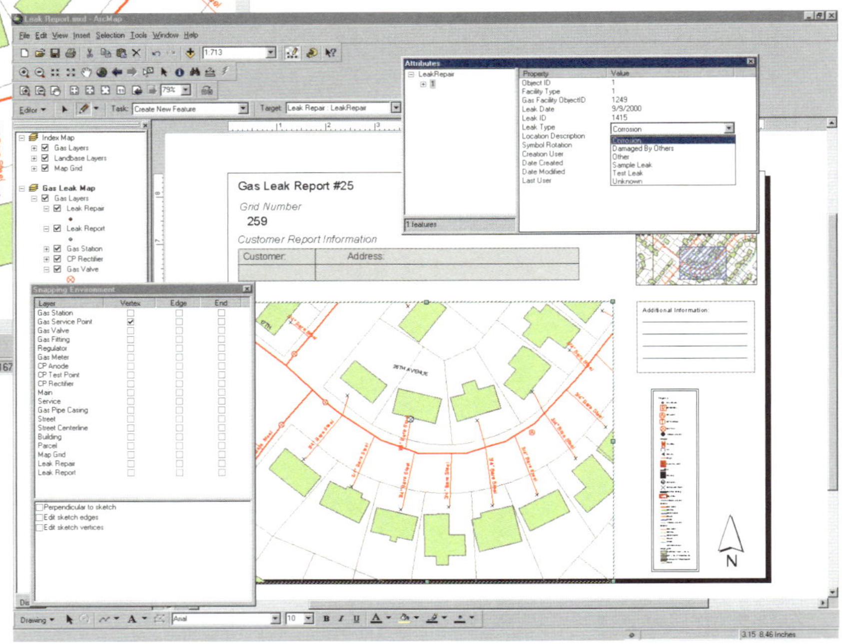

When you are preparing a map, you can edit features directly using layout view.

Tools for editing and managing topologies

ArcMap provides tools to edit features that have topological relationships defined in a geodatabase or in a map topology. ArcView seats are limited to editing map topology, a simpler, temporary form of topology that allows shared parts of features to be simultaneously edited.

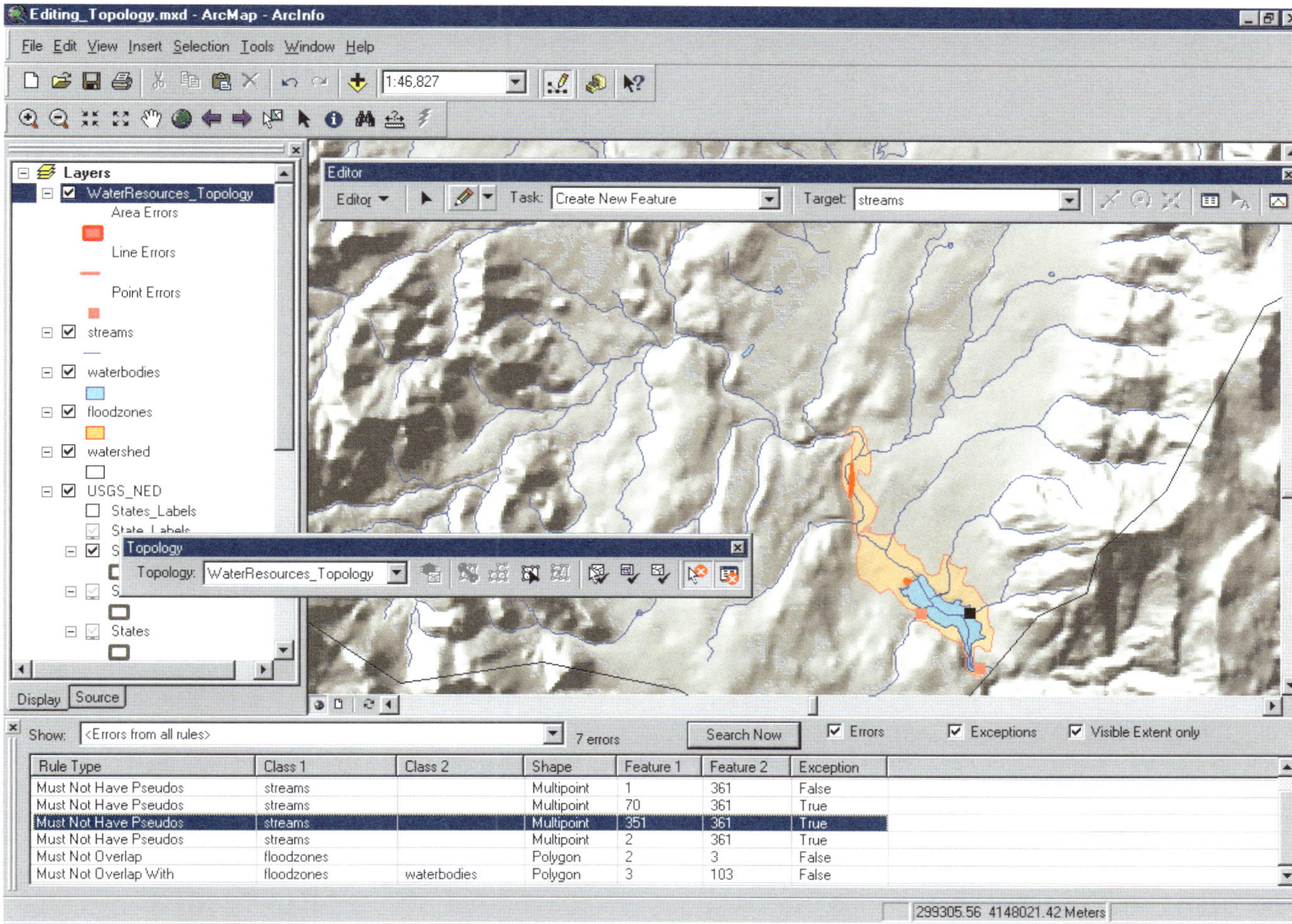

Tools for editing and managing networks in a geodatabase

ArcMap provides tools to edit geometric networks stored in a geodatabase.

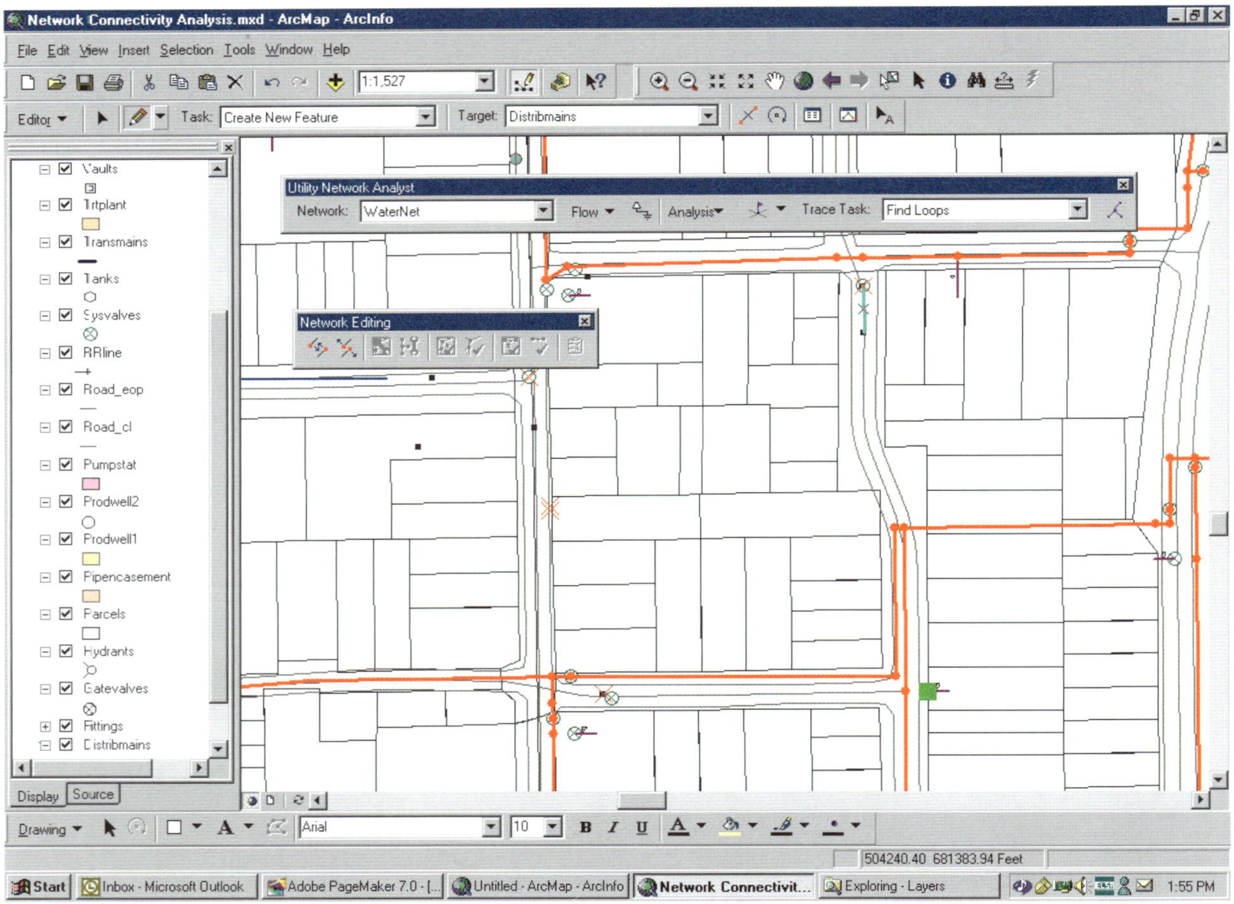

Context menus and shortcut keys for increased productivity

ArcMap contains numerous context menus and shortcut keys to help you create and edit features quickly.

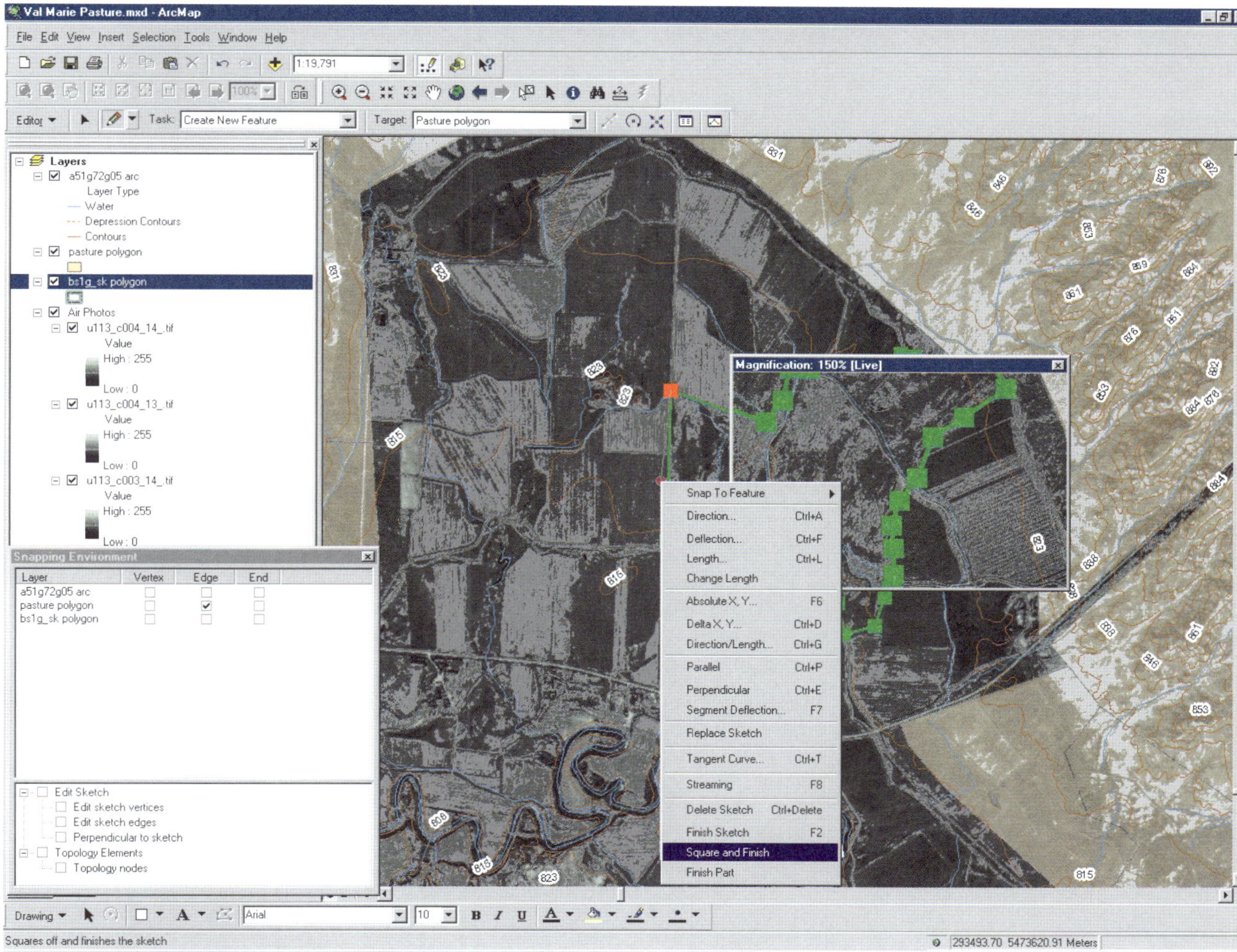

Use the Sketch tool context menu and shortcut keys to access advanced feature creation tools.

Tools for rubber sheeting, adjusting, and edgematching feature data

ArcMap provides tools to transform, adjust, rubber sheet, and edgematch feature data from different sources.

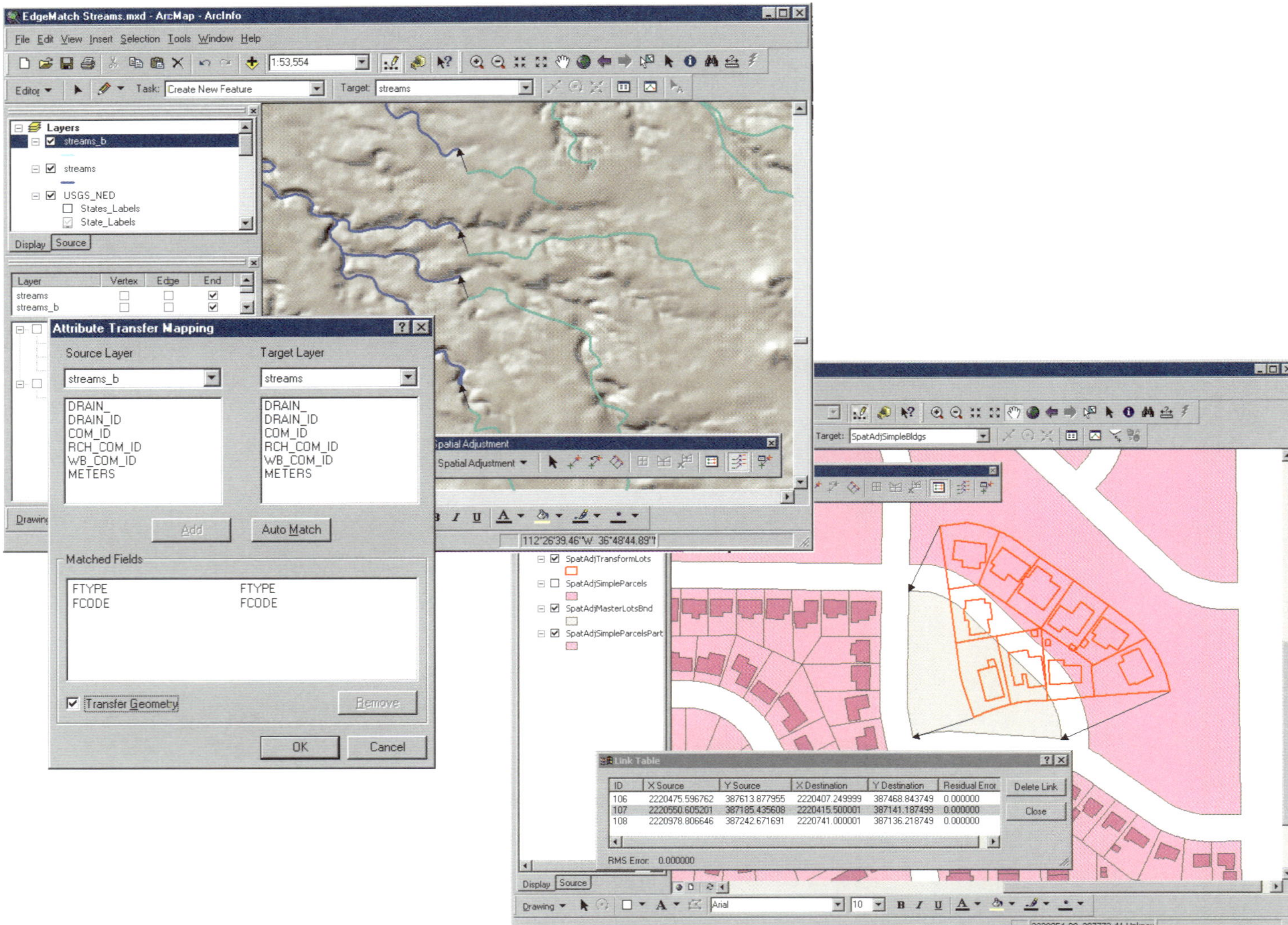

Multiuser editing with version management and conflict detection

If you have several users that need to edit the same data at the same time, ArcMap can help you manage versions of your ArcSDE geodatabase.

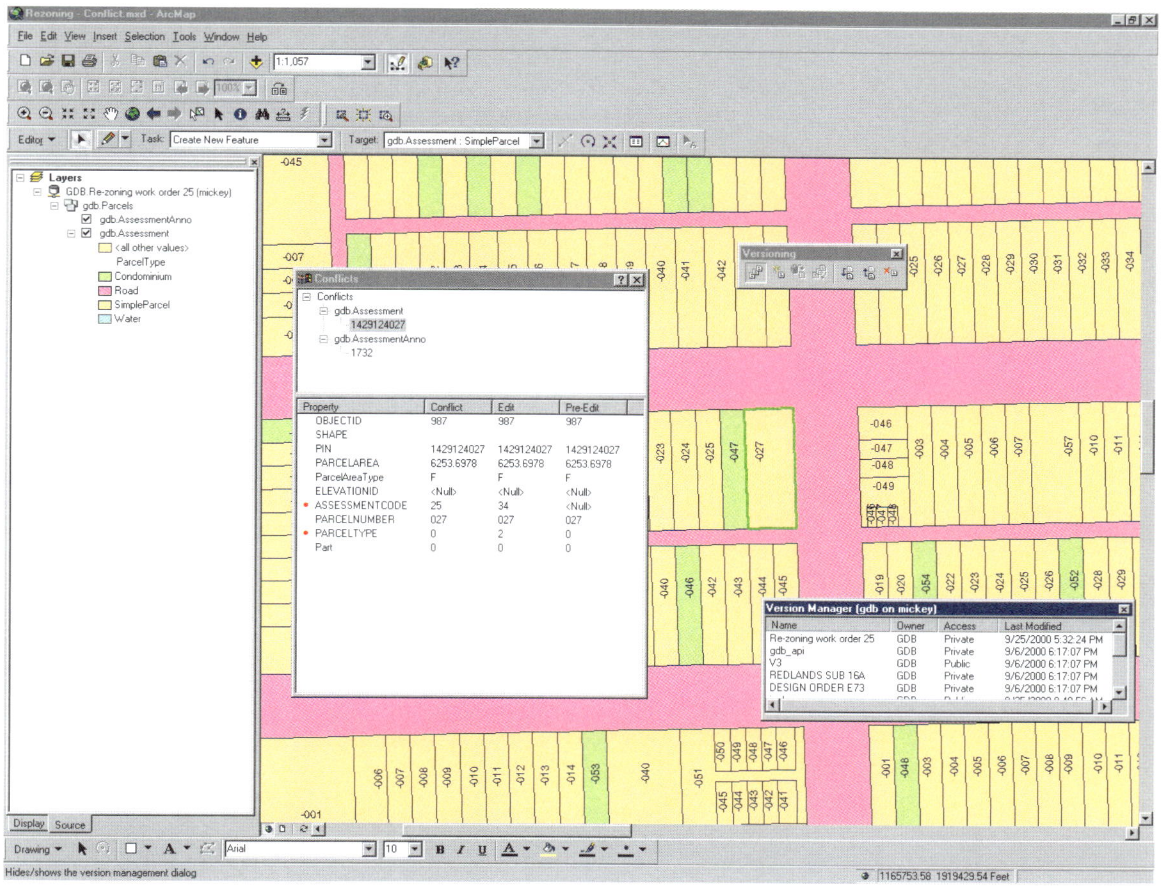

Sophisticated version management tools in ArcMap will help you maintain a multiuser editing environment.

Remote editing of data checked out from your versioned geodatabase

If you have people who need to work with part of your geodatabase away from your network, ArcEditor and ArcInfo seats allow you to check out features to a personal geodatabase, edit them in the field, and check them back into the master geodatabase upon their return.

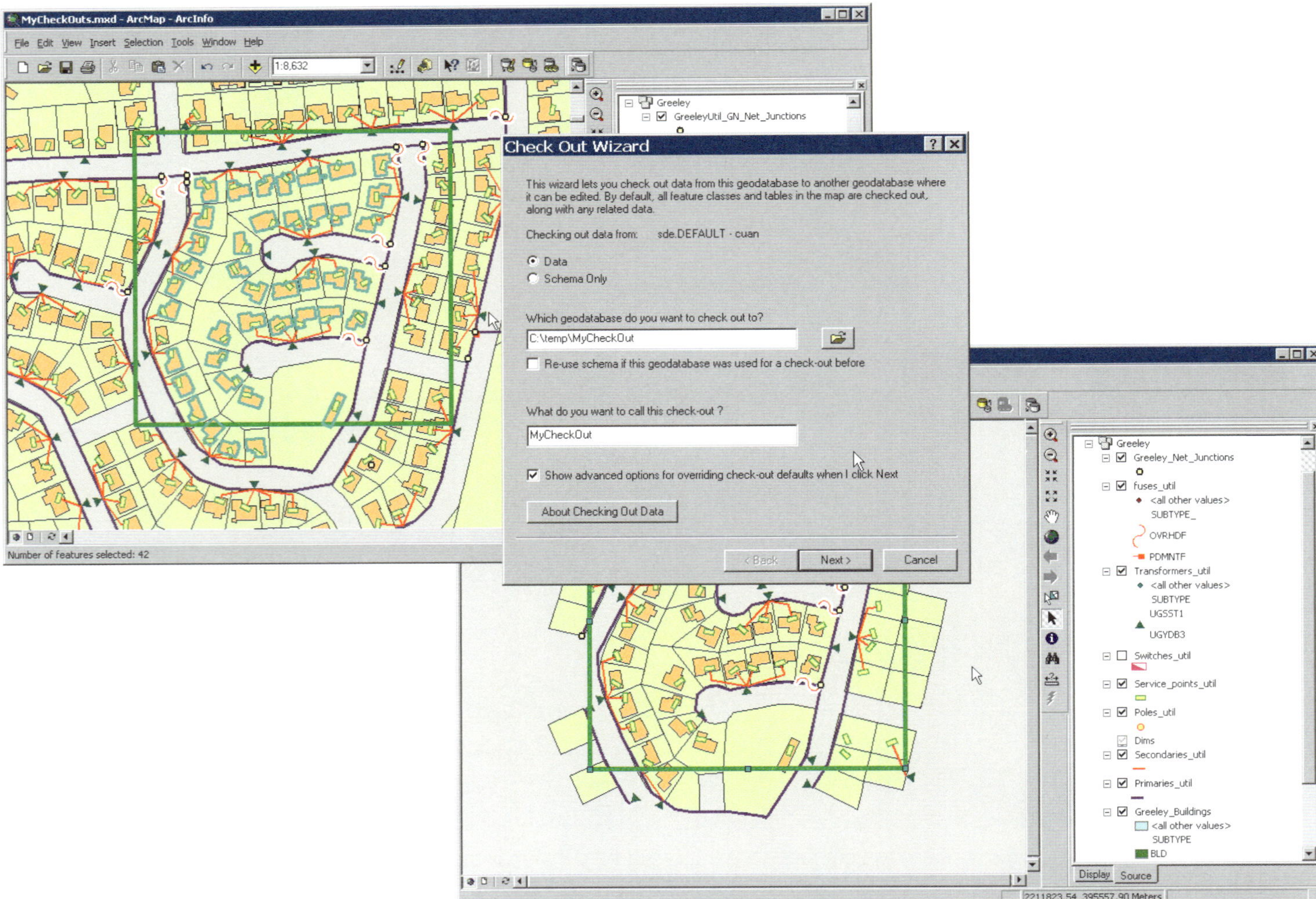

Editing in projected space

If you've collected data from a variety of sources, chances are that not all layers contain the same coordinate system information. Using ArcMap, you can set the coordinate system for a data frame. As you add layers to your map, they are automatically transformed to that projection. That means that you can edit the shapes and attributes of a layer regardless of the coordinate system it was stored in.

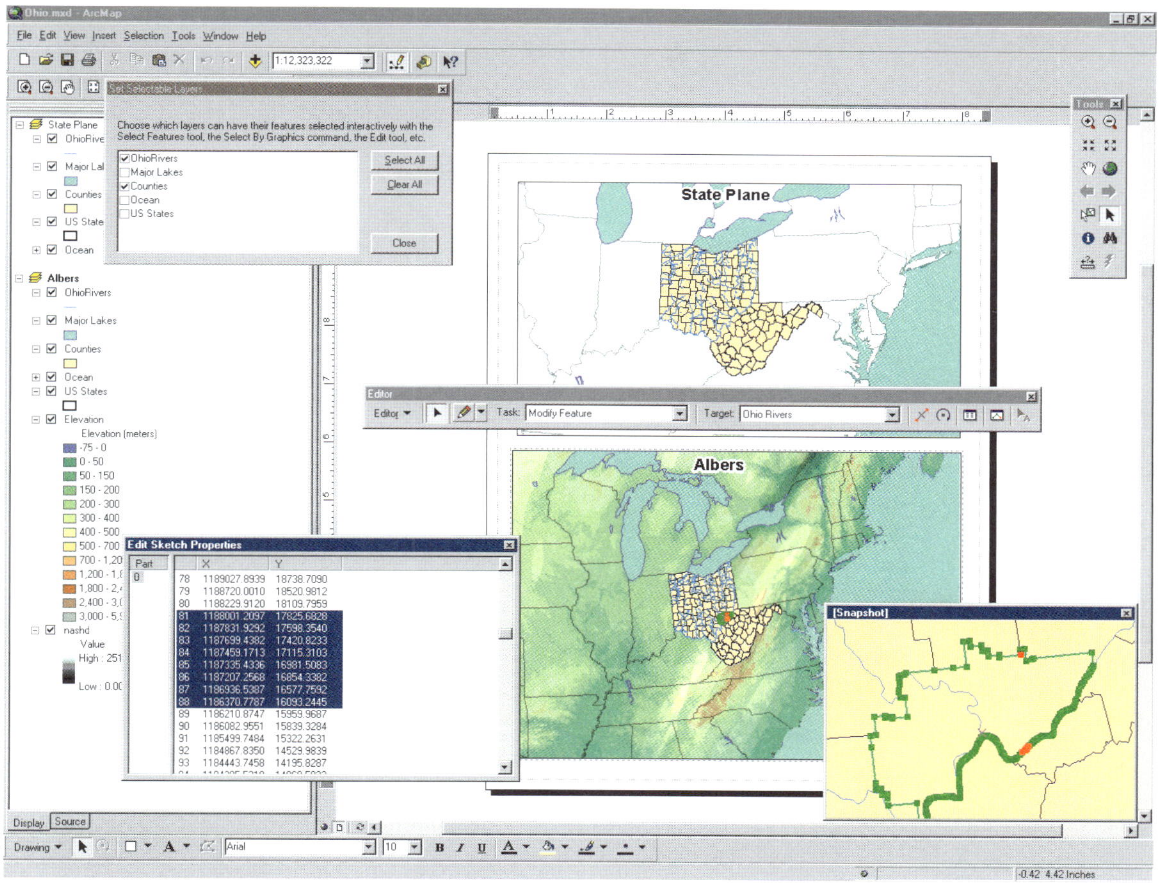

ArcMap has project-on-the-fly capabilities that let you edit layers in the coordinate system that are most important to you without having to transform your data.

Tips on learning how to edit in ArcMap

If you're new to GIS, remember that you don't have to learn everything about editing in ArcMap to get immediate results. To learn how to edit your GIS data, see the *Geodatabase Workbook*. ArcMap comes with the data used in the tutorial, so you can follow along step by step at your computer. You can also read the tutorial without using your computer.

Finding answers to your questions

Like most people, your goal is to complete your tasks while investing a minimum amount of time and effort in learning how to use software. You want intuitive, easy-to-use software that gives you immediate results, without having to read pages of documentation. However, when you do have a question, you want the answer quickly so that you can complete your task. That's what this book is all about—getting you the answers you need, when you need them.

This book describes editing tasks—from basic to advanced—that you'll perform with ArcMap. Although you can read this book from start to finish, you'll likely use it more as a reference. When you want to know how to do a particular task, such as creating a new feature, just look it up in the table of contents or index. What you'll find is a concise, step-by-step description of how to complete the task. Some chapters also include detailed information that you can read if you want to learn more about the concepts behind the tasks. You may also refer to the glossary in this book if you come across any unfamiliar GIS terms or need to refresh your memory.

About this book

This book is designed to introduce editing in ArcMap and its capabilities. If you have never used a GIS before or feel you need to refresh your knowledge, please take some time to read *Getting Started with ArcGIS*, which you received in your ArcGIS package. It is not necessary to do so to continue with this book, but you should use it as a reference if you encounter tasks with which you are unfamiliar.

Getting help on your computer

In addition to this book, use the ArcGIS Desktop Help system to learn how to use ArcMap. To learn how to use the ArcGIS Desktop Help system, see *Using ArcMap*.

Contacting ESRI

If you need to contact ESRI for technical support, see the product registration and support card you received with ArcMap or refer to 'Contacting Technical Support' in the 'Getting more help' section of the ArcGIS Desktop Help system. You can also visit ESRI on the Web at *www.esri.com* and *support.esri.com* for more information about ArcMap and ArcGIS.

ESRI education solutions

ESRI provides educational opportunities related to geographic information science, GIS applications, and technology. You can choose among instructor-led courses, Web-based courses, and self-study workbooks to find educational solutions that fit your learning style. For more information go to *www.esri.com/education*.

Editing basics

2

IN THIS CHAPTER

- **An overview of the editing process**
- **Exploring the Editor toolbars and keyboard shortcuts**
- **Adding editing toolbars**
- **Upgrading a geodatabase**
- **Adding the data you want to edit**
- **Starting and stopping an edit session**
- **Managing the edit cache**
- **Changing the options of the task menu**
- **Selecting, moving, copying and pasting, and deleting features**
- **Setting the number of decimal places used for reporting measurements**

In addition to mapmaking and map analysis, ArcMap is also the application for creating and editing your spatial databases. ArcMap has tools to edit shapefiles and geodatabase feature datasets.

This chapter provides an introduction on how to edit in ArcMap and describes the basic tasks you need to know before you can start to create and edit spatial data. For instance, this chapter shows you how to perform such tasks as adding the Editor toolbar; adding other editing toolbars, such as the Edit cache, Advanced Editing, Spatial Adjustment, Topology, and Network Editing toolbars; starting and stopping an edit session; and selecting features.

An overview of the editing process

The following is a general overview of how to use ArcMap and the Editor toolbar to edit your data. Each of the following steps is outlined in detail in this chapter or other chapters in this book.

1. Start ArcMap.
2. Create a new map or open an existing one.

3. Add the data you want to edit to your map.

   If there are no existing layers for the feature classes you want to edit, you can create them using ArcCatalog™. For more information on creating a feature layer, see *Using ArcCatalog*.

4. Add the Editor toolbar to ArcMap.

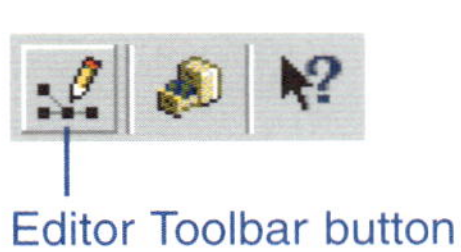

5. Choose Start Editing from the Editor menu.

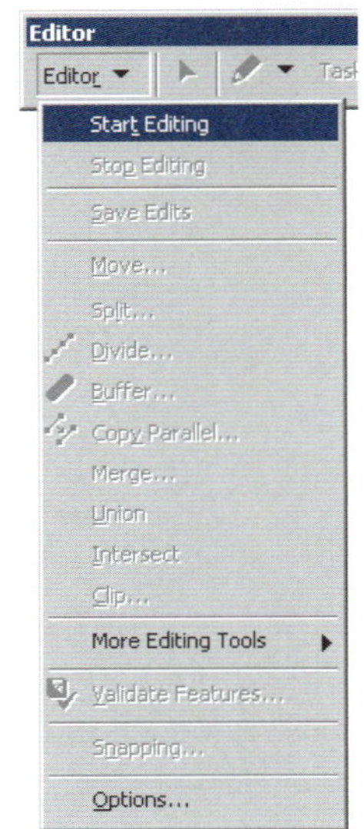

6. Create or modify features and/or their attributes.

7. Choose Stop Editing from the Editor menu and click Yes when prompted to save your edits.

There is no need to save the map—all edits made to the database will automatically be reflected the next time you open the map.

The Editor toolbar

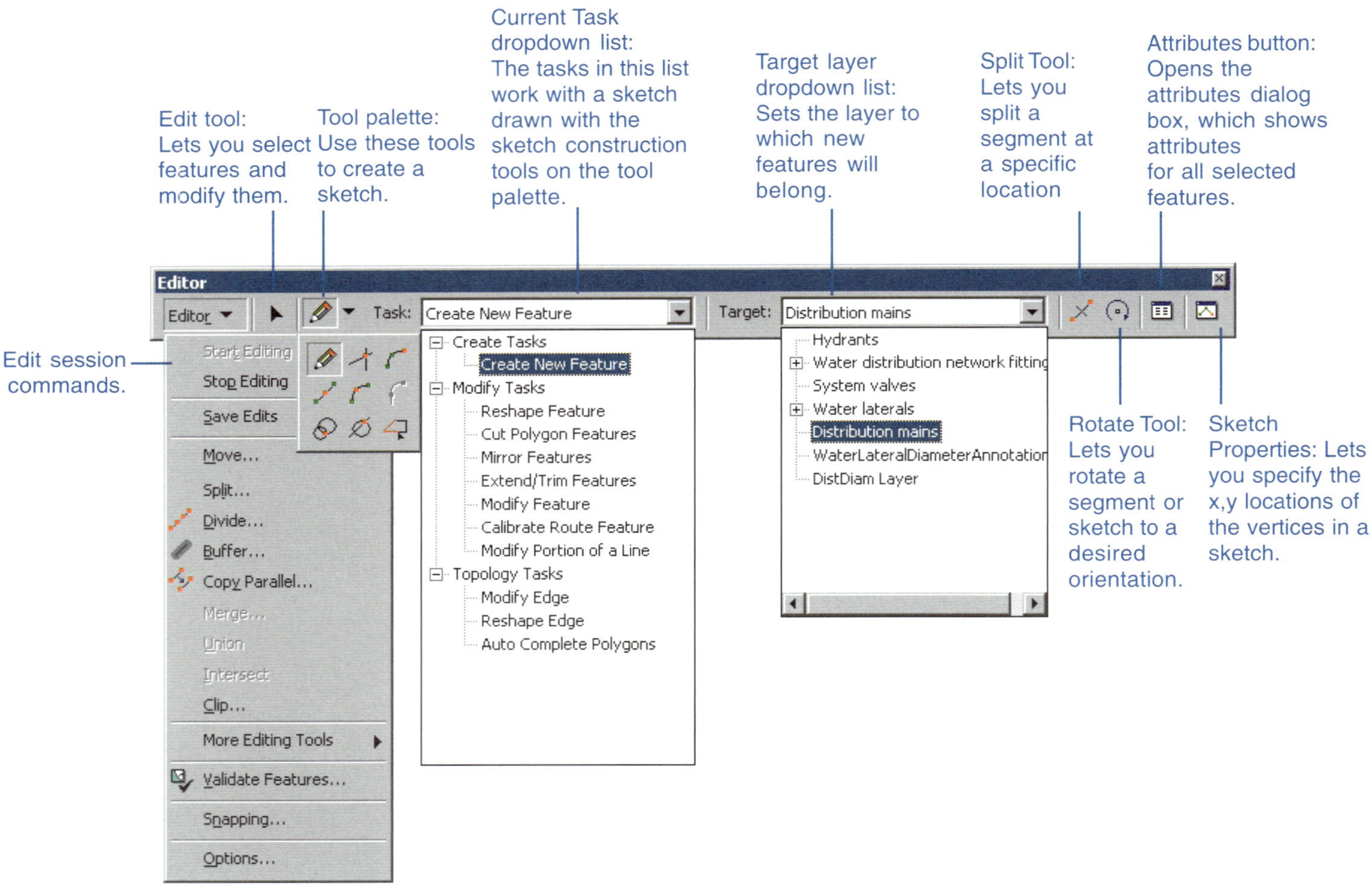

Exploring the Editor toolbar

This section shows you how editing in ArcMap helps you complete the tasks that you need to do. You'll learn about the types of data you can edit as well as the basics of creating and modifying features and their attributes.

The structure of vector datasets

ArcMap provides a common editing environment for features stored in geodatabase feature datasets and shapefiles.

When you edit data with ArcMap, you edit feature classes (collections of features) that the layers on your map represent.

Editing the feature classes lets you edit the actual data source, not just the representation on the map.

A feature class is a collection of the same type of features, for example, a collection of points or a collection of polygons.

A dataset is a collection of feature classes that share the same spatial reference. A dataset might be a collection of land base

Comparing the structure of vector datasets

	Geodatabase	Shapefile
Collections of datasets	A *geodatabase* is a collection of feature datasets.	A shapefile folder is a collection of shapefiles.
Datasets	A *feature dataset* is a collection of feature classes.	A *shapefile* has one shapefile feature class.
Collections of features	A *feature class* is a collection of features of the same type.	A shapefile feature class is a collection of shapefile features.
Features	Point, multipoint, polyline, polygon, annotation, and network.	Point, multipoint, line, and polygon.
Topology	Geodatabase datasets may contain topologies or a Geometric Network.	Map Topology may be used to integrate and edit shapefile feature classes.

feature classes or a collection of utility feature classes. Shapefiles are an exception; they do not hold a collection of feature classes but only one shapefile feature class.

A collection of feature datasets is stored in a geodatabase. Shapefiles are stored in a shapefile folder. Although you may add multiple collections of datasets to your map (geodatabases, ArcInfo workspaces, and shapefile folders), you can only edit feature classes within one collection at a time. Coverage feature classes can't be edited with ArcMap.

What is a sketch and how does it work with a task?

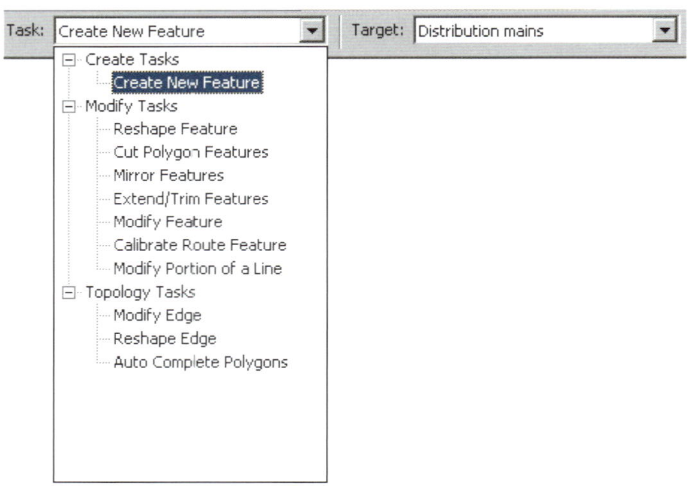

Current Task dropdown list

A *sketch* is a shape you draw that performs various tasks when editing, such as adding new features, modifying features, and

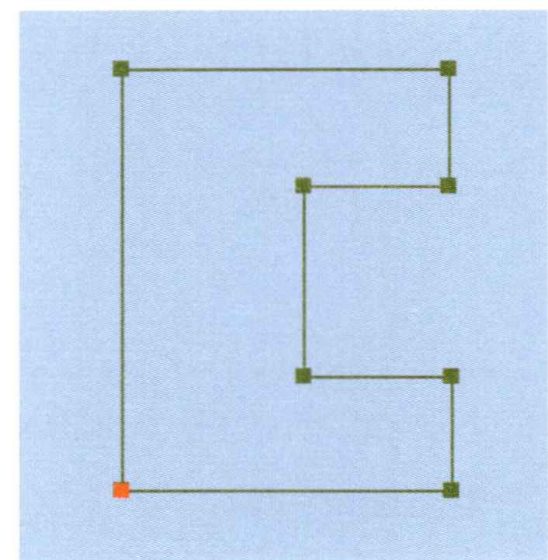

Building as sketch

Building as feature

reshaping features. Tasks are listed in the Current Task dropdown list. You must create a sketch in order to complete a task.

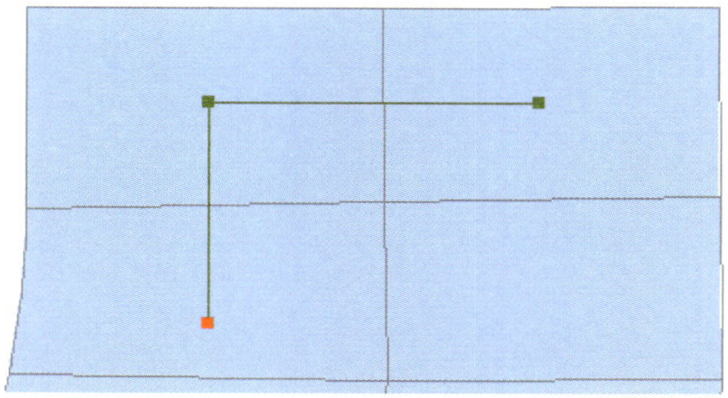

Sketch intersects parcels to be selected.

Parcels intersected by the sketch are now selected.

For instance, the Create New Feature task uses a sketch you create to make the new feature.

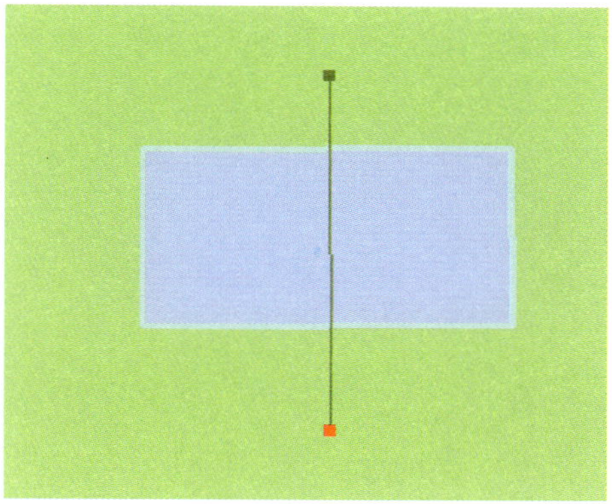

Sketch showing where the polygon is to be cut.

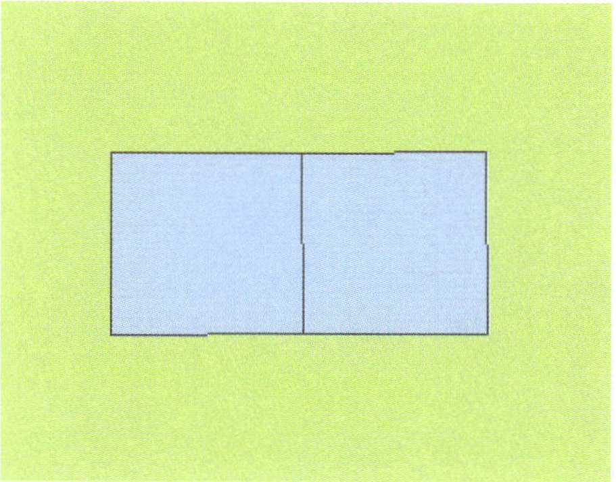

Polygon divided into two features where the sketch was drawn.

The Select Features Using a Line task uses a sketch you create to select features; the features the line intersects are selected.

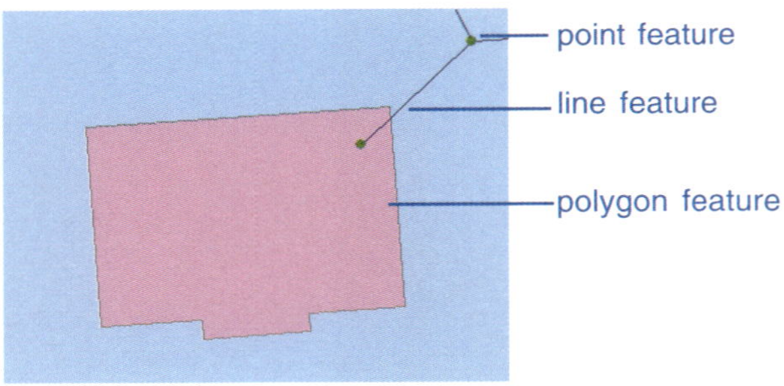

The Cut Polygon Features task uses a line sketch you draw to cut a polygon.

Creating new features

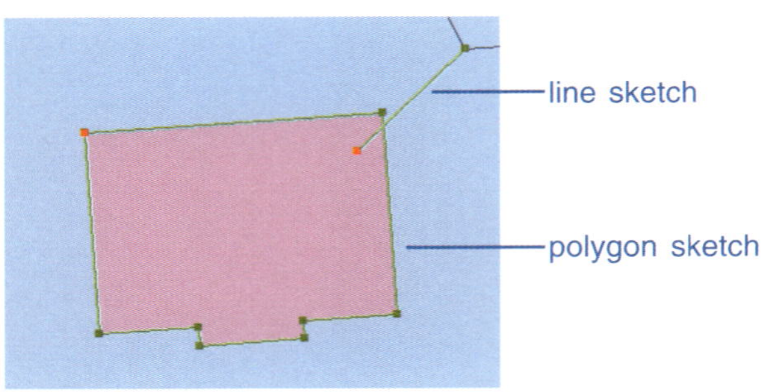

You can create three main types of features with the Editor toolbar: points, lines, and polygons.

To create a line or polygon, you must first create a sketch. A sketch's shape is composed of all the vertices and segments of the feature. *Vertices* are the points at which the sketch changes direction, such as corners, and *segments* are the lines that connect the vertices.

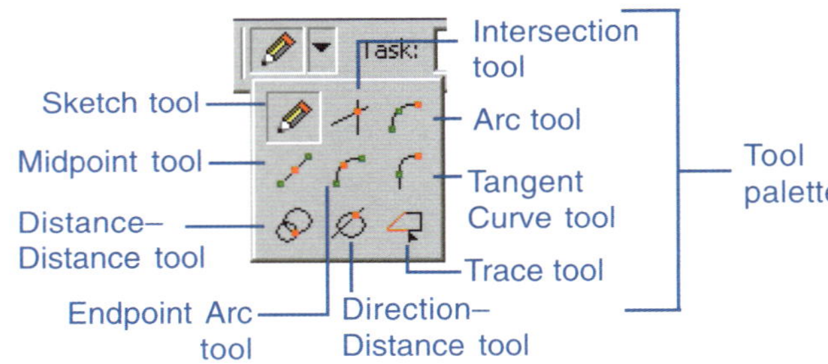

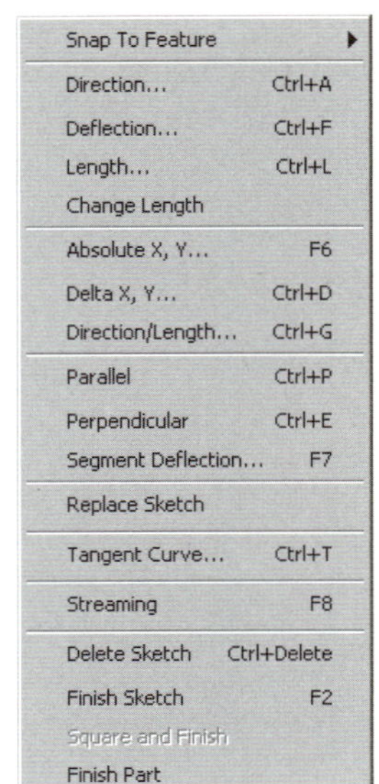

Sketch tool context menu

You can create a sketch by creating the vertices and segments that make up the features. Vertices are marked in green, with the last vertex added marked in red.

The Sketch tool is the tool you use most often to create a sketch. It has an accompanying context menu that helps you place vertices and segments more accurately. The Arc tool, the Distance–Distance tool, and the Intersection tool—located with the Sketch tool on the tool palette—also help you create vertices and segments using other construction methods.

When you're creating a new feature, the target layer determines in which layer a new feature will belong. The Target layer dropdown list contains the names of all the layers in the datasets with which you're working. Subtypes are also listed, if applicable. For instance, if you set the target layer to Buildings: Commercial, any features you create will be part of the Commercial subtype of the Buildings layer.

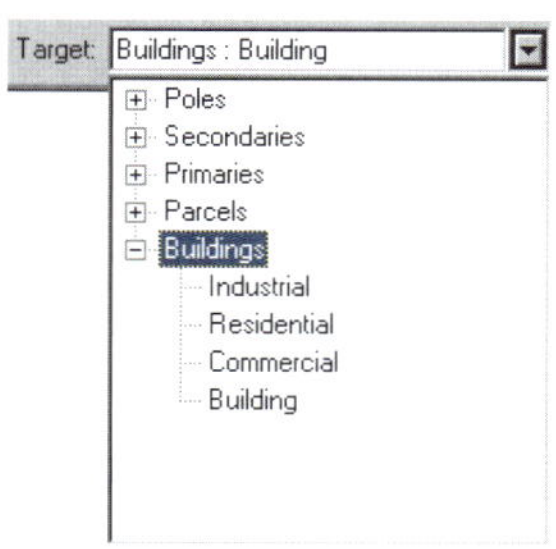

Target layer dropdown list

You must set the target layer whenever you're creating new features—whether you're creating them with the Sketch tool, by copying and pasting, or by buffering another feature.

Modifying features

For every feature on the map, there is an alternate form, a sketch. In the same way that you must create a sketch to create a feature, to modify a feature you must modify its sketch. Because the vertices are visible in a sketch, you can edit the feature in detail; you can move the vertices, delete them, or add new ones using the Sketch context menu.

Besides editing a feature by working with its sketch, you can also use another sketch you create to modify the feature for certain tasks. An example of this type of task is Cut Polygon Features, where a sketch you construct is used to divide one polygon into two.

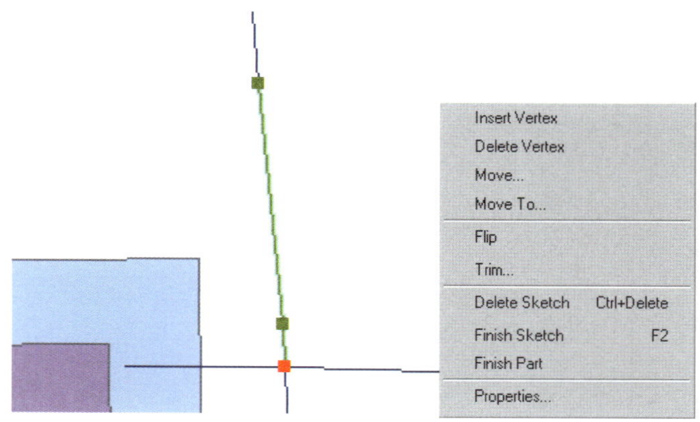

When you edit a feature's sketch, you edit its vertices using the Sketch context menu.

Simple modifications to features, such as moving, copying, or deleting, can be made by selecting the feature and choosing the appropriate tool or command.

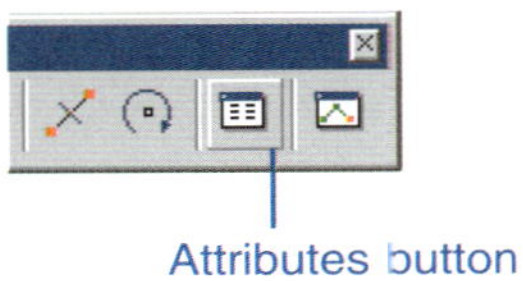

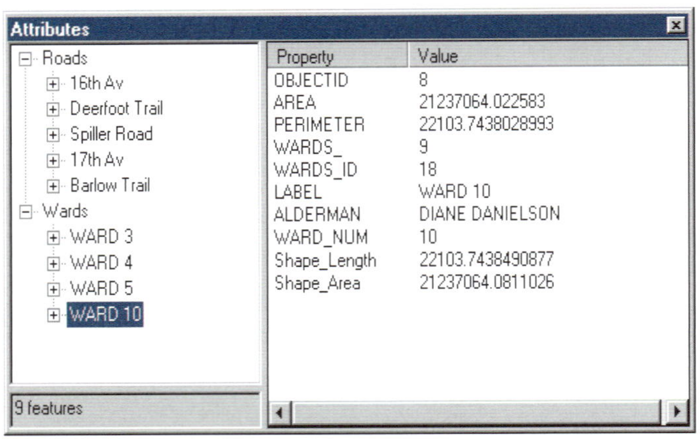

Attributes dialog box

Editing attributes

Attributes can be created or edited in the Attributes dialog box. After selecting the features whose attributes you want to edit, click the Attributes button to see the dialog box.

Editor keyboard shortcuts

There are keyboard shortcuts associated with a number of the editing tools and commands. You can use keyboard shortcuts to make editing quicker and more efficient.

Sketch tool

Direction	Ctrl + A
Deflection	Ctrl + F
Length	Ctrl + L
Absolute X,Y	F6
Delta X,Y	Ctrl + D
Direction/Length	Ctrl + G
Parallel	Ctrl + P
Perpendicular	Ctrl + E
Segment Deflection	F7
Tangent Curve	Ctrl + T
Streaming	F8
Delete Sketch	Ctrl + Delete
Finish Sketch	F2
Show Vertices	V
Show Tolerance	T
Zoom In	Z
Zoom Out	Y
Pan	C
Suspend Snapping	Spacebar
Add/Remove Selection	Shift
Move Selection Anchor	Ctrl

Trace tool

Change Side	Tab
Erase Trace	Esc
Offset	O

Endpoint Arc tool

Radius	R

Direction–Distance tool

Direction	D
Distance	D

Edit tools

Select Anchor	Ctrl + Left mouse click
Copy	Ctrl + C
Paste	Ctrl + V
UndoCreate	Ctrl + Z
Next Feature	N

Topology Edit tool

Select Node	Shift + N
Select Edge	Shift + E
Move Selection Anchor	Ctrl + Left mouse click
Split/Move Node	S + Left mouse click

Validate tool

Validate Area within sketch	Ctrl

Scale/Rotate

Insert second anchor	S

The Advanced Editing toolbar

While the Advanced Editing toolbar is available with ArcEditor and ArcInfo, it is not a feature available with ArcView.

Fillet tool:

Creates a fillet curve between two line features.

Extend tool:

Extends a line to touch another line.

Traverse tool:

Adds to the edit sketch using COGO descriptions.

Explode tool:

Splits a multipart feature into its component features.

Copy Features tool:

Creates a copy in the Editor Target Layer of selected features where you click on the map.

Smooth tool:

Smoothes the shape of a selected feature by converting its segments a series of smoothed line segments.

Trim tool:

Trims a line that crosses a selected line.

Proportion tool:

Divides a line into a number of segments of specified length, allocating any error between the new segments.

Inverse tool:

Adds COGO description of a feature to the feature attribute table.

Generalize tool:

Simplifies the shape of a selected feature.

Adding editing toolbars

Before editing geographic feature data within ArcMap, you must first add the Editor toolbar.

Tip

Adding the Editor toolbar from the Tools menu

You can also add the Editor toolbar from the Tools menu. Click Tools and click the Editor Toolbar button.

Tip

Adding the Editor toolbar from the View menu

You can also add the Editor toolbar by clicking the View menu, pointing to Toolbars, and checking Editor.

Tip

Adding the Editor toolbar using the Customize dialog box

Click the Tools menu and click Customize. Click the Toolbars tab and check Editor.

Adding the Editor toolbar

1. Start ArcMap.
2. Click the Editor Toolbar button to display the Editor toolbar.
3. Click the toolbar's title bar and drag it to the top of the ArcMap application window.

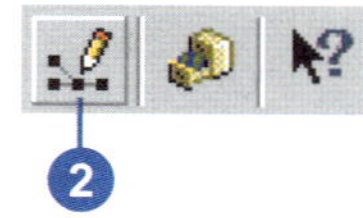

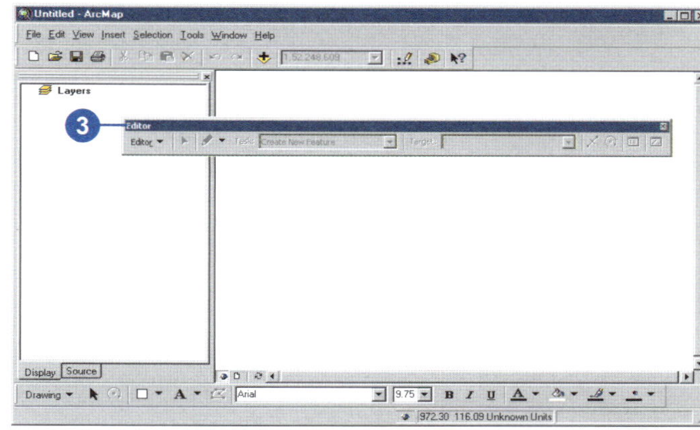

Adding other editing toolbars

1. Click Editor, point to More Editing Tools, and click the toolbar that you want to add.

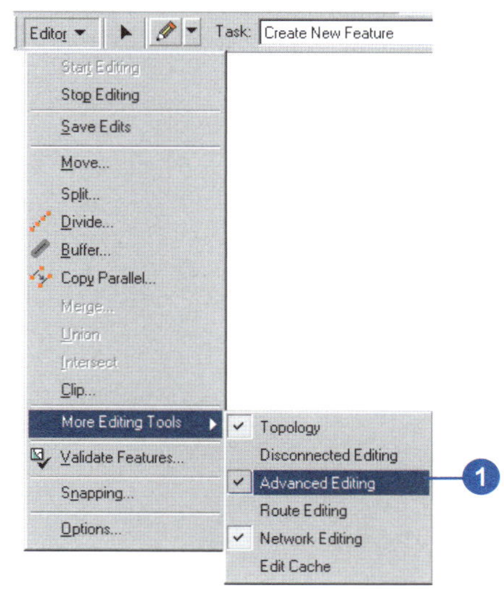

Upgrading a geodatabase

Geodatabases built using previous versions of ArcGIS do not support some of the newer functions of ArcGIS.

If your geodatabase was developed using a previous version of ArcGIS, you may wish to upgrade your geodatabase.

Tip

Creating a backup copy of your geodatabase

Bear in mind that once a geodatabase is upgraded, previous versions of ArcGIS can view, but cannot edit, the geodatabase. For this reason, you may wish to make a copy of the geodatabase and upgrade the copy, thus leaving you with both an original and an upgraded geodatabase.

1. Start ArcCatalog.
2. Right-click the geodatabase you want to upgrade and click Properties.
3. Click the General tab.
4. Click Upgrade Personal Geodatabase.
5. Click OK.

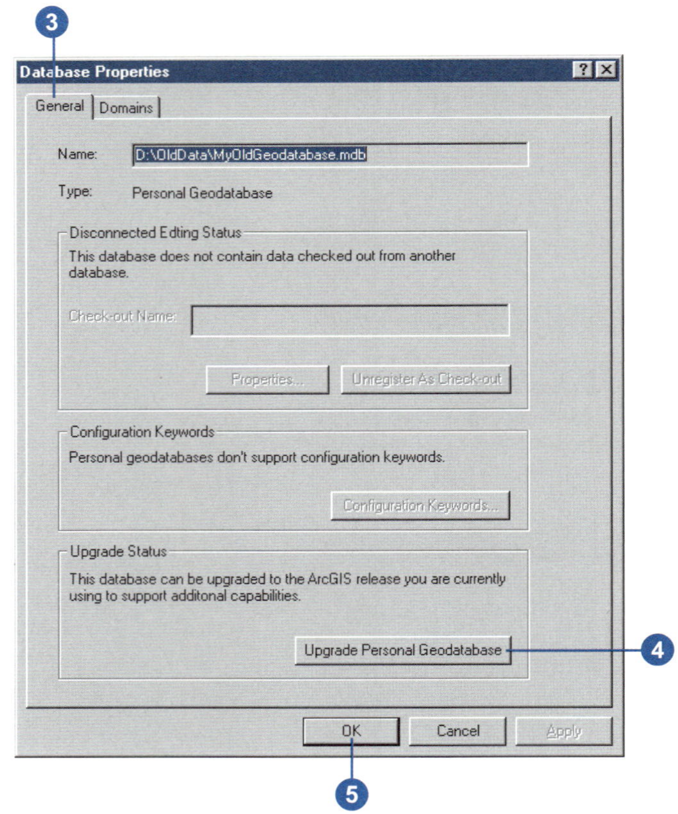

Adding the data you want to edit

Before you can start editing, you must add the data you want to edit to your map. In ArcMap, you can edit feature data in shapefiles or geodatabases.

Tip

Stopping the drawing of data

You can stop the drawing process without clearing the map by pressing the Esc key.

Tip

Loading data from a geodatabase

You can import features from a geodatabase into a layer on your map using the Load Objects command. For more information, see Building a Geodatabase.

1. Start ArcMap.
2. Click the Add Data button.
3. Navigate to the location of your data and click Add.

   The data is added to your map.

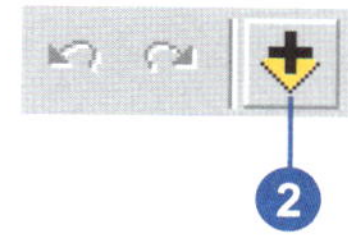

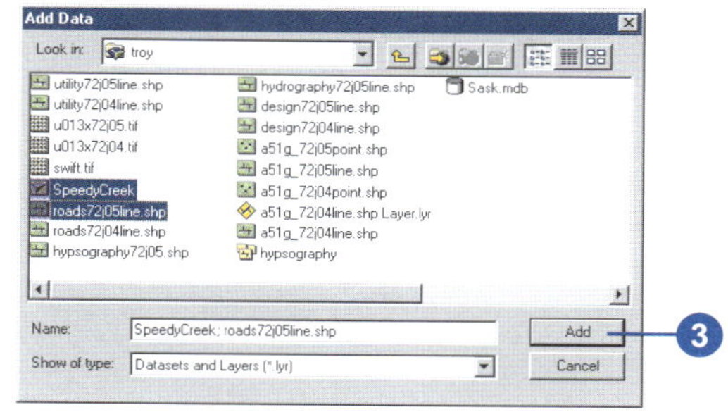

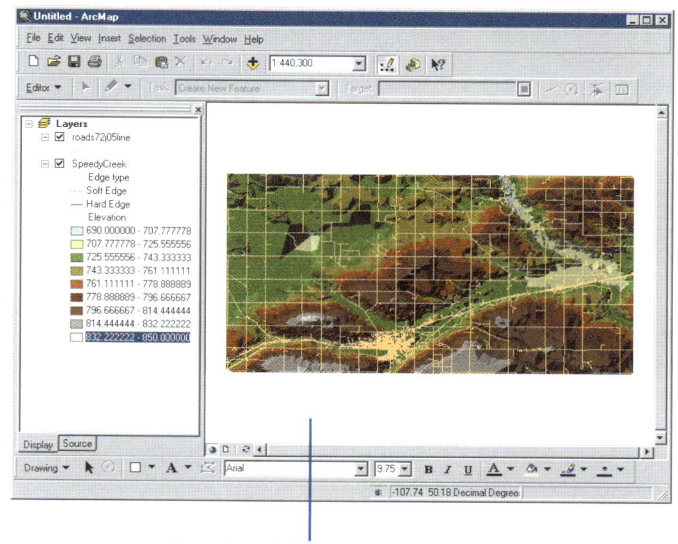

Data is added to the map.

Starting and stopping an edit session

All editing takes place within an *edit session*. To begin, choose Start Editing from the Editor menu. The edits you make are immediately visible on your map but are not saved to the database until you choose to do so.

If you're working with large amounts of data, you can speed up the editing and selection of features by creating an edit cache. An *edit cache* holds the features visible in the current map extent in memory on your local machine. An edit cache results in faster editing because ArcMap doesn't have to retrieve data from the server. You can create an edit cache by clicking the Build Edit Cache ►

Tip

Editing a map with more than one collection of datasets

You can only edit one collection of datasets at a time. These can be geodatabases or folders containing a collection of shapefiles. If your map contains more than one collection, when you choose Start Editing you will be prompted to choose which one you want to edit.

Starting an edit session

1. Start ArcMap and add the Editor toolbar.
2. Click Editor and click Start Editing.

   The Editor toolbar is now active.

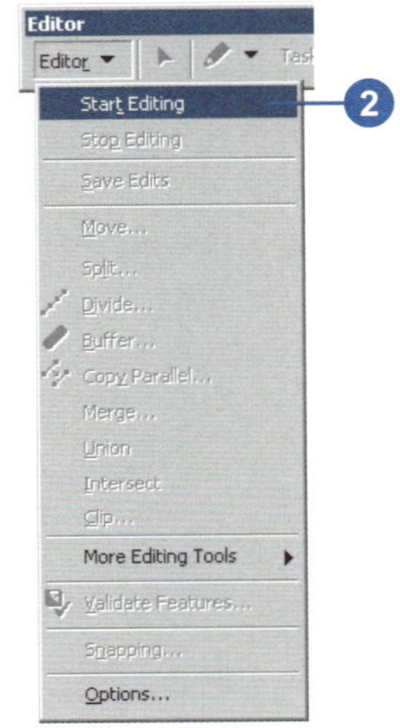

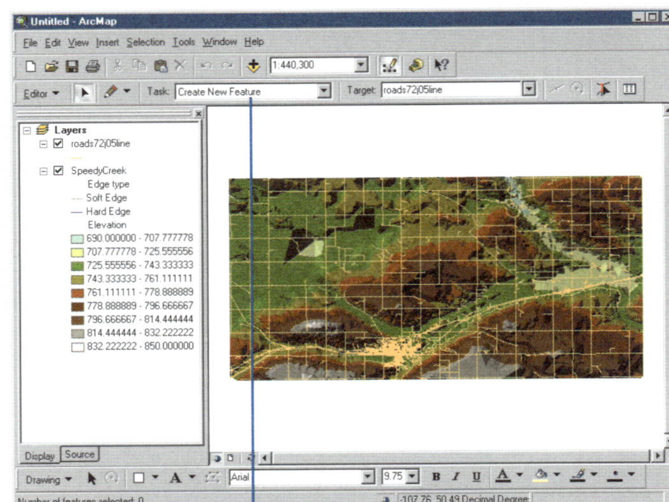

The Editor toolbar is now active.

command on the Edit Cache toolbar.

When you're finished editing, you can save any changes you've made or quit editing without saving. You can also save the edits you've made at any time by choosing Save Edits from the Editor menu.

Tip

Zooming to your edit cache extent

You can quickly return to your edit cache extent at any time in your edit session. Click the Zoom to Edit Cache button on the Edit Cache toolbar.

Creating an edit cache

1. Add the data you want to edit.
2. Click the Zoom In button on the Tools toolbar.
3. Zoom in to the area on the map that you want to edit.
4. Click the Build Edit Cache button on the Edit Cache toolbar.

   The features visible in the current extent are held in memory locally.

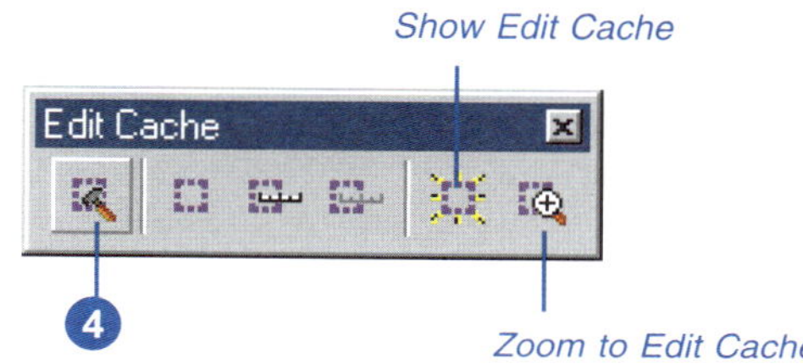

Saving your edits in the middle of an edit session

1. Click Editor.
2. Click Save Edits.

   Any edits you have made are saved to the database.

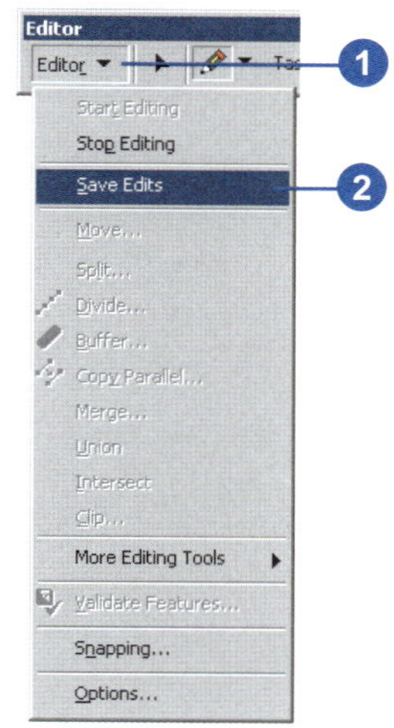

Tip

Editing a map with more than one data frame

If your map contains more than one data frame, you will be editing the data frame that is active when you choose Start Editing. To edit a different data frame, you must choose Stop Editing, then choose Start Editing with the desired data frame active.

For a discussion of data frames, see Using ArcMap.

Tip

Editing in layout view

You can also edit data in a map that you're preparing. Click the View menu and click Layout View. For more information about working in layout view, see Using ArcMap.

Stopping an edit session

1. Click Editor and click Stop Editing.
2. To save changes, click Yes. To quit without saving, click No.

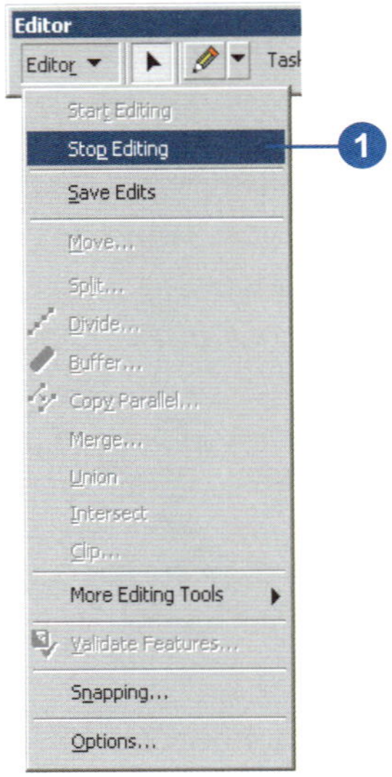

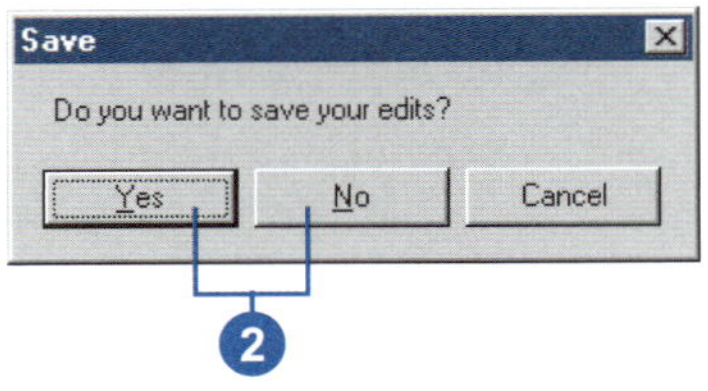

Managing the edit cache

Using the edit cache when editing data stored in an ArcSDE geodatabase will greatly improve performance. Since the edit cache caches features on the client, it cuts down on the number of queries that the client needs to execute on the server during editing.

This improves the performance of your edit session and also reduces the load on the server itself in a multiuser environment. When editing network features or when using the shared geometry editing tools on ArcSDE geodatabases, you should always use the edit cache.

ArcMap has tools to build and help you work with the edit cache. These tools are found on the Edit Cache toolbar.

Tip

Adding the toolbar

You can also add the toolbar by clicking the View menu, pointing to Toolbars, then clicking Edit Cache or clicking the Editor context menu under More Editing Tools.

Adding the Edit cache toolbar

1. Right-click the Main menu.
2. Click Edit Cache.
3. Dock the toolbar to the ArcMap window. Now each time you start ArcMap, the toolbar will be displayed.

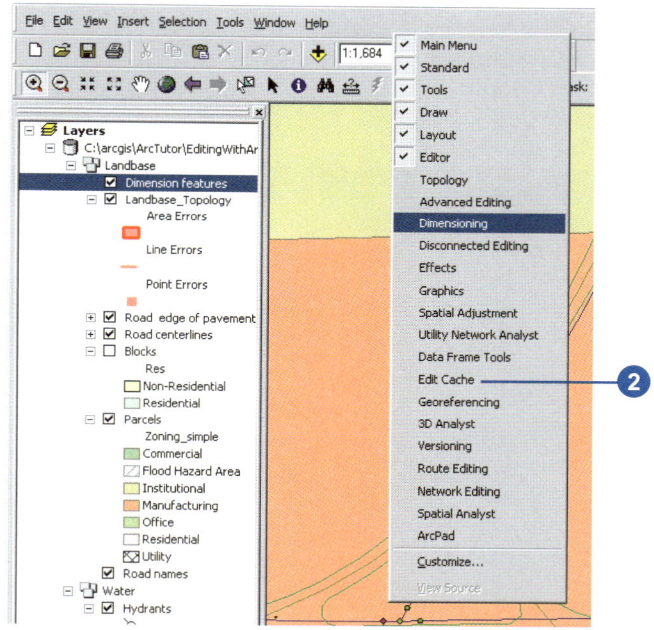

Using the edit cache

You can't build or use an edit cache unless you have started editing.

Edit cache tools

Click the Show Edit Cache command on the Edit Cache toolbar to display the extent of the current edit cache. This command will only be enabled if your map's extent intersects the extent of the current edit cache.

Click the Zoom To Edit Cache command on the Edit Cache toolbar to zoom to the extent of the current edit cache.

The Show Edit Cache and the Zoom To Edit Cache commands are enabled only when you have built an edit cache.

Auto-cache builds a new edit cache for the current extent each time you pan or zoom. This is convenient when you do not know the exact bounds of the area you want to cache, but it can have a large performance cost. It is a good idea to set a maximum scale for autocaching, so you do not retrieve every feature when you pan or zoom at the full extent of the data.

Read-only features

In an ArcSDE geodatabase, the features that you do not have write privileges to are not cached, nor are features in feature classes that are not registered as versioned.

Building an edit cache

1. Pan and/or zoom to the area of your map where you will be performing your edits.
2. Click the Build Edit Cache command on the Edit Cache toolbar.

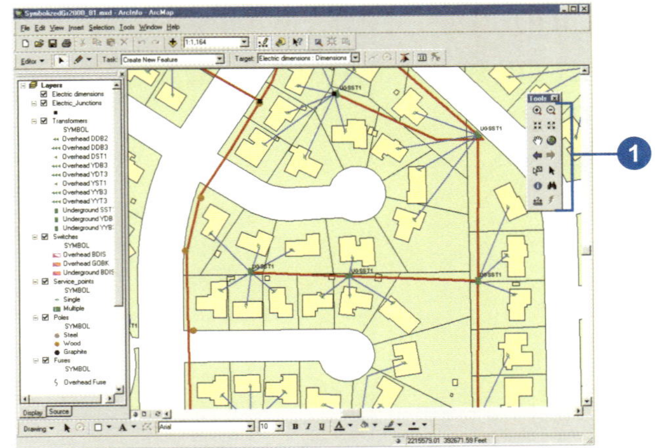

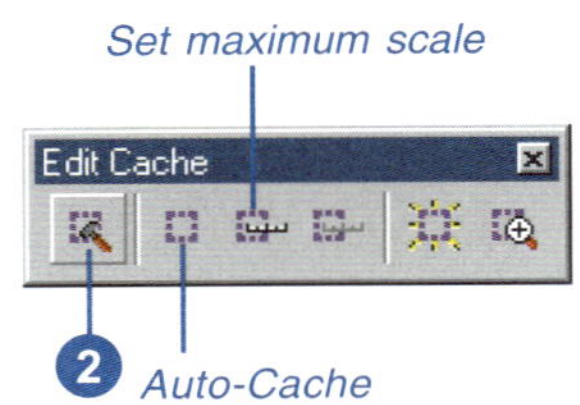

Using auto-cache

1. Zoom out just beyond the scale at which you'll be editing.
2. Click Set Auto-Cache Scale on the Edit Cache toolbar.
3. Click the Auto-Cache command on the Edit Cache toolbar.

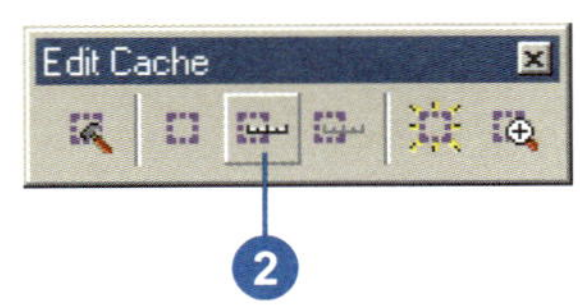

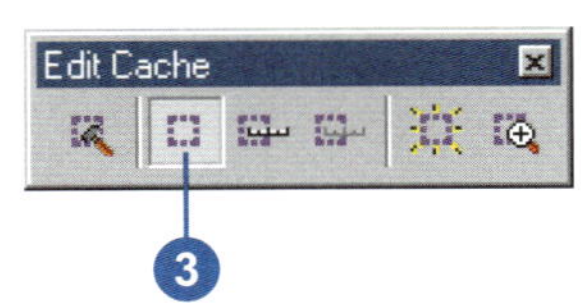

Changing the options of the task menu

Many times your project may call for specific editing tasks to be used more than others. Task options can be added to or removed from the task dropdown menu. This will simplify your editing interface and facilitate your selection process.

Modifying the task dropdown menu

1. Click Editor and click Options.
2. Click the Edit Tasks tab.
3. To remove tasks, select the task to be removed and click Remove.
4. To add tasks, click Add.
5. On the Add Tasks dialog box, select the tasks to add and click OK.

   The task dropdown menu will be modified according to the changes made.

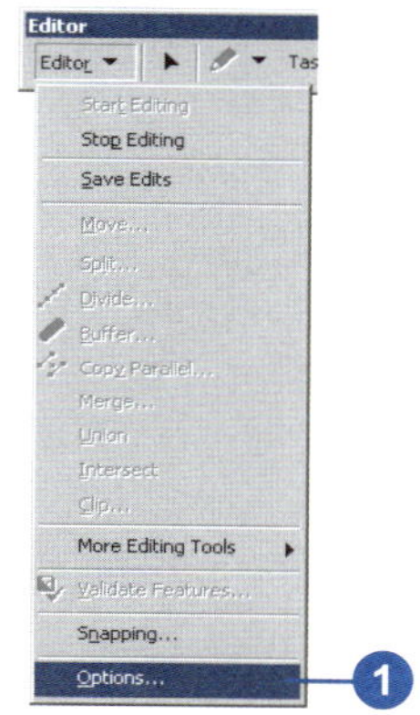

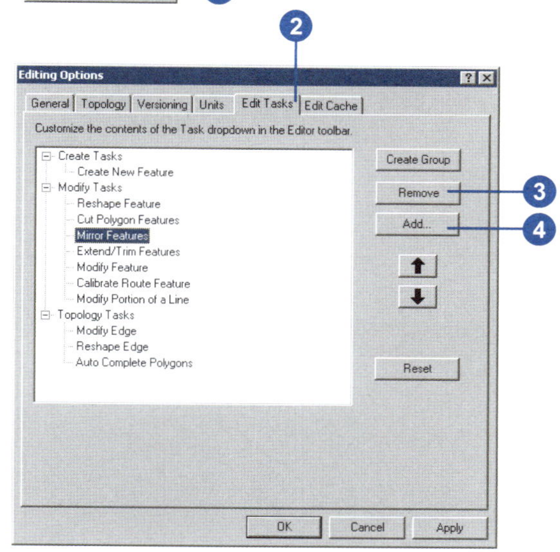

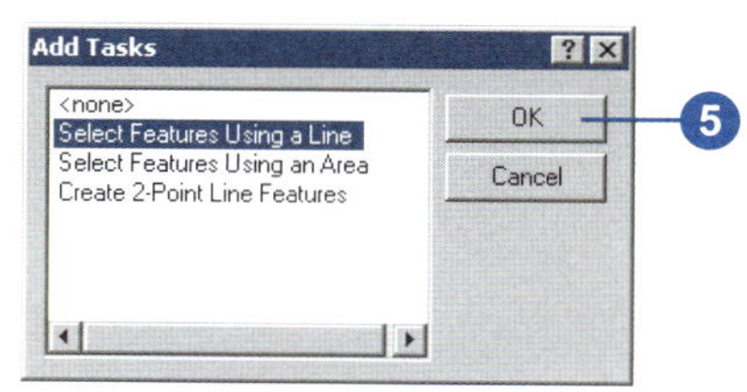

Selecting features

Selecting features identifies the features on which you want to perform certain operations. For example, before you move, delete, or copy a feature, you must select it. You must also select features before you can view their attributes.

You can select features in several different ways, either by clicking them with the Edit tool or by creating a line or a polygon that intersects the features you want to select. The number of features selected is shown immediately after you make the selection, in the lower-left corner of the ArcMap window. ►

Tip

Selecting more than one feature

To select more than one feature, hold down the Shift key while you click the features. You can also use the Edit tool to drag a box around a group of features.

Tip

Removing features from the selection

To remove features from the selection set, hold down the Shift key while you click the features.

Selecting features using the Edit tool

1. Click the Edit tool.
2. Move the pointer over a feature and click.

   The selected feature is highlighted.

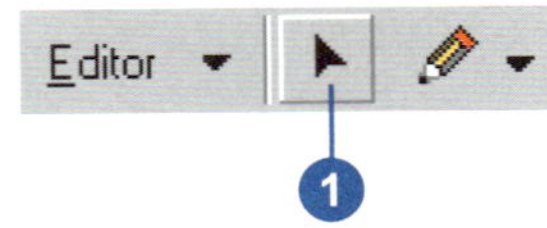

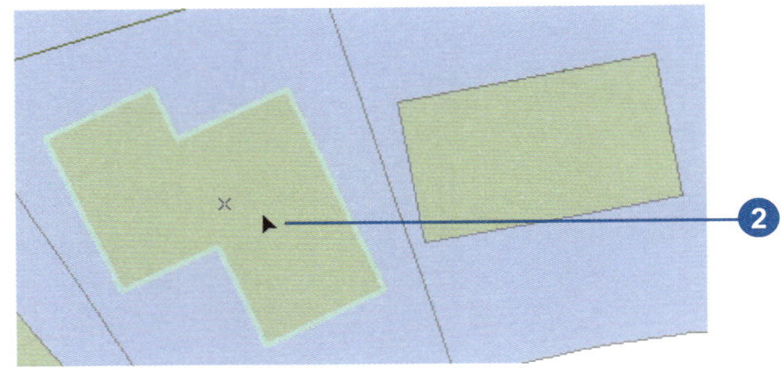

The selected building is highlighted.

The small "x" located in the center of the selected features is the *selection anchor*. The selection anchor is used when rotating features, moving features using snapping, and scaling features.

You can choose which layers you can select by choosing Set Selectable Layers from the Selection menu and using the Selectable Layers list.

For example, suppose you wanted to select a large number of buildings by drawing a box around them but selected a parcel by mistake as you drew the selection box. To avoid this, you might uncheck the Parcels layer in the Selectable Layers list so that parcels cannot be selected.

Tip

Adding options to the task dropdown menu

Tasks on the task dropdown menu can be added or removed from the selection on the Edit Options dialog box under the Edit Tasks tab. For further details, see 'Modifying the task dropdown menu' in this chapter.

See Also

For more information on creating a line, see Chapter 3, 'Creating new features'.

Selecting features using a line

1. Click the Task dropdown arrow and click Select Features Using a Line.
2. Click the tool palette dropdown arrow and click the Sketch tool or any of the other construction tools in the tool palette.
3. Construct a line that intersects the features you want to select.

   The features that the line intersects are now selected.

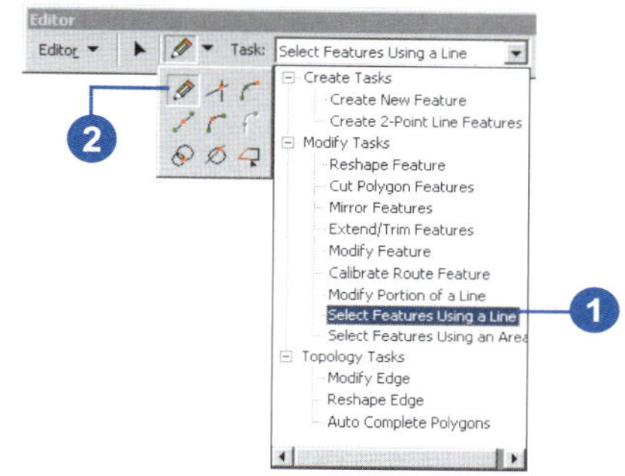

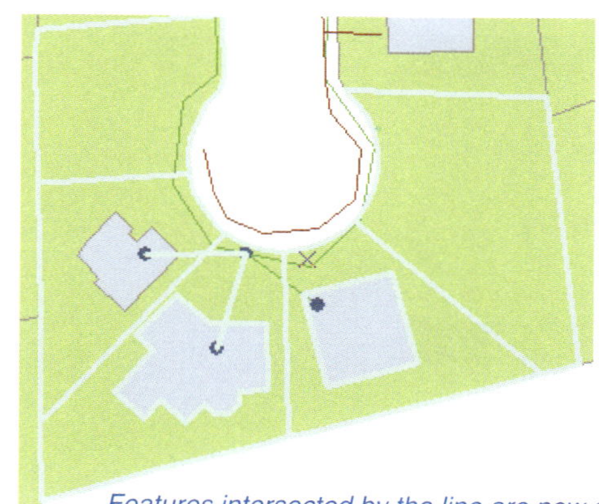

Features intersected by the line are now selected.

Tip

Use the Selection menu for more control over your selection

The Selection menu gives you more tools with which to make a selection, such as adding to the current selection, selecting all features onscreen, or creating an SQL statement.

Tip

Pan and zoom while you select features

You can pan and zoom while selecting features without having to change tools. Hold the Ctrl key and press Z to zoom in, X to zoom out, or C to pan.

See Also

For more information on creating a polygon, see Chapter 3, 'Creating new features'.

See Also

For more information on the selection anchor, see 'Moving features' in this chapter and Chapter 7, 'Editing existing features'.

See Also

For more information on selecting features in ArcMap, see Using ArcMap.

Selecting features using a polygon

1. Click the Task dropdown arrow and click Select Features Using an Area.
2. Click the tool palette dropdown arrow and click the Sketch tool or any of the other construction tools in the tool palette.
3. Construct a polygon that intersects the features you want to select.

   The features that intersect with the polygon you created are now selected.

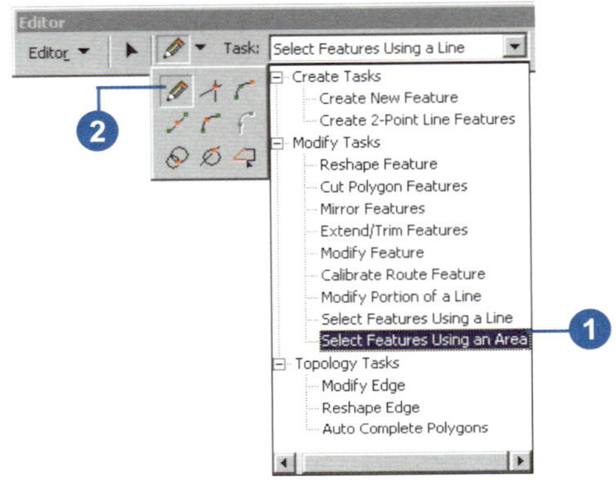

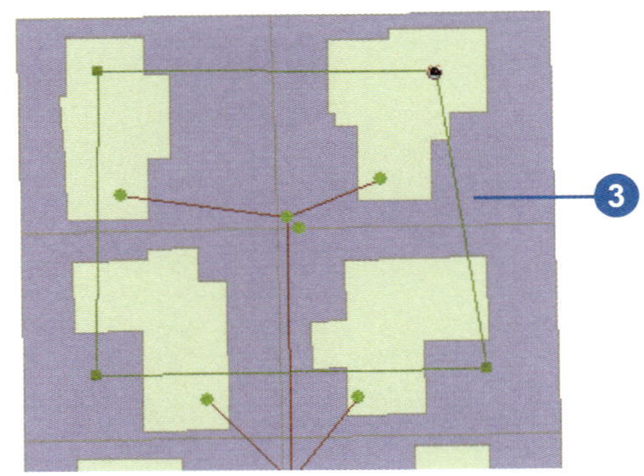

Features that intersect with the area are now selected.

Tip

Adding the Selection tab to the ArcMap table of contents

You can quickly change the selectable layers from the ArcMap table of contents if you add the Selections tab. You can add the Selections tab to the table of contents from the TOC tab of the ArcMap Tools Options dialog box.

Making a layer selectable

1. Click Selection and click Set Selectable Layers.
2. Click the check boxes next to the layer names you want to be able to select. Uncheck the boxes next to the names you don't want to be able to select.

   Layers whose names are unchecked are still visible in your map but cannot be selected.

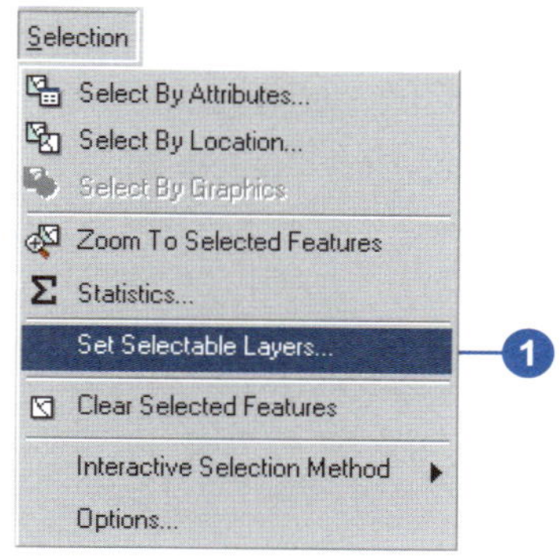

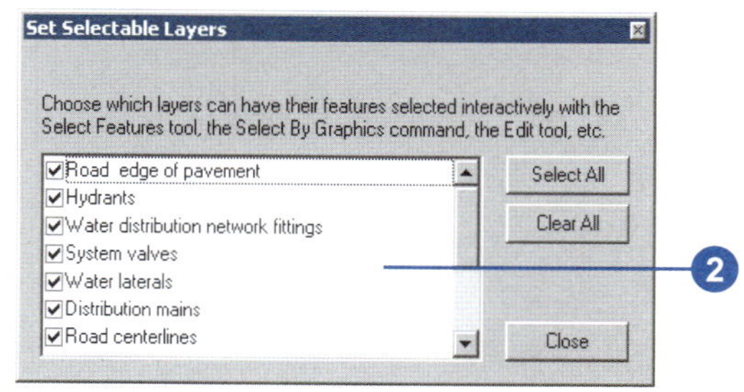

Moving features

You can move features in three different ways: by dragging; by specifying delta x,y coordinates; or by rotating.

Dragging is the easiest way to move a feature. Use this method when you have a general idea of where you want to move the feature.

Specify delta x,y coordinates when you want to move a feature to a precise location. ArcMap uses the current location of the selected feature or features as the origin (0,0) and moves them from that location according to the coordinates you specify. ►

Tip

Avoiding accidental moves

The Sticky move tolerance allows you to set a minimum number of pixels your pointer must move on the screen before a selected feature is moved. This can be a useful way to prevent features from being accidentally moved small distances when they are clicked with the Edit tool. You can set the sticky move tolerance on the General tab of the Editing Options dialog box.

Dragging a feature

1. Click the Edit tool.
2. Click the feature or features you want to move.
3. Click and drag the feature or features to the desired location.

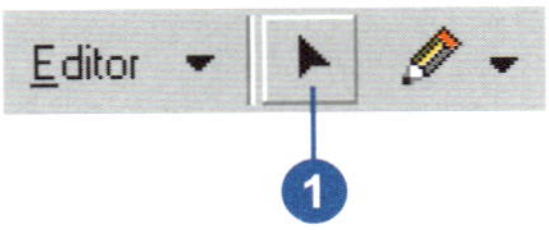

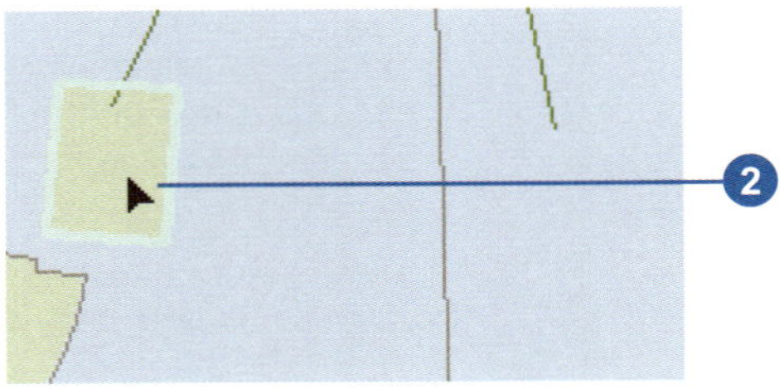

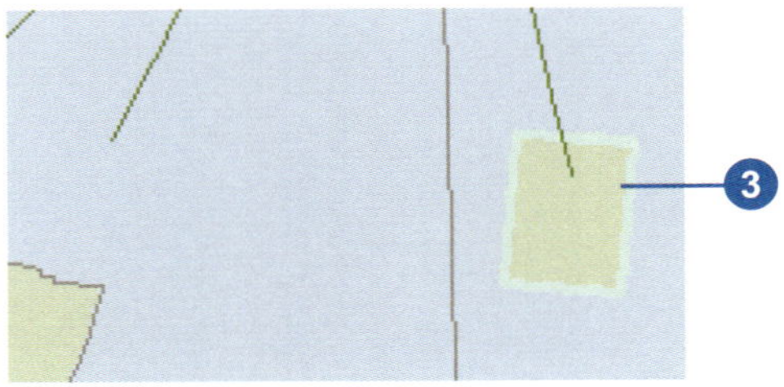

The selected building is dragged to a new location.

The coordinates are measured in map units. The graphic below illustrates the change in location when delta x,y coordinates of 2,3 are specified for a building.

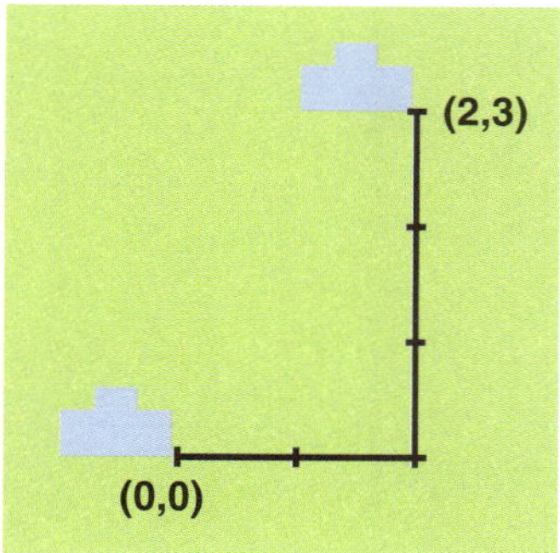

In the example above, the building is moved up and to the right, as positive coordinate values are specified. To move the building left or down, you would specify negative values.

You can rotate features in ArcMap using the Rotate tool. After selecting the features, drag the mouse pointer so that the features rotate to the desired position. Features rotate around the selection anchor, the small x located in the center of selected features.

If you want to move a feature to a precise location in relation to another feature, you can use the snapping environment. For example, you can move a ►

Moving a feature relative to its current location

1. Click the Edit tool.
2. Click the feature or features you want to move.
3. Click Editor and click Move. ►

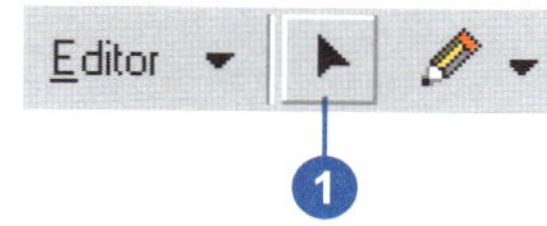

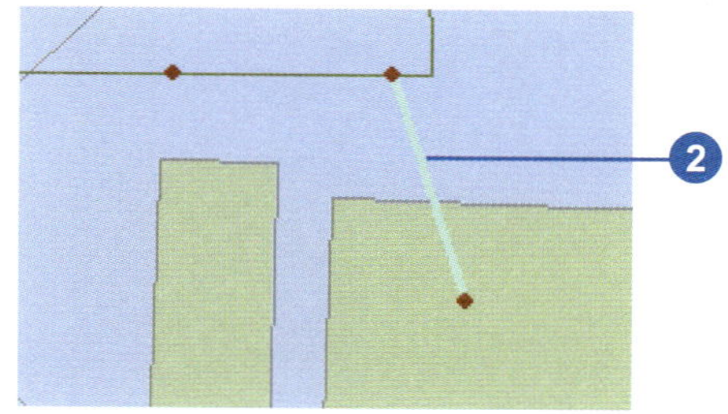

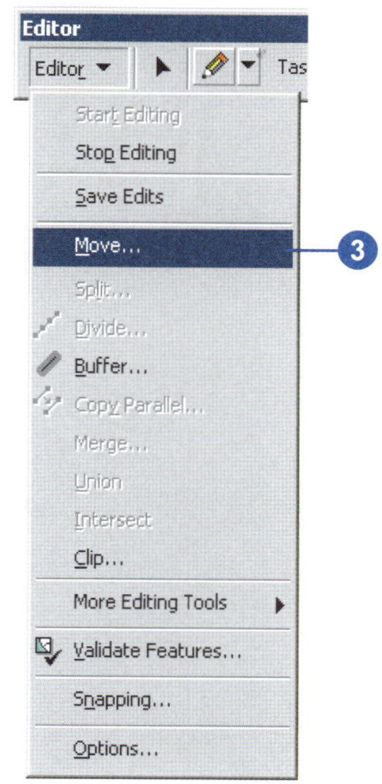

parcel and have one of its corners jump, or snap, precisely to a corner of another parcel. Simply move the parcel's selection anchor to its corner vertex after setting the appropriate snapping properties. Then move the parcel toward its new location until the selection anchor snaps to the corner vertex of the other parcel. Snapping is discussed in Chapter 3, 'Creating new features'.

Tip

Moving the selection anchor

To move the selection anchor, move the pointer over it, press the Ctrl key, and drag the selection anchor to the desired location.

Tip

Undoing a move

You can undo any edit you make to a feature by clicking the Undo button on the ArcMap Standard toolbar.

4. Type the desired coordinates and press Enter.

   The feature is moved according to the specified coordinates.

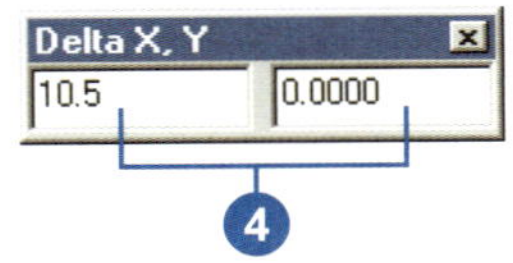

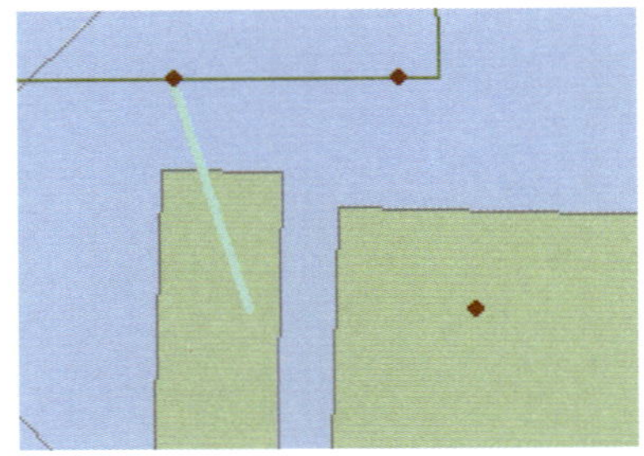

The feature is moved 10.5 map units to the left.

Tip

Rotating by degree

To specify the exact amount of rotation, click the Rotate tool, press A, and type the number of degrees. A positive number rotates the feature in the clockwise direction, a negative number in the counter-clockwise direction.

Tip

Rotating with snapping

If you want to rotate a feature until a point on it snaps to a feature specified in the current snapping environment, add an Auxiliary selection anchor and drag it to the part of the feature that you want snapped to another feature. You can add an Auxiliary selection anchor by pressing the S key when using the Rotate tool. The Auxiliary selection anchor also works with the Scale tool.

Tip

Rotating a point's symbology

If your data already has a field that contains the rotation angle for each point symbol, you can use ArcMap to rotate the symbology.

Right-click the point layer name in the map's table of contents and click Properties. Click the Symbology tab. Click the Advanced button, then click Rotation. From the dropdown list, choose the field that contains the rotation angle. Click the option that describes how you want that angle calculated.

Rotating a feature

1. Click the Edit tool.
2. Click the feature or features you want to rotate.
3. Click the Rotate tool.
4. Click anywhere on the map and drag the pointer to rotate the feature to the desired position.

   The feature or features rotate around the selection anchor. You can drag the selection anchor to a new location to change the center of rotation.

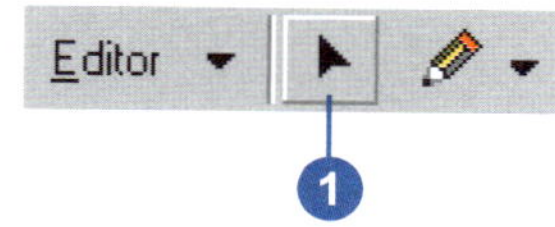

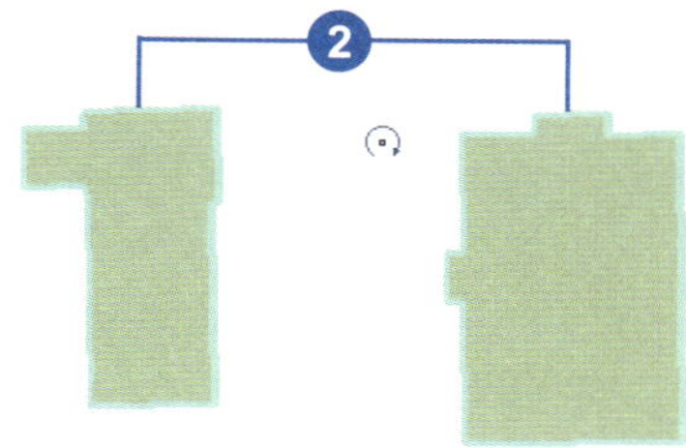

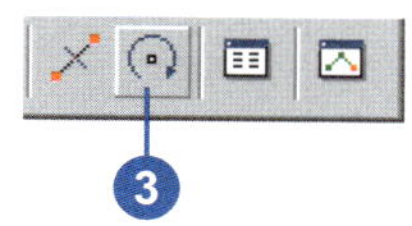

The selected features are rotated.

Copying and pasting features

To copy an existing feature, use the tools on the ArcMap Standard toolbar. From the Target layer dropdown list, choose the layer containing the type of features you want the new feature to be—for example, a building.

You can copy a feature and paste it as part of another layer, but it must be the same type of layer (point, line, or polygon) as the one from which you copied. There is one exception to this rule—you can copy polygons into a line layer.

Attributes from the original feature are only copied to the new feature if you are copying and pasting within the same layer.

See Also

For more information on attributes, see Chapter 9, 'Editing attributes'. You can also see Using ArcMap.

1. Click the Target layer dropdown arrow and click the layer containing the type of features you want the new features to be.
2. Click the Edit tool.
3. Click the feature or features you want to copy.
4. Click the Copy button.
5. Click the Paste button.

   The feature is pasted on top of the original feature.

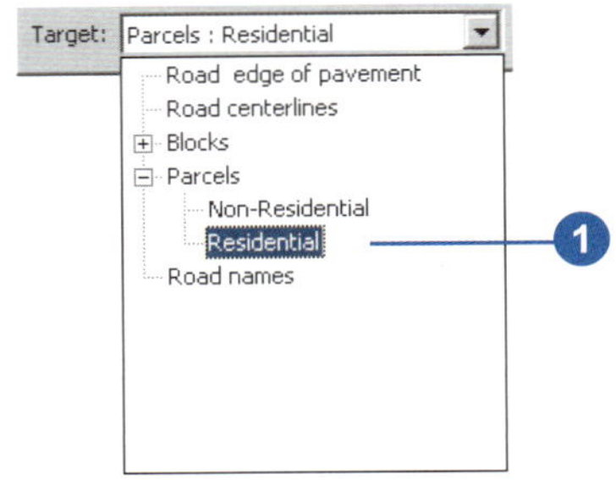

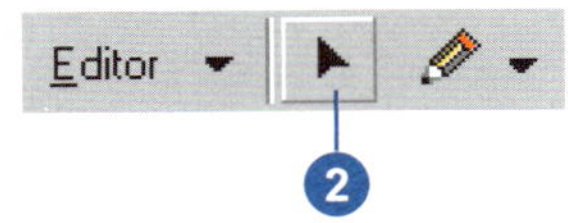

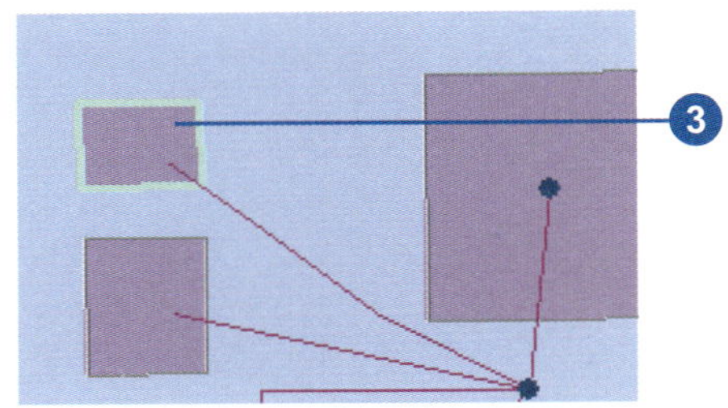

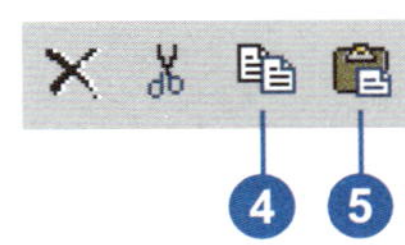

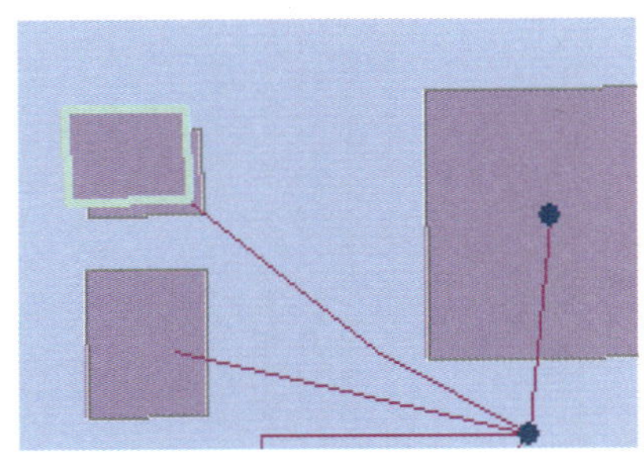

The selected feature is pasted on top of the original feature.

Deleting features

To delete a feature from the map and from the database, use the Delete button on the ArcMap Standard toolbar.

Tip

Deleting features using the Delete key

You can also press the Delete key on the keyboard to remove selected features.

1. Click the Edit tool.
2. Click the feature or features you want to delete.
3. Click the Delete button.

   The selected features are deleted.

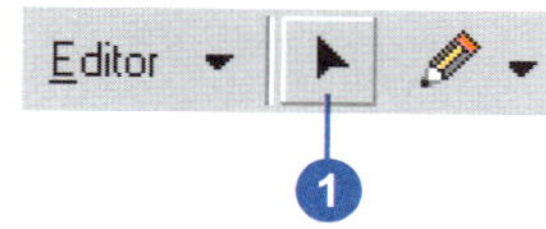

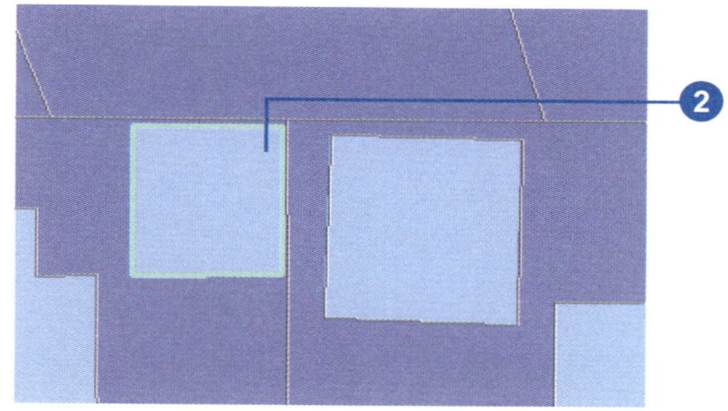

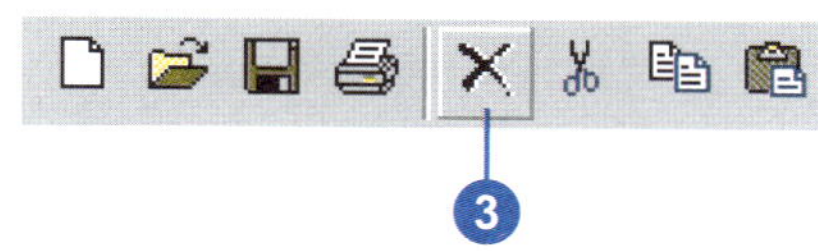

The selected building is deleted.

Setting the number of decimal places used for reporting measurements

When creating or editing a feature with the Sketch tool, you can use the Sketch tool context menu to view such measurements as the distance between two vertices, the angle between two segments, or the current coordinate location of the pointer.

By default, ArcMap displays these measurements using three decimal places. However, you can easily change the number of decimal places displayed. After you set the number of decimal places, ArcMap will report all measurements using that number of decimal places.

1. Click Editor and click Options.
2. Click the General tab.
3. Type the number of decimal places you want to use.
4. Click OK.

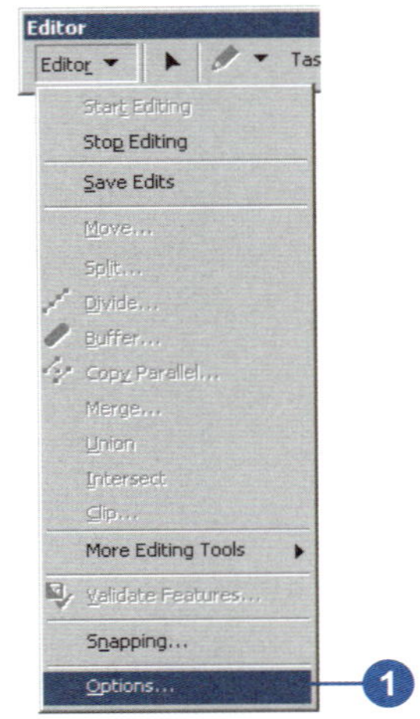

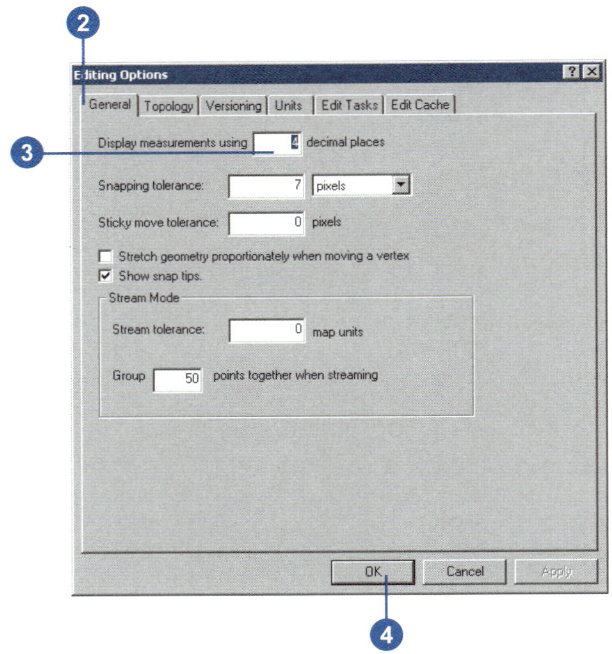

Creating new features

3

IN THIS CHAPTER

- **Creating new point, line, and polygon features**
- **Creating segments using directions and lengths or angles**
- **Creating segments that are circular arc curves**
- **Copying, filleting, extending, and trimming lines**
- **Proportionally dividing a line**
- **Creating features from a traverse**
- **Getting COGO attributes of features**
- **Exploding multipart features**
- **Generalizing and smoothing features**
- **Using the snapping environment**
- **Setting direction type and units**

It's easy to create a wide array of new features using the editing sketch construction tools. To begin, simply specify the layer in which you want to create the new feature. Then, use the appropriate tool to digitize the vertices of the feature.

With the editing tools you can create new point, line, or polygon features for many practical purposes. Using the Sketch tool and its accompanying context menu, you can add a water main perpendicular to an existing water main in a subdivision. The Distance–Distance tool lets you create a land parcel that begins 55 meters from one corner of an existing lot and ends 40 meters from another lot corner. Creating a cul-de-sac is simplified using the Arc tool to create a circular curve. With the Intersection tool, you can add a parcel to a subdivision by establishing a corner vertex using segments of an adjoining parcel.

Using the editing tools, you can create a variety of features by constructing segments at specific angles and of specific lengths. You can create features that are parallel or perpendicular to other features. You can also create multipoint features, such as a system of oil wells, and multipart features, such as a group of islands that form a country or state.

These are just a few examples of how you can use ArcMap to easily and accurately create new features for your database.

How to create a new feature

To create a new feature using ArcMap, you create an edit sketch. A *sketch* is a shape that you draw by digitizing vertices. You can use a sketch to complete various tasks; these tasks are listed in the Task dropdown list shown below.

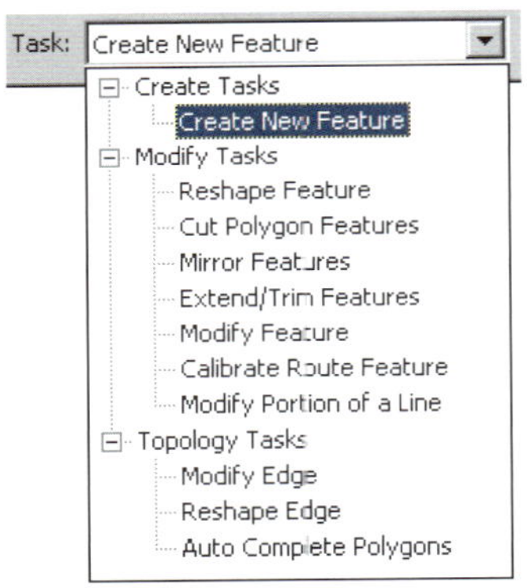

Task dropdown list

Tasks you can complete with a sketch include creating new features, modifying features, extending or trimming features, and reshaping features.

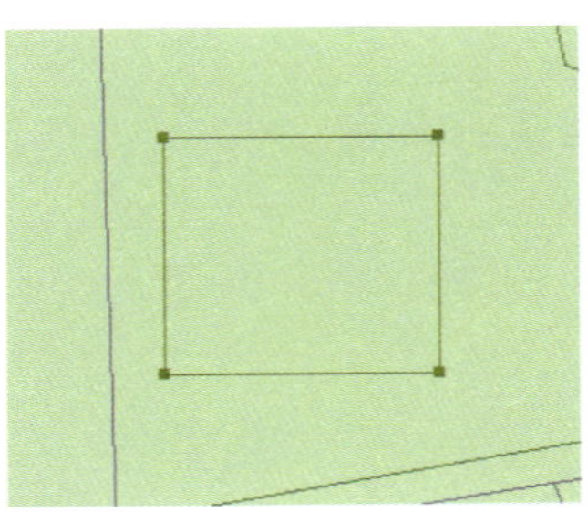

Building as sketch

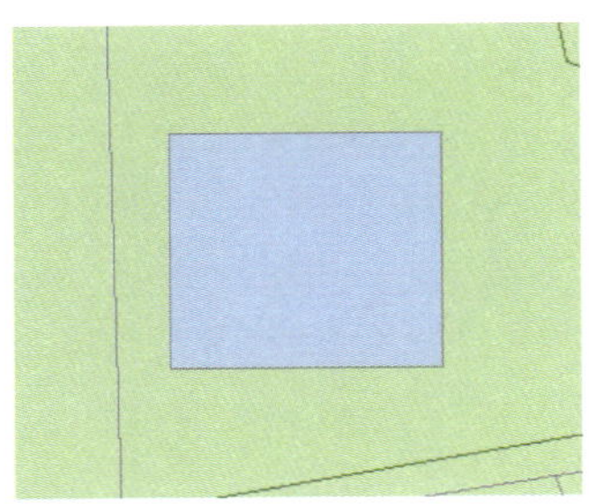

Building as feature

This chapter focuses on using sketches to create new features. When the current task setting is Create New Feature, the shape you create becomes the new feature.

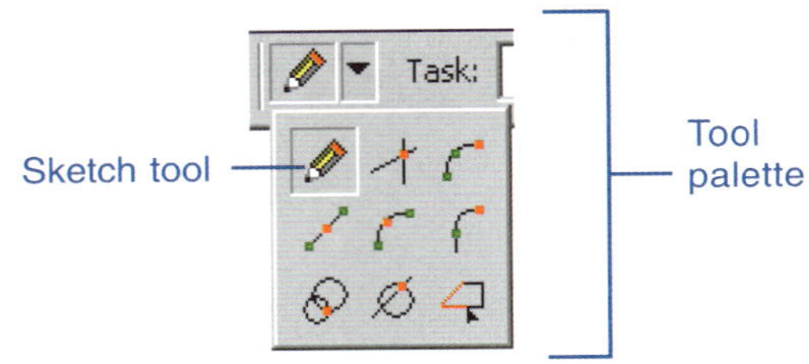

A sketch is composed of *vertices*—the points at which the sketch changes direction, such as corners—and *segments*—the lines that connect the vertices. You create a sketch using the Sketch tool located on the tool palette.

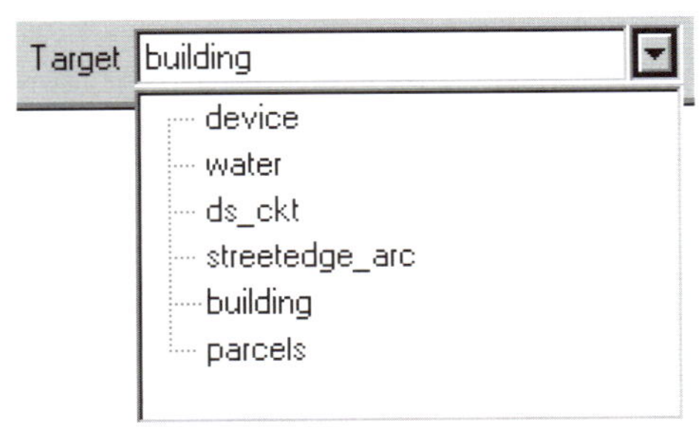

Target layer dropdown list

The type of feature you create is determined by the setting of the Target layer dropdown list. This list contains the names of all the layers in the datasets with which you're working.

Choose the layer to which you want to add new features before you start to create them.

To create point features, click once on the map. To create line or polygon features (see the example below), use the Sketch tool to click on the map to digitize the vertices that make up that feature.

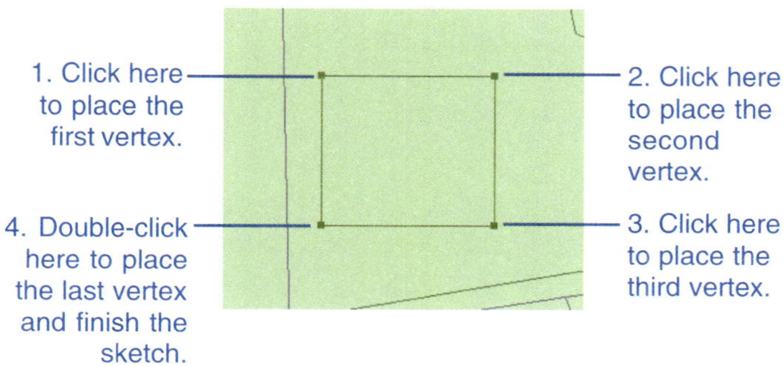

To create the last vertex and finish the sketch, double-click with the mouse. After you finish the sketch, ArcMap adds the final segment of the sketch and the sketch turns into a feature.

Of course, you won't always be able to place vertices or segments interactively. When you're using the Sketch tool, you can see a menu called the Sketch tool context menu. You can access this menu when you right-click the mouse away from the sketch you're creating. The menu has choices to help you place

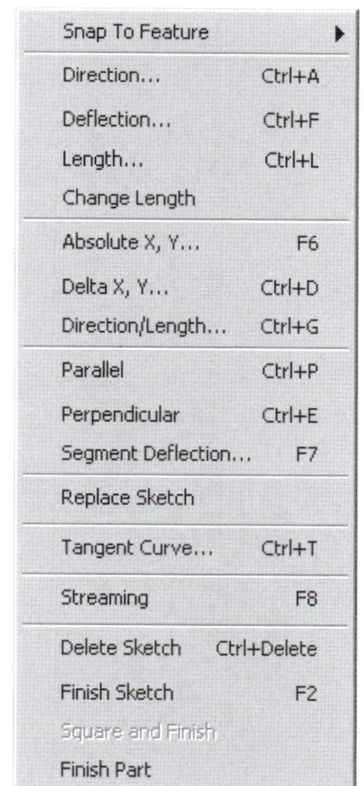

Sketch tool context menu

the vertices and segments exactly where you want them. For example, you can set a segment to be a certain length or angle, or create a vertex at a specific x,y coordinate location.

All the tools on the tool palette help you create a sketch. Three tools use more specific construction methods to create either points or vertices: the Distance–Distance tool, Direction–Distance tool, and the Intersection tool.

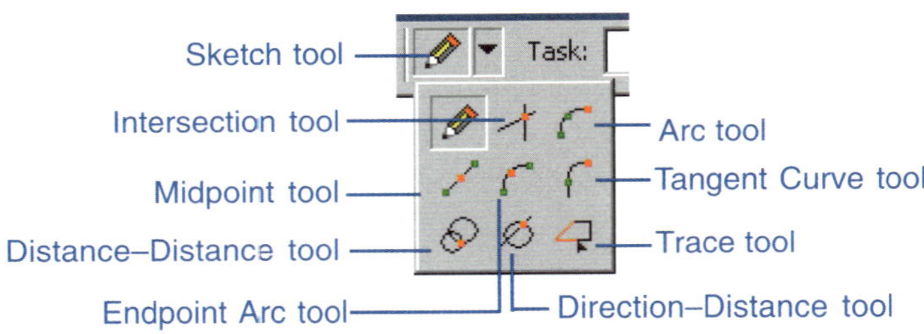

The Distance–Distance tool lets you create a point or vertex at the intersection of two distances from two other points. You might use this tool to place a new electrical primary based on field measurements. Suppose you know that the next point for the primary is 50 feet from one building corner and 75 feet from another.

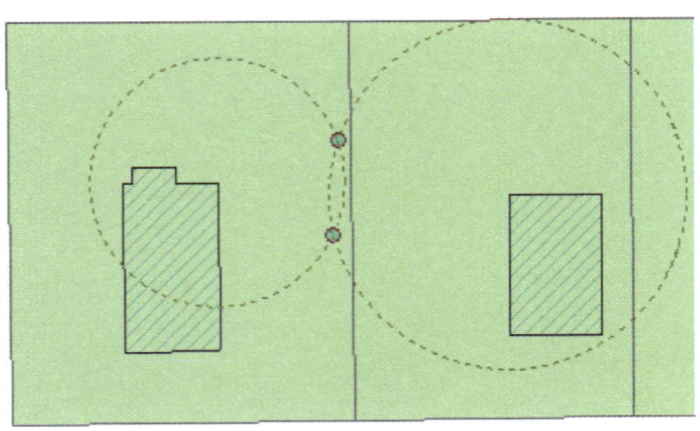

The Distance–Distance tool allows you to choose one of two intersection points of two circles; the size of the circles is determined by the radius you set.

The Distance–Distance tool creates two circles based on these distances and finds two possible intersection points where the primary can be placed.

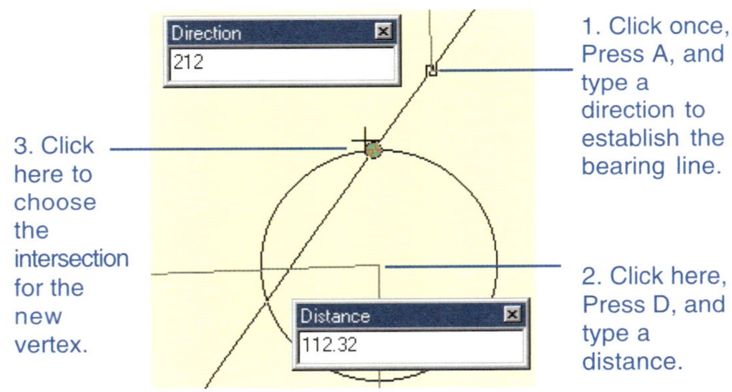

The Direction–Distance tool, like the Distance–Distance tool, allows you to create a vertex using a distance from a known point, plus information from another point. Instead of using a distance from the second point, the Direction–Distance tool uses a direction from a known point to define a bearing line. You can choose which of the intersection points will be the new vertex.

You could use the Direction–Distance tool to place a tree location point based on field notes stating that the tree is on a bearing of 212 degrees from a fence corner and 112 feet from the northeast corner of a building.

The Intersection tool creates a point or vertex at the place where two segments would intersect if extended far enough.

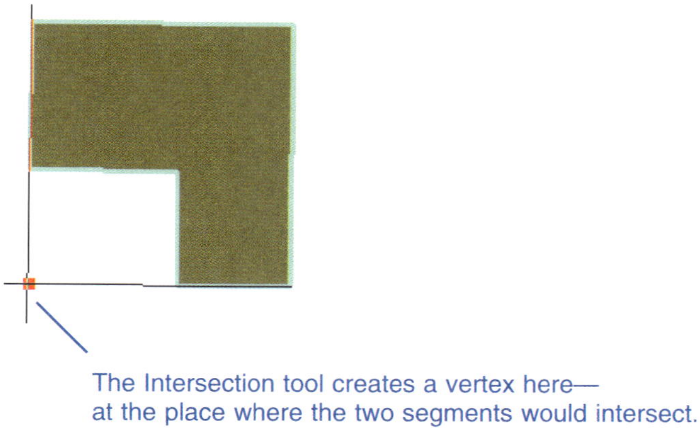

Suppose you want to create a parking lot adjoining an L-shaped building. The outer corner of the lot should be located at the point where the two outermost walls of the building would intersect if they were extended. You could use the Intersection tool to find this implied intersection point and create the corner vertex of the lot.

The Midpoint tool lets you define the location of the next vertex by clicking two points—the new vertex is placed at the midpoint of the line between these points.

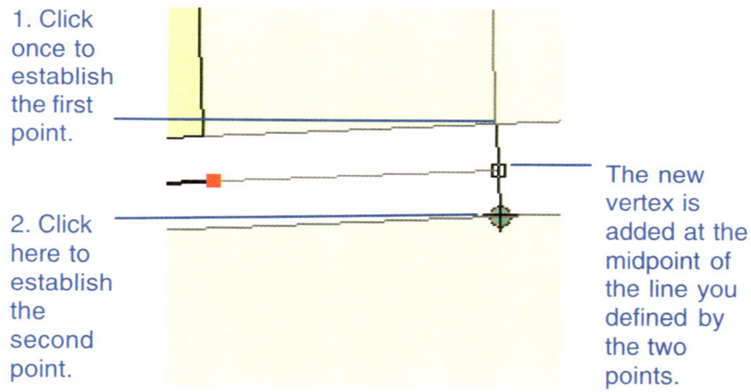

You might use the Midpoint tool to place the next vertex of a street centerline midway between the parcels on either side of the street.

The Arc tool helps you create a segment that is a circular arc.

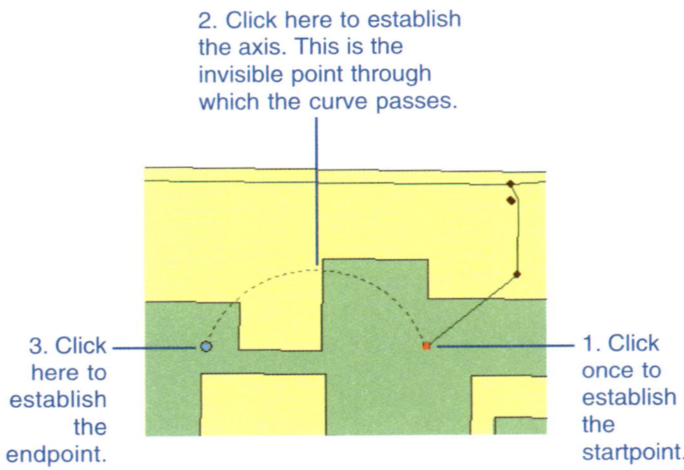

The Endpoint Arc tool, like the Arc tool, helps you create a segment that is a circular arc. The Endpoint Arc tool allows you to specify the start and endpoints of the curve, then define a radius for the curve.

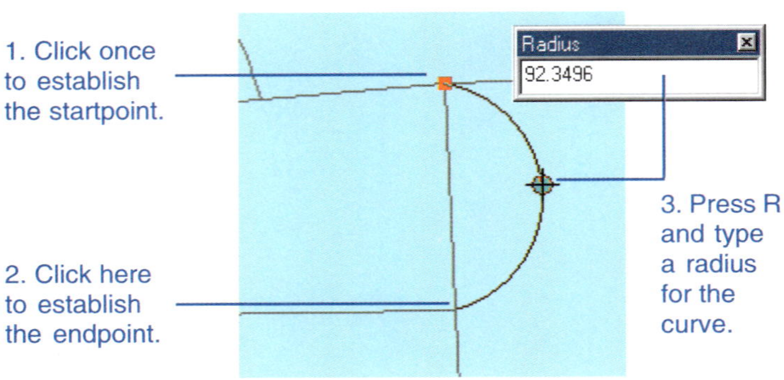

The Tangent Curve tool helps you create a segment that is a circular arc. This tool adds a segment that is tangential to the previously sketched segment. The Tangent Curve tool can only be used if you have already sketched a segment using one of the other sketch tools.

Once a line segment has been sketched, the Tangent Curve tool will draw a circular arc off of the previous line segment.

The Trace tool helps you create segments that follow along existing segments. Suppose you want to add a new road casing feature that is offset 15 feet from the front of a parcel subdivision.

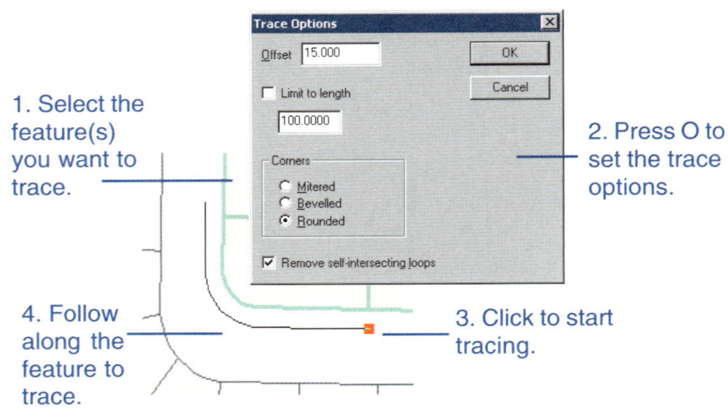

You could use the Trace tool to trace along the existing line features instead of typing the angle and length of each segment.

You can use any combination of the following methods for creating vertices or segments to create a new line or polygon feature:

- Sketch tool
- Sketch tool context menu
- Distance–Distance tool
- Direction–Distance tool
- Intersection tool
- Midpoint tool
- Arc tool
- Endpoint Arc tool
- Tangent Curve tool
- Trace tool

ArcMap has another context menu—the Sketch context menu—that works more directly with the sketch as a whole. With this menu, you can add, move, or delete vertices; switch the direction of the sketch; reduce its length; or display the properties of the sketch shape.

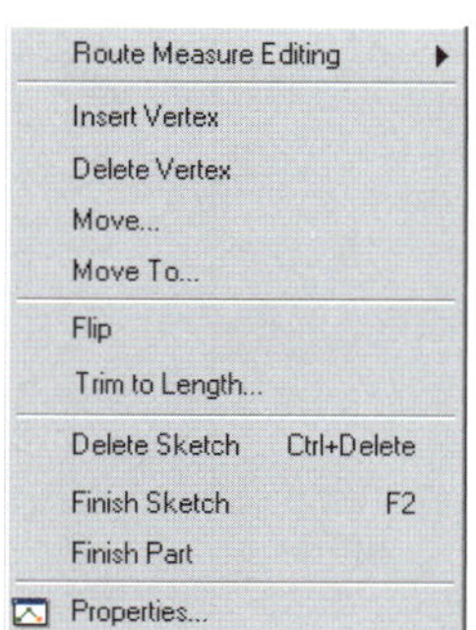

Sketch context menu

From the properties dialog box, you can remove parts from a multipart feature, remove many vertices in one operation, add points, and/or modify x,y values as well as m- and z-values. The Sketch context menu is available when you right-click while the pointer is positioned over any part of the sketch using any tool. It differs from the Sketch tool context menu, which you can access only when working with the Sketch tool and when you right-click away from your sketch.

Creating point features and vertices

You can think of vertices as being much the same as point features, except that vertices are connected by segments and make up line or polygon features.

Point features and vertices are created using the same methods. The Target layer setting determines whether you're creating a point feature or a vertex that is part of a line or polygon sketch.

You can create point features or vertices of a sketch in several different ways:

- By digitizing freehand with the Sketch tool (you can also use the snapping environment to help)
- By using Absolute X, Y or Delta X, Y on the Sketch tool context menu ▸

Tip

The snapping environment can help you create points and vertices

The snapping environment can help you create points or vertices at more exact locations relative to other features. For more information, see 'Using the snapping environment' in this chapter.

Creating a point or vertex by digitizing

1. Click the Current Task dropdown arrow and click Create New Feature.
2. Click the Target Layer dropdown list and click a point layer.
3. Click the tool palette dropdown arrow and click the Sketch tool.
4. Click on the map to create the point.

   The point is created on your map and marked as selected.

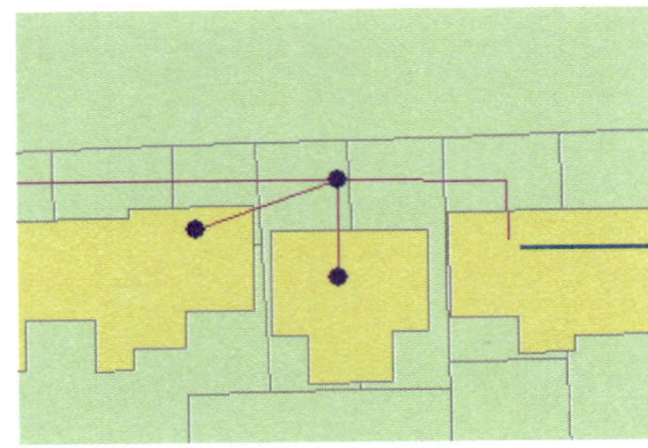

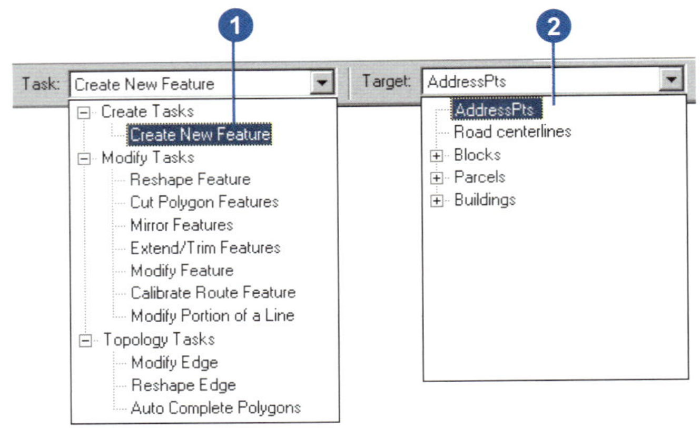

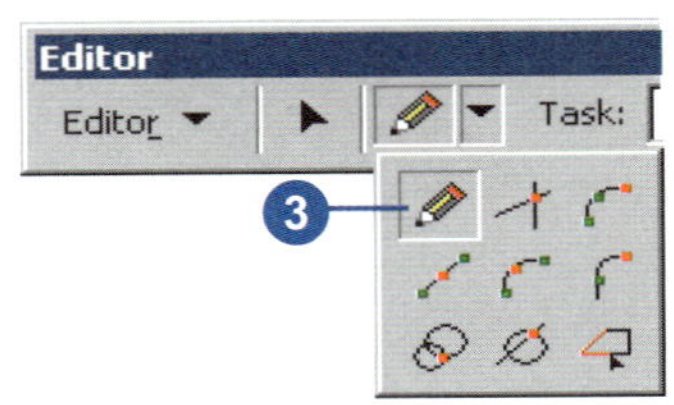

- By using the Distance–Distance, Direction–Distance, Midpoint, or Intersection tools

To create new features, you must have an existing layer to which you want to add them. If you do not, you can create one using ArcCatalog. For more information on creating a feature layer, see *Using ArcCatalog*.

To digitize freehand, simply click the Sketch tool and click on the map.

Absolute X, Y on the Sketch tool context menu lets you create a point or vertex at a specific location using the map's coordinate system. You might use Absolute X, Y to create a pole in a utility database if you have the x,y coordinates of a pole from using a global positioning system (GPS) unit. ►

Tip

Shortcut for Absolute X, Y

After clicking the Sketch tool, you can press F6 to set the x,y coordinates.

Tip

Closing the Sketch tool context menu

You can close the Sketch tool context menu by pressing the Esc key.

Creating a point or vertex using the coordinate system of the map (Absolute X, Y)

1. Click the tool palette dropdown arrow and click the Sketch tool.
2. Right-click anywhere on the map and click Absolute X, Y.
3. Type the coordinates and press Enter.

   A vertex or point is created at the specified coordinates.

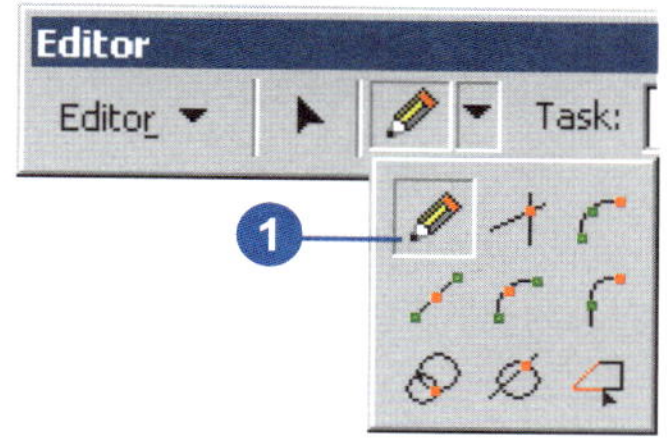

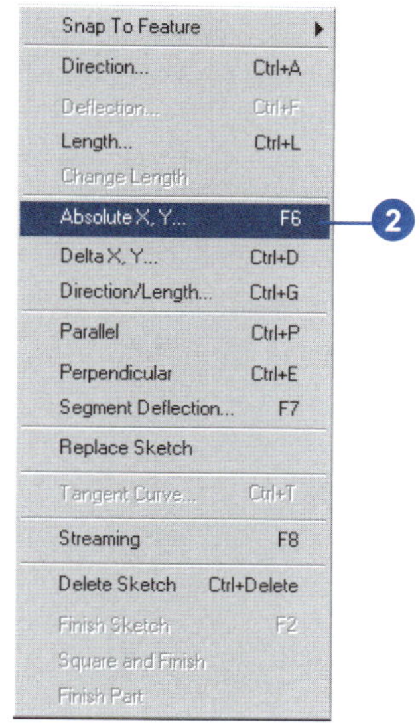

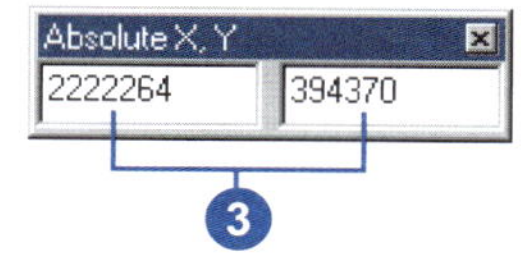

A point is created at (2222264,394370).

Delta X, Y on the Sketch tool context menu lets you create a vertex using the last vertex in the sketch as the origin. You can think of it as another way of measuring angle and length from a point already on the map.

For example, just as the red point in the diagram below can be measured at a distance of 20 feet from the last point at an angle of 53 degrees, it can also be measured in coordinates measured from the last point.

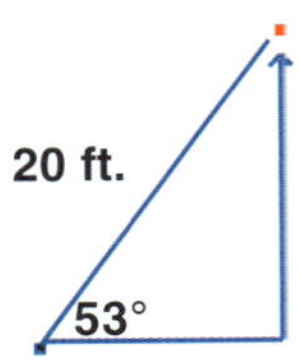

Point measured using an angle and length

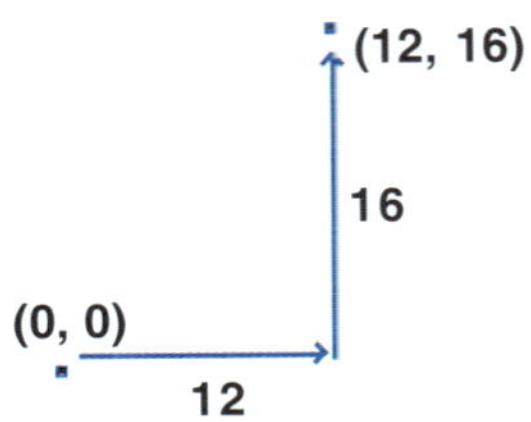

Same point measured using delta x,y coordinates

Tip

Shortcut for Delta X, Y

After clicking the Sketch tool, you can press Ctrl + D to set the delta x,y coordinates.

Creating a vertex relative to the location of the last vertex (Delta X,Y)

1. Click the tool palette dropdown arrow and click the Sketch tool after creating at least one vertex.
2. Right-click away from the vertex or sketch and click Delta X, Y.
3. Type the coordinates and press Enter.

   A vertex is created at the specified coordinates.

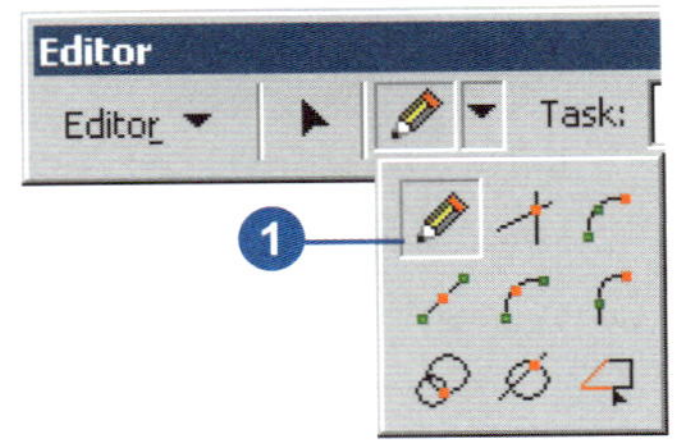

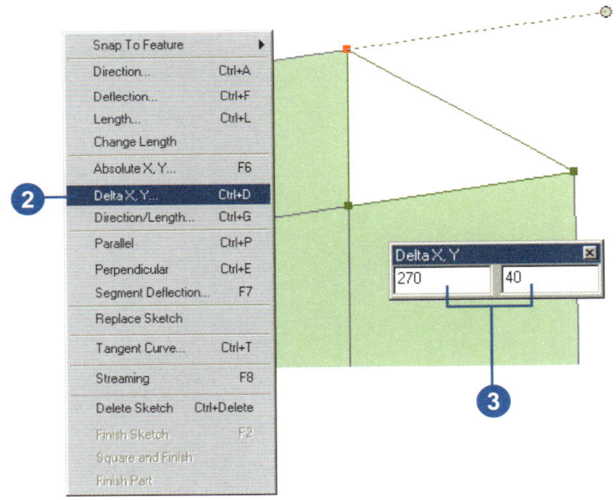

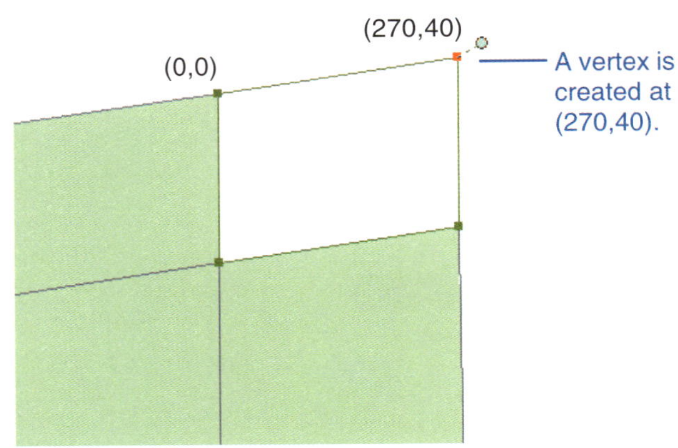

The Distance–Distance tool offers another way to create a point or vertex at a specific location. Suppose you want to create a pole feature. If you don't have the exact coordinate location but know that it is at the intersection of 50 map units from the corner of one building and 70 map units from the corner of another, you can use the Distance–Distance tool to place the point. The Distance–Distance tool lets you create a point or vertex at the intersection of two distances from two other points. ►

Tip

Undoing and redoing a vertex

You can undo any vertex you create by clicking the Undo button on the ArcMap Standard toolbar. Click the Redo button if you want to readd the vertex.

Creating a point or vertex using the Distance–Distance tool

1. Click the tool palette dropdown arrow and click the Distance–Distance tool.
2. Click once to establish the centerpoint of the first circle and press the D key on the keyboard.
3. Type the radius length for the first circle and press Enter.

   A circle is created with the specified radius. ►

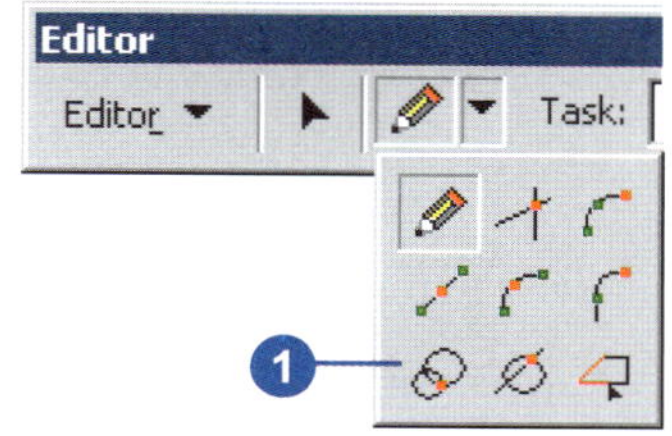

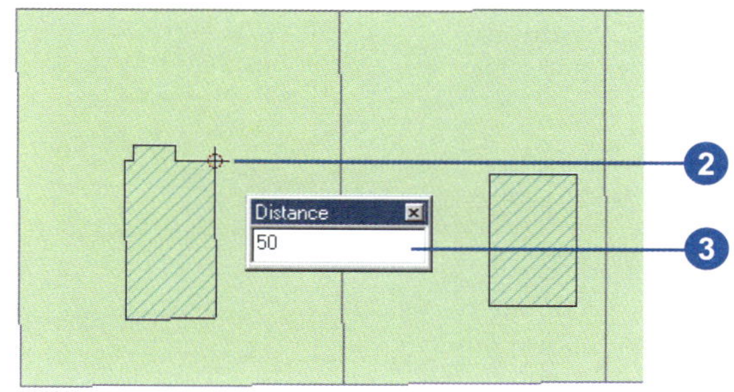

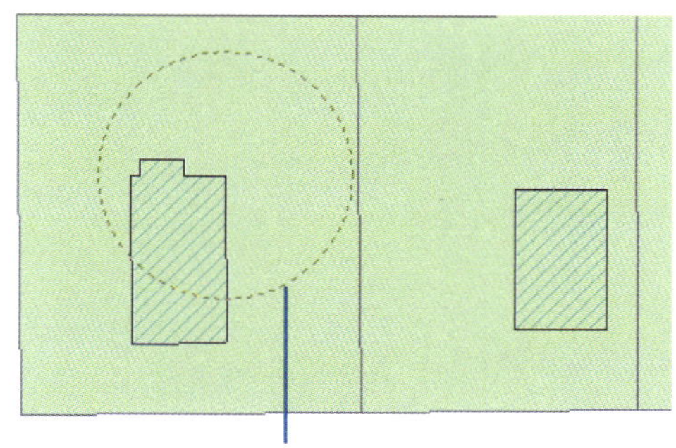

A circle with a 50-map unit radius is created.

As shown in the example, you'd create one circle with the centerpoint on the corner of the first building and a radius of 50 map units. You'd create another circle with the centerpoint on the corner of the other building and a radius of 70 map units. The Distance–Distance tool calculates the two locations where the radii of the circles intersect.

Tip

Choosing an intersection point

Press Tab to alternate between the two points of intersection and press Enter to create the point.

4. Click once to establish the centerpoint of the second circle and press the D key on the keyboard.
5. Type the radius length for the second circle and press Enter.

   A second circle is created with the specified radius. The two locations where the radii of the circles intersect are highlighted when you move the pointer over them.
6. Position the pointer over the location you want and click.

   A vertex or point is added to your map.

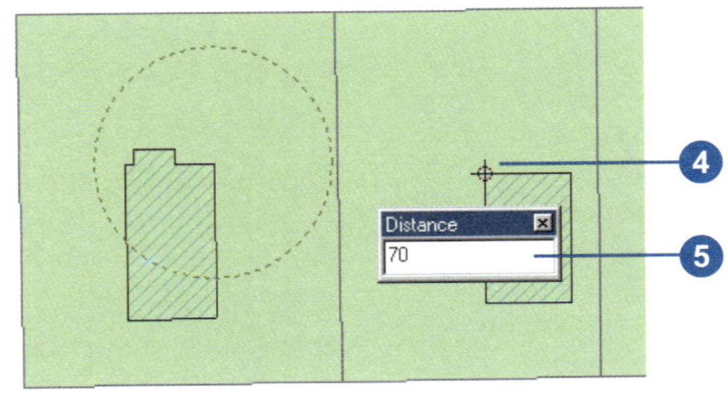

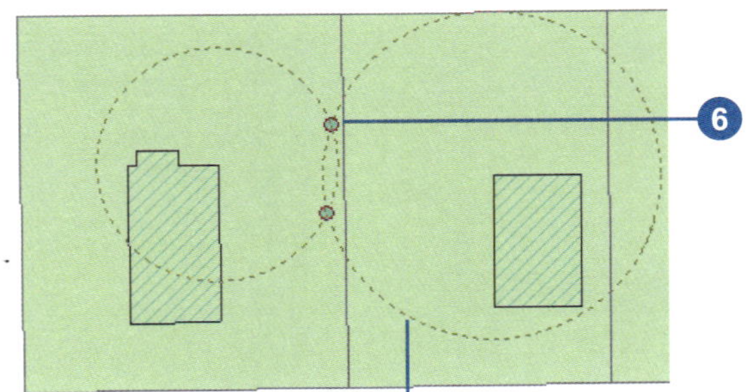

A circle with a 70-map unit radius is created.

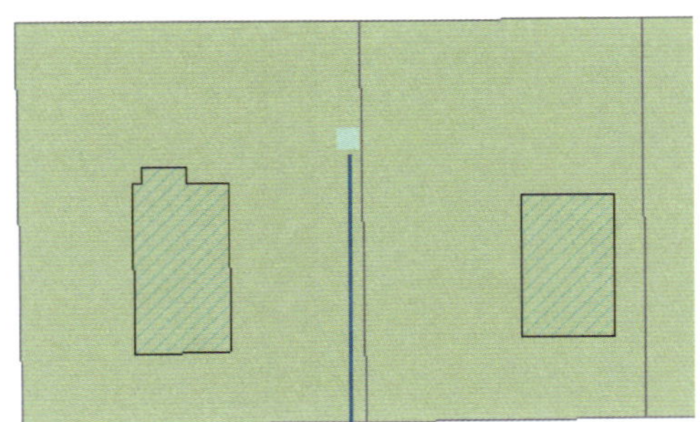

A point is created at one of two places where the radii of the circles intersect.

The Midpoint tool is ideal for placing a vertex directly between two known features. For example, you may want to place the road center line directly between two parcels or place an additional power pole directly between two preexisting power poles.

Tip

Pan and zoom while adding points

Hold down the Control key and press Z to zoom out, X to zoom in, or C to pan the display when using any sketch tool.

Creating a point or vertex using the Midpoint tool

1. Click the tool palette dropdown arrow and click the Midpoint tool.
2. Click once to establish the first of two points. The new vertex will be created between this point and the next point you click.
3. Click the second point.

   A vertex or point is placed at the midpoint of the line between the two points you clicked.

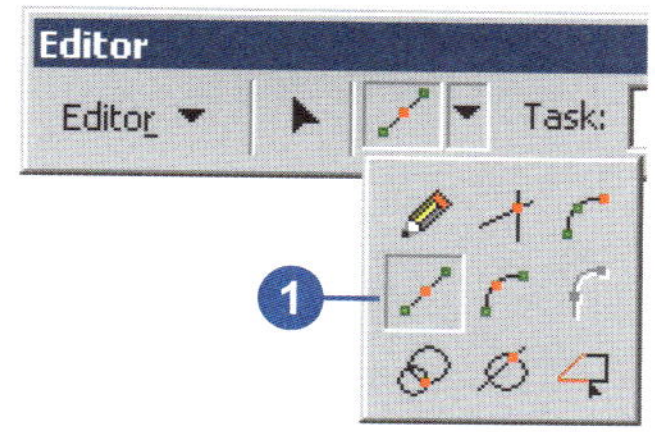

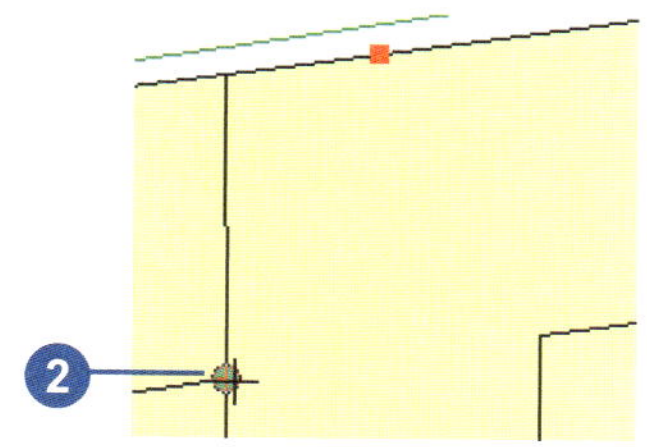

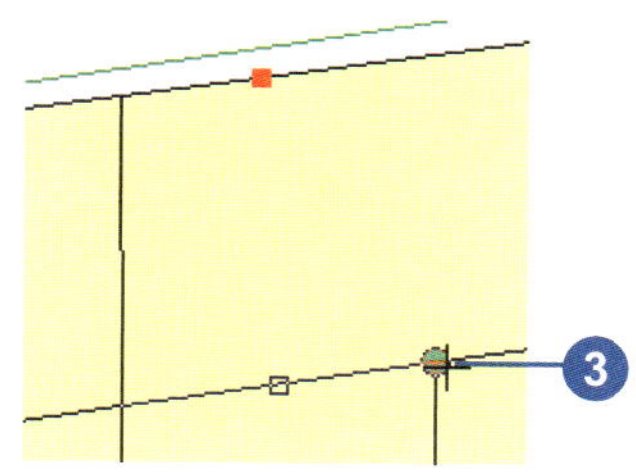

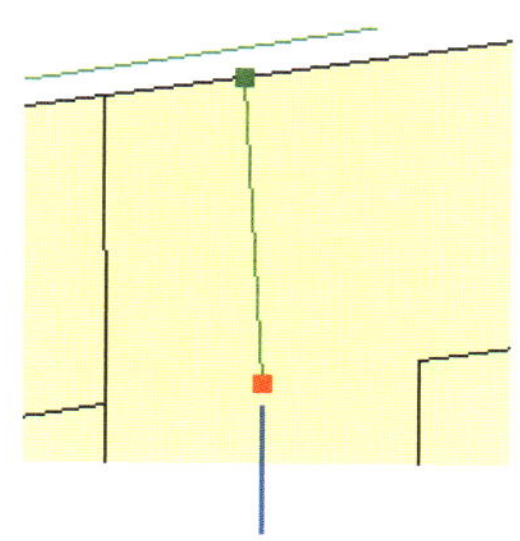

A point is created at the midpoint along the line defined by the two points you clicked.

The Intersection tool creates a point or vertex at the implied intersection of two segments. Implied means that the segments don't have to actually intersect on the map. In the example, suppose you want to create a new parcel. One corner of the parcel must be placed at the implied intersection of two segments of an adjoining parcel. You can use the Intersection tool to find this implied intersection point and create the corner vertex of the new parcel.

Creating a point or vertex using the Intersection tool

1. Click the tool palette dropdown arrow and click the Intersection tool.

   The pointer turns into crosshairs.

2. Position the crosshairs over the first segment and click.

3. Position the crosshairs over the second segment and click.

   A vertex or point is added at the implied intersection of the two segments.

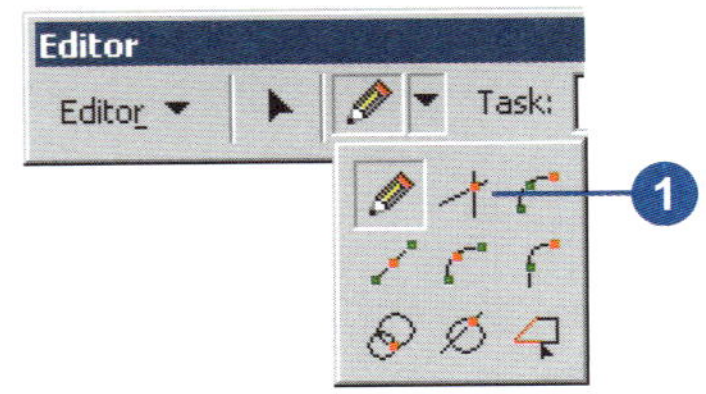

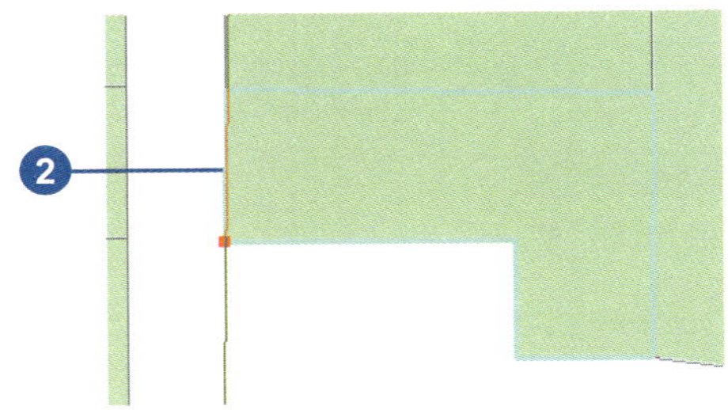

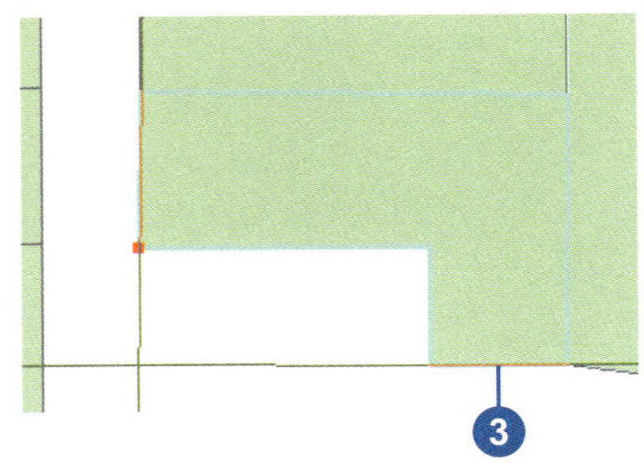

A vertex is added at the place where the two segments would intersect if extended.

Often a vertex location is only known according to the direction and the distance from two known features. For example, the location of a tree on a lot could be located at a particular direction from the particular corner of the lot and a given number of feet from another corner. In this situation, the Direction–Distance tool will place a vertex at the specified location.

As illustrated in this example, the direction is specified from one corner of the parcel. ▸

Creating a point or vertex using the Direction–Distance tool

1. Click the tool palette dropdown arrow and click the Direction–Distance tool.
2. Click a point to specify from where the direction to the next vertex was measured.
3. Move the pointer to get the approximate direction toward the next vertex.

   As you move the pointer the direction from the point you clicked to, the pointer is displayed in the lower-left corner of the ArcMap window.

   Press the D key and type the direction to the vertex from the point you clicked, then press Enter.

   You can also click again on the map to set the direction.
4. Click a point to specify from where the distance to the next vertex was measured. ▸

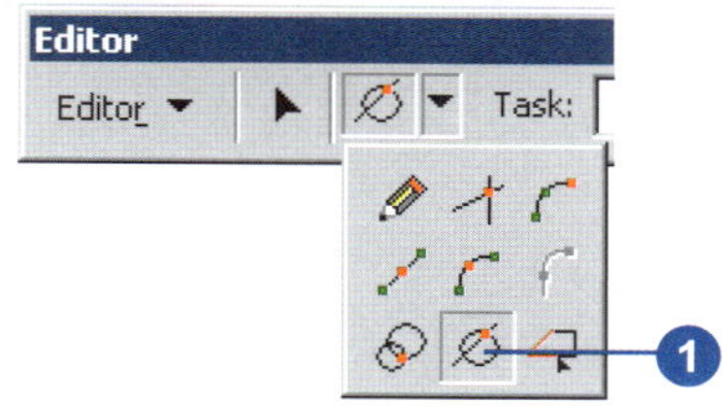

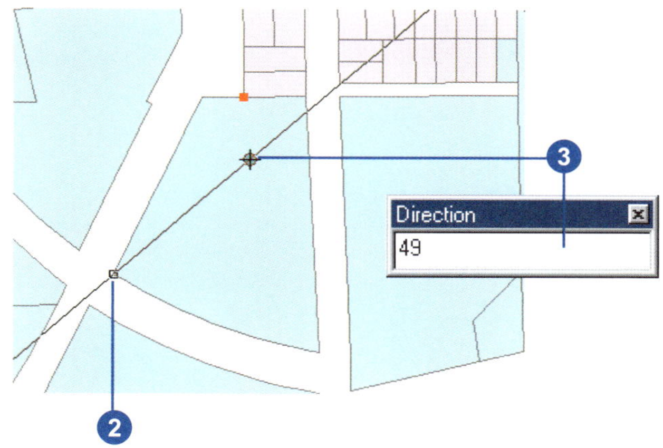

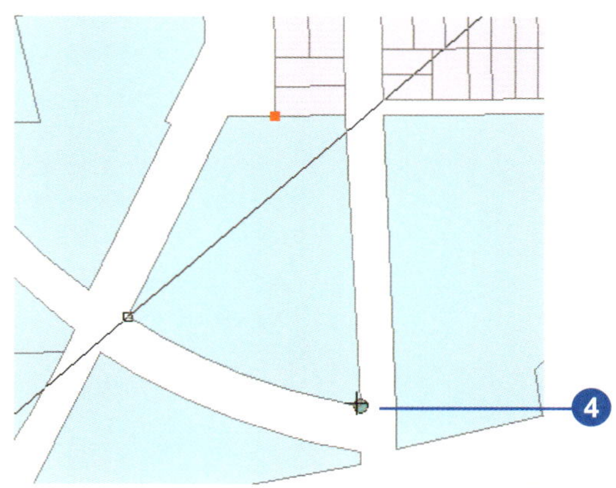

The distance is specified from another corner of the parcel, and the locations where both the direction and distance criteria are met, a potential vertex is placed. By clicking one of the two vertices, the new vertex is placed.

Tip

Determining map units

With many of these sketch tools you must specify a distance in map units. The map units are specified under the Coordinate System tab of the Data Frame Properties dialog box. This is found by clicking View on the main menu, then clicking Data Frame Properties.

5. Move the pointer to get the approximate distance from the point you clicked to the next vertex.

   As the cursor is moved, a circle, centered at the second point you clicked, is dragged out to intersect the direction line you just defined. The radius of the circle is displayed in the lower-left corner.

   Optionally, you can press the D key and type a distance, in map units, to the next vertex from the point you specified. Pressing the Enter key will set the size of the circle.

   The new vertex will be at one of the intesection points of this circle and the direction line.

6. Click the intersection of the circle and the direction line which corresponds to the position of the new vertex.

   The intersection point nearest the cursor will be highlighted and, once clicked, the new vertex is placed.

   Optionally, you can use the Tab key to switch between the two intersection points and press Enter to select one of them.

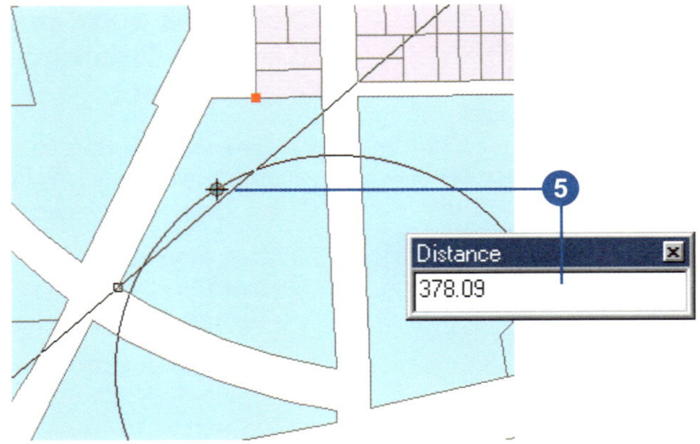

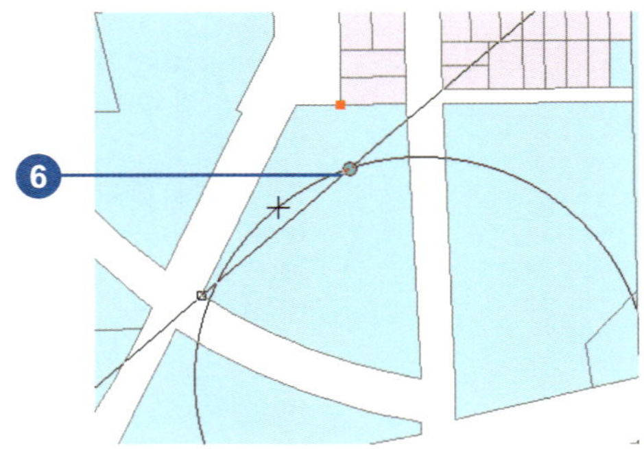

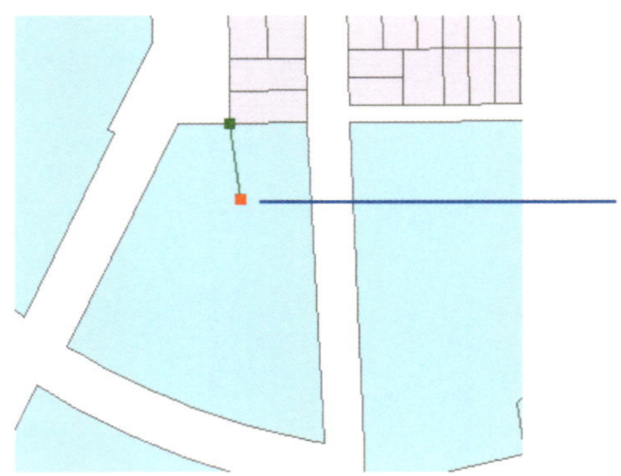

A vertex is added at the intersection you chose. It is at the angle you specified from the first point and the distance you specified from the second point.

You can also create a *multipoint feature*, a feature that consists of more than one point but only references one set of attributes in the database. For example, a system of oil wells could be created as a multipoint feature; the database references a single set of attributes for the main well and the multiple well holes in the system.

Tip

Creating multipoint layers

When creating multipoint features, your target layer must also be a multipoint feature class.

Creating a multipoint feature

1. Click the Task dropdown arrow and click Create New Feature.
2. Click the Target layer dropdown arrow and click a multipoint layer.
3. Click the tool palette dropdown arrow and click the Sketch tool.
4. Click the map to create parts of the multipoint feature.
5. Right-click anywhere on the map when you have created the last point of the multipoint feature and click Finish Sketch. ▸

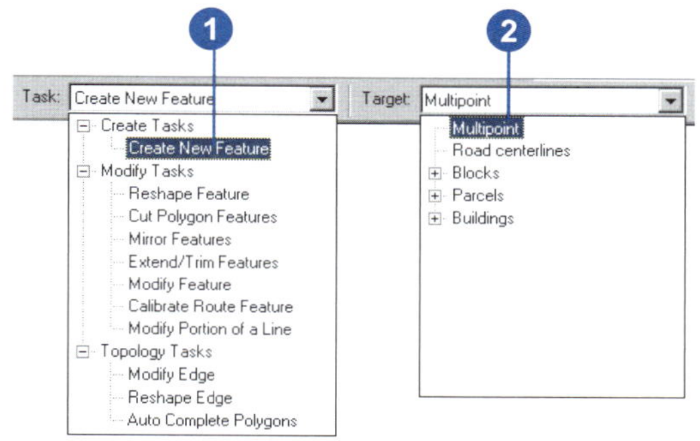

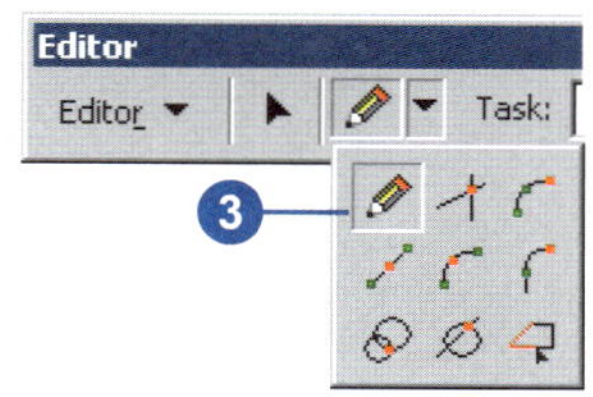

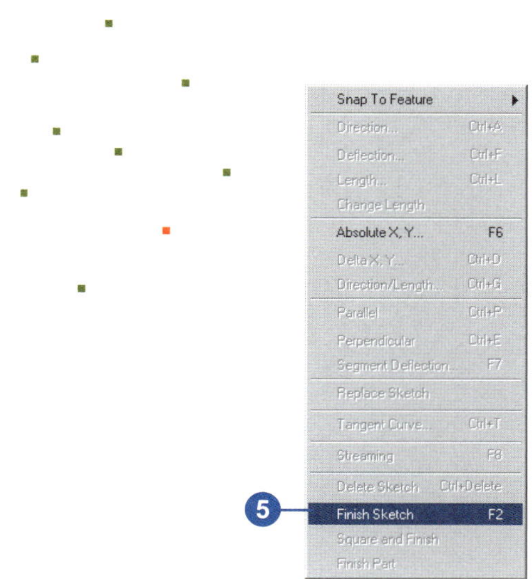

Tip

Modifying existing features

Double-click a feature or select single features and change the task in the Editor toolbar to Modify Feature. Use the Sketch Properties dialog box to edit vertices. You can open the sketch properties by clicking the Sketch Properties button on the Editor toolbar.

Now when you click one part of the multipoint feature to select it, all points are automatically selected because they all belong to one multipoint feature.

All points of the feature are selected.

Creating lines and polygons

You can create lines or polygons by digitizing the vertices that make up the feature. For example, to create a square building, you would digitize the four corners. Using any combination of methods for creating vertices or segments.

By selecting in the target list a feature class that is represented by lines or polygons, the editor tools will create the corresponding feature type.

Tip

Shortcut for finishing the sketch

You can double-click the last vertex of the feature to finish the sketch. Or, press F2 when you've finished creating the sketch.

Creating a line or polygon feature by digitizing

1. Click the Task dropdown arrow and click Create New Feature.
2. Click the Target layer dropdown arrow and click a line or polygon layer.
3. Click the tool palette dropdown arrow and click the Sketch tool.
4. Click the map to digitize the feature's vertices.
5. Right-click anywhere on the map when finished and click Finish Sketch. ▸

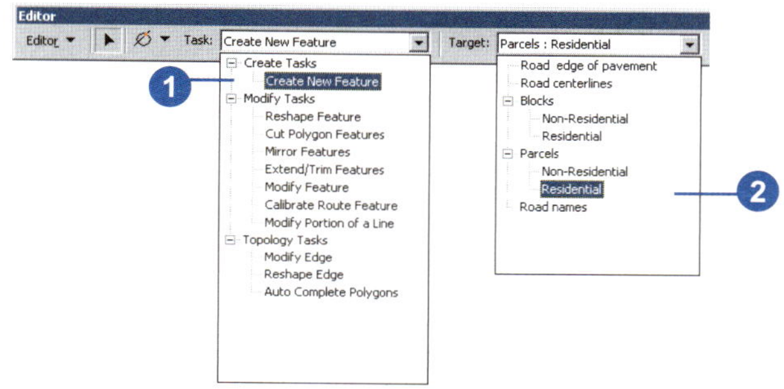

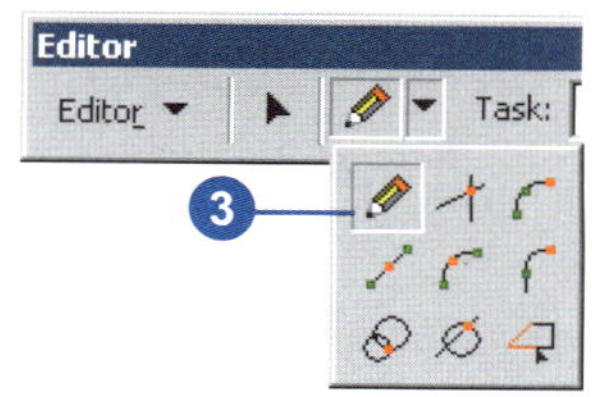

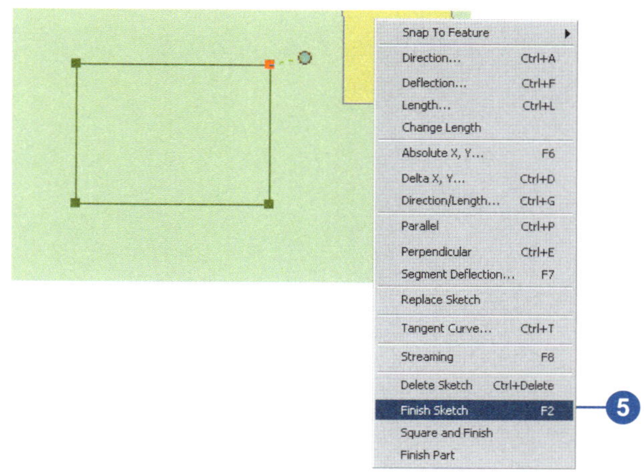

Tip

Deleting a vertex

To delete a single vertex from a sketch, center the pointer over the vertex, right-click, then click Delete Vertex.

Tip

Deleting the sketch

To delete the entire sketch of the feature you are creating, position the pointer over any part of the map, right-click, and click Delete Sketch. You can also delete a sketch by pressing Ctrl + Delete.

If the sketch has already been finished, select the feature with the Edit tool, right-click, and click Delete Sketch.

The line or polygon is created on your map.

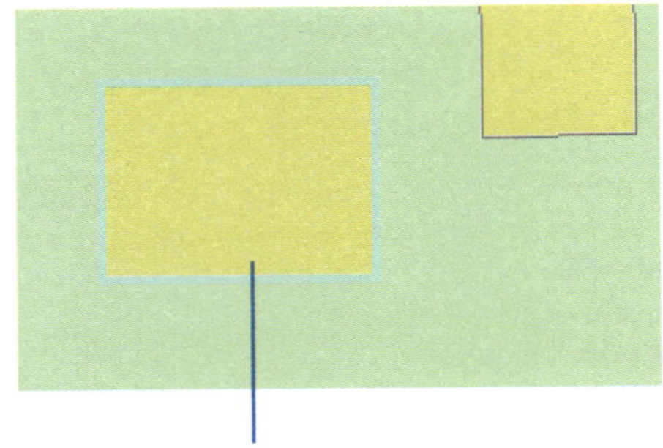

A new feature is created on your map.

ArcMap also provides a way to create a *multipart feature*, a feature that is composed of more than one physical part but only references one set of attributes in the database. For example, the State of Hawaii could be considered a multipart feature. Although composed of many islands, it would be recorded as one feature. A multipart feature can only share vertices, not edges.

Tip

Replace sketch

You can add the shape of a line or polygon feature to the sketch by right-clicking over the feature with the Sketch tool and clicking Replace Sketch. The sketch will contain the shape of the feature you clicked over. Using the Edit tool, the sketch can be dragged and dropped anywhere on the map.

Tip

Undoing and redoing a vertex

You can undo the last vertex you created by clicking the Undo button on the ArcMap Standard toolbar. Click the button again to undo the second-to-last vertex you created, and so on. Click the Redo button if you want to readd the vertex.

Creating a multipart line or polygon

1. Create a line or polygon feature.
2. Right-click anywhere on the map when you have finished creating the first part of the feature and click Finish Part.
3. Create the next part of the feature.
4. Right-click anywhere on the map when you have finished the last part of the feature and click Finish Sketch. ▸

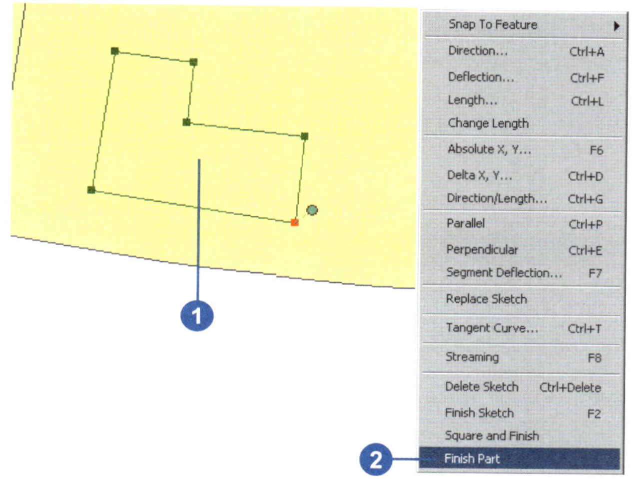

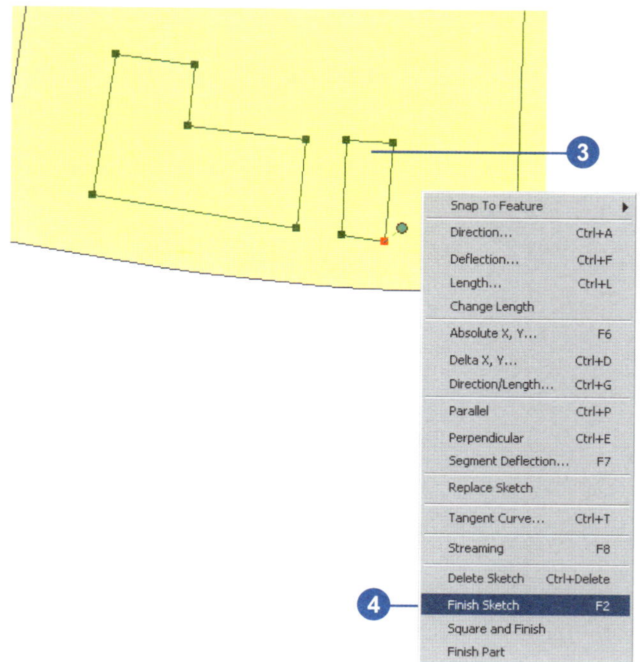

Tip

Shortcut for finishing the sketch

You can double-click the last vertex of the new feature to finish the sketch.

Tip

Shortcut for finishing a part

When creating a multipart sketch, you can hold down the Shift key and double-click on the last vertex of a part to finish it.

Now when you click one part of the feature to select it, all parts are automatically selected because they all belong to one multipart feature.

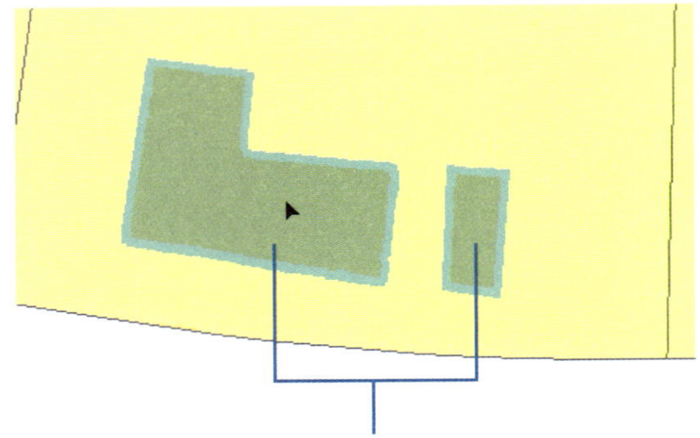

Both parts of the feature are selected.

The Square and Finish command on the Sketch tool context menu is a way of completing a polygon. It finishes a polygon by adding two new segments at 90-degree angles. Square and Finish saves you time and ensures precision when creating square-cornered buildings.

Squaring a polygon or polyline

1. Click the Current Task dropdown arrow and click Create New Feature.
2. Click the Target layer dropdown arrow and click a polygon or polyline layer.
3. Click the tool palette dropdown arrow and click the Sketch tool.
4. Digitize at least two segments.
5. Right-click anywhere away from the sketch and click Square and Finish. ▶

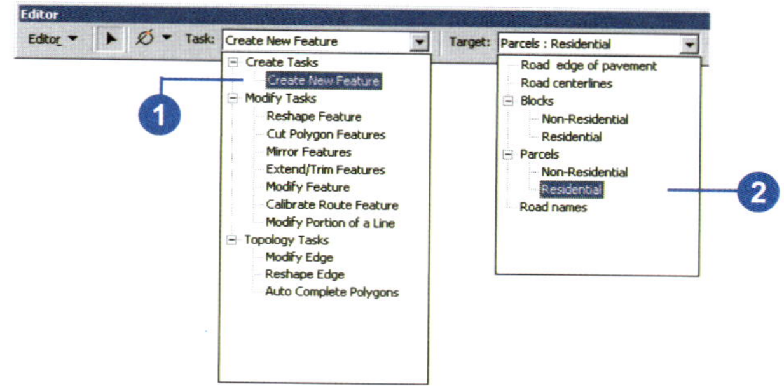

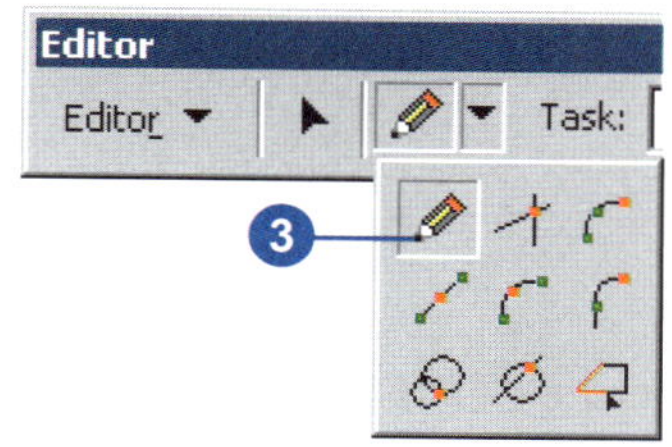

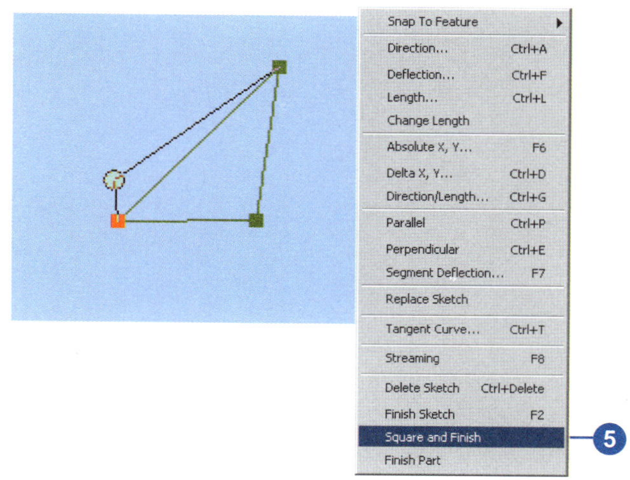

Tip

Streaming

You can also create lines and polygons with the mouse using stream mode digitizing (streaming). For more information, see Chapter 5, 'Using a digitizer'.

Tip

Adding attributes to created features

To add attributes to the newly created features, with the Edit tool, right-click the feature and click Attributes. The Attributes dialog box will open, which allows you to modify the attributes.

The angles from the first vertex and the last vertex are squared. A new vertex is added, and the sketch is finished where the resulting segments intersect.

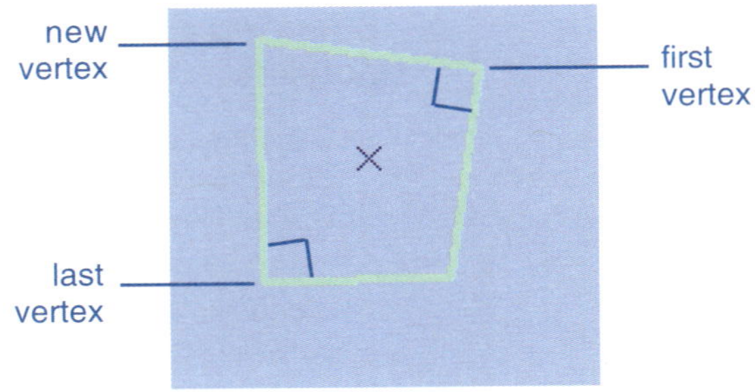

Creating segments using directions and lengths

The edit tools help you create segments in specific directions, measured either using the map coordinate system (Direction) or from the last segment (Deflection).

The Direction command uses east as 0 degrees and measures positive angles counterclockwise by default. For example, a 90-degree angle represents north and a 180-degree angle represents west. You can change the direction measuring system and angular units on the Units tab of the Editing Options dialog box.

The Deflection command uses the last segment as 0 degrees and calculates the angle you specify from there. Positive values are calculated in a counterclockwise direction from the existing segment, while negative values are calculated clockwise. ▸

Tip

Shortcut for direction angle

After clicking the Sketch tool and creating at least one vertex, you can press Ctrl + A to set the direction angle.

Creating a segment using an angle and a length

1. Click the tool palette dropdown arrow and click the Sketch tool after creating at least one vertex.
2. Right-click away from the sketch and click Direction.
3. Type the direction and press Enter.

   The segment is constrained to the specified direction. ▸

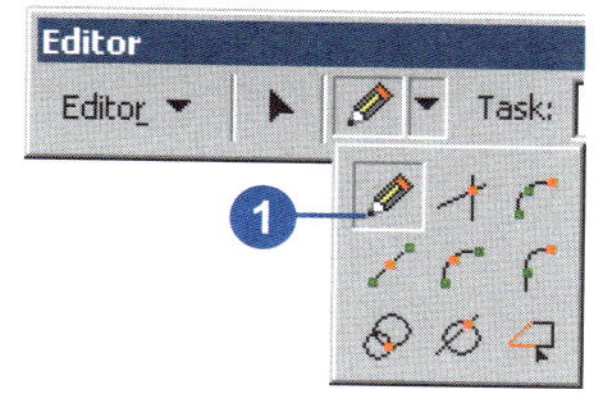

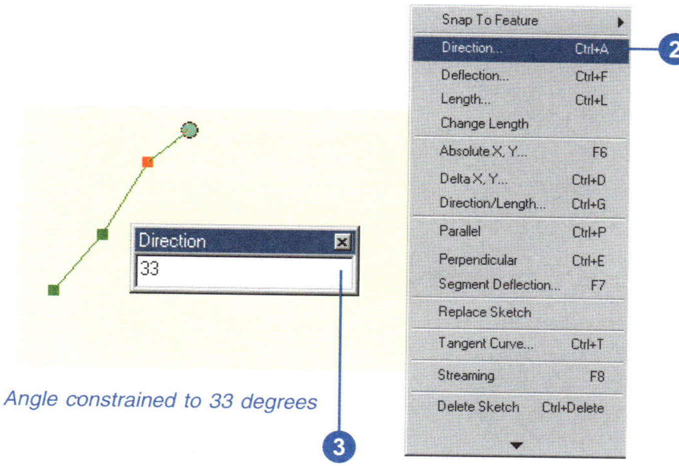

Angle constrained to 33 degrees

You might use Deflection to create the bent end of a water or gas line at a 33-degree angle to a house.

Both the Direction and Deflection commands constrain the angle of the segment. For example, if you type 45 as the Direction, the segment will be constrained to a 45-degree angle one way and a 225-degree angle the other.

Use the Length command to specify the length of a segment you're creating.

Tip

Shortcut for length

After clicking the Sketch tool and creating at least one vertex, you can press Ctrl + L to set the length.

Tip

Changing the length of a segment

If you want to change the length of a segment you have already created, you can use Change Length on the Sketch tool context menu. This undoes the last vertex while keeping the direction constraint.

4. Right-click anywhere on the map and click Length.
5. Type the length and press Enter.

   The vertex that makes the segment the desired angle and length is created.

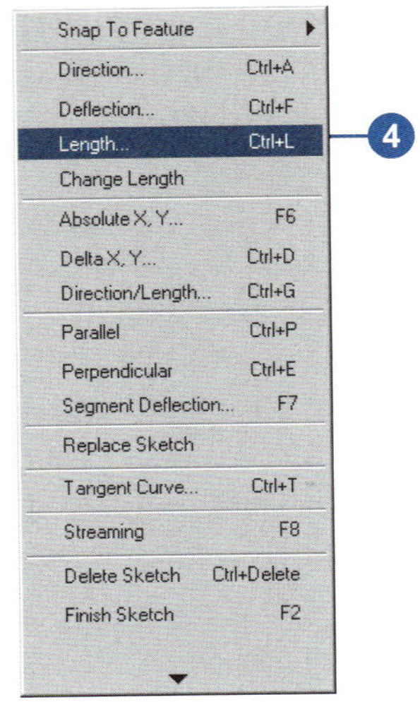

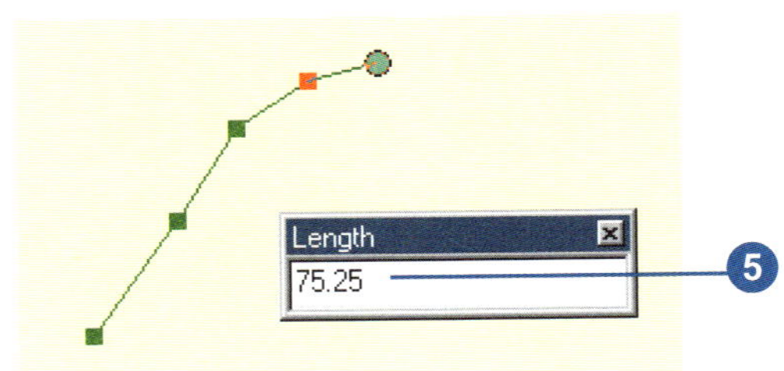

The vertex that makes the segment the desired angle and length is created.

Tip

Shortcut for deflection

After clicking the Sketch tool and creating at least one vertex, you can press Ctrl + F to set the deflection angle.

Creating a segment at an angle from the last segment (deflection)

1. Click the tool palette dropdown arrow and click the Sketch tool after creating at least one vertex for the new segment.
2. Right-click away from the sketch.
3. Click Deflection.
4. Type the desired angle from the last segment and press Enter. ►

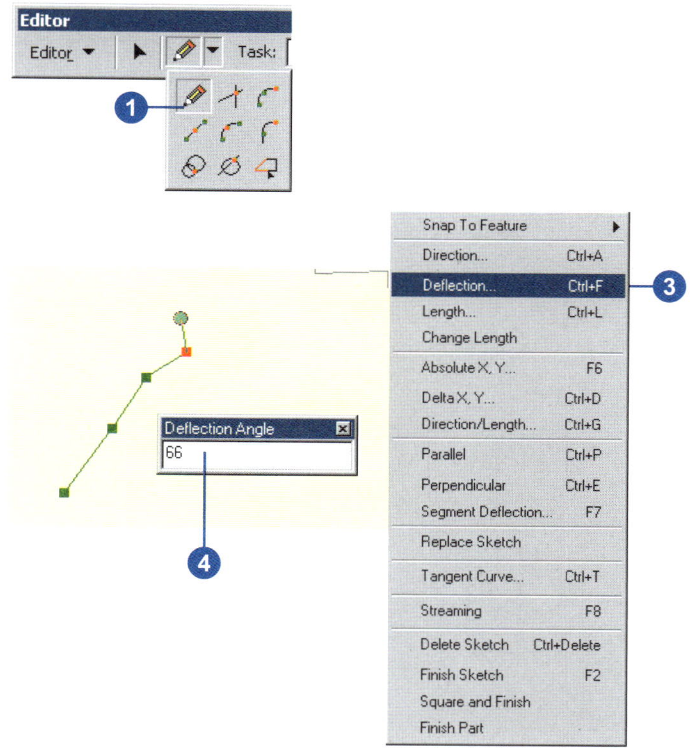

The segment is constrained to the specified angle.

5. Click once to digitize the endpoint of the segment or choose Length from the Sketch tool context menu.

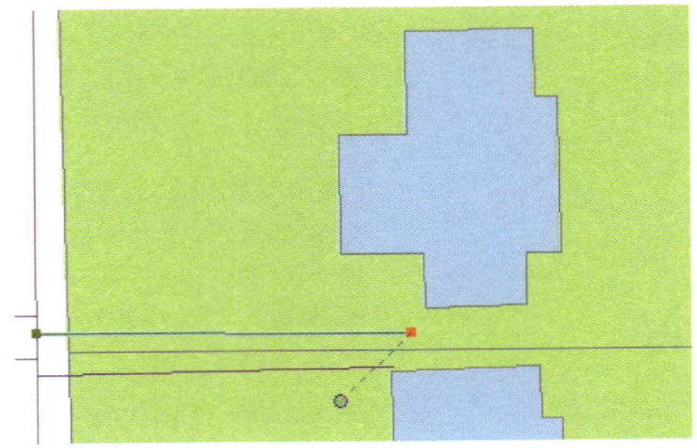

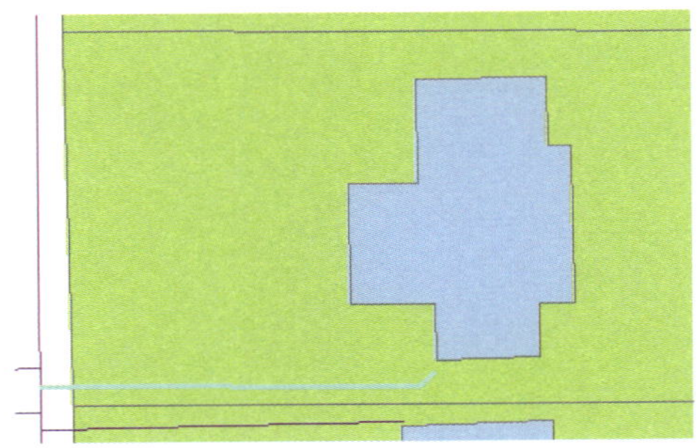

Creating segments using angles from existing segments

Three commands on the Sketch tool context menu—Segment Deflection, Parallel, and Perpendicular—help create segments with angles relative to segments that already exist.

The Segment Deflection command lets you create a segment at an angle relative to any existing segment. While Deflection creates a segment at a specific angle from the last segment in the sketch you're creating, Segment Deflection lets you choose a segment in an existing feature.

As with the Deflection command, the segment you work from with Segment Deflection is 0 degrees, and the deflection ▶

Tip

Shortcut for segment deflection

After clicking the Sketch tool, creating at least one vertex, and positioning the pointer over the segment from which you want to be the specific angle, you can press F7 to set the angle.

Creating a segment at an angle from any other segment

1. Click the tool palette dropdown arrow and click the Sketch tool after creating at least one vertex.
2. Position the pointer over the segment of a feature class from which you want to create a segment and right-click with the mouse.
3. Click Segment Deflection.
4. Type the desired angle from the segment you chose and press Enter. ▶

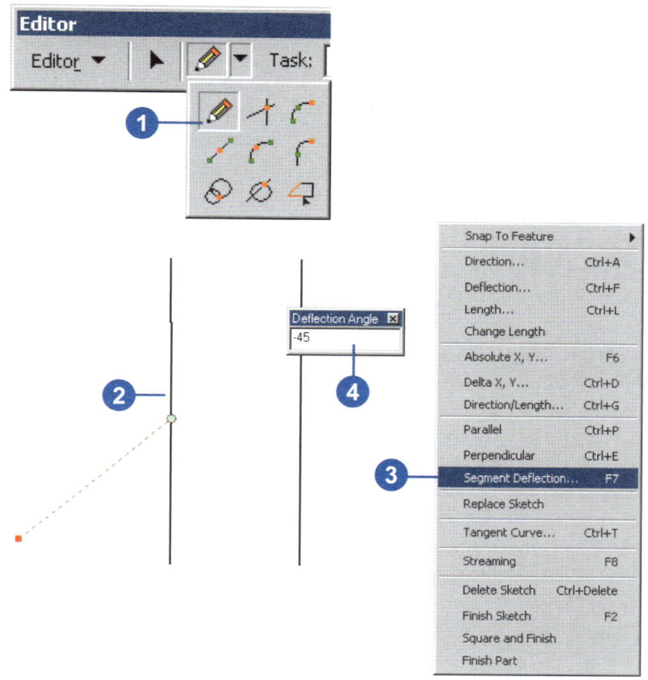

angle you specify for the new segment is calculated from there. Positive values are calculated in a counterclockwise direction from the existing segment, while negative values are calculated clockwise. The example given shows a cross street created at a -45-degree angle to the existing streets.

Tip

Using only positive values with segment deflection

If you wish to work only with positive angle values, convert negative angles to positive angles by adding 180 to the negative value. For example, a -45-degree angle, measured clockwise, becomes a 135-degree angle, measured counterclockwise.

The segment is constrained to the specified angle.

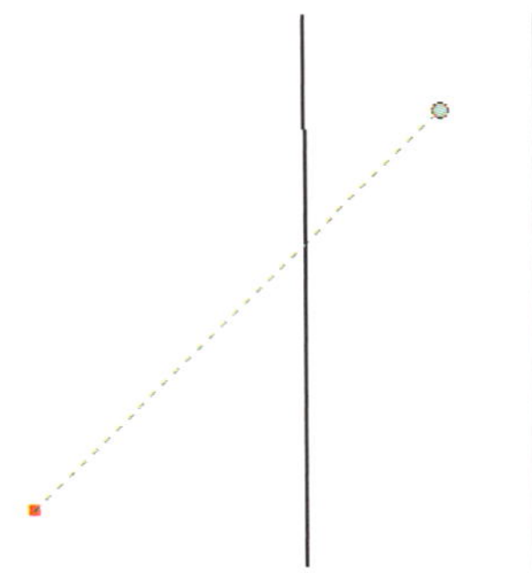

5. Click once to digitize the endpoint of the segment or choose Length from the Sketch tool context menu.

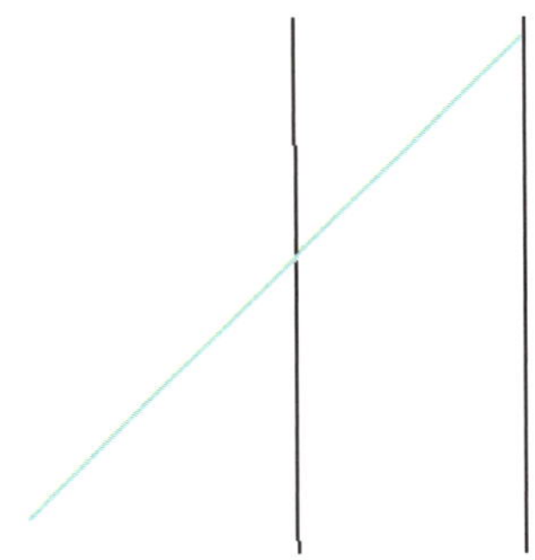

The Parallel command on the Sketch tool context menu constrains a segment to be parallel to any segment you choose. For instance, you might use this command to create a gas main line parallel to the street.

The Perpendicular command on the Sketch tool context menu constrains a segment to be perpendicular to an existing segment. You might use this command to place a service line perpendicular to the main line.

Tip

Shortcut for parallel

After clicking the Sketch tool, creating at least one vertex, and positioning the pointer over the segment to which the new segment will be parallel, you can press Ctrl + P to make the segment parallel.

Creating a segment parallel to another segment

1. Click the tool palette dropdown arrow and click the Sketch tool after creating at least one vertex.
2. Position the pointer over the segment to which the new segment will be parallel and right-click.
3. Click Parallel.

   The segment is constrained to be parallel to the specified segment.
4. Click once to digitize the endpoint of the segment or choose Length from the Sketch tool context menu.

Shortcut for perpendicular

After clicking the Sketch tool, creating at least one vertex, and positioning the pointer over the segment to which the new segment will be perpendicular, you can press Ctrl + E to make the segment perpendicular.

Creating a segment perpendicular to another segment

1. Click the tool palette dropdown arrow and click the Sketch tool after creating at least one vertex.
2. Position the pointer over the segment to which the new segment will be perpendicular and right-click with the mouse.
3. Click Perpendicular.

   The segment is constrained to be perpendicular to the specified segment.
4. Click once to digitize the endpoint of the segment or choose Length from the Sketch tool context menu.

Creating segments that are circular arc curves

When creating features, it is often necessary to create a circular arc. Instead of being made of numerous vertices, a circular arc has only two vertices as endpoints. ArcMap offers four ways to create a segment that is a circular arc. These include the Arc tool, the Endpoint Arc tool, the Tangent Curve tool, and the Tangent Curve command.

First, you can create a circular arc using the Arc tool. You might use the Arc tool to digitize a cul-de-sac using an aerial photo image as a backdrop.

A circular arc can also be created using the Endpoint Arc tool. This tool allows you to place the vertices for both ends of the arc and adjust the radius.

The Tangent Arc tool creates a circular arc based on the previously sketched segment.

You can also create a circular arc using the Tangent Curve command on the Sketch tool context menu. You can use the Tangent Curve command to ▸

Creating a segment that is a circular arc using the Arc tool

1. Click the tool palette dropdown arrow and click the Arc tool.
2. Click once to establish the starting point of the arc.

   A vertex is created.
3. Click once to establish the axis of the arc.

   This is the invisible point through which the curve passes.
4. Click once again to establish the endpoint of the arc.

   A segment that is a true curve is created.

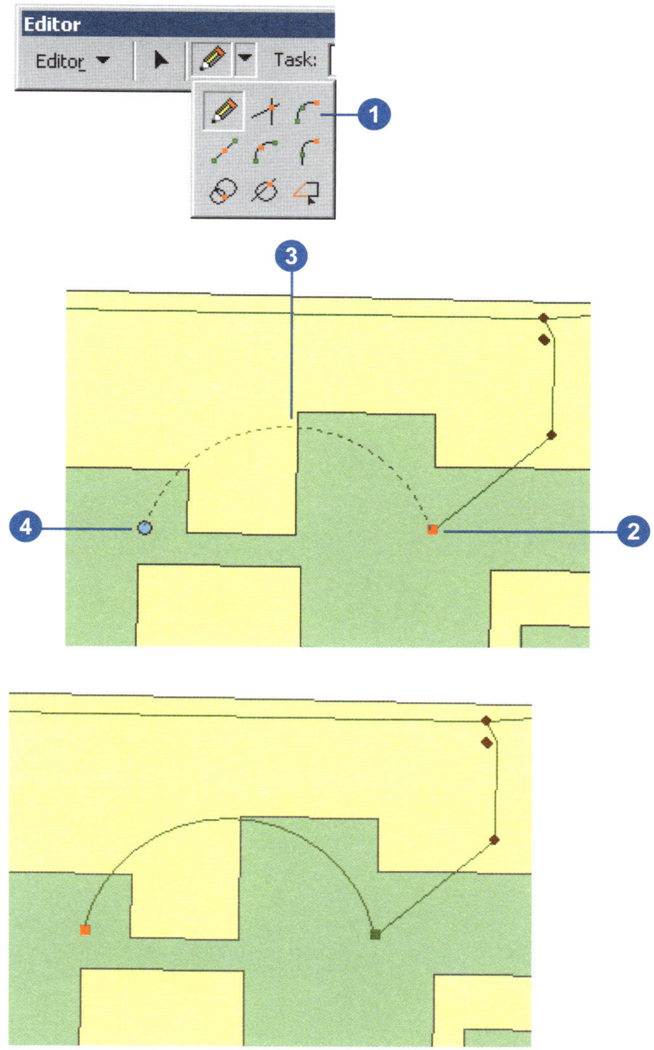

A segment that is a true curve is created.

add a circular arc to an existing segment. For example, you might use this command to add a curved segment to extend a centerline along a curved road.

When you create a tangent curve, you must specify two parameters for the curve from the following options: arc length, chord, radius, or delta angle. You must also specify whether you want to create the curve to the right of the line or to the left of the line, according to the direction in which the line was drawn. The curve is created from the last vertex of the existing segment based on the parameters you defined.

If you choose chord length and radius to construct the curve, there are two possible solutions: the major and minor portions of the circle.

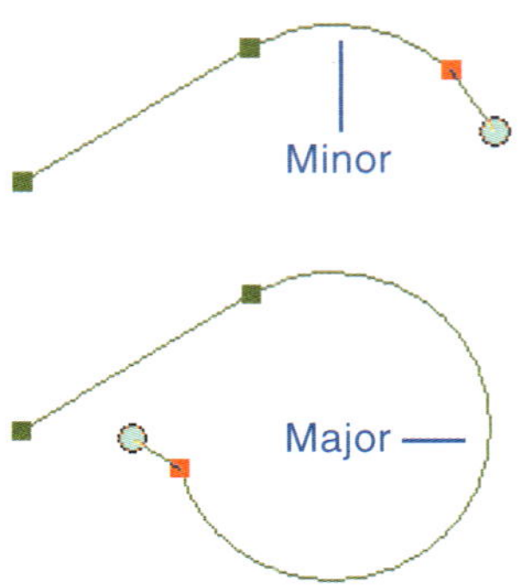

Creating a segment that is a circular arc using the Endpoint Arc sketch tool

1. Click the tool palette dropdown arrow and click the Endpoint Arc tool.
2. Click the starting point of the arc.
3. Click the endpoint of the arc.
4. Move the pointer to get the approximate radius for the curve.

   Press the R key and type the radius for the curve.

   You can also click again on the map to set the radius.

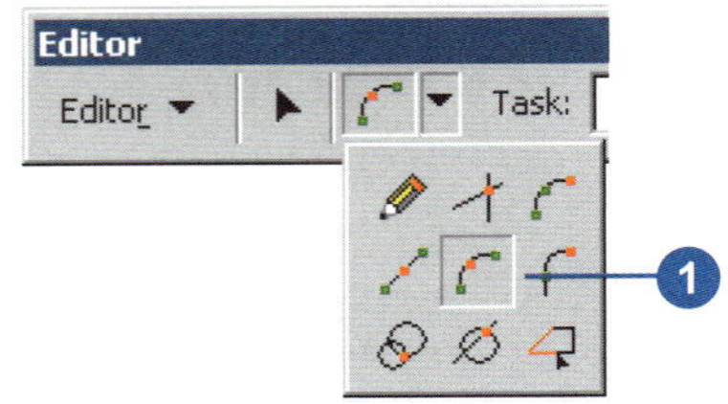

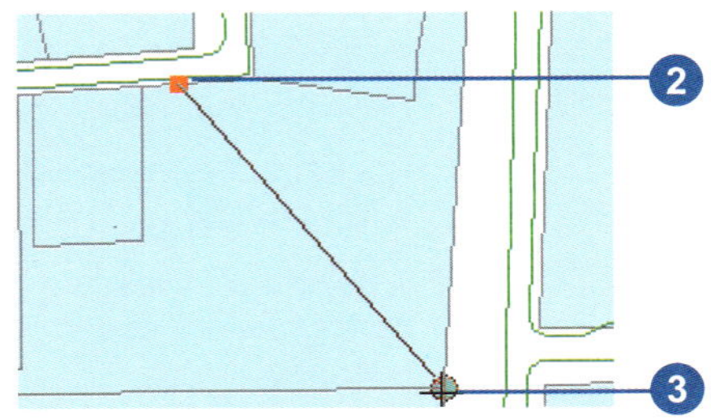

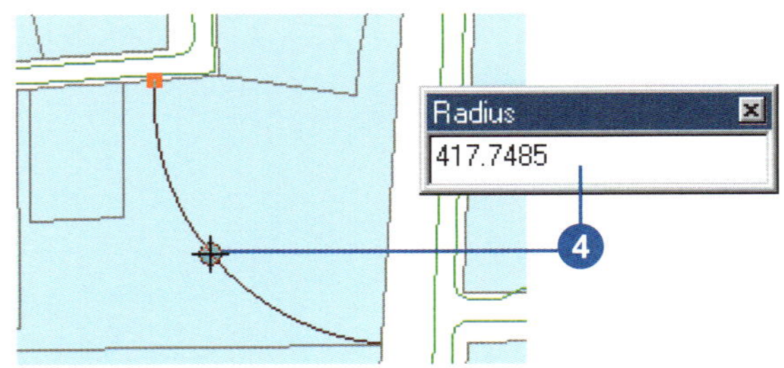

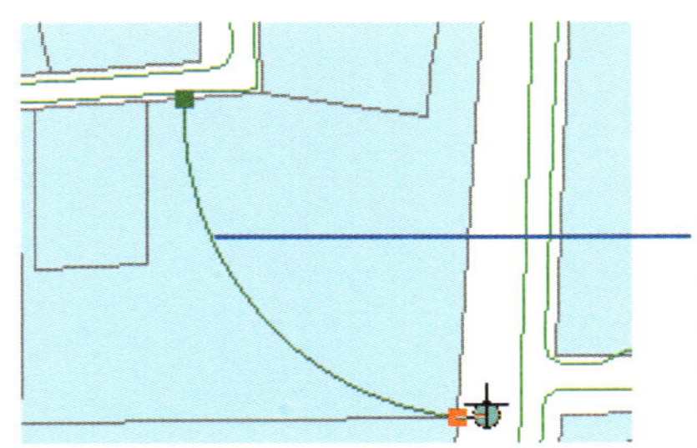

A circular arc with the radius you specified is created between the two points you clicked.

The Minor check box will appear at the bottom of the Tangent Curve dialog box. Check it to construct the minor portion of the circle.

Tip

Differentiating between the Arc tools

The Arc tool: Creates a circular arc from three points that lie on the circular arc.

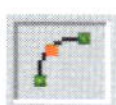

The Endpoint Arc tool: Creates a circular arc from two points on the circle and from the radius.

The Tangent Curve Tool: Creates an arc tangent to the previous segment in the sketch.

Creating a segment that is a circular arc using the Tangent Curve tool

1. Click the tool palette dropdown arrow and click the Tangent Curve tool once a segment has been created using one of the other sketch tools.

   As the cursor is moved, the arc will bend and change length to remain tangent to the previous segment.

2. Click again to place the endpoint of the arc.

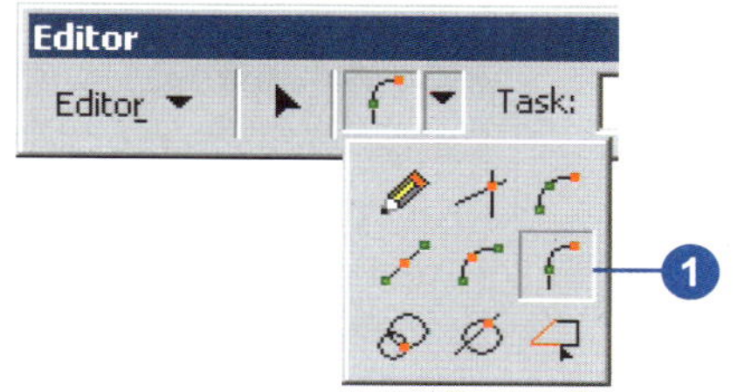

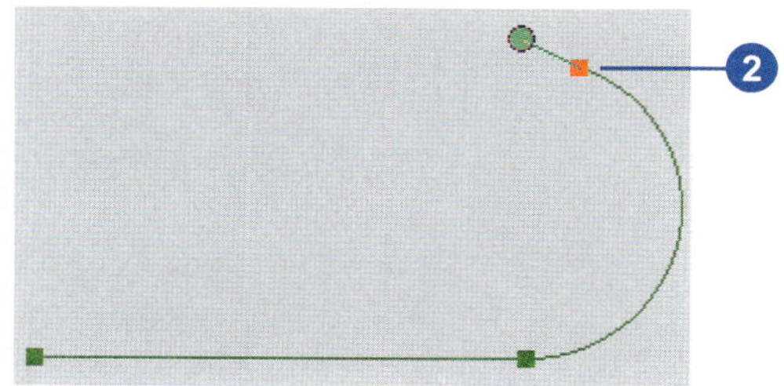

Creating a segment that is a circular arc using the Tangent Curve command

1. Click the tool palette dropdown arrow and click the Sketch tool after creating at least one segment.
2. Right-click anywhere on the map and click Tangent Curve.
3. Click the dropdown arrows and click two parameters by which you want to define the curve.
4. Type the appropriate values for the parameters (distance in map units for arc length, chord, and radius; degrees for delta angle).
5. Click Left to create the tangent curve to the left of the segment. Click Right to create the curve to the right.
6. Press Enter. ►

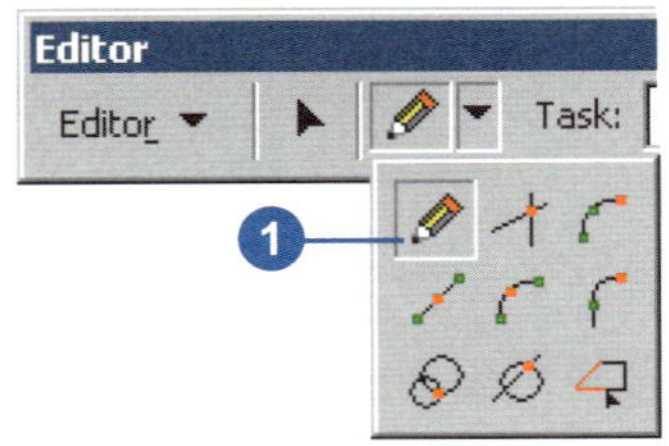

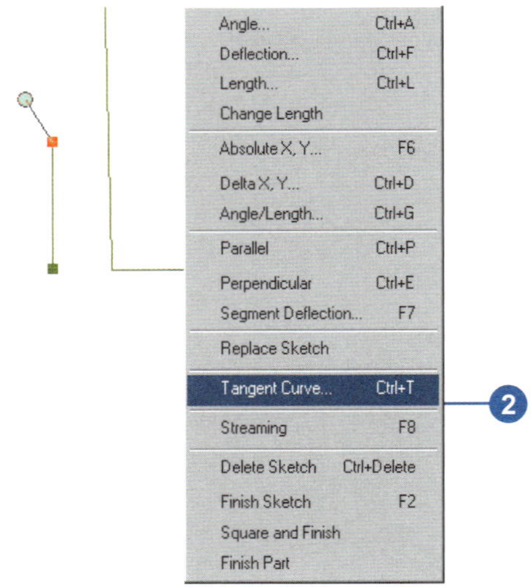

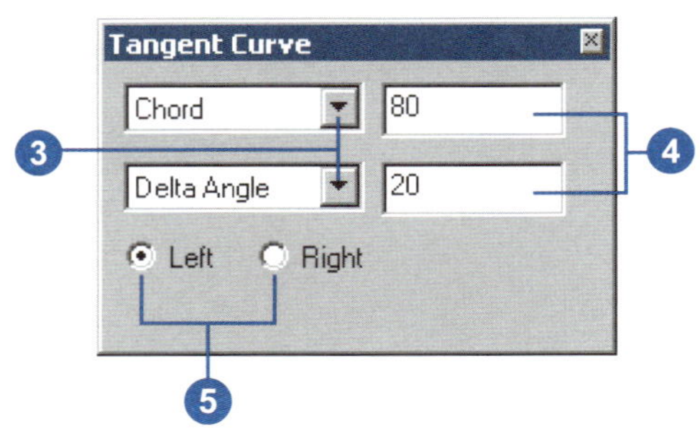

A segment that is a true curve is created from the last vertex of the segment according to the parameters you specified.

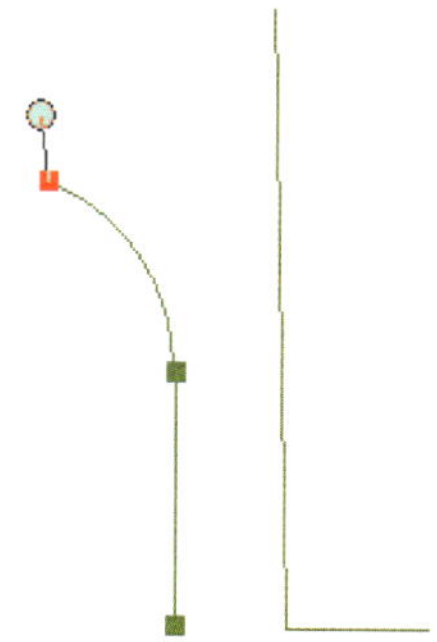

A circular arc curve with a chord length of 80 map units and a delta angle of 20 degrees is created to the left of the last vertex.

Creating segments by tracing features

You can create segments by tracing over the segments of selected features using the Trace tool.

Suppose you want to create a new water main that is offset seven meters from the parcel boundaries. Using the Trace tool, you can create new segments in the sketch that are at the same angle as the selected parcel boundaries yet constructed at an offset value of seven meters.

Tip

Backing up a trace

If you traced too far or have traced the wrong direction, move the mouse backwards over what you have traced. If you have clicked to stop the trace, click Undo to remove all vertices added during the trace.

Tip

Canceling a trace

Press Esc to quickly cancel a trace.

Tip

Finishing the sketch

When you're finished tracing, double-click to finish the sketch.

Creating segments by tracing features

1. Click the Edit tool.
2. Select the feature or features you want to trace.
3. Click the tool palette dropdown arrow and click the Trace tool.
4. Press O to open the Trace Properties dialog box.

   The Trace Properties dialog box lets you specify many of the properties of the trace segment.
5. Type an offset value. If you want to trace directly on top of existing features, enter a value of 0.
6. Optionally, you can limit the length of the trace segment.
7. Click the corresponding radio button to specify type of corners.

   Close the Trace Options dialog box by clicking OK.
8. Click to start tracing.
9. Click to stop tracing.

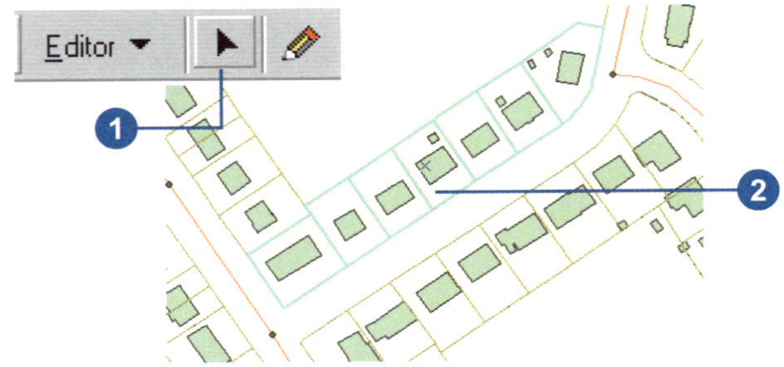

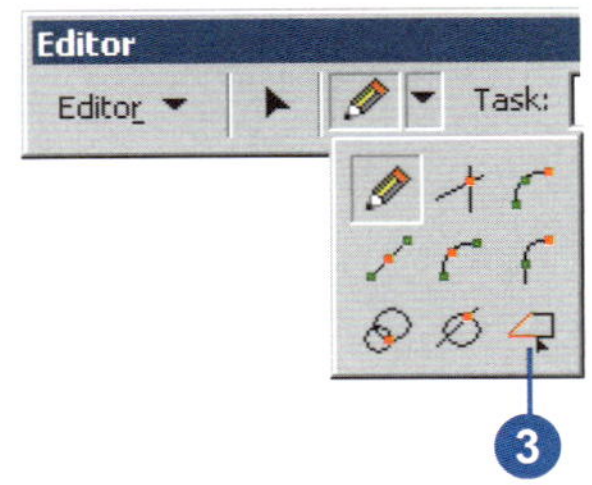

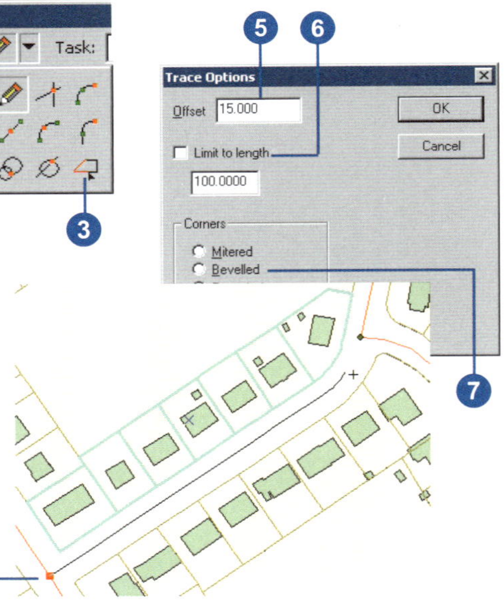

Duplicating features with the Copy Feature tool

You can quickly create a duplicate of a selected feature or features by using the Copy Feature tool. The Copy Feature tool is located on the Advanced Editing toolbar. Only the geometry, not the attributes, of the selected feature are copied to the Target layer.

If the Target feature class or subtype has default values, feature-linked annotation, or connectivity rules defined in a geodatabase, the copy will have the default values and behavior appropriate for the Target layer. Attributes without default values will have a null value.

Tip

Using the Advanced Editing toolbar

The Advanced Editing toolbar has a more complex step of editing features. This toolbar is only available with ArcEditor and ArcInfo licenses.

Tip

Opening the Advanced Editing toolbar

The Advanced Editing toolbar is found on the Editor context menu under More Editing Tools.

1. Click the Edit tool.
2. Select the features that you want to copy.
3. Click the Copy Feature Tool.
4. Click the place where you want a copy of the feature to be placed.

   A copy of the geometry of the selected feature is created at the location you clicked. The feature is created in the Editor's current Target layer.

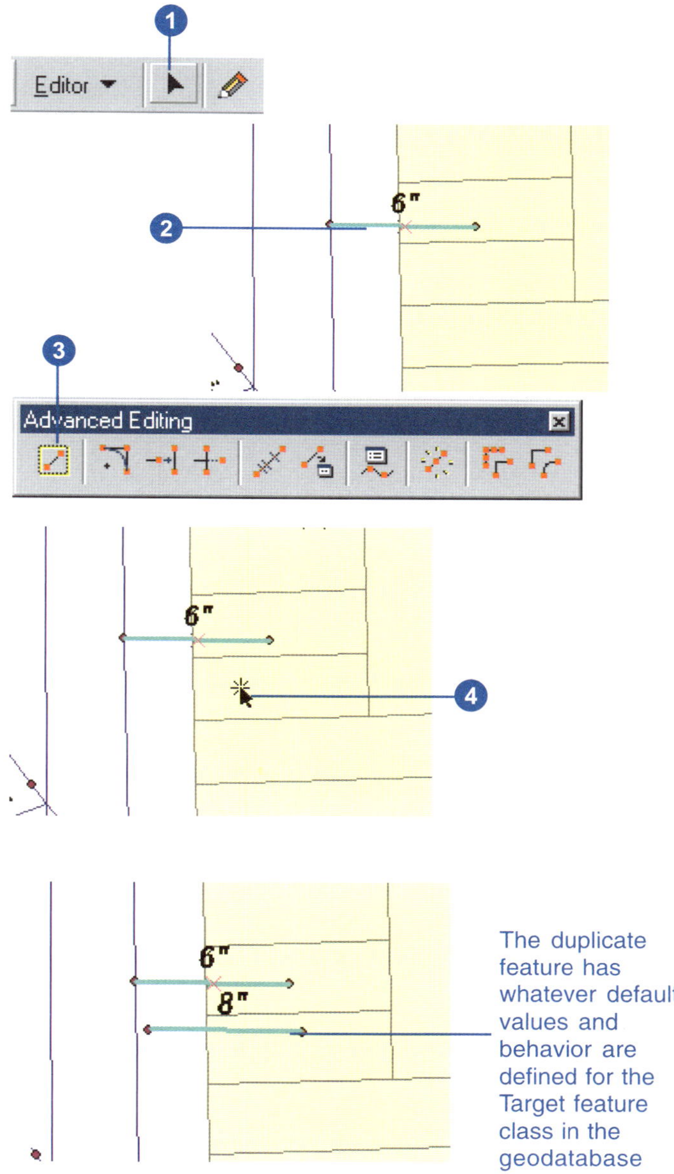

Creating a fillet curve between two lines

Fillets are segments of a circular arc that are often used to connect two intersecting lines. Fillets are used to create smoothly curving connections between lines, such as edge of pavement lines at street intersections, or rounded corners on parcel features.

Tip

Shortcut for fillet radius

After clicking the Fillet tool, you can set a default fillet radius on the Fillet Options dialog box by pressing the R key. If a default radius is set, the fillet curve will be constructed immediately with that radius when you click the second line of those you want to fillet.

Tip

Radius units

The Fillet Options will use current data frame coordinate system units for the fillet radius, unless you specify another unit of measure when you type the fillet radius.

Tip

Placement of the fillet

If you specify a fixed radius, the fillet curve is created on the side of the second line where the pointer was when you clicked.

1. Click the Fillet tool.
2. Click first one line then the other to specify which lines you want to construct the fillet between.

   The radius of the fillet curve will change as you drag the pointer away from the first line you clicked.

   When the curve looks like it has the right radius, you can click to finish the curve. If you want the curve to have a specific radius, you can set the radius from the Fillet Options window.
3. Press the R key to set the Fillet Options.
4. Optionally, check Trim existing segments if you want to remove the segments outside of the curve's radius.
5. Optionally, check Fixed radius if you want to specify a radius for the curve.
6. Type a radius for the curve and press Enter, or click OK.
7. Click the map to indicate the quadrant of the intersection of the lines where the fillet will be created.

   A new fillet curve joins the two selected lines. The extra line segments outside the curve are trimmed off, if you checked Trim existing segments.

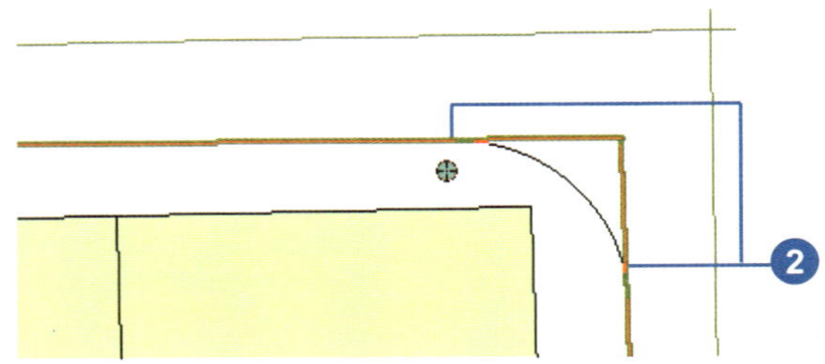

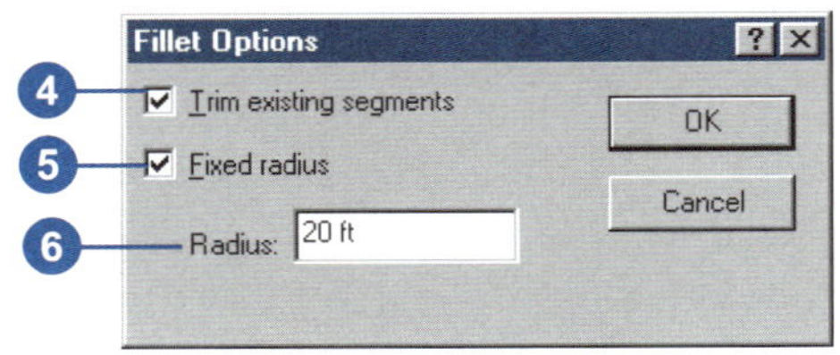

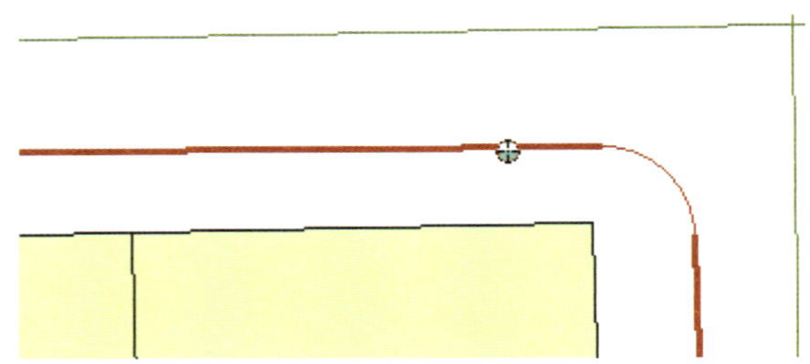

Extending a line

You can extend a line segment that's a little too short so that it touches another line segment. The Extend tool on the Advanced Editing toolbar lets you click a line feature and extend it to another selected line feature.

1. Click the Edit tool.
2. Select the line segment to which you want to extend a line.
3. Click the Extend tool.
4. Click the end of the feature that you want to extend.

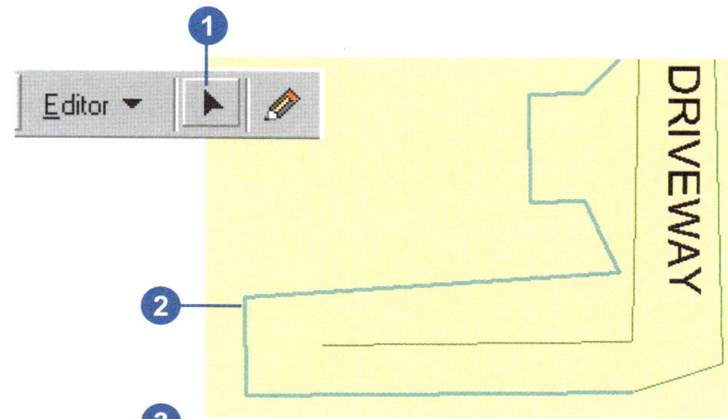

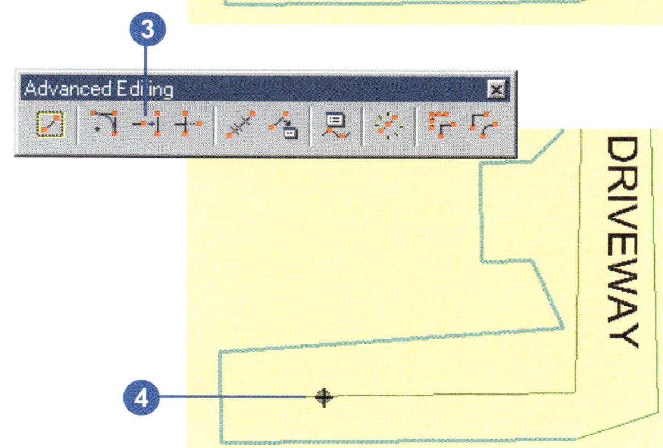

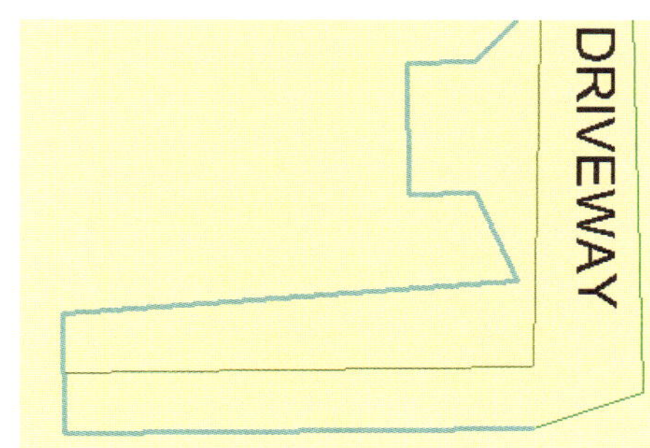

The line you clicked is extended to the selected line.

Trimming a line

You can trim off the part of a line that crosses another line segment. The Trim tool on the Advanced Editing toolbar lets you click one side of a line feature that crosses a selected line in order to trim it off.

1. Click the Edit tool.
2. Select the line segment at which you want to trim a line.
3. Click the Trim tool.
4. Click the end of the feature that you want to trim.

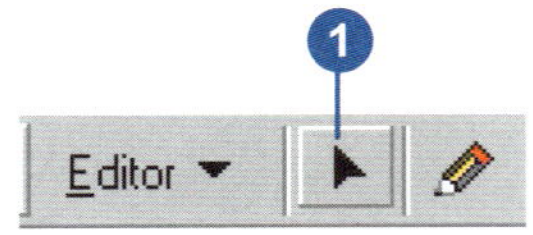

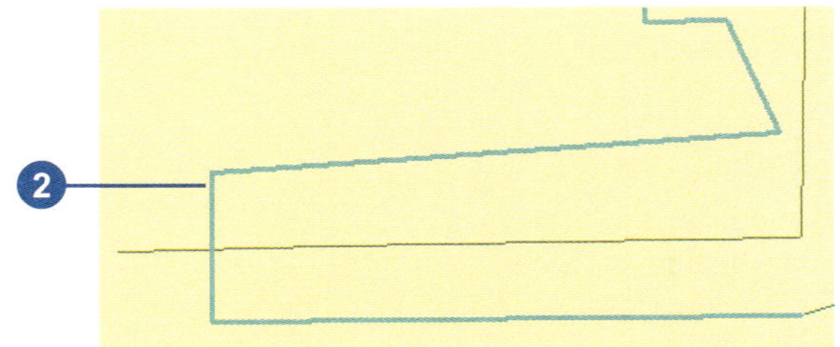

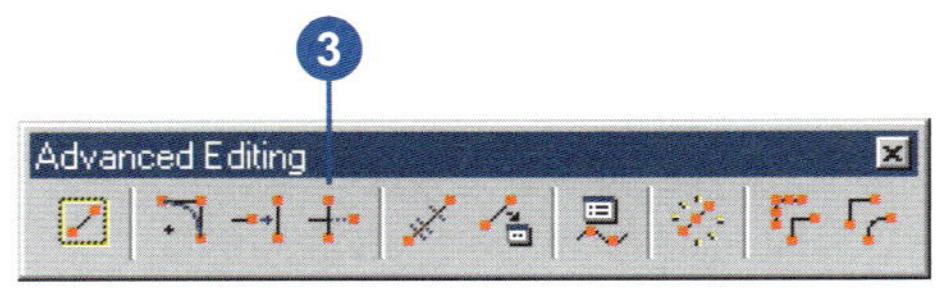

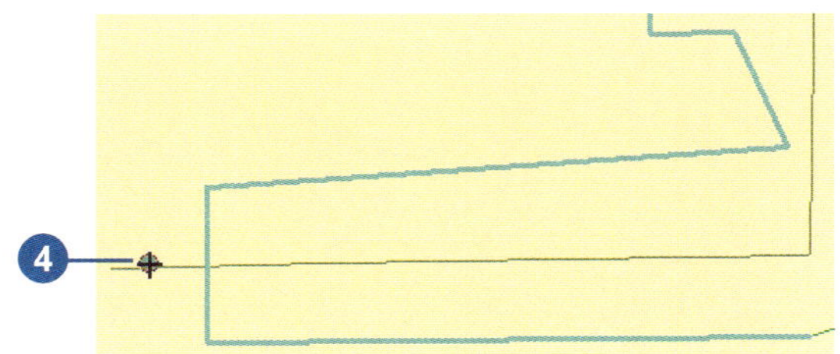

The line you clicked is trimmed at the selected line.

Proportionally dividing a line

One common coordinate geometry editing task is to divide an existing line feature into a number of segments of a specified length.

Sometimes the task can be complicated by measuring errors. For example, the length of the line in the geographic information system (GIS) might not be exactly equal to the total length of the segments, as measured in the field. The Proportion tool allows you to divide a selected line into a number of segments and allocate the difference between the length of the line feature and the total length of the segments between all of the new segments.

Suppose you have a line that you want to divide into segments that you can snap to when creating new parcel corners. The length of the line feature in the GIS is 320.38 ft. The lot plan you've been given shows that the lot corners are at 111.78 ft, 70.43 ft, and 138.65 ft along this line. Unfortunately, there is a difference of 0.48 ft between the measurements and the feature length. ▸

1. Click the Edit tool.
2. Select the line that you want to proportionally divide.
3. Click the Proportion tool. ▸

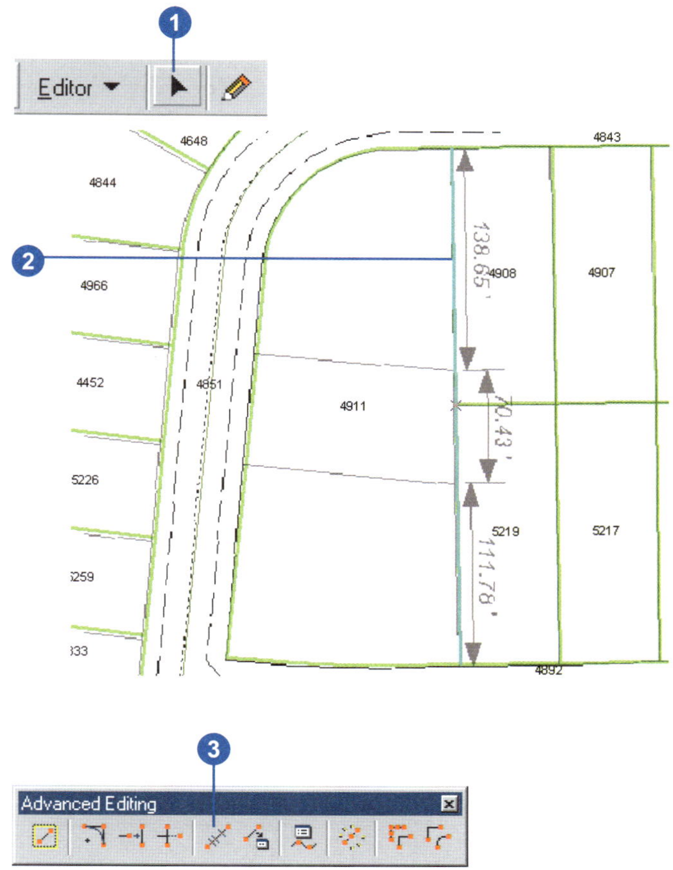

The Proportion dialog box lets you enter the lengths of the segments. As you do it reports the length of the selected feature, the sum of the lengths of the segments, the amount that is left over, and the relative error, expressed as the ratio between the leftover and the original feature length.

The Proportion tool creates new proportioned features of 111.61 ft, 70.32 ft, and 138.44 ft, dividing the difference between the features.

If the line feature that you are splitting has an attribute named Distance the new line features will have the values that you typed as the attributed length, and the Shape_length field will show the true length.

4. Type the lengths of the segments into which to divide the line.
5. Optionally, click Reverse to switch the orientation of the line if the arrows indicating the orientation of the feature on the map are the reverse of the order in which you entered the segments.
6. Click OK.

   The line is split proportionally to the length of the measured segments. Any difference between the line length and the sum of the length of the segments is allocated proportionately to the new lines' length.

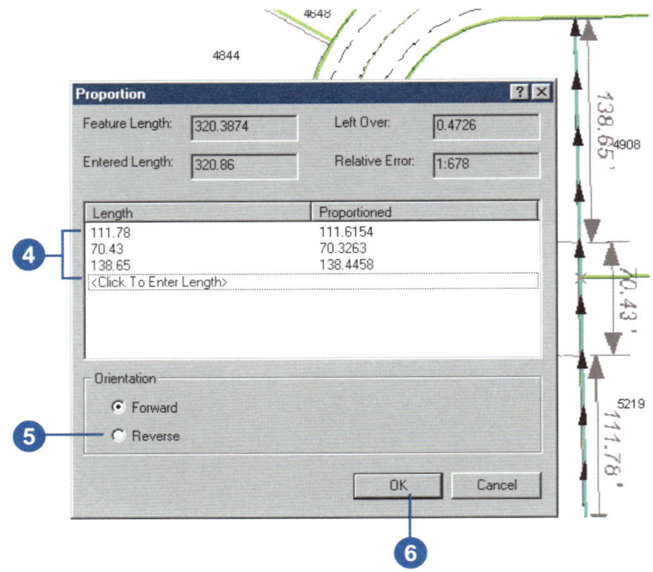

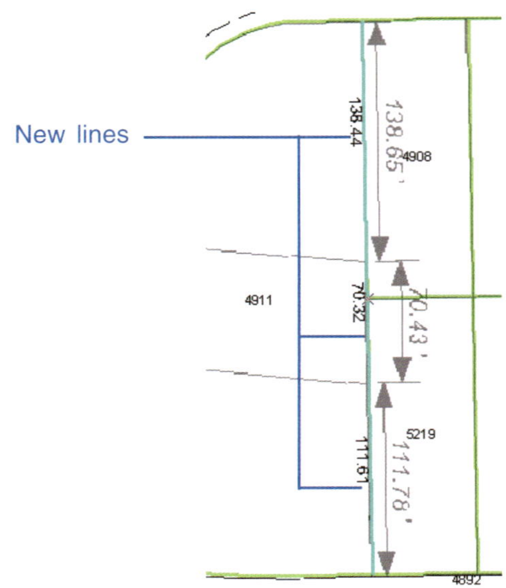

Getting a COGO description of a feature with the Inverse tool

You can get the COGO description of a feature using the Inverse tool. The Inverse tool will calculate the direction and length of the feature as well as curve parameters, if required, and populate appropriate attributes on the feature.

In order to use the Inverse tool, the feature must have the following attributes as text fields: Direction, Distance, Radius, Delta, Tangent, Arclength, and Side.

The feature must be either a straight line or circular arc. It typically only has two vertices but more are allowed as long as the feature is consistently straight or consistently curved. The current direction type and angular units are used when populating the attributes.

Tip

Changing the COGO attribute names

The names of the COGO attributes can be changed by using the Advanced ArcMap settings utility.

1. Click the Edit tool.
2. Select the feature that you want to generate a COGO description for.
3. Click the Inverse tool.

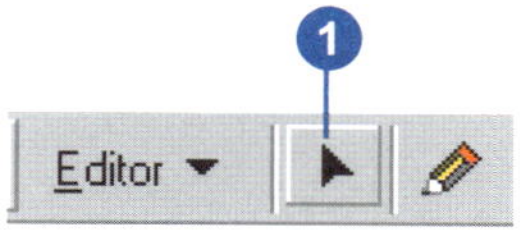

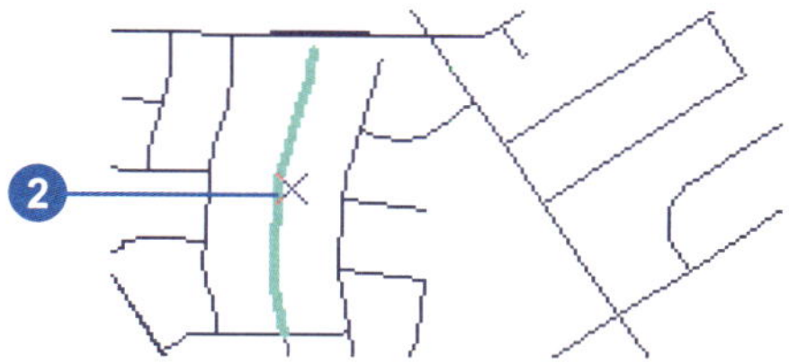

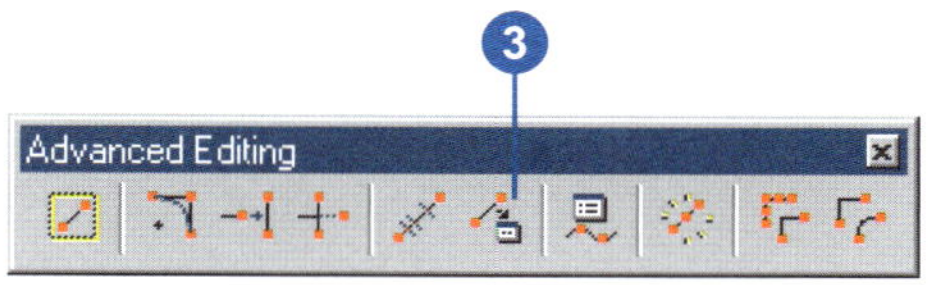

Creating edit sketch geometry with the Traverse tool

Another common coordinate geometry editing task is to create a line or polygon edge using a set of survey measurements collected in the field. The data may be in the form of directions and distances, angles and distances, curves, or tangent curves, measured from a known point. These are collectively known as a traverse.

The Traverse tool lets you create edit sketch geometry from a wide variety of traverse measurements. ►

Tip

Interactive start point

You can click the Interactive Start Point Selection tool to set a start point by clicking on the map.

Tip

Starting from a sketch

If you have already started an edit sketch when you click the Traverse tool, the start point will be the last vertex of the edit sketch.

Starting a traverse from a known coordinate

1. Click the Traverse tool.
2. Click Edit, to the right of the Start box.
3. Type X and Y values for the starting coordinate.
4. Click OK.

   The start point of the traverse is set in the Start Point text box. The next vertex of the edit sketch will be placed on the measurements you specify from this location.

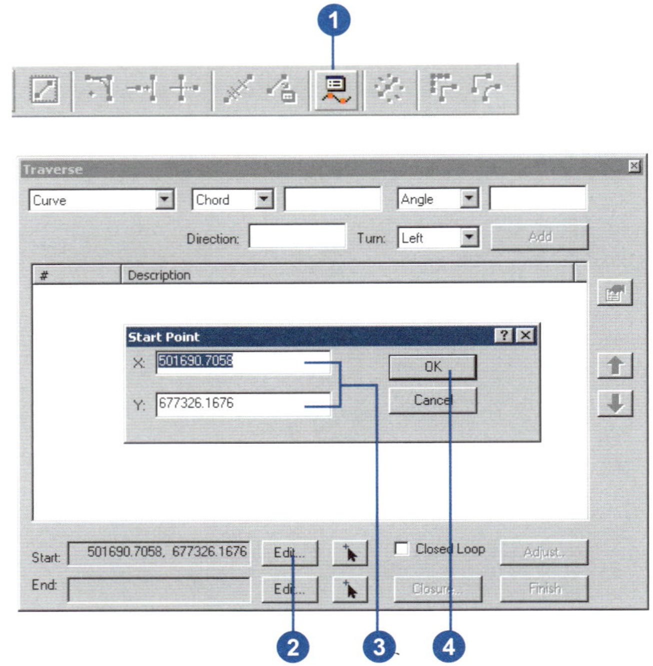

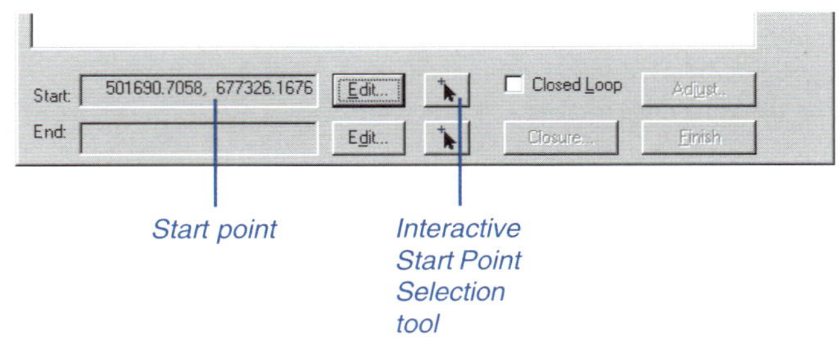

Once you've started a traverse you can add segments or curves to the edit sketch using Direction–Distance, Angle–Distance, Curve, and Tangent Curve methods.

The Traverse tool adds each segment as a line in the Traverse course table and to the edit sketch. You can select each course segment by clicking it in the table—the corresponding segment will flash on the map.

Tip

Duplicating segments

You can insert a duplicate of a segment by right-clicking it in the table and clicking Insert.

Tip

Reordering segments

You can reorder a segment by selecting it in the table and clicking the up and down arrows.

Tip

Removing segments

You can remove a segment by selecting it in the table and clicking Remove.

Adding a segment to a traverse using a Direction–Distance course

1. Click the course type dropdown arrow and click Direction–Distance.
2. Type a direction.

   The Traverse tool uses degrees measured counterclockwise from east by default. You can change the angular measuring system and units on the Editing Options dialog box, Units tab.
3. Type a distance.

   The Traverse tool uses the data frame's coordinate system units of measurement by default. You can change these units by modifying the coordinate system description, but not during an edit session.
4. Click Add.

   The course is added to the course table, and the segment is added to the edit sketch.

   You can continue to add segments using any of the traverse methods.

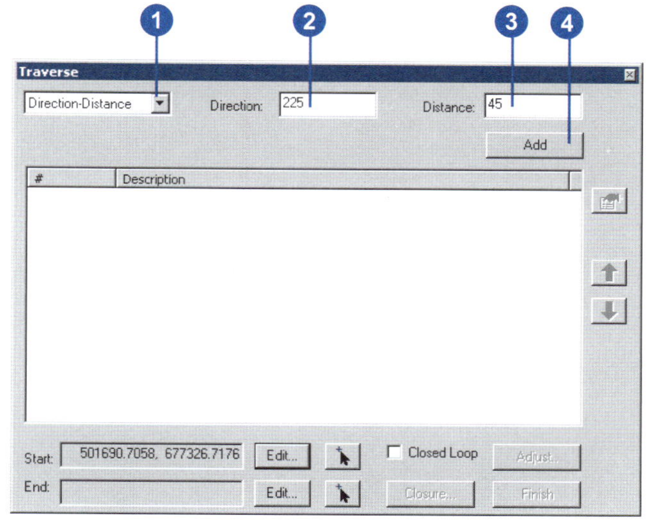

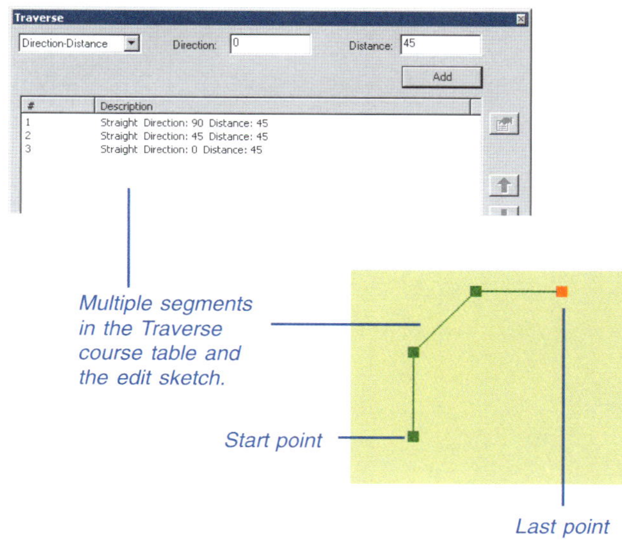

Tip

Limitations to first course

The first course in your traverse table cannot be based on an Angle Distance or Tangent Curve because these techniques are calculated from the direction of the previous course.

Adding a segment to a traverse using an Angle–Distance course

1. Click the course type dropdown arrow and click Angle–Distance.
2. Type an angle.

   The Traverse tool uses degrees measured counter-clockwise from the previous course of the traverse. You can change the angular measuring system and units on the Editing Options dialog box, Units tab.
3. Type a distance.

   The Traverse tool uses the data frame's coordinate system units of measurement by default. You can change these units by modifying the coordinate system description, but not during an edit session.
4. Click Add.

   The course is added to the course table, and the segment is added to the edit sketch.

   You can continue to add segments using any of the traverse methods.

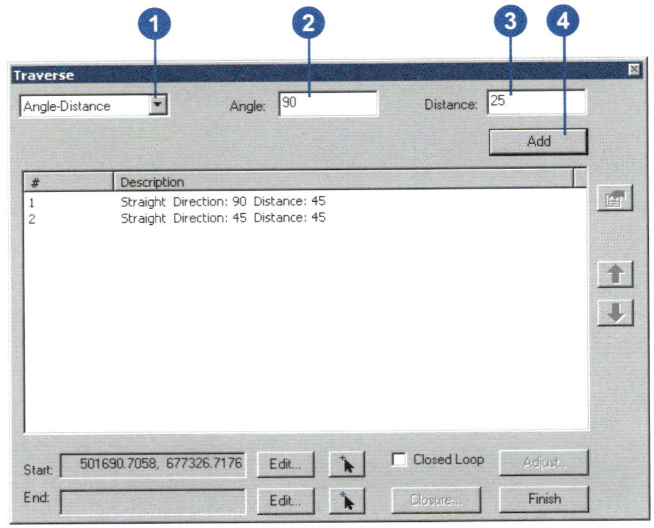

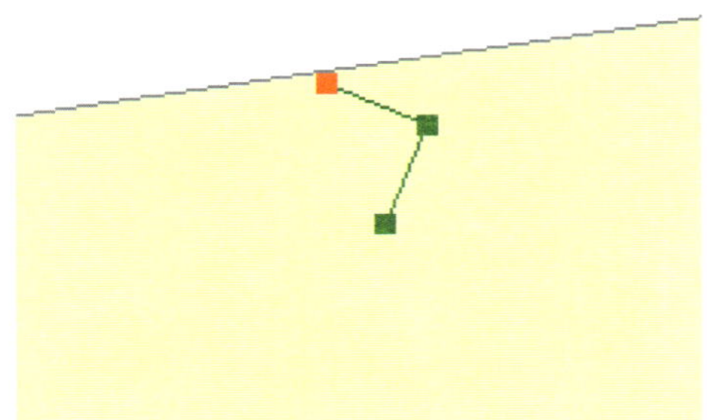

Tip

Productivity with the Traverse tool

You can use the keyboard to enter and manage courses in the traverse. Enter can be pressed whenever there is enough information to add a course. The Tab key and Shift + Tab can be used to navigate between the fields. For dropdown lists, the first letter of the choice can be used as a shortcut or you can use the Up/Down Arrow keys to make your selection.

Adding a segment to a traverse using a Curve course

1. Click the course type dropdown arrow and click Curve.
2. Click the dropdown arrows and choose two parameters to use in determining the curve.
3. Type the appropriate values for the parameters. Distance values will use map units.
4. Type a chord direction for the curve.
5. Click the dropdown arrow to choose whether the curve will be to the right or left.
6. Click Add.

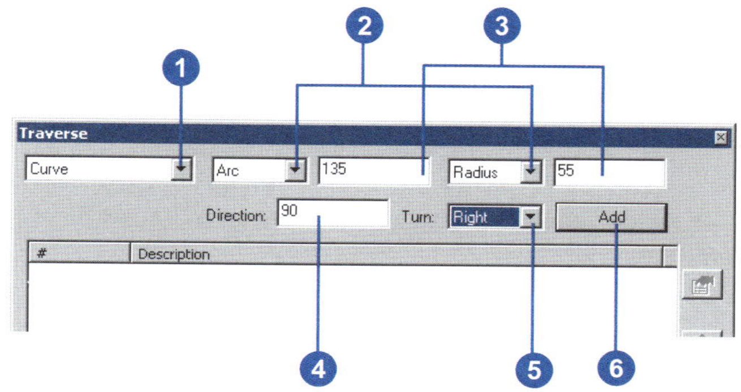

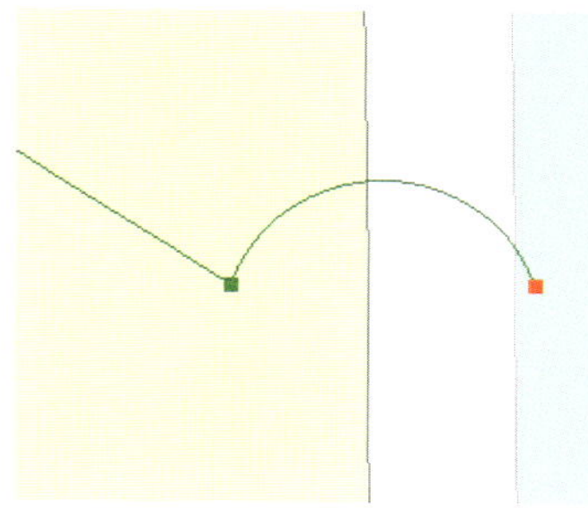

Adding a segment to a traverse using a Tangent Curve course

1. Click the course type dropdown arrow and click Tangent Curve.
2. Click the dropdown arrows and choose two parameters to use in determining the curve.
3. Type the appropriate values for the parameters. Distance values will use map units.
4. Click the dropdown arrow to choose whether the curve will be to the right or left.
5. Click Add.

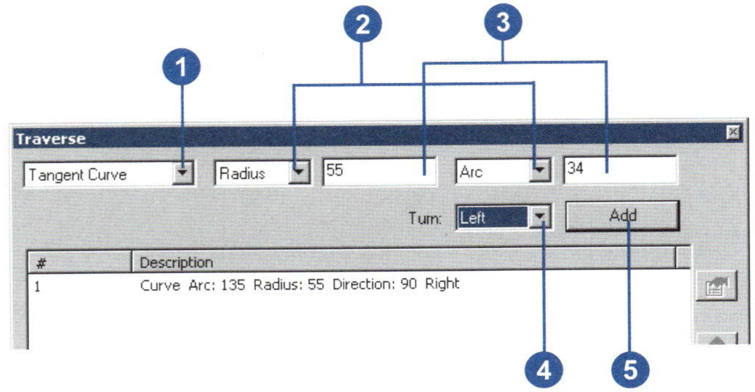

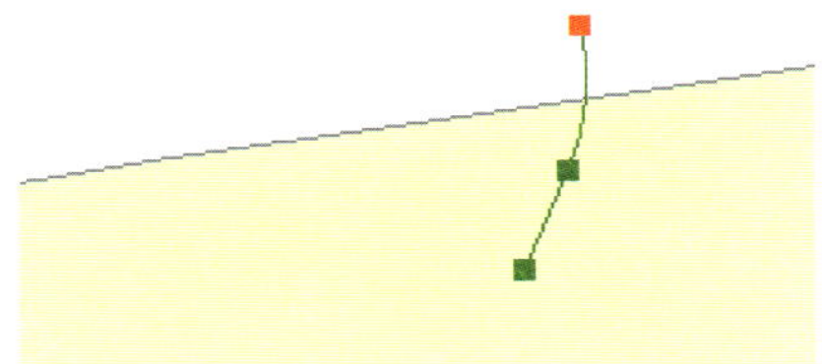

Creating two-point line features to populate COGO attributes

You can easily create features using the Traverse tool to populate COGO attributes in a feature class. Using the Create 2-Point Line Features task allows you to not only create the feature geometry with the Traverse tool, it also allows you to save each course in the traverse as a COGO two-point line feature.

To create two-point line features using this process, the target layer must have the following attributes as text fields: Direction, Distance, Radius, Delta, Tangent, Arclength, and Side.

Tip

Adding options to the Task dropdown menu

Tasks on the Task dropdown menu can be added or removed from the selection on the Edit Options dialog box under the Edit Tasks tab. For further details, see Chapter 2, 'Editing basics'.

1. On the Task dropdown menu, select Create 2-Point Line Features.
2. On the Editor toolbar, set your target layer to a layer that contains COGO attributes.
3. Using the Traverse tool, specify the courses of the traverse.
4. When you finish the traverse, the new line features will be saved as COGO two-point line features.

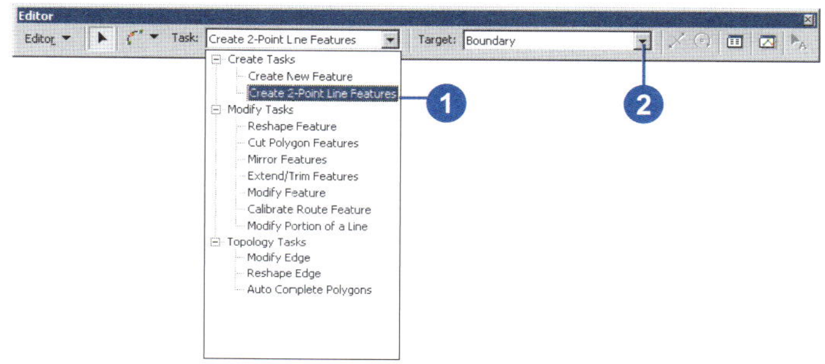

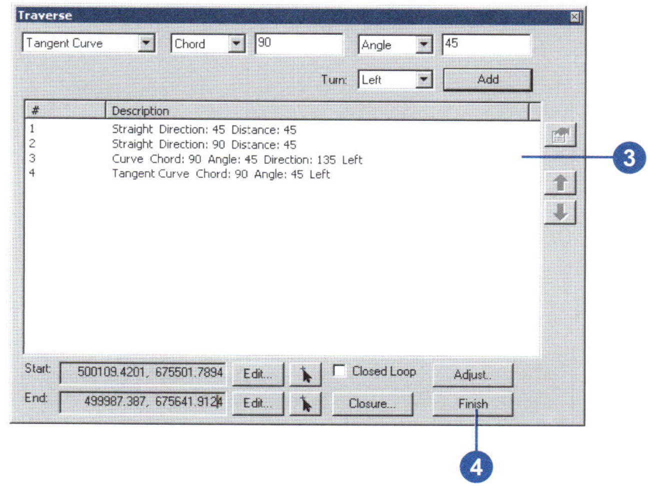

Tip

Opening the Course properties dialog box

You can open the properties for the Course dialog box using three different methods. You can double-click the segment in the course list, select the segment and click Properties on the Traverse dialog box, or right-click the course and select Properties.

Modifying a segment in a traverse

1. Select the segment in the course list.
2. Click the Properties button.
3. In the Course properties, adjust the values as required.
4. Click OK.

   The traverse and the sketch are updated.

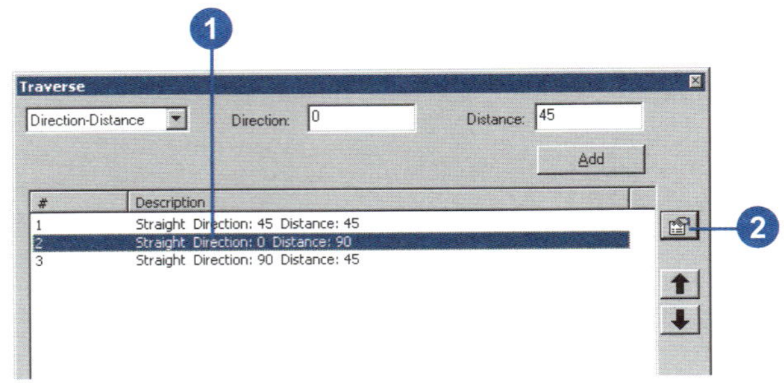

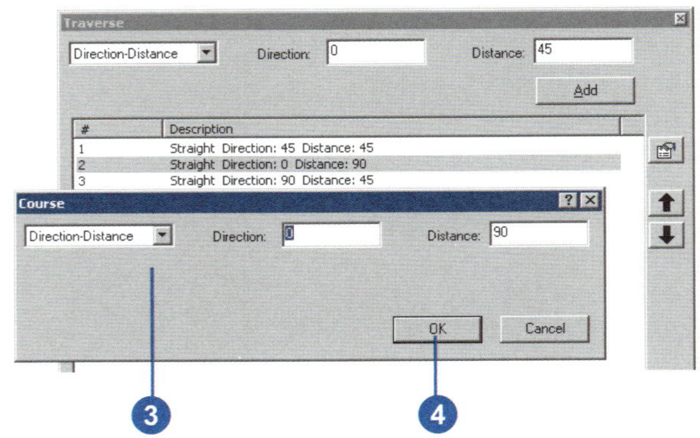

Obtaining a closure report

A traverse must always start from a known coordinate. It can also end at a known coordinate. When this occurs a difference can be calculated between the specified endpoint and the traversed endpoint.

A closure report is a summary of the difference between the endpoint coordinate of a traverse and the calculated endpoint.

Tip

Interactive endpoint

Click on the Endpoint Selection tool to set the endpoint by clicking on the map.

Tip

Starting and ending at the same point

Checking the Closed Loop check box sets the beginning and ending points the same.

1. Click Edit to the right of the End box.
2. Type the x,y coordinates for the ending point.
3. Click OK.
4. Click Closure.

   The closure report lists the following:

- Number of courses
- Total length of the traverse
- The specified and calculated endpoints
- The difference of misclosure in both x,y and direction/distance values
- Relative error that is a ratio of the distance misclosure over the total length

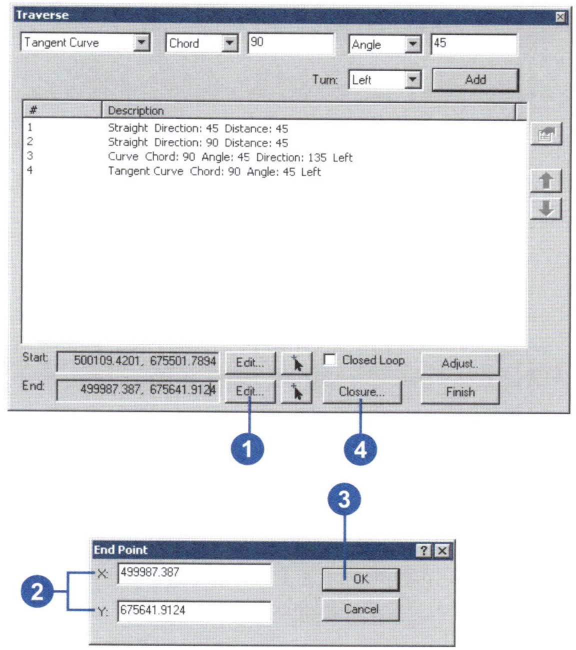

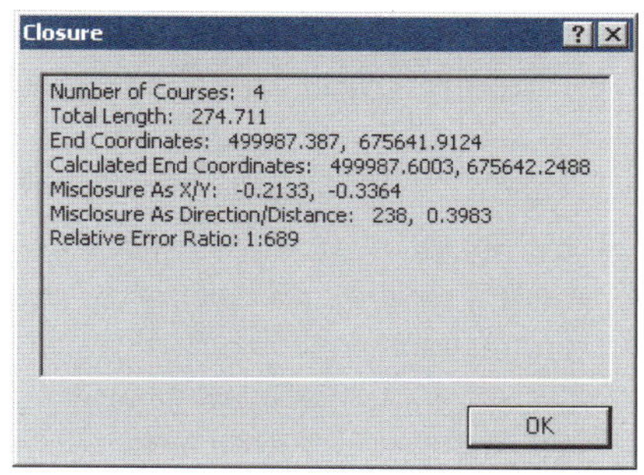

Techniques for adjusting a traverse

Often when generating a traverse, the coordinates of an ending destination are known. You have already discovered that ArcMap provides a method to specify this endpoint and determine the difference between the traverse endpoint and the desired ending point. This difference is know as the misclosure.

ArcMap also provides three different techniques for adjusting the traverse to eliminate misclosure. Each of these adjusting techniques vary in the amount of adjustment of the direction and distance of the individual courses of the traverse. These techniques include compass correction, transit correction, and Crandall correction.

The compass correction technique specifies that the misclosure, or difference in x and y between the resulting endpoint and the desired endpoint, are equally distributed among the individual two-point arcs and curves that make up the traverse. This is done by adjusting the location and distance of each arc proportional to the difference in closure. The compass correction technique is the most often used to resolve errors in misclosure. It assumes that the errors are related to both errors in the direction measurements as well as the distance measurements. Thus, the corrections are reflected in each distance and direction value. This technique is also known as the Bowditch rule.

Much like the compass correction technique, the transit correction method specifies that the misclosure is equally distributed among the individual two-point arcs and curves that make up the traverse. However, this technique favors the direction measurements over the distance measurements. In determining the location change required of each arc, the proportion assigned to each arc is proportional to the total x or y values of all the arcs. This results in changes that will affect both the direction and the distance of each arc, but will alter the distance to a greater extent.

The Crandall correction technique is used when the direction values are assumed to be precise and accurate and that any misclosure is due solely to errors in distance measurements. This adjustment will preserve all of the direction measurements and will alter only the distance measurements to eliminate the closure error.

With this array of correction techniques, you will be able to not only correct the errors in the traverse, but you will also be able to place greater or lesser value on specific characteristics of the traverse data.

Adjusting a traverse

1. Ensure that an endpoint is set.
2. Click Adjust.
3. Choose an adjustment method from the dropdown menu.
4. Optionally, you can save the adjusted values in a text file by clicking Save Report.
5. Click Accept.

   The traverse will be adjusted based on the adjustment method chosen and feature created.

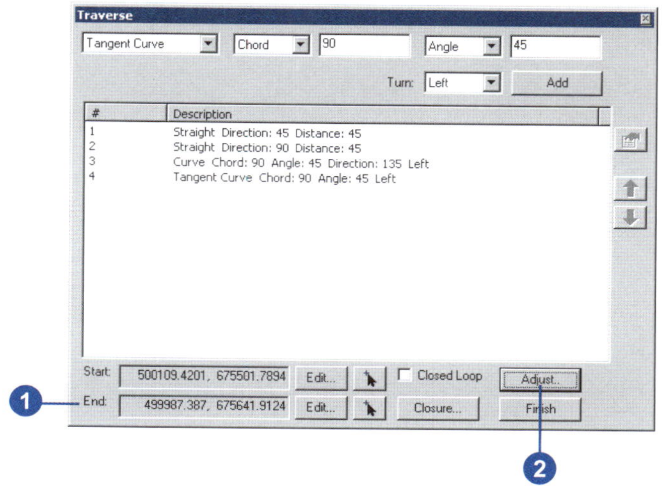

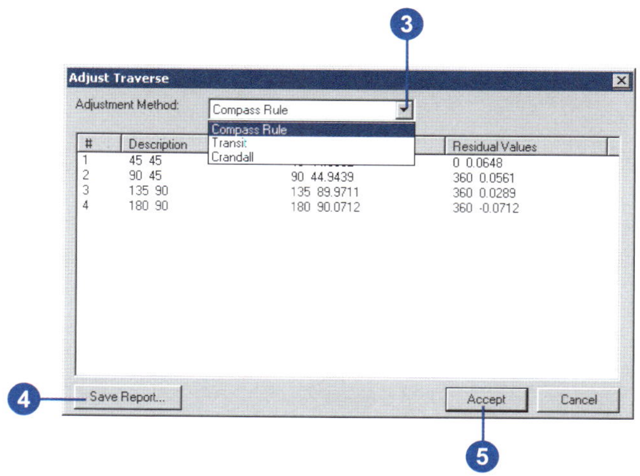

Saving a traverse

You can save a traverse to a text file to share with others or for your own use later.

1. Right-click the Traverse dialog box and click Save Traverse.
2. Navigate to the place where you want to save the traverse.
3. Type a name for the traverse.
4. Click Save.

   The traverse information is saved to the text file.

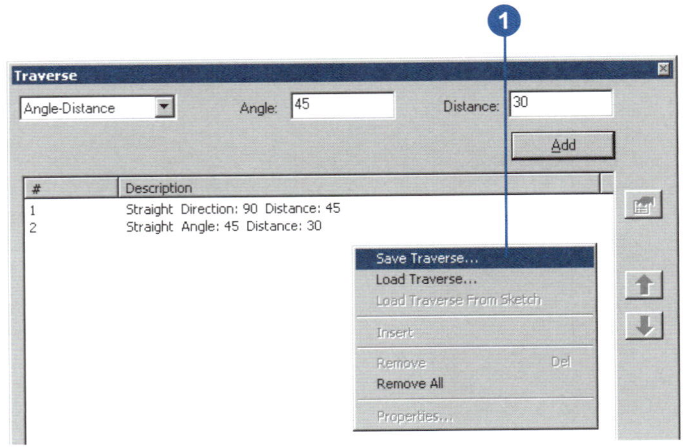

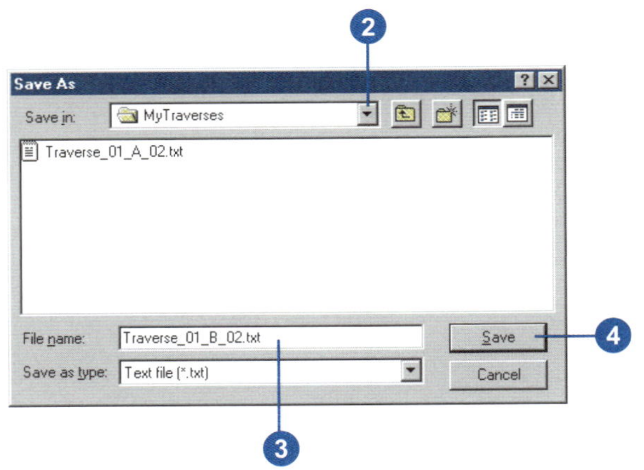

Loading a traverse

You can load a traverse that has been saved to a text file.

1. Right-click the Traverse dialog box and click Load Traverse.
2. Navigate to the traverse text file.
3. Click the traverse you want to open.
4. Click Open.

   The traverse information is loaded from the text file.

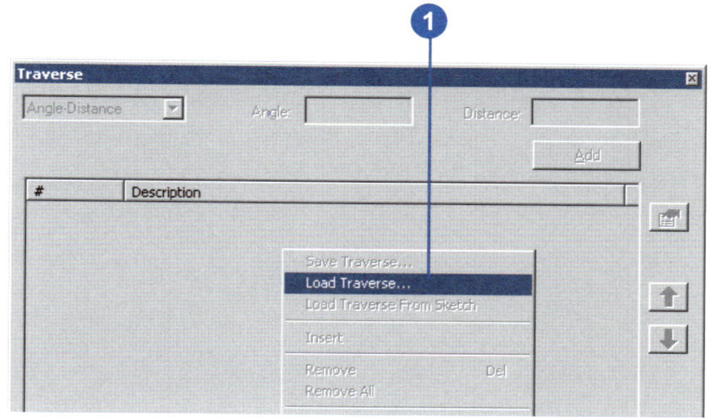

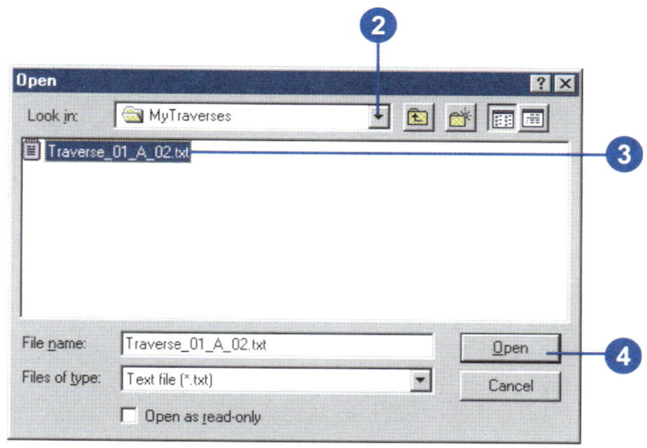

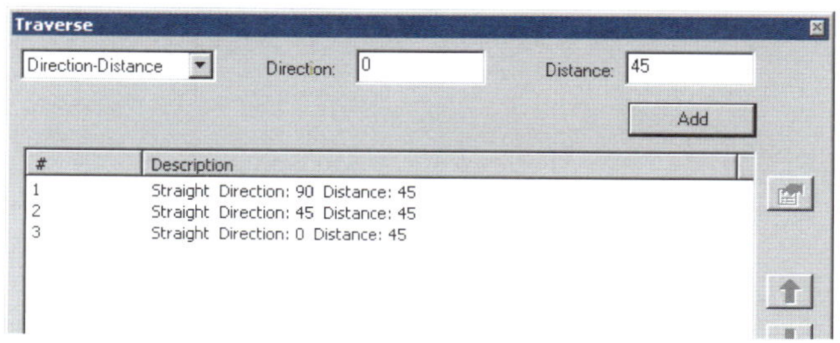

Exploding a multipart feature

You can use the Explode tool to separate a multipart feature into its component features.

1. Click the Edit tool.
2. Select the multipart feature that you want to Explode.
3. Click the Explode tool.

   The parts of the multipart feature become independent features.

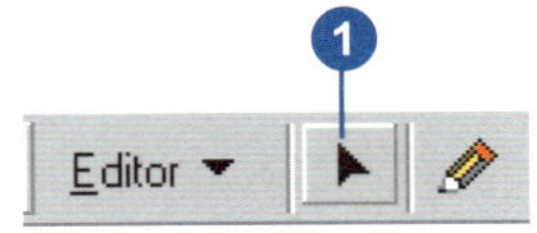

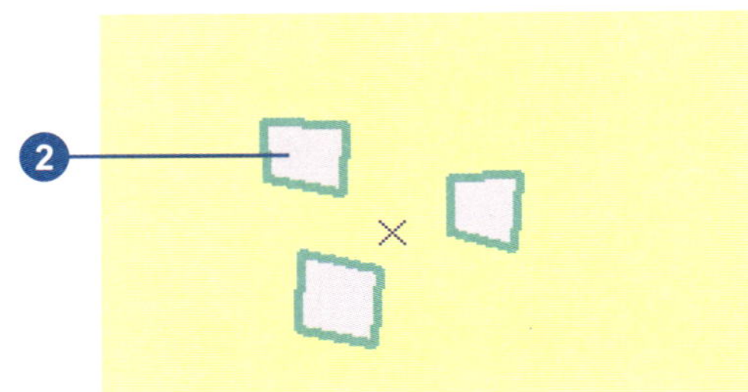

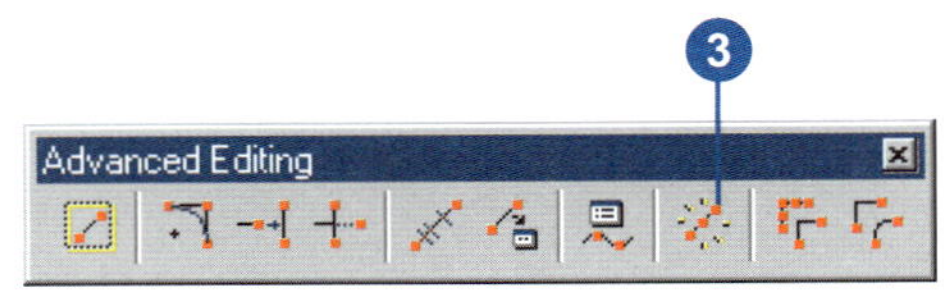

Generalizing a feature

You can use the Generalize tool to simplify the shape of features. The Generalize tool uses the Douglas–Poiker algorithm to simplify the input geometry of the selected feature.

The degree to which the geometry is simplified depends on the Maximum allowable offset, which limits how far the output geometry can be from the input geometry.

For features composed of linear segments, the output vertices will be a subset of the original feature vertices.

If you use the Generalize tool on a true curve, the output will be a series of straight line segments. The vertices may fall on all parts of the original curve, not just the vertices. The output of the Generalize tool on lines with nonlinear curves may have more vertices than the original curve, but all of the segments will be straight.

1. Click the Edit tool.
2. Select the feature that you want to Generalize.
3. Click the Generalize tool.
4. Type the Maximum allowable offset.

   The Maximum allowable offset is the maximum distance any part of the output geometry can be from the input geometry, in map units.
5. Click OK.

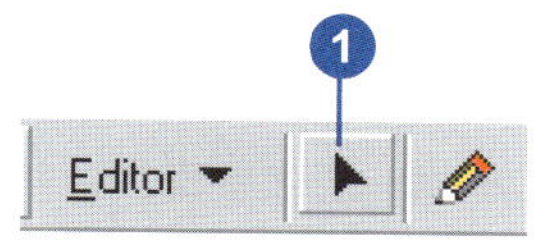

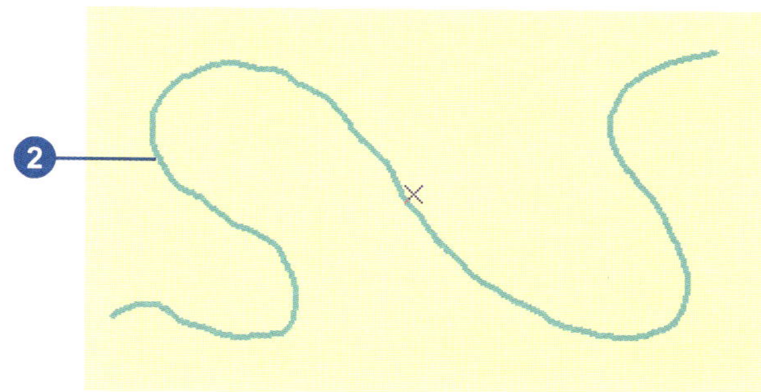

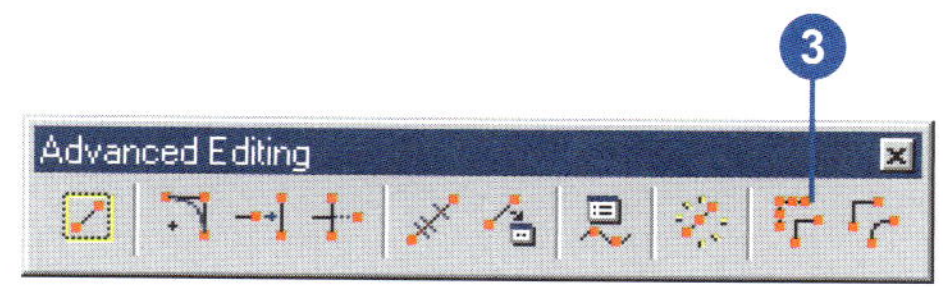

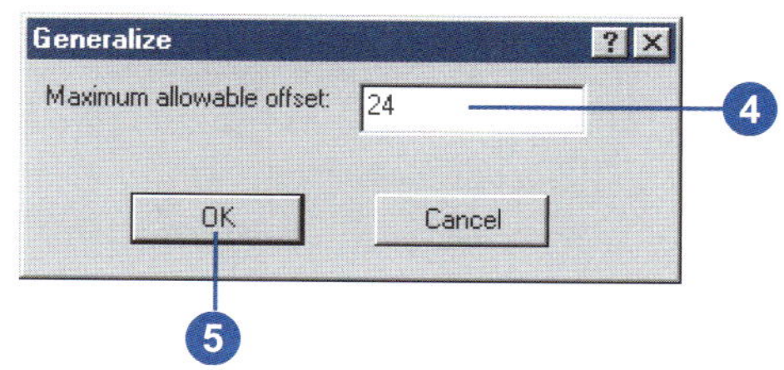

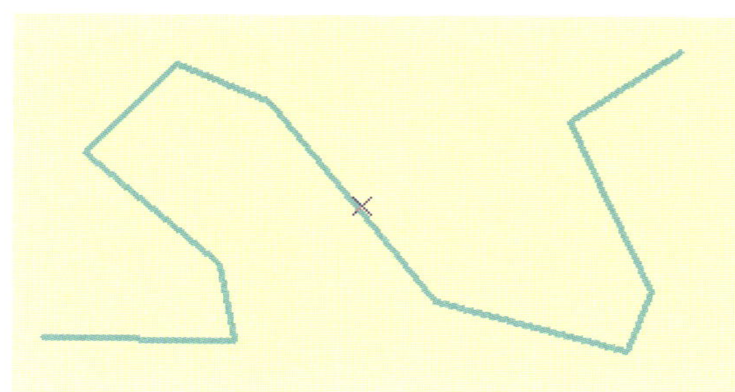

Smoothing a feature

You can use the Smooth tool to smooth the straight edges and angular corners of a feature. The feature geometry is replaced by a series of smoothed line segments.

The degree to which the geometry is smoothed depends on the Maximum allowable offset, which limits how far the output geometry can be from the input geometry.

1. Click the Edit tool.
2. Select the feature that you want to Smooth.
3. Click the Smooth tool.
4. Type the Maximum allowable offset.

   The Maximum allowable offset is the maximum distance any part of the output geometry can be from the input geometry, in map units.
5. Click OK.

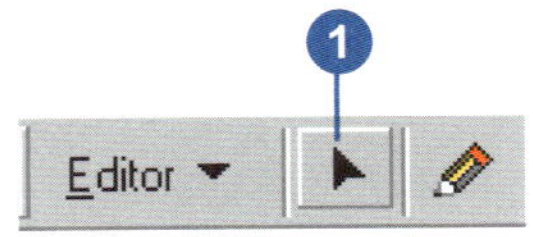

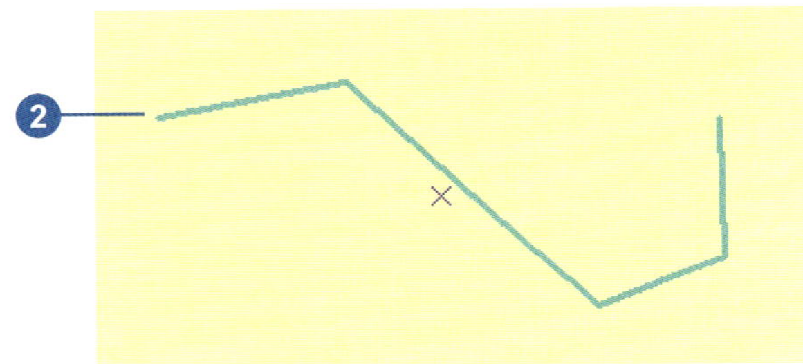

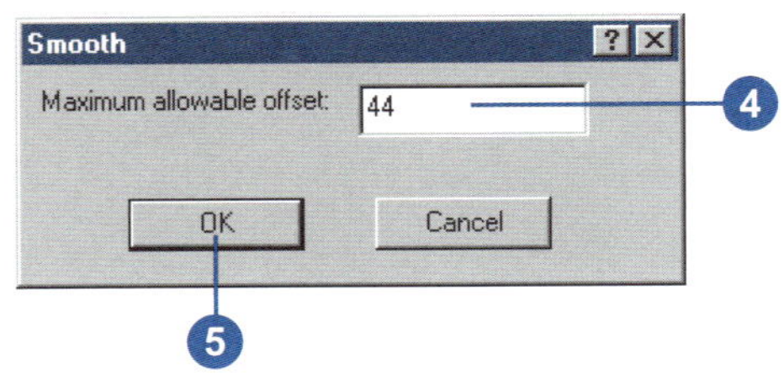

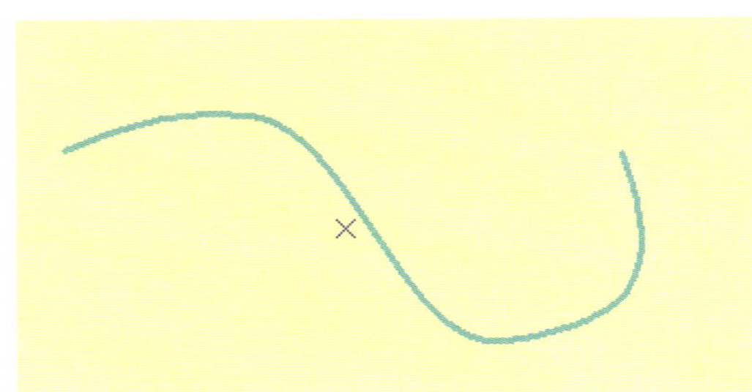

The Snapping Environment window

You can keep the window open as you work—any changes in settings are effective immediately. Click the Close button when you are finished.

The layers in your map document are listed here. Set the snapping priority—the order in which snapping will occur by layer—by dragging the layer names to new locations.

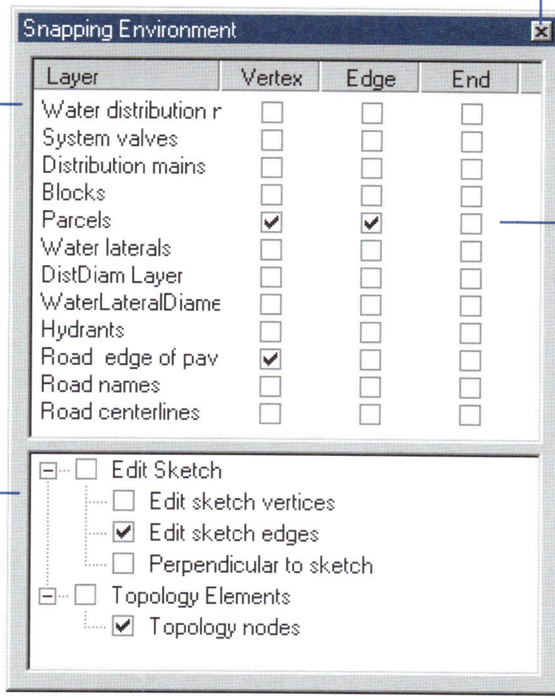

Check the type of snapping properties you want for each layer.

The bottom portion of the window shows snapping properties that work with a sketch or with elements of a Topology.

Types of snapping properties

When you use the snapping environment to create or place a new feature in an exact location relative to other features, you must choose to which part of existing features—vertex, edge, or endpoint—you want your feature to snap. These choices are called layer *snapping properties*. You can also specify snapping properties for the edit sketch and for topology elements; these are called sketch and topology snapping properties. You can set all three types of snapping properties using the Snapping Environment window. The following table briefly explains each of the layer snapping and sketch snapping properties.

Layer snapping properties	**Sketch snapping properties**	**Topology snapping properties**
Vertex  Snaps to each vertex of the features in that layer.	Perpendicular to sketch 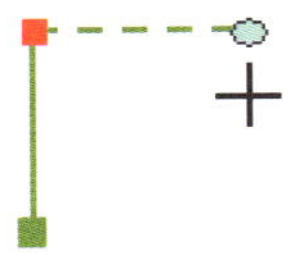 Lets you create a segment that will be perpendicular to the previous.	Topology nodes 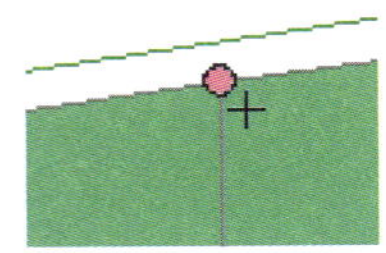 Snaps to the nodes in the topology.
Edge 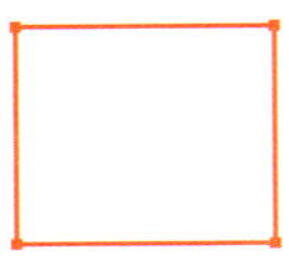 Snaps to the entire outline—both segments and vertices—of each feature in that layer.	Edit sketch edges 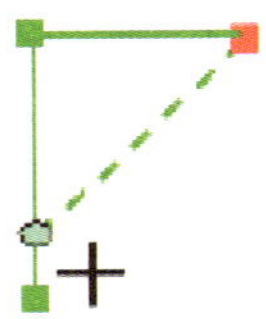 Snaps to the entire outline—both segments and vertices—of the sketch.	
Midpoint Snaps to the midpoint between the two end vertices of the segment.	Edit sketch vertices 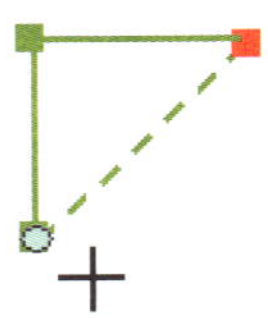 Snaps to the vertices of the sketch.	
Endpoint Snaps to the first vertex and the last vertex in a line feature.		

Using the snapping environment

The *snapping environment* can help you establish exact locations in relation to other features. Suppose you're creating a new segment of primary that begins from an existing transformer; you want to ensure that the vertex of the primary connects precisely to the transformer.

The snapping environment makes this type of task accurate and easy. Setting the snapping environment involves setting a snapping tolerance, snapping properties, and a snapping priority. ►

Tip

Showing snap tips

You can choose to display the layer name or target being snapped to. Check Show snap tips on the General tab of the Editing Options dialog box. A small text box will appear when snapping, which identifies the layer you have snapped to.

Tip

Viewing the snapping tolerance

To see the current snapping tolerance area, hold down the T key while using the Sketch tool.

Setting the snapping tolerance

1. Click Editor and click Options.
2. Click the General tab.
3. Click the Snapping tolerance dropdown arrow and click the type of measurement unit you want to use for the snapping tolerance—pixels or map units.
4. Type the desired number of measurement units in the Snapping tolerance text box.
5. Click OK.

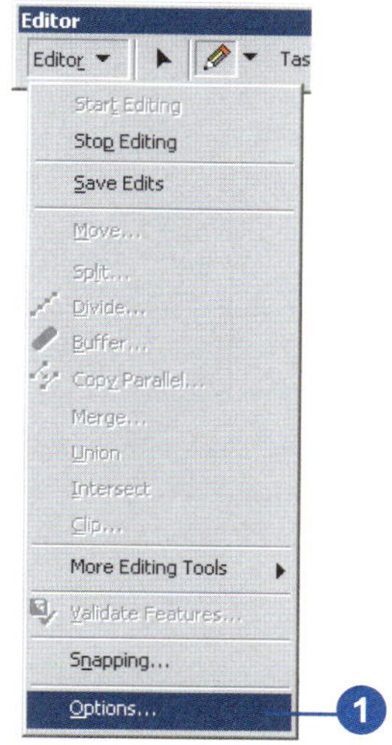

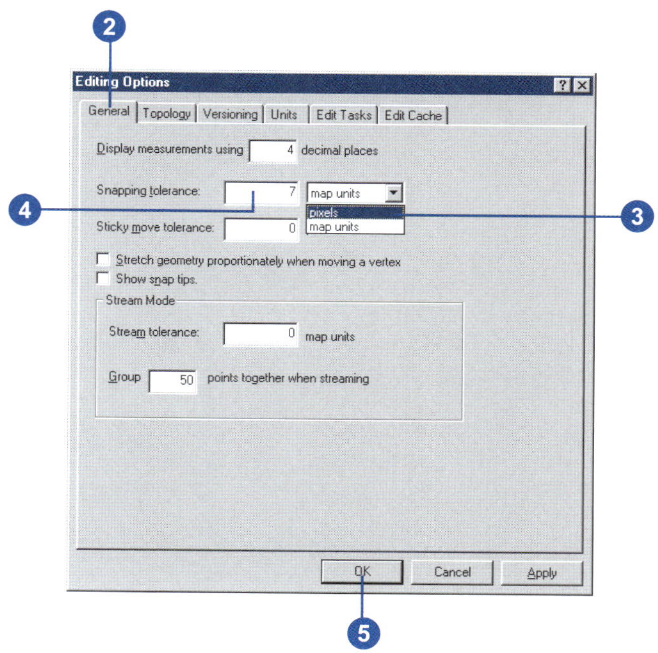

The *snapping tolerance* is the distance within which the pointer or a feature is snapped to another location. If the location being snapped to—vertex, edge, or endpoint—is within the distance you set, the pointer automatically snaps (jumps) to the location.

The circle around the pointer in the graphics below represents the snapping tolerance. When the location being snapped to (orange point) is outside the snapping tolerance, the snapping location (blue dot) stays with the pointer (top graphic). When the location being snapped to is inside the snapping tolerance, the snapping location moves away from the pointer and snaps to the target location (bottom graphic).

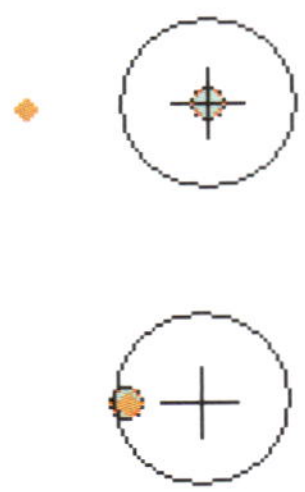

You can choose the part of the feature—vertex, edge, or endpoint—to which you want your new feature to snap by ▸

Setting snapping properties

1. Click Editor and click Snapping.

   The Snapping Environment window appears.

2. Check the snapping properties you want.

   The snapping properties are effective as soon as they are checked or unchecked.

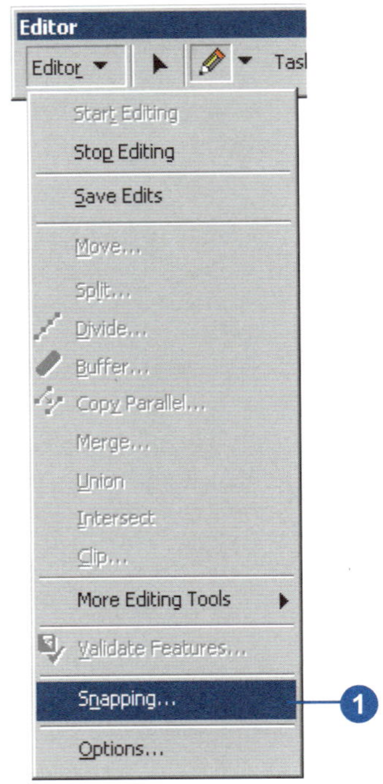

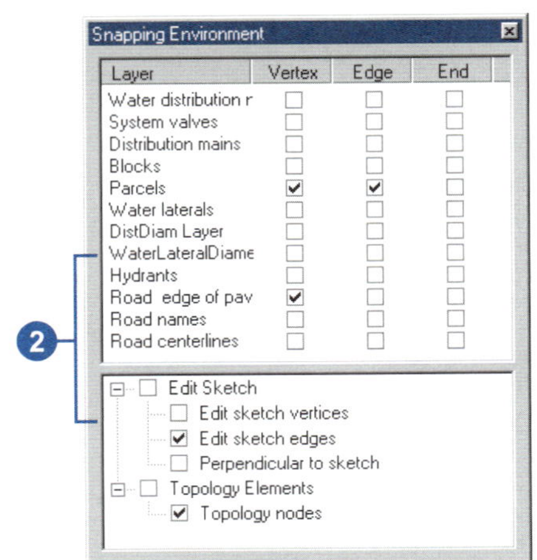

setting the layer *snapping properties*. For example, if you want your new feature—a segment of primary—to snap to the vertex of an existing transformer in the transformers layer, you would check the box under Vertex and next to the transformers layer in the Snapping Environment window. When the pointer comes within the snapping tolerance of the transformer, the first vertex of the primary snaps to the vertex of the transformer.

You can also set the *snapping priority* for layers on your map. The order of layers listed in the Snapping Environment window determines the order in which snapping will occur. Snapping occurs first in the layer at the top of the list, then in each consecutive layer down the list. You can easily change the snapping priority by dragging the layer names to new locations.

Tip

Sketch and topology snapping properties

You can set snapping properties that apply specifically to the edit sketch and to topology elements in the Snapping Environment dialog box; these are located at the bottom of the Snapping Environment window. For more information, see 'Types of snapping properties' in this chapter.

Setting the snapping priority

1. Click Editor and click Snapping.

   The Snapping Environment window appears.

2. Click and drag the layer names to arrange them in the order in which you want snapping to occur. (The first layer in the list will be snapped to first.)

   The snapping priorities you set are effective immediately.

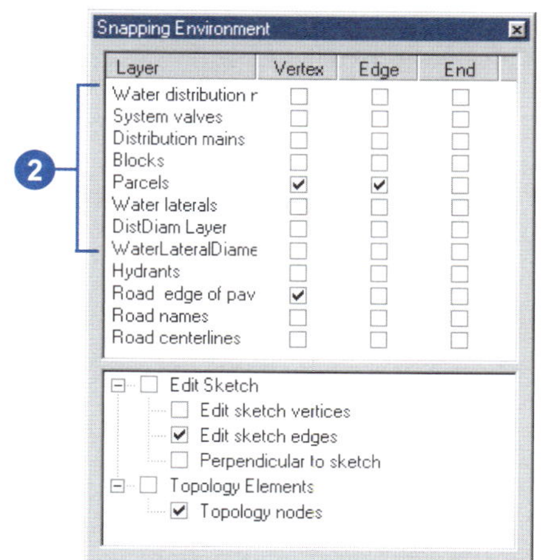

The Snap to Feature command

Apart from the rules set in the Snapping Environment, you can also snap on the fly to specific features. This is done through the Snap to Feature command of the sketch context menu.

By specifying a feature and the part of the feature to which you wish to snap, your next vertex will automatically be placed regardless of the Snapping Environment settings.

Snapping to a specific feature

1. Click the tool palette dropdown arrow and click the Sketch tool.
2. Right-click the feature to which you want to snap your next vertex.
3. On the dropdown menu, choose Snap to Feature.
4. Click the part of the feature to which you want to snap the next vertex.

   The vertex will be placed at the nearest location that matches your selection.

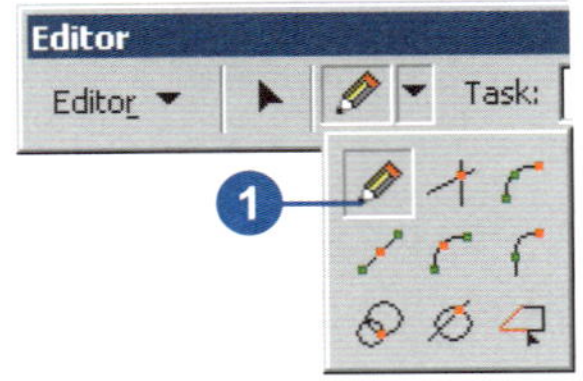

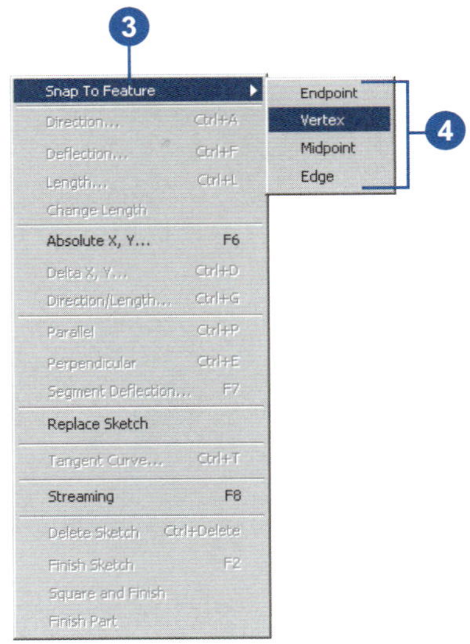

Setting the direction measuring system and units for editing tools

Some of the ArcMap editing tools allow you to enter an angle, direction, or deflection when constructing an edit sketch. These include the Direction–Distance tool and Traverse tool as well as several of the commands on the sketch context menu. You can change the direction measuring system and units these tools use on the Units tab of the Editing Options dialog box. When you change the direction measuring system and units, the editing tools will all recognize inputs in the new system and units.

Direction measuring systems

You can choose from the following direction measuring systems: North Azimuth, South Azimuth, Quadrant Bearing, and Polar. By default, the tools accept angular measurements in the Polar direction measuring system.

Polar angles are measured counterclockwise from the positive x-axis.

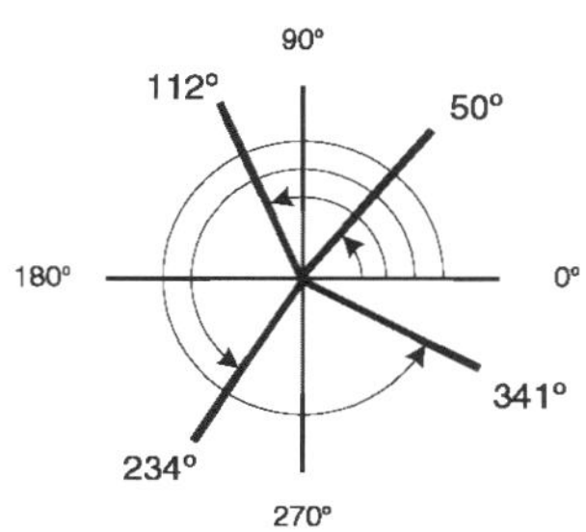

In the *North Azimuth* system the azimuth of a line is the horizontal angle measured from a meridian to the line, measured in the clockwise direction from north.

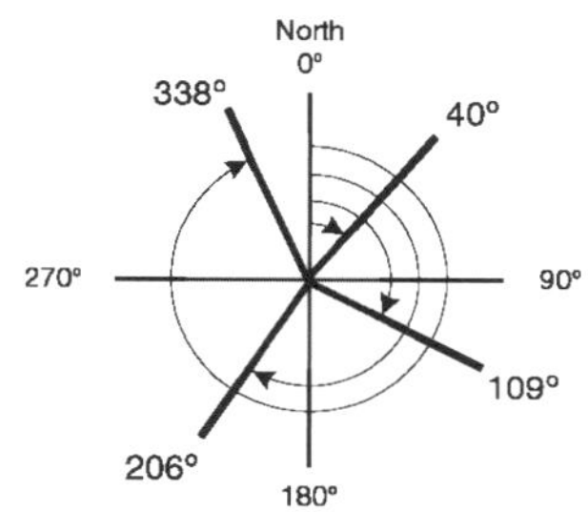

In the *South Azimuth* system the angles are measured clockwise from south.

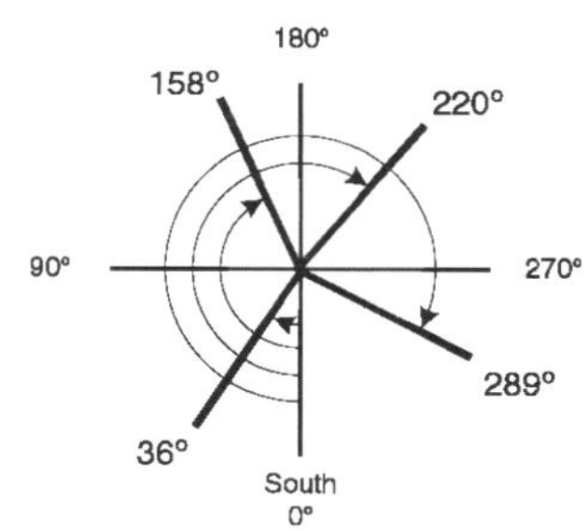

In the *Quadrant Bearing* system the bearing of a line is measured as an angle from the reference meridian, either the north or the south, and measured toward the east or the west. Bearings in the Quadrant Bearing system are written as a meridian, an angle, and a direction. For example, a bearing of N 25 W defines an angle 25 degrees west measured from north. A bearing S 18 E defines an angle 18 degrees east measured from the south.

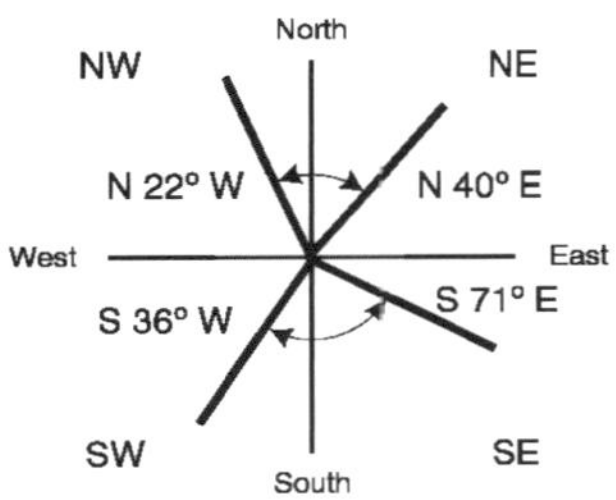

Valid input formats for Quadrant Bearing measurements include:

- [NS] dd.dddd [EW] where the first letter is an N or S, indicating the meridian of origin, and the last letter is an E or W, indicating which direction the angle is toward.
- dd.dddd-[1234] where the second to last character is a - (dash) and the last digit indicates the quadrant that the bearing is in. The quadrants are numbered 1 - NE, 2 - SE, 3 - SW, 4 - NW.

Direction measuring units

The editing tools use decimal degrees as their default units of angular measure. You can choose from the following direction measurement units: decimal degrees, degrees/minutes/seconds, radians, gradians, and gons.

Degrees are the standard unit of angular measurement, where one degree represents 1/360 of a circle, and fractions of a degree are represented as decimal values.

Degrees Minutes Seconds also uses the degree, but fractions of a degree are represented in minutes and seconds, where one minute equals 1/60 of a degree, and one second equals 1/60 of a minute.

Valid input formats for degrees/minutes/seconds values include:

- dd-mm-ss.ss
- dd.mmssss
- dd^mm'ss.ss"

Radians are the Standard International (SI) unit of plane angular measure. There are 2pi, approximately 6.28318, radians in a circle. One radian is equivalent to about 57.296 degrees. The length of a circular arc with an angle of one radian is equal to the radius of the arc.

Gradians are a unit of angular measure where the right angle is divided into 100 parts. One gradian equals 1/400 of a circle.

Gons are the same as gradians. One gon equals 1/400 of a circle. The term gon is primarily used in German, Swedish, and other northern European languages where the word grad means degree.

Setting the direction type and angular units

You can set the direction measurement system and the units with which you measure angles from the Units tab of the Editing Options dialog box. All of the editing tools that accept angular measurements will interpret angular measurements using the direction type and units that you specify here.

Setting the direction type

1. Click Editor and click Options.
2. Click the Units tab.
3. Click the Direction Type dropdown list and choose a direction measuring system.
4. Click OK.

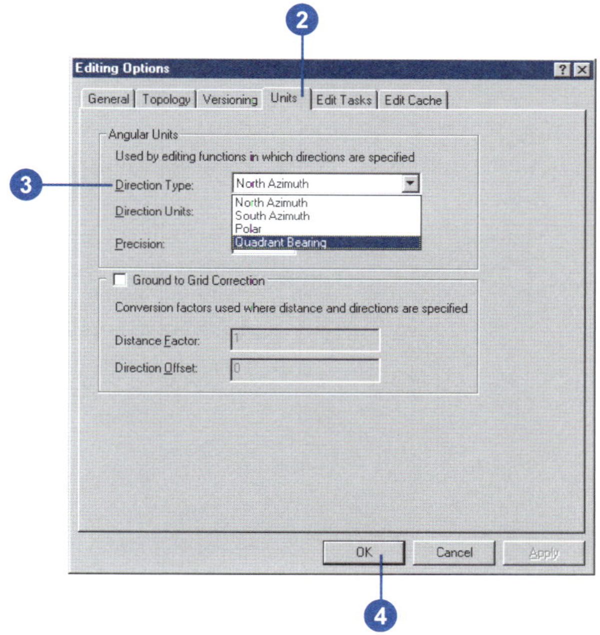

Setting the direction units

1. Click Editor and click Options.
2. Click the Units tab.
3. Click the Direction Units dropdown list and click the type of direction measurement unit you want to use.
4. Set Precision to specify the number of decimal places used when displaying angles and directions.
5. Click OK.

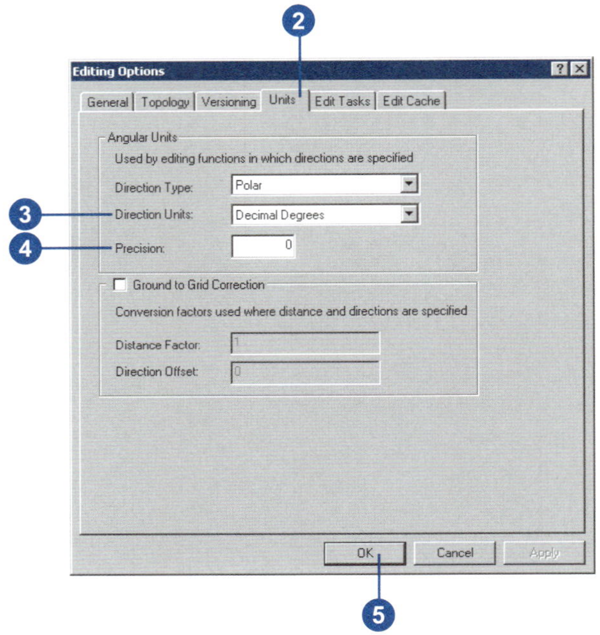

Setting the ground to grid conversion

Sometimes when creating features using the coordinate geometry tools in ArcMap, you need to convert angles and distances measured in the field so they match the coordinate system of your data. For example, if you are given distances measured in ground units at a high elevation, the distances will need to be adjusted slightly to fit your GIS dataset's coordinate system, in which distances are assumed to have been measured at a given ellipsoid—or sea level—surface. The distance conversion is applied as a scale factor to distances you type into the coordinate geometry editing tools.

Sometimes you need to correct angular measurements taken in the field as you enter them in your GIS. For example, suppose you are creating features using coordinate geometry from a set of measurements on a surveyor's field notes. The surveyor's angular measurements were based on True North—defined using the astronomical meridian. Your GIS dataset has a coordinate system where north differs by 0 degrees, 2 minutes, and 3 seconds from true north. Instead of going through the field notes and manually converting the measurements, you can set a direction offset to automatically correct the angles as you type them.

Direction and distance corrections can be calculated by measuring directions and distances between control points in the GIS and comparing them to directions and distances measured on the ground, as shown on a survey plan.

Using a ground to grid correction

You can set distance and direction conversion factors to allow you to correct for differences between your GIS coordinate system and the angles and distances a surveyor may have measured in the field.

Setting a ground to grid distance conversion factor

1. Click Editor and click Options.
2. Click the Units tab.
3. Check Ground to Grid Correction.
4. Type a conversion factor to convert distances measured in the field (ground distance) to distances in the GIS (grid distance).
5. Click OK.

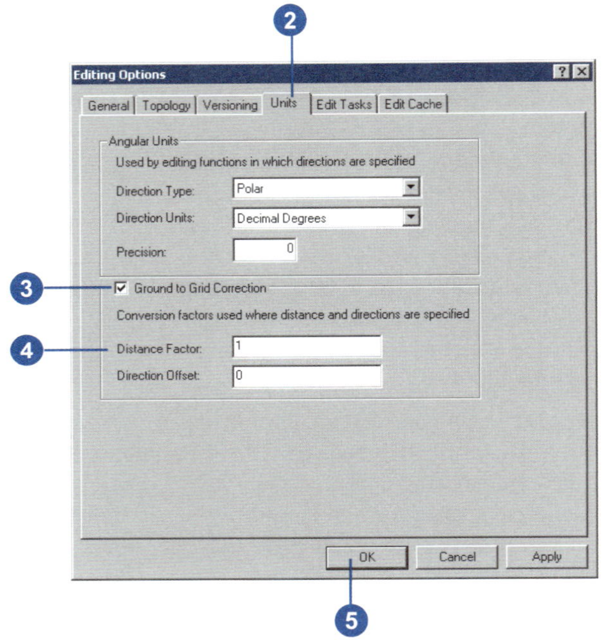

Setting a ground to grid direction offset

1. Click Editor and click Options.
2. Click the Units tab.
3. Check Ground to Grid Correction.
4. Type the desired number of measurement units in the Direction Offset text box.
5. Click OK.

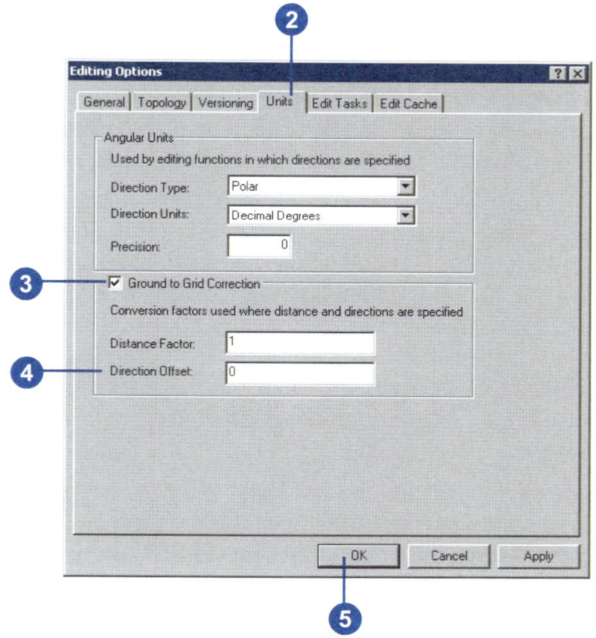

Setting the distance units for tools

Sometimes when creating features using the editing tools in ArcMap you need to enter data that was recorded in different distance units than the coordinate system of your data.

For example, suppose your data is in a State Plane coordinate system and the linear units are U.S. Survey Foot (1 Foot_US = 0.3048006096 m). You are given measurements in International Feet (1 Foot = 0.3048 m). Rather than convert all of the measurements, you can type the abbreviation for International Foot, ft, after the measurements and the tools will convert the distance correctly. You could also change the dataset coordinate system definition to use International Feet, and type the distances without a unit suffix.

Whenever you are typing a distance into one of the editing tools, you have the option to specify the linear units to use or to simply type a number, which the tool will interpret as being in the dataset's coordinate system units.

Distance Units	Abbrev.	Meters per Unit	Description
Metric Units			
Kilometer	km	1,000	1,000 meters exactly
Meter	m	1	International meter
Millimeter	mm	0.001	1/1000 meter exactly
Imperial or international units			
Foot	ft	0.3048	Standard foot used in the United States. Also known as international foot or imperial foot that was used in most non-U.S. countries before the metric system.

Distance Units	Abbrev.	Meters per Unit	Description
Imperial or international units (Cont'd)			
Mile	mi	1,609.344	Also referred to as a statute mile, equal to 5,280 international feet.
Nautical mile	nm	1,852	The nautical mile is a unit of distance used primarily in sea and in aviation. The nautical mile is defined to be the average distance on the earth's surface represented by one minute of latitude. In 1929, the nautical mile was defined to be exactly 1,852 meters or 6,076.11549 feet, a distance known as the international nautical mile.
Chain	ch	20.1168	66 international feet
Yard	yd	0.9144	Three international feet
Rod	rd	5.0292	1/4 chain or 16.5 international feet
Link	lk	0.201168	1/100 international chain or 66/100 international feet
Inch	in	0.0254	1/12 of an international foot

Distance Units	Abbrev.	Meters per Unit	Description
		U.S. survey units	
Survey foot	ftUS	0.304800-6096	The U.S. survey foot is used in the State Plane Coordinate Systems. In the United States, fundamental survey units such as rods, chains, statute miles, acres, sections, and townships all depend on the U.S. survey foot. An exact conversion to meters can be accomplished by multiplying U.S. survey feet by the fraction 1200/3937.
Survey mile	miUS	1609.347-2186944	5,280 survey feet
Survey chain	chUS	20.11684-02337	66 survey feet
Survey rod	rdUS	5.029210-0584	1/4 survey chain
Survey link	lkUS	0.201168-4023	1/100 survey chain
Survey yard	ydUS	0.914401-8288	3 survey feet

Editing topology

4

IN THIS CHAPTER

- **An introduction to topology**
- **Topology rules, errors, and exceptions**
- **Editing features in a topology**
- **Adding the topology toolbar**
- **Creating a map topology**
- **Editing shared geometry**
- **Rebuilding the topology cache**
- **Using the edit sketch to make topology elements**
- **Changing symbology for errors and topology features**
- **Validating edits to a topology**
- **Creating new features from existing features**

In addition to simple features, ArcMap lets you edit collections of features related by a topology. In a geodatabase, a topology contains rules that define how features share space.

Topologies are useful for maintaining high-quality spatial data by ensuring that your features conform to simple rules. When you set up a geodatabase with a topology, you specify which feature classes participate in the topology and define rules that control when and how features can share geometry. For example, in a geodatabase of states and coastlines, you might have a rule that states polygons cannot overlap each other and another rule that the coastline must coincide with the boundaries of the states. These rules would help you ensure, when updating the state boundaries to match the coastline, that you don't inadvertently create places where the states overlap, or where a state's coastal edge doesn't match the shape of the coastline.

ArcMap allows you to simultaneously edit multiple features from layers that share geometry. Only those features that participate in the topology will be affected by these edits. For instance, you can update lot lines that define a parcel, and update the corresponding parcel polygon feature in your geodatabase. Similarly, you could move a road centerline and, at the same time, update all of the bus routes that follow that road. In this respect, editing topologies is a bit like editing features in a geometric network. However topologies offer many more possible ways that polygon, arc, and point features can be related than networks do. ArcMap provides some new tools for editing and managing topologies. With ArcView licensed seats of ArcMap you can edit shared geometry by creating a map topology.

What is topology?

Topology has historically been viewed as a spatial data structure used primarily to ensure that the associated data forms a consistent and clean topological fabric. With advances in object-oriented GIS development, an alternative view of topology has evolved. The geodatabase supports an approach to modeling geography that integrates the behavior of different feature types and supports different types of key relationships. In this context, topology is a collection of rules and relationships that, coupled with a set of editing tools and techniques, enables the geodatabase to more accurately model geometric relationships found in the world.

Topology, considered from the feature behavior perspective, allows a more flexible set of geometric relationships to be modeled than the data structure perspective. It also allows topological relationships to exist between more discrete types of features within a feature dataset. In this alternative view, topology may still be employed to ensure that the data forms a clean and consistent topological fabric; but also, more broadly, it is used to ensure that the features obey the key geometric rules defined for their role in the database.

Why use topology?

Topology is used most fundamentally to ensure data quality and to allow your geodatabase to more realistically represent geographic features. A geodatabase provides a framework within which features can have behavior, such as subtypes, default values, attribute domains, validation rules, and structured relationships to tables or other features. This behavior enables you to more accurately model the world and maintain referential integrity between objects in the geodatabase. Topology may be considered an extension of this framework for behavior that allows you to control the geometric relationships between features and to maintain their geometric integrity. Unlike other feature behavior, topology rules are managed at the level of the topology and dataset, not for individual feature classes.

How do I work with topology?

Different people work with topology in different ways, depending upon their role in an organization and its GIS design and management work flow.

Initially, creating a topology requires a geodatabase designer. A topology organizes the spatial relationships between features in a set of feature classes. The designer analyzes an organization's data modeling needs, identifies the key topological relationships required in the geodatabase, and defines the rules that will constrain different features' topological relationships.

Once the participating feature classes have been added to the topology and the rules defined, the topology is validated. Data quality managers use the topology tools to analyze, visualize, report, and, where necessary, repair the spatial integrity of the database after it is initially created as well as after editing. Topology provides these users with a set of validation rules for the topologically related features. It also provides a set of editing tools that let users find and fix integrity violations.

As the geodatabase is used and maintained, new features are added and existing features are modified. Data editors update features in the geodatabase and use the topology tools to construct and maintain relationships between features, within the constraints imposed by the database designer. Depending on the work flow of the organization, the topology may be validated after each edit session or on a schedule.

Topology basics for data editors

Topologies store three sets of parameter rules, ranks, and a cluster tolerance. When editing a geodatabase, you will not typically need to modify these parameters, but you will need to be aware of them, especially the rules.

Topologies also maintain a feature layer that stores *dirty areas*, *errors*, and *exceptions*. You use these to maintain the quality of data in your topology.

The sections that follow describe each of these parameters and concepts in more detail.

Rules

Rules define the permissible spatial relationships between features. The rules you define for a topology control the relationships of features allowed within a feature class, between features in different feature classes, or between subtypes of features.

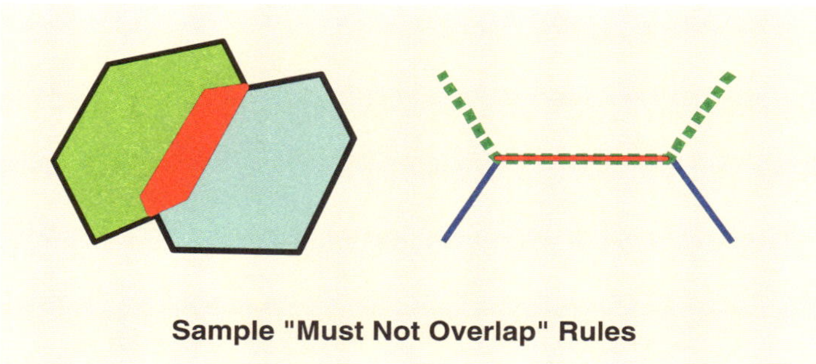

Example of a "Must Not Overlap" rule applied to polygons and lines. The red polygon and line mark the places where the rule is violated. These are stored in the topology as error features. Such rules can apply to features within the same feature class, to pairs of feature classes, or to subtypes of features.

The initial validation of the topology checks all of the features against all of the rules. This initial check can take some time, but subsequent checks are performed only on the areas that have been edited—the dirty areas.

Cluster tolerance

The *cluster tolerance* defines how close vertices must be to each other in order to be considered coincident and limits the distance features can move during *validation*. The cluster tolerance is the minimum distance between vertices of features that are not coincident. Vertices that fall within the cluster tolerance are defined as coincident and are snapped together. The cluster tolerance is typically a very small actual distance to minimize the movement of correctly placed features.

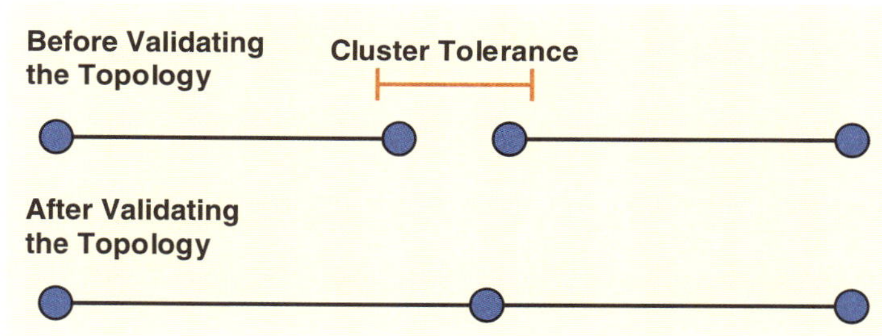

When you validate a topology, features within the cluster tolerance are snapped together.

Ranks

Ranks control which features may be moved to other features when snapping the topology together during validation. The ranks you specify for feature classes in the topology control which feature classes will be moved when snapping coincident vertices during the initial validation of the topology as well as subsequent validations.

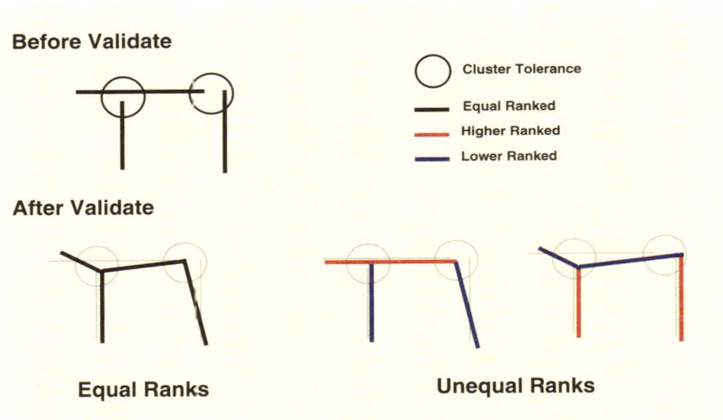

When you validate a topology, the ranks of the feature classes in the topology control how features are snapped together. Lower-ranking features snap to higher-ranking features. Equally ranked features snap to the geometric average of their position.

When different feature classes have different levels of intrinsic reliability, such as when one was collected by survey or differential global positioning system (GPS) and another was digitized from less accurate source material or collected with uncorrected GPS, ranks can allow you to ensure that reliably placed vertices are not snapped to the location of less reliable vertices. Lower-ranked features' vertices will be snapped to the location of higher-ranked vertices, if they fall within the cluster tolerance. The location of equally ranked vertices are geometrically averaged when they fall within the cluster tolerance.

Feature layers maintained by a topology

Instead of storing topological information for all features, the topology discovers those relationships when the information is requested, such as when you are editing using the Topology Edit tool. The topology stores some feature layers that let it efficiently track the places where the topology may have been violated during editing—dirty areas—and features that were found to violate topology rules after validation—error features. Certain errors may be acceptable, in which case the error features are marked and stored as exceptions.

Dirty areas

Dirty areas are areas that have been edited, updated, or affected by the addition or deletion of features. Dirty areas allow the topology to limit the area that must be checked for topology errors during topology validation. Dirty areas track the places where topology rules may have been violated during editing. This allows selected parts, rather than the whole extent of the topology, to be validated after editing.

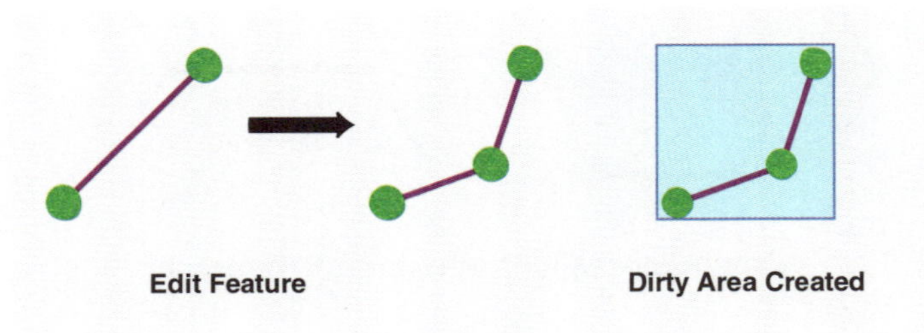

When you edit features in a topology, the topology creates a dirty area to mark the area that should be checked for violations of the topology rules.

Dirty areas are created when:

- A feature is created or deleted
- A feature's geometry is modified
- A feature's subtype is changed
- Versions are reconciled
- The topology properties are modified

Dirty areas are stored in the topology as a single feature, with each new dirty area joined with the existing dirty area, and each area that has been validated removed from the dirty area.

Errors and exceptions

Errors and exceptions are stored as features in the topology layer and allow you to render and manage the cases in which features do not obey the rules of the topology. Error features record where topological errors were discovered during validation. Certain errors may be acceptable, in which case the error features can be marked as exceptions.

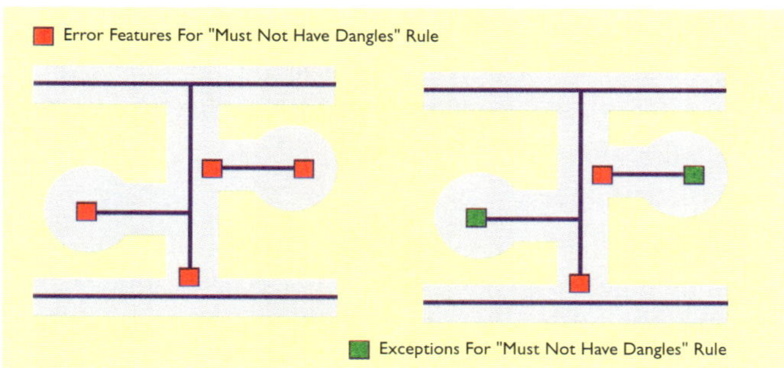

When you validate a topology, features that violate the rules are marked as error features. You can edit the features to fix the errors, or you can mark the errors as exceptions. In this example, the street line features cannot have dangles, which are endpoints that do not connect to other street features. Because cul-de-sac streets are a legitimate exception to this rule, they may be marked as exceptions in the topology. The remaining errors should be fixed by editing the street features.

ArcMap and ArcCatalog allow you to create a report of the total number of errors and exceptions for the feature classes in your topology. You can use the report of the number of error features as a measure of the data quality of a topological dataset. The error inspector in ArcMap lets you select different types of errors and zoom to individual errors. You can correct topology errors by editing the features that violate the topology's rules. After you validate the edits, the error is deleted from the topology.

Topology rules

Many topology rules can be imposed on features in a geodatabase. A well-designed geodatabase will have only those topology rules that define key spatial relationships needed by an organization.

Some topology rules govern the relationships of features within a given feature class, while others govern the relationships between features in two different feature classes. Topology rules can also be defined between subtypes of features in one or another feature class. For example, a topology rule can require street features to be connected to other street features at both ends, except in the case of streets belonging to the cul-de-sac or dead-end subtypes.

Some of the key topological rules that you may encounter are discussed in the following pages.

Polygon rules

Must Not Overlap

This rule requires that the interior of polygons in the feature class not overlap. The polygons can share edges or vertices. This rule is used when an area cannot belong to two or more polygons. It is useful for modeling administrative boundaries, such as ZIP Codes or voting districts, and mutually exclusive area classifications, such as land cover or landform type.

Must Not Have Gaps

This rule requires that polygons not have voids within themselves or between adjacent polygons. Polygons can share edges, vertices, or interior areas. Polygons can also be completely disconnected. This rule is used when polygons or blocks of contiguous polygons should not have empty spaces within them. It is useful for modeling landownership, as in a parcel fabric, where a given area is completely allotted to various polygons, but where external areas—roadways, for example—are not modeled in the same feature class.

Must Not Overlap With

This rule requires that the interior of polygons in one feature class must not overlap with the interior of polygons in another feature class. Polygons of the two feature classes can share edges or vertices or be completely disjointed. This rule is used when an area cannot belong to two separate feature classes. It is useful for combining two mutually exclusive systems of area classification, such as zoning and water-body type, where areas defined within the zoning class cannot also be defined in the water-body class and vice versa.

Must Be Covered By Feature Class Of

This rule requires that a polygon in one feature class must share all of its area with polygons in another feature class. An area in the first feature class that is not covered by polygons from the other feature class is an error. This rule is used when an area of one type, such as a state, should be completely covered by areas of another type, such as counties.

Must Cover Each Other

This rule requires that the polygons of one feature class must share all of their area with the polygons of another feature class. Polygons may share edges or vertices. Any area defined in either feature class that is not shared with the other is an error. This rule is used when two systems of classification are used for the same geographic area and any given point defined in one system must also be defined in the other. One such case occurs with nested hierarchical datasets, such as census blocks and block groups or small watersheds and large drainage basins. The rule can also be applied to nonhierarchically related polygon feature classes, such as soil type and slope class.

Must Be Covered By

This rule requires that polygons of one feature class must be contained within polygons of another feature class. Polygons may share edges or vertices. Any area defined in the contained feature class must be covered by an area in the covering feature class. This rule is used when area features of a given type must be located within features of another type. This rule is useful when modeling areas that are subsets of a larger surrounding area, such as management units within forests or blocks within block groups.

Boundary Must Be Covered By

This rule requires that boundaries of polygon features must be covered by lines in another features class. This rule is used when area features need to have line features that mark the boundaries of the areas. This is usually when the areas have one set of attributes and their boundaries have other attributes. For example, parcels might be stored in the geodatabase along with their boundaries. Each parcel might be defined by one or more line features that store information about their length or the date surveyed, and every parcel should exactly match its boundaries.

Area Boundary Must Be Covered By Boundary Of

This rule requires that boundaries of polygon features in one feature class be covered by boundaries of polygon features in another feature class. This is useful when polygon features in one feature class, such as subdivisions, are composed of multiple polygons in another class, such as parcels, and the shared boundaries must be aligned.

Contains Point

This rule requires that a polygon in one feature class contain at least one point from another feature class. Points must be within the polygon, not on the boundary. This is useful when every polygon should have at least one associated point, such as when parcels must have an address point.

Line rules

Must Not Overlap

This rule requires that lines not overlap with lines in the same feature class. This rule is used where line segments should not be duplicated—for example, in a stream feature class. Lines can cross or intersect but cannot share segments.

Must Not Intersect

This rule requires that line features from the same feature class not cross or overlap each other. Lines can share endpoints. This rule is used for contour lines that should never cross each other or in cases where the intersection of lines should only occur at endpoints, such as street segments and intersections.

Must Not Have Dangles

This rule requires that a line feature must touch lines from the same feature class at both endpoints. An endpoint that is not connected to another line is called a dangle. This rule is used when line features must form closed loops, such as when they are defining the boundaries of polygon features. It may also be used in cases where lines typically connect to other lines, as with streets. In this case, exceptions can be used where the rule is occasionally violated, as with cul-de-sac or dead-end street segments.

Must Not Have Pseudonodes

This rule requires that a line connect to at least two other lines at each endpoint. Lines that connect to one other line, or to themselves, are said to have pseudonodes. This rule is used where line features must form closed loops, such as when they define the boundaries of polygons or when line features logically must connect to two other line features at each end, as with segments in a stream network, with exceptions being marked for the originating ends of first-order streams.

Must Not Intersect Or Touch Interior

This rule requires that a line in one feature class must only touch other lines of the same feature class at endpoints. Any line segment in which features overlap, or any intersection not at an endpoint, is an error. This rule is useful where lines must only be connected at endpoints, such as in the case of lot lines, which must split (only connect to the endpoints of) back lot lines and which cannot overlap each other.

Must Not Overlap With

This rule requires that a line from one feature class not overlap with line features in another feature class. This rule is used when line features cannot share the same space—for example, roads must not overlap with railroads or depression subtypes of contour lines cannot overlap with other contour lines.

Must Be Covered By Feature Class Of

This rule requires that lines from one feature class must be covered by the lines in another feature class. This is useful for modeling logically different but spatially coincident lines, such as routes and streets. A bus route feature class must not depart from the streets defined in the street feature class.

Must Be Covered By Boundary Of

This rule requires that lines be covered by the boundaries of area features. This is useful for modeling lines, such as lot lines, that must coincide with the edge of polygon features, such as lots.

Endpoint Must Be Covered By

This rule requires that the endpoints of line features must be covered by point features in another feature class. This is useful for modeling cases where a fitting must connect two pipes or a street intersection must be found at the junction of two streets.

Must Not Self Overlap

This rule requires that line features not overlap themselves. They can cross or touch themselves, but must not have coincident segments. This rule is useful for features such as streets, where segments might touch, in a loop, but where the same street should not follow the same course twice.

Must Not Self Intersect

This rule requires that line features not cross or overlap themselves. This rule is useful for lines, such as contour lines, that cannot cross themselves.

Must Be Single Part

This rule requires that lines must have only one part. This rule is useful where line features, such as highways, may not have multiple parts.

Point rules

Must Be Covered By Boundary Of

This rule requires that points fall on the boundaries of area features. This is useful when the point features help support the boundary system, such as boundary markers, which must be found on the edges of certain areas.

Must Be Properly Inside Polygons

This rule requires that points fall within area features. This is useful when the point features are related to polygons, such as wells and well pads or address points and parcels.

Must Be Covered By Endpoint Of

This rule requires that points in one feature class must be covered by the endpoints of lines in another feature class. This rule is similar to the line rule 'Endpoint Must Be Covered By' except that, in cases where the rule is violated, it is the point feature that is marked as an error, rather than the line. Boundary corner markers might be constrained to be covered by the endpoints of boundary lines.

Must Be Covered By Line

This rule requires that points in one feature class must be covered by lines in another feature class. It does not constrain the covering portion of the line to be an endpoint. This rule is useful for points that fall along a set of lines, such as highway signs along highways.

Topology errors and exceptions

Topology rules may represent an ideal situation, but geodatabases are flexible enough to handle exceptions to the rules found in real-world data. Violations of topology rules are initially stored as errors in the topology, but where appropriate, you can mark them as exceptions. Exceptions are thereafter ignored, though you can return them to error status if you decide that they are actually errors and that the features should be modified to comply with the topology rules.

Exceptions are a normal part of the data creation and update process. An assessor's geodatabase might have a topology rule requiring that building features not cross parcel lines as a quality control for the building digitizing effort. This rule might be true for 90 percent of the features in the city, but it could be violated by some high-density housing and commercial buildings.

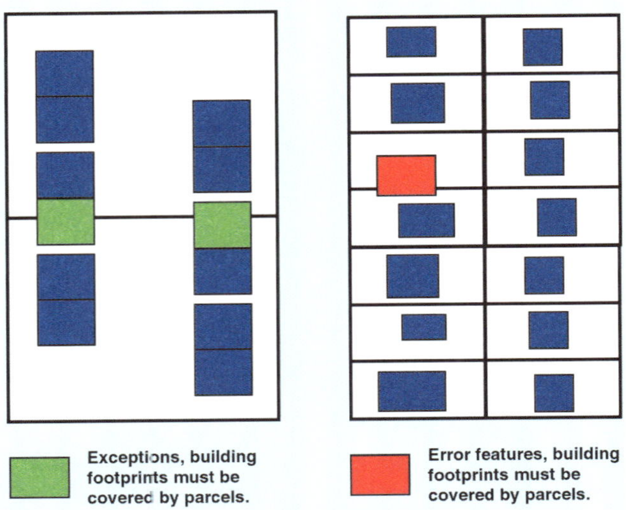

If you create a condominium building feature that crosses parcel boundaries, it will be discovered as an error when you validate your edits, but you can mark it as a legitimate exception to the rule. Similarly, a street database for a city might have a rule that centerlines must connect at both ends to other centerlines. This rule would normally ensure that street segments are correctly snapped to other street segments when they are edited. However, at the boundaries of the city you might not have street data. Here, the external ends of streets might not snap to other centerlines. These cases could be marked as exceptions, and you would still be able to use the rule to find cases where streets were incorrectly digitized or edited.

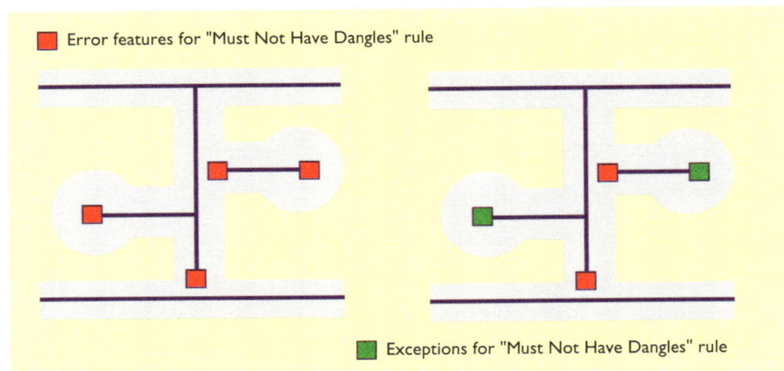

Topology errors can be fixed quickly using the Fix Topology Error tool. This tool allows you to select a topology error and choose from a number of fixes that have been predefined for that error type. You can also use the tool to get more information about the rule that has been violated or mark the error as an exception.

Geometric elements of a topology

When you create a topology, you specify the feature classes that participate in the topology. These feature classes may contain point, line, or polygon features. In the topology, the geometric relationships are between the parts of the features rather than the features themselves. Polygons in a topology have edges that define the boundary of the polygons, nodes where edges intersect, and vertices, which define the shape of the edges.

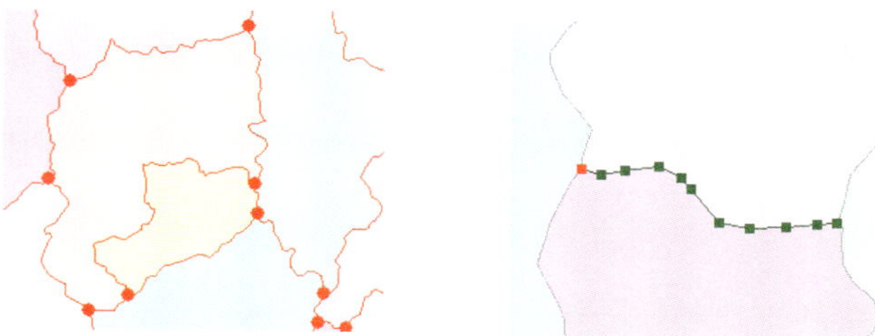

Polygon features share edges and nodes, in red. Vertices define the shape of the edges, in green.

Similarly, line features are made up of an edge, at least two nodes that define the endpoints of the edge, and vertices that define the shape of the edge. Point features behave as nodes when they are coincident with other features in a topology.

Line features share edges and nodes, in red. Vertices define the shape of the edges, in green.

When features in the topology have parts that intersect or overlap, the edges and nodes that define these parts are shared.

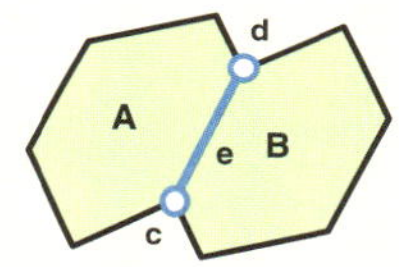

Polygons A and B have shared nodes c and d and shared edge e.

Lines A and B have endpoint nodes c, d, and e.
Lines A and B share node e.

You can use the Topology Edit tool to move nodes and whole edges that are shared between features or to move the vertices that define the shape of shared edges.

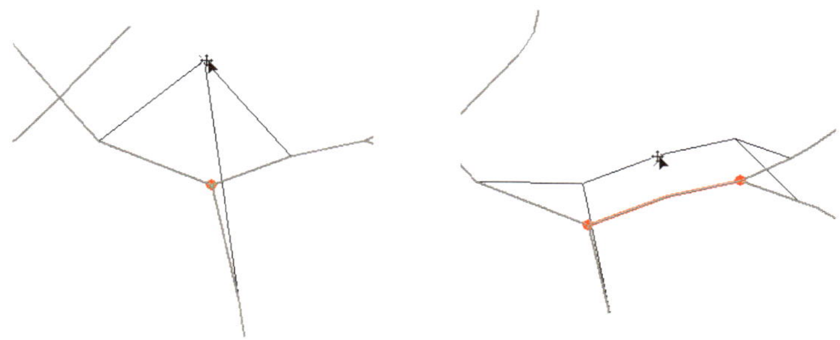

Moving a node stretches the connected edges so they stay connected. Moving an edge with its endpoint nodes also stretches the edges attached to the nodes.

When you move nodes or vertices, you can choose whether you want the segment between the vertex and the closest vertex to be stretched or whether you want the whole edge to be proportionately stretched.

You can choose whether to stretch the feature geometry proportionally when moving a vertex, or whether to just stretch the segments between the vertex and the next vertices.

You can also temporarily add new topology nodes to split edges. This simply splits the edge for the topology; it does not break the feature into two features. This can be useful when you want to move one part of an edge without affecting other parts of the edge or when you want to create a new node to snap to.

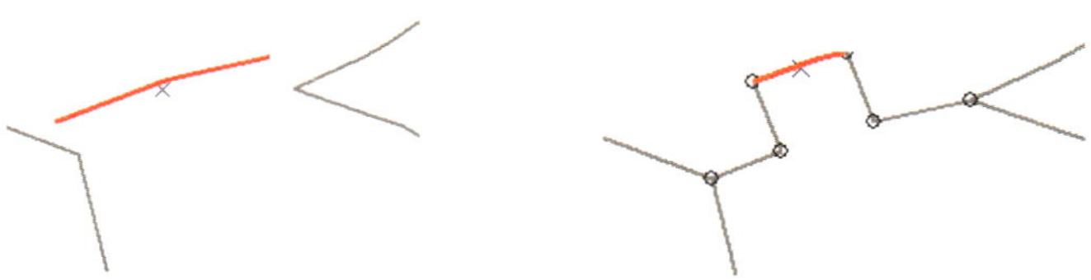

Moving an edge that defines a single feature without selecting its nodes moves the whole edge without maintaining connections to the nodes. Moving a split edge that is a part of a single feature does maintain connectivity within the feature.

You can find out which features share a given topology element and control whether or not the geometry should be considered shared with the Show Shared Features tool.

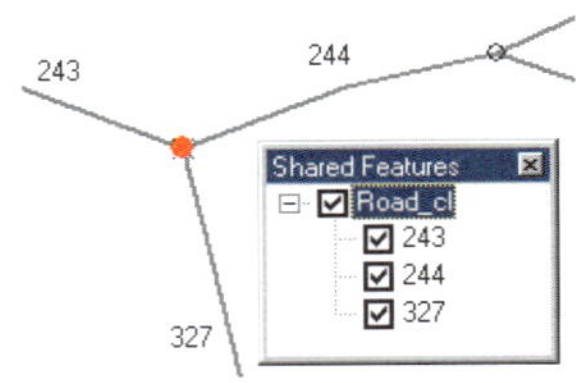

If two or more features share an edge or node, you can use the Show Shared Features tool to turn off geometry sharing for one or more of the features. Changes that you make to that topology edge or node with the Topology Edit tool will only affect the features for which the geometry is still shared.

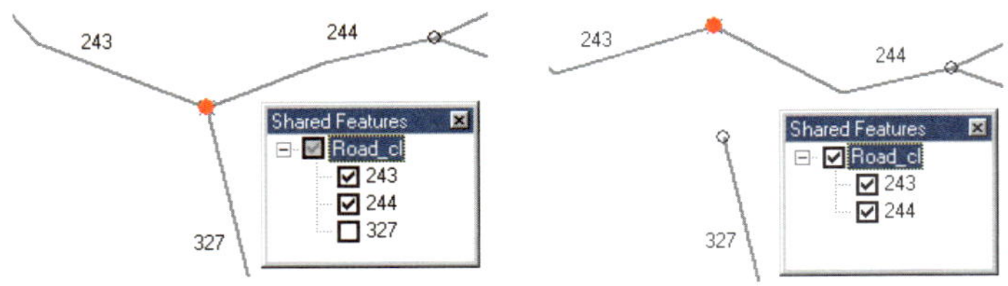

Uncheck a feature with the Shared Features tool in order to make a shared topology element independent of that feature.

Editing features in a topology

Editing features that participate in a topology is similar to editing simple features—in fact, you can use the same sketch tools to create new features that participate in a topology that you would use for features that do not. When you want to modify a feature that shares edges or nodes with other features in the topology, you can use the Topology Edit tool.

When editing topological features, you often have a choice of several ways of doing something. For example, suppose you manage a forest and there are two polygon feature classes, Forest and Stand, in your forest dataset.

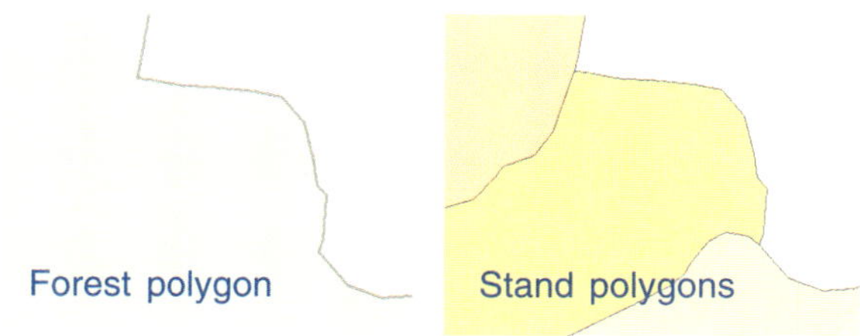

The Forest and Stand polygons are related by three topology rules: Stand must not overlap, Forest must cover Stand, and Stand must cover Forest. These rules prevent Stand polygons from overlapping, since no area can be in two stands at once. They also prevent stands from extending outside of the official boundary of the whole forest, or the forest boundary from covering an area not in a stand.

Suppose you are editing the Stand feature class and need to change the boundary between two of the stands. You could start editing; use the Topology Edit tool to select the shared edge; double-click the edge to edit its vertices; then add, remove, and move vertices along that edge to shape it to fit the new boundary.

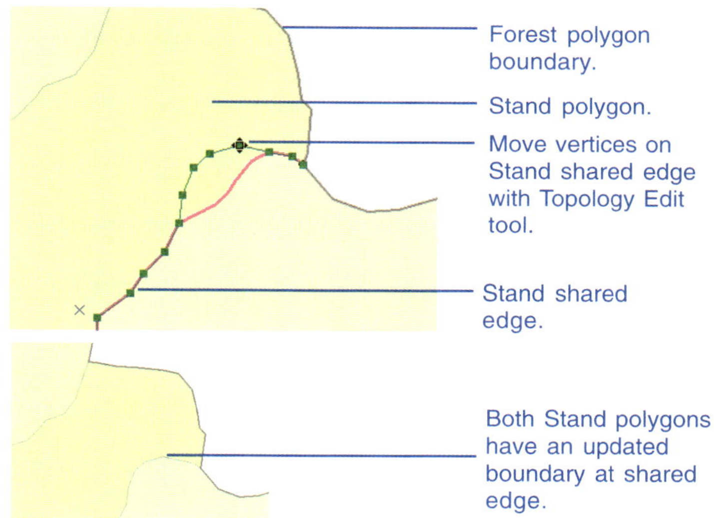

Use the Topology Edit tool to edit an edge shared by two polygons from the same feature class.

You could also use the Topology Edit tool to modify edges shared by the Stand and Forest polygons. For example, suppose the actual boundary of the forest has been determined to be 150 meters east and 20 meters north of the corner where two stand features meet at the edge of the forest. You could use the Topology Edit tool to select the topology node at this intersection of features and move it to the correct location.

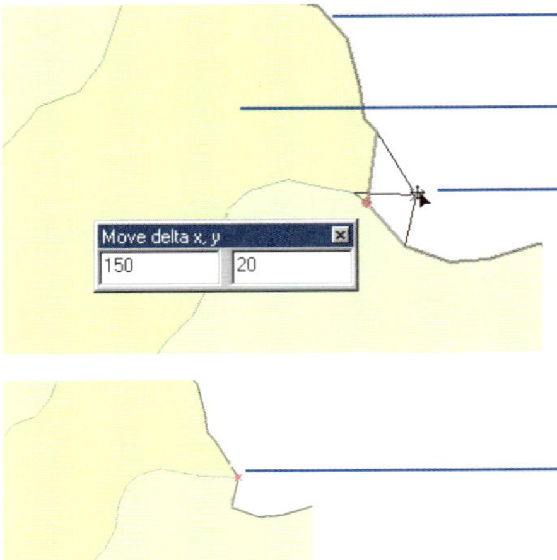

Use the Topology Edit tool to edit a topology node shared by two polygons from the same feature class and a third polygon from another feature class. You can move the node freehand, move it relative to its current position, or move it to an absolute location. When the node moves, the edges connected to it in both the Forest and Stand polygon feature classes are stretched to stay connected to it.

You can also use the regular editing tools to edit individual features that participate in a topology. When you edit topologically related features using the nontopological editing tools, you are only modifying one feature at a time. If this feature shares geometry with other features, the shared geometry is not updated. If the edits create a violation of the topology rules, you can use the Error Inspector to find the error and the Fix Topology Error tool to fix the error. There are several predefined ways that you can fix a given type of topology error. The Fix Topology Error tool allows you to right-click an error and choose which fix to use for the error.

1) Edit a feature with an editing tool.

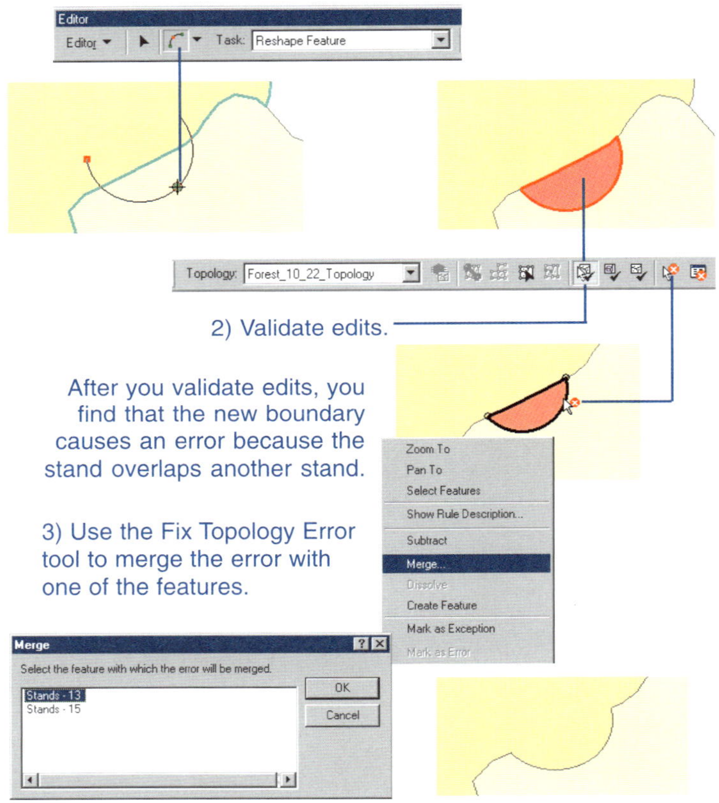

You can also edit features with the nontopological Edit tool. If the features participate in a topology rule, you can validate the topology to identify errors that your edits may have caused. If the edit created an error, you can use the Fix Topology Error tool to fix it using one of the predefined topology error fixes. The shared geometry is updated by the fix. You can also apply edits to shared edges using an edit sketch with the Modify Edge and Reshape Edge topology edit tasks.

The Modify Edge and Reshape Edge topology edit tasks allow you to update a selected shared edge using an edit sketch.

1. Reshape an edge using Reshape Edge editing task and the edit sketch tools.

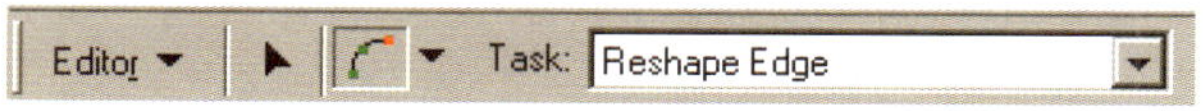

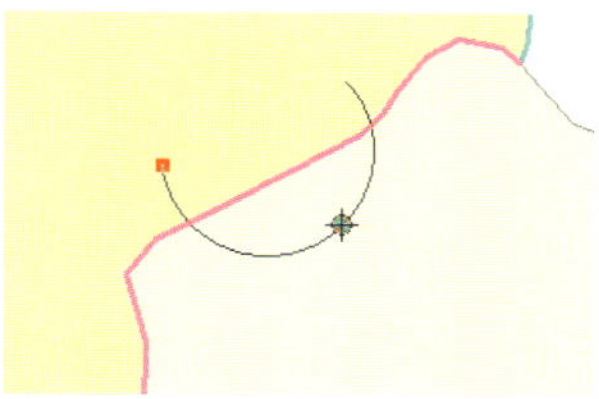

2. Finish the sketch. All features that share the edge are updated.

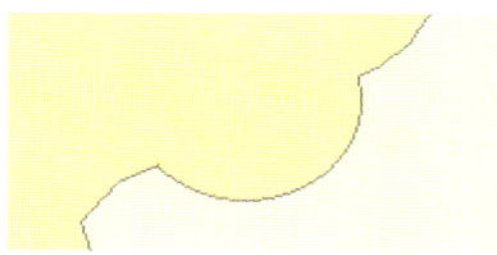

You can use the editor tools and the Reshape Edge and Modify Edge edit tasks to simultaneously edit several features that share an edge.

Correcting topology errors

There are a couple of ways to correct topology errors once you've discovered them. You can select the error on the map with the Fix Topology Error tool, or select the error from within the Error Inspector and apply one of the fixes listed in the context menu for that error type.

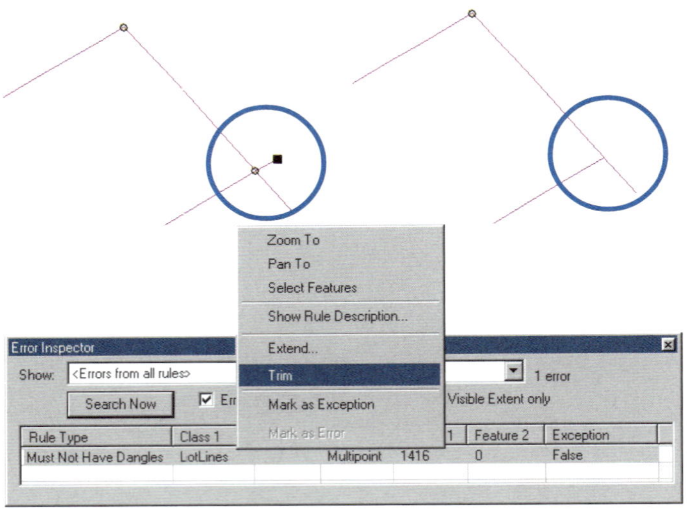

Different error types have different predefined fixes available for them. For example, a dangling line can be trimmed or extended to another line. Errors caused by overlapping polygons can be merged into one of the polygons, subtracted from both, or turned into a separate new polygon feature.

Errors caused by violations of the "must be covered by" rules can be fixed by creating a new feature or deleting a feature.

Making new features with topology tools

There are several ways that you can use topology tools to make new features from existing ones. In ArcCatalog you can create a whole new polygon feature class from an input set of line features. In ArcMap you can construct new polygon features from the intersection of selected existing line and polygon features or create new line features by splitting selected line features where they cross each other.

Creating polygons from lines in ArcCatalog

In ArcCatalog, the Polygon Feature Class From Lines tool takes one or more existing line or polygon feature classes in a feature dataset and creates new polygon features from the closed shapes that are defined by the intersection of all of the lines or polygon edges.

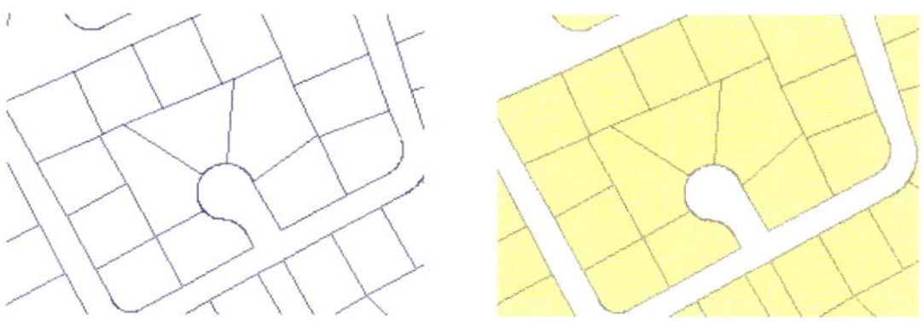

You have the option to use a point feature class to supply attributes for the new polygons. If a point falls within one of the new polygons, the polygon is assigned the point's attributes.

Creating features in ArcMap

In ArcMap, the Construct Features tool takes selected features from one or more feature classes and creates new features in the target feature class.

The tool uses the input geometries of the selected features to construct polygons or lines following polygon boundaries, depending upon the geometry of the target feature class.

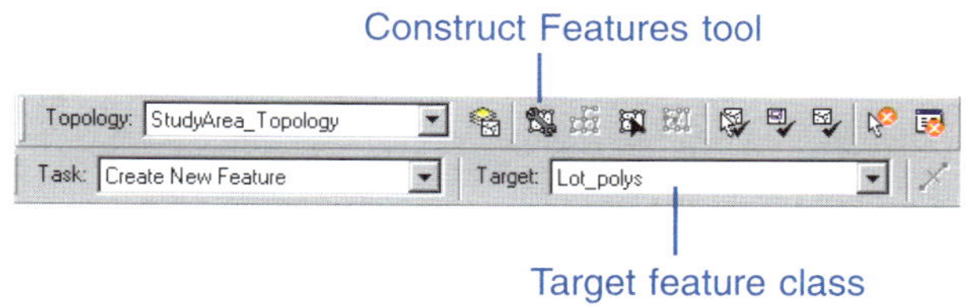

Selected lines can be turned into a polygon in the target feature class.

You can use this tool to build parcel polygons from selected lot line features or lake shorelines from selected lake polygon features.

Selected polygon features can be turned into line features in the target feature class.

You can also use the Planarize tool to create separate line features from selected touching or crossing line features. This can be useful when you have nontopological linework that has been spaghetti digitized or imported from a computer-aided design (CAD) drawing.

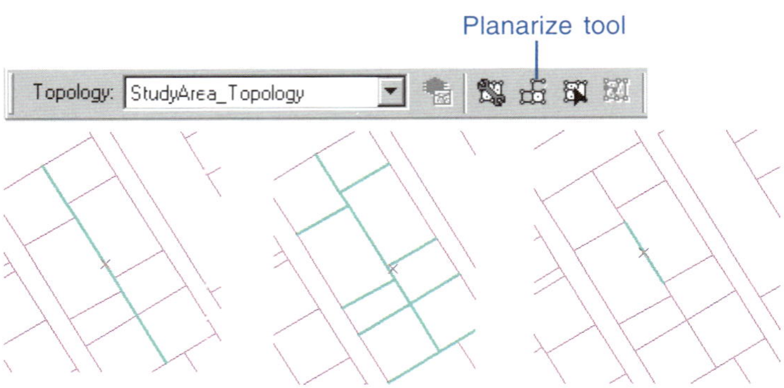

Selected line features can be split into separate features where they touch or cross other selected line features.

Adding the topology toolbar

The topology toolbar contains tools that you can use to create a map topology as well as tools that you can use to work with map and geodatabase topologies.

The topology tools operate within an edit session, so you will need to start editing before any of the topology tools are available.

1. Click Editor, point to More Editing Tools, and click Topology.

   The Topology toolbar appears.

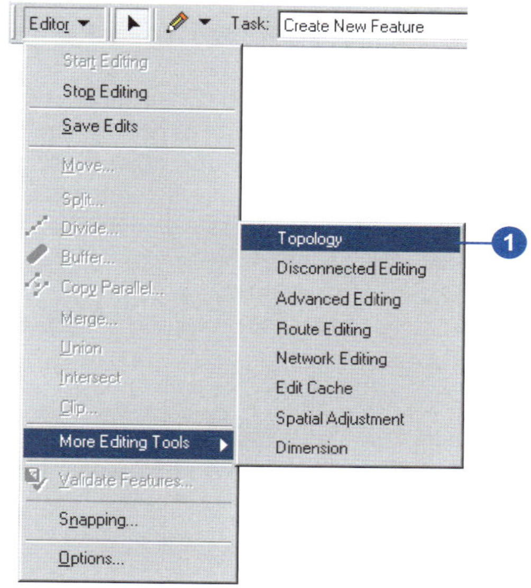

Map topology concepts

A *map topology* is a simple topology that you can impose upon simple features on a map during an edit session. A map topology allows you to simultaneously edit simple features that overlap or touch each other using the tools on the topology toolbar. The features can be in one or more feature classes and may have different geometries. Line features and the outlines of polygon features become topological edges when you create a map topology. Point features, the endpoints of lines, and the places where edges intersect become nodes.

There are two steps to creating a map topology. First, you specify which feature classes on the map will participate in the topology; then you specify a cluster tolerance. The cluster tolerance is a distance within which features will be coincident.

After you create a map topology, you can use the Topology Edit tool to edit the edges and nodes shared by the features. Editing an edge or node shared by two or more different features results in each feature being modified. This lets you move a border to update two forest polygons or move a corner vertex and update several parcel polygons and a few lot boundaries at the same time.

A map topology can be applied to simple features in a shapefile or to simple feature classes in a geodatabase. The feature classes that participate in the map topology must be in the same folder or geodatabase. A map topology cannot be applied to feature classes that participate in a geometric network.

Although they cannot edit geodatabase topologies, ArcView seats of ArcMap can be used to edit a map topology.

A map topology creates topological relationships between the parts of features that are coincident. You can choose the distance, or *cluster tolerance*, that defines how close together edges and vertices must be in order to be considered coincident. You can also specify the feature classes that you want to participate in the map topology.

You do not specify any topology rules for a map topology. All edges or vertices of features in the map topology that fall within the cluster tolerance are considered to be topologically shared. You edit shared edges and vertices in a map topology in the same way, and with the same tools, as you would edit a geodatabase topology. Since there are no topology rules, there is no need to validate a map topology, and there is no creation of error features.

At the geometry level, topologies are about simple relationships, such as coincidence, covering, and crossing, between the geometric primitives that make up features. While all simple feature class geometries—point, line, polygon—may participate in topologies, internally, the types of geometry that are acted on when editing a topology are:

- Edges—line segments that define lines or polygons.
- Nodes—points at the end of an edge.
- Pseudonodes—a node connecting only two edges or a logical split defined in the topology cache while editing. Pseudonodes of the latter sort become a vertex after editing.

When you create a map topology, the cluster tolerance that you specify is used to determine which parts of the features are coincident and which edges and nodes in the topology are shared. The cluster tolerance is typically a very small actual ground distance. Setting large cluster tolerances can result in features being collapsed or distorted when vertices within a given feature snap together.

Creating a map topology

Once the data that you want to create a map topology for is on the map and you've started an edit session, you can create a map topology.

1. On the Editor toolbar, click Editor and click Start Editing.
2. Click the source folder or geodatabase that contains the data that you want to edit.
3. Click OK.
4. On the Topology toolbar, click the Map topology button. ▸

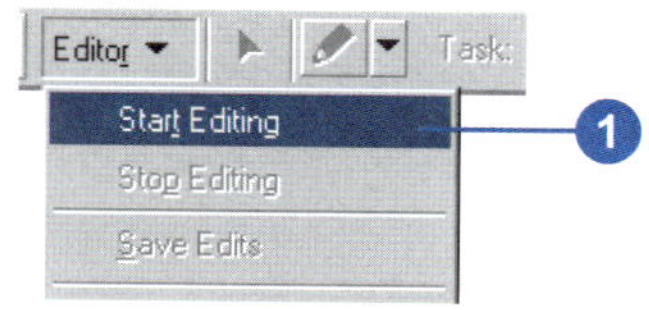

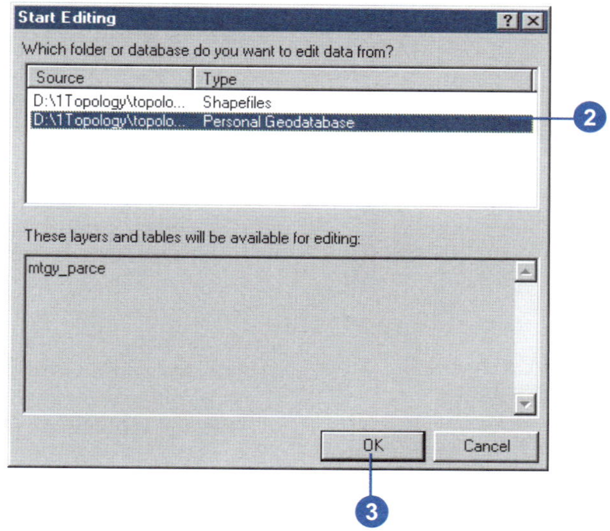

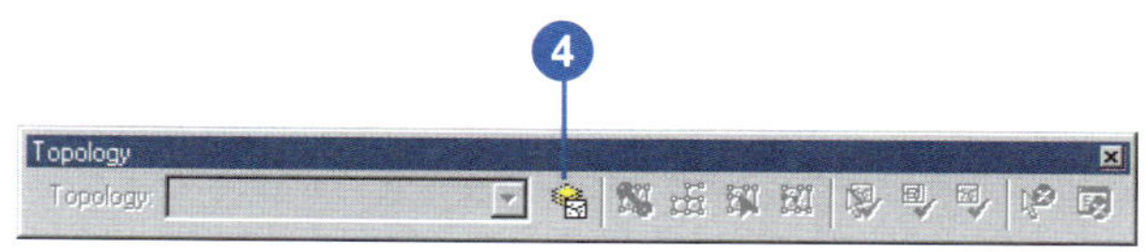

5. Check the feature classes that will participate in the map topology.

   Annotation, dimension, and relationship classes as well as feature classes in a geometric network or geodatabase topology, cannot be added to a map topology.

6. Optionally, set a cluster tolerance for the map topology.

   The default cluster tolerance is the minimum possible cluster tolerance. Increasing the cluster tolerance may cause more features to be snapped together and considered coincident, but this may reduce the spatial accuracy of your data.

7. Click OK.

8. Click the Topology Edit tool and click on the features you want to edit using the map topology.

   The map topology is created for the features that are visible in the current display extent.

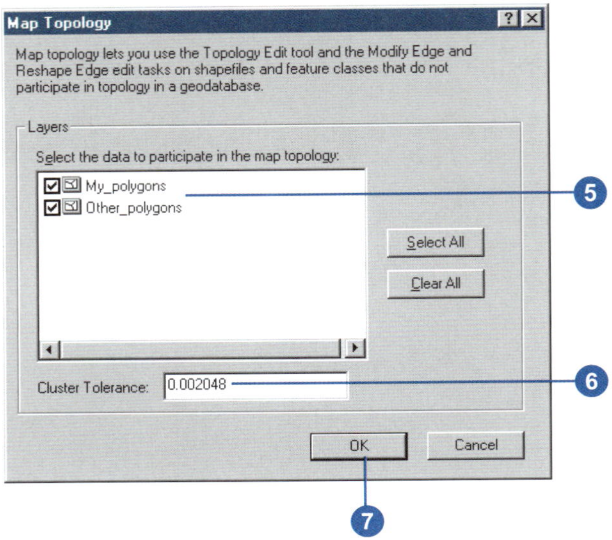

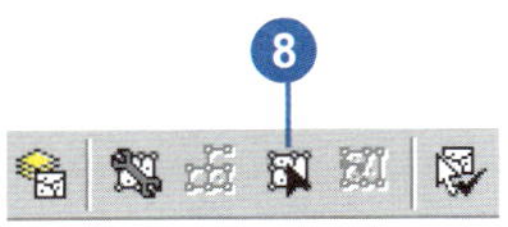

Editing shared geometry

The Topology Edit tool allows you to select and modify edges and nodes that may be shared by more than one feature. It also allows you to select and move the individual vertices that define the shape of edges. When you move vertices, edges, or nodes with the Topology Edit tool, all of the features that share the node or edge are updated.

Tip

Adding to a topology selection

Pressing the Shift key while selecting a node or edge adds that node or edge to the currently selected topology elements.

Tip

Selecting only nodes

You can select nodes by holding the N key while clicking the node or while dragging a box around the node with the Topology Edit tool.

Tip

Selecting only edges

You can select edges by holding the E key while clicking the edge or while dragging a box around the edge with the Topology Edit tool.

Selecting a node

1. Click the Topology Edit tool on the Topology toolbar.
2. Click the node that you want to select.

   You can ensure that edges are not selected by holding the N key while selecting the node. Another easy way to select a node is to drag a rectangle around it, while holding the N key.

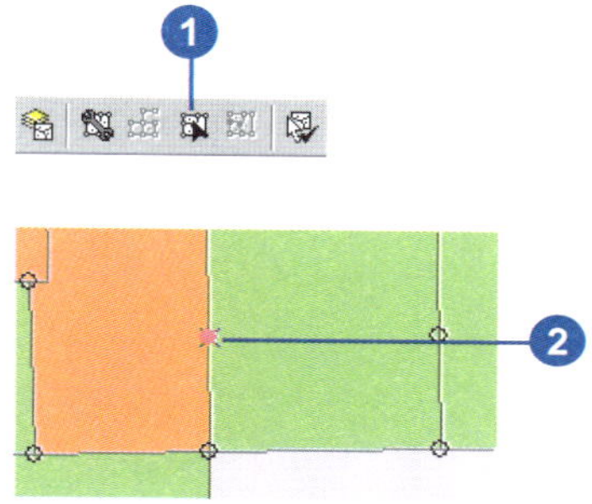

Selecting an edge

1. Click the Topology Edit tool on the Topology toolbar.
2. Click the edge that you want to select.

   You can ensure that nodes are not selected by holding the E key while selecting the node.

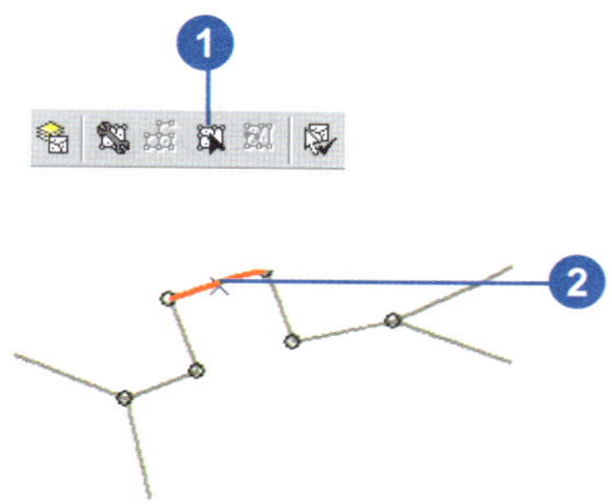

Moving a vertex on a shared edge

1. Click the Topology Edit tool on the Topology toolbar.
2. Double-click the edge that you want to move the vertices of.
3. Click and drag the vertex that you want to move.

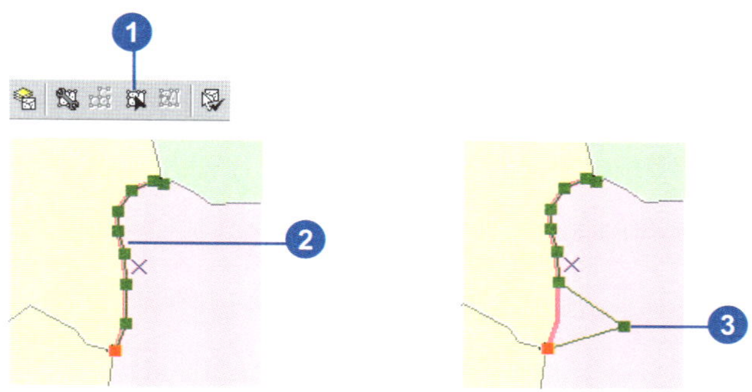

When you move a node in a topology, all of the edges that connect to it are stretched to stay connected to the node. When you move an edge, edge segments stretch to maintain the connection of shared endpoint nodes to their previous location. Sometimes you may want to move a node and a connected edge without stretching the other connected edge. Moving a shared endpoint node of an edge requires you to temporarily split the topological relationship between the node and the other shared edges, then reestablish it. This is known as a Split-move of the node. You'll need to snap the node to an edge or to another topology node.

Tip

Stretching geometry proportionately

You can choose to have the entire geometry of features stretch proportionately when you move a vertex or a topology node. On the Editor Options dialog box, click the General tab, then click the check box to Stretch geometry proportionately when moving a vertex. This overrides the default behavior of stretching the segments between the moved vertex and adjacent vertices.

Moving a node

1. Click the Topology Edit tool on the Topology toolbar.

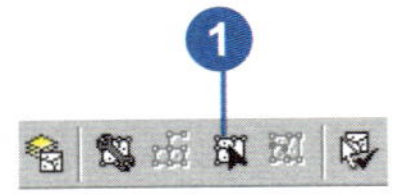

2. Click the node that you want to move.

   You can ensure that edges are not selected by holding the N key while selecting the node. Another easy way to select a node is to drag a rectangle around it, while holding the N key.

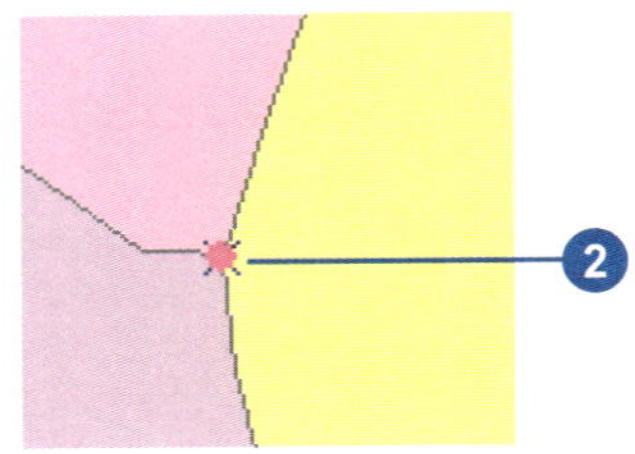

3. Click and drag the node that you want to move.

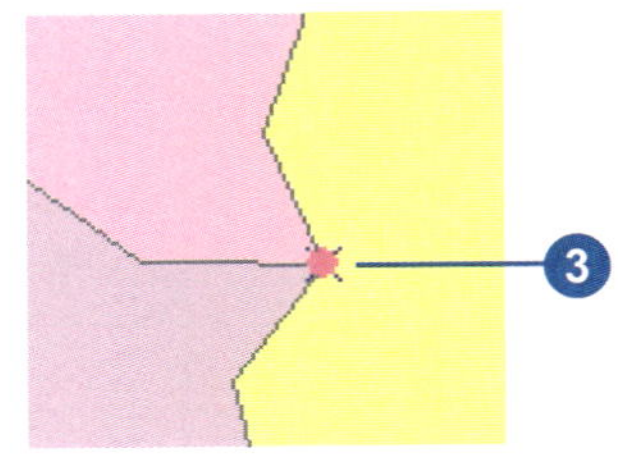

Moving an edge

1. Click the Topology Edit tool on the Topology toolbar.

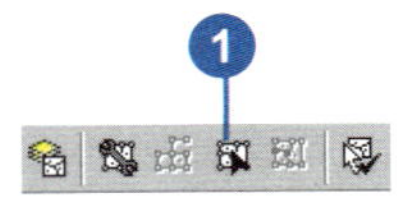

2. Click the edge that you want to move.

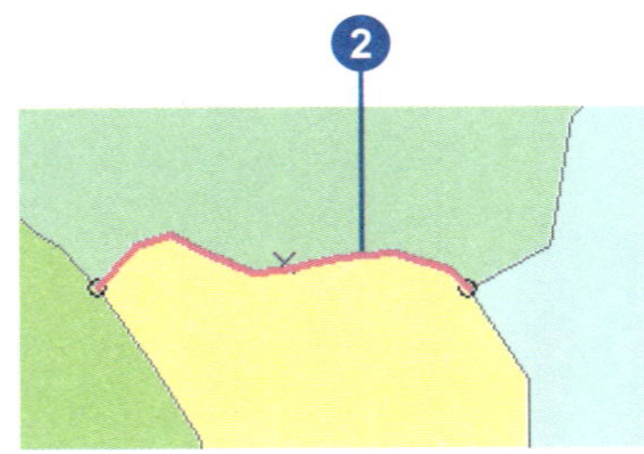

3. Click and drag the edge to a new location. Edge segments stretch to connect the edge's endpoint nodes to their previous positions, where they are shared.

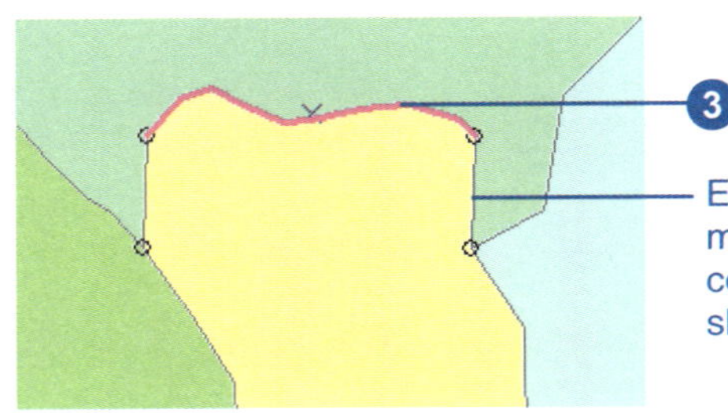

Moving a topology element by a given X and Y distance

1. Click the Topology Edit tool on the Topology toolbar.
2. Click the node or edge that you want to move.
3. Right-click and click Move.
4. Type an X and a Y distance to move the topology element relative to its current location and press Enter.

   If you are moving an edge, or more than one topology element, the Move command moves the selection anchor to the specified location and moves the topology element to maintain its position relative to the selection anchor.

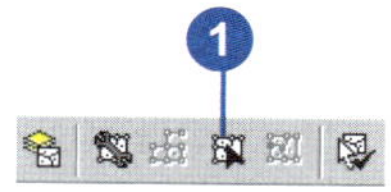

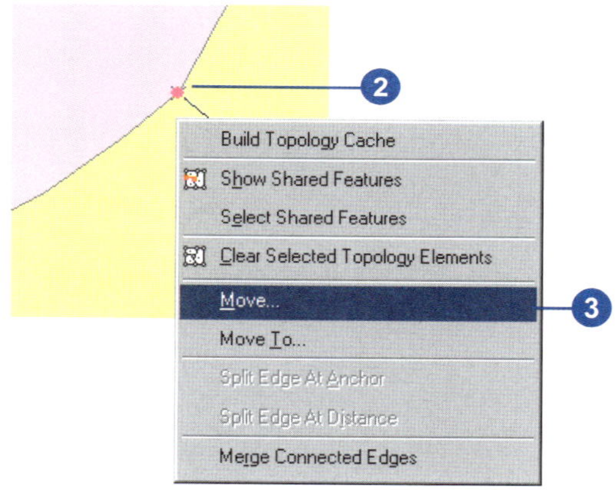

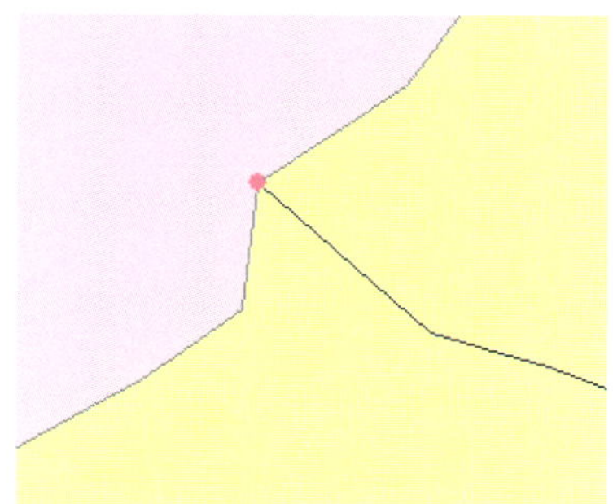

Moving a topology element to a given location

1. Click the Topology Edit tool on the Topology toolbar.
2. Click the node or edge that you want to move.
3. Right-click and click Move To.
4. Type the absolute X and Y coordinate to which you want to move the topology element and press Enter.

   If you are moving an edge, or more than one topology element, the Move To command moves the selection anchor to the specified location and moves the topology element to maintain its position relative to the selection anchor.

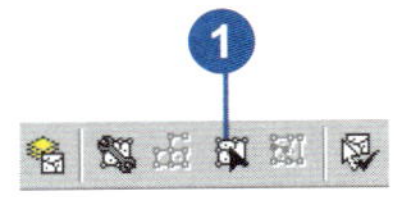

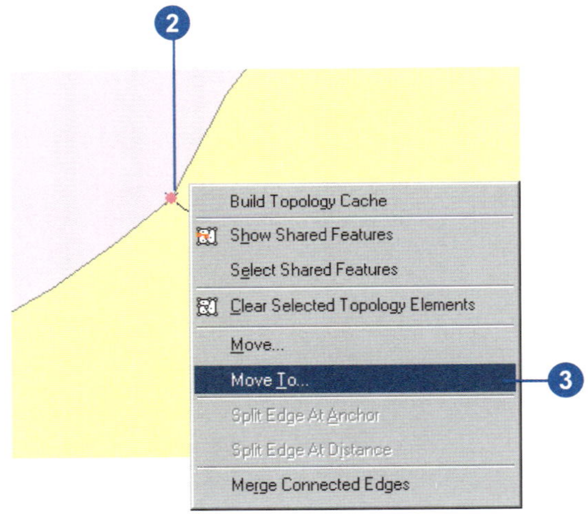

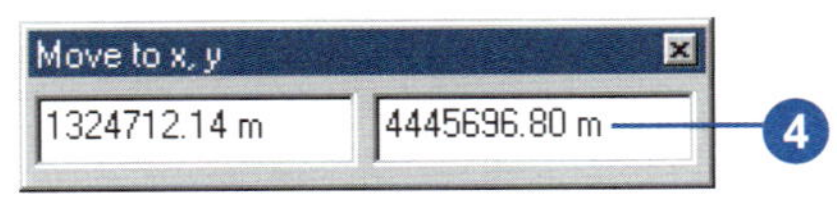

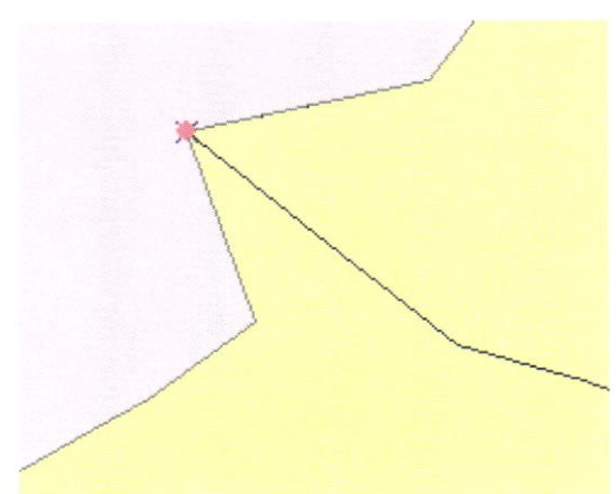

Splitting an edge with the selection anchor

1. Click the Topology Edit tool on the Topology toolbar.
2. Click the edge that you want to split.
3. Hold the Ctrl key and click and drag the selection anchor to the place where you want to split the edge.
4. Right-click and click Split Edge At Anchor.

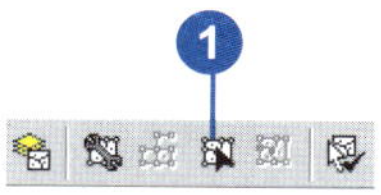

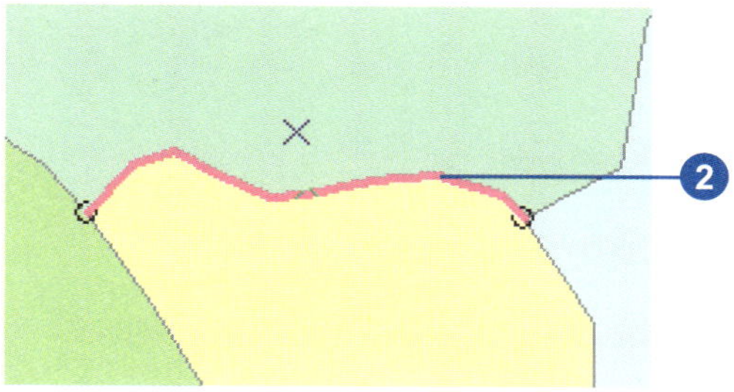

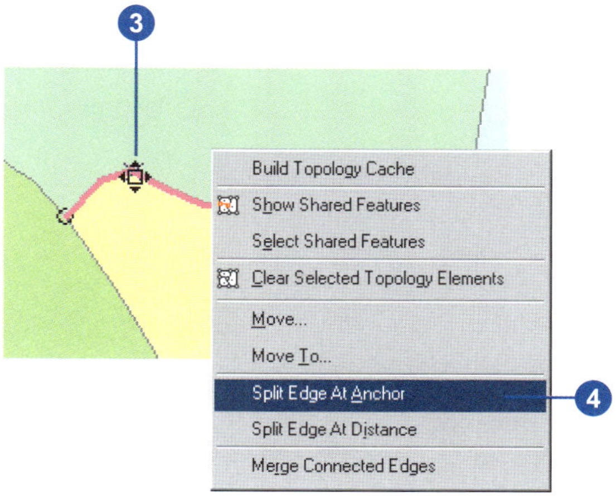

Splitting an edge at a distance from an endpoint

1. Click the Topology Edit tool on the Topology toolbar.
2. Click the edge that you want to split.
3. Right-click and click Split Edge At Distance.

   Arrows appear along the edge to indicate the edge direction. The edge can be split at a distance or a percentage of its length, measured from the start point or the endpoint of the edge. ►

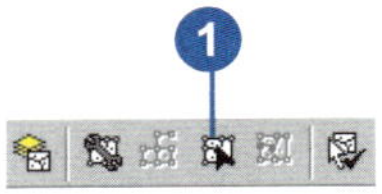

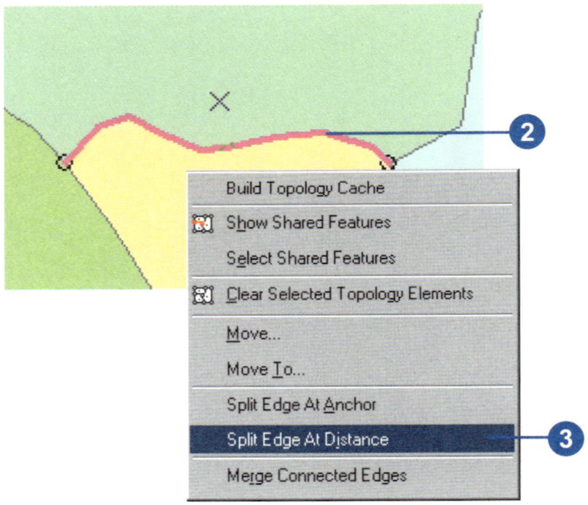

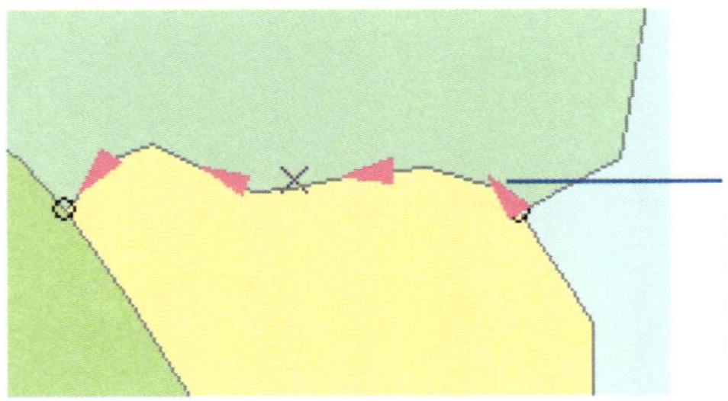

4. Optionally, click the button to specify a percentage of the edge length.
5. Type a distance or a percentage if you chose to make the split at a percentage of the edge length.
6. Optionally, click the button to indicate that you want the distance to be measured from the endpoint of the edge.
7. Click OK.

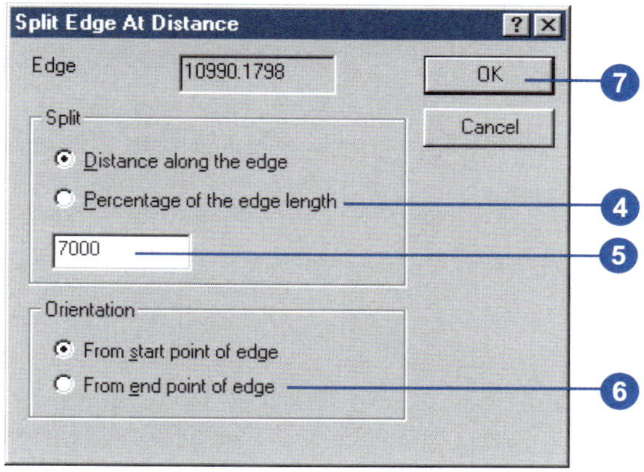

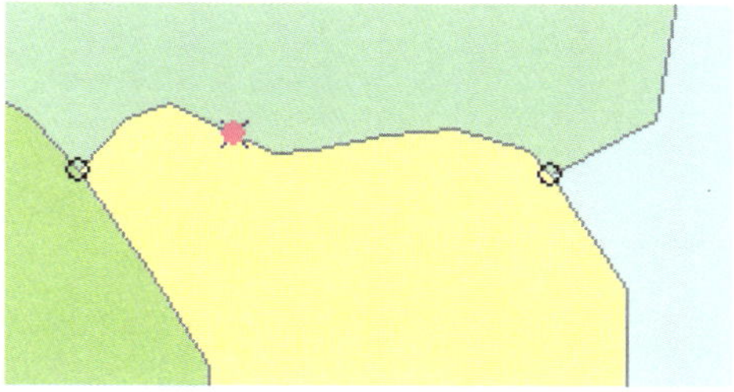

Tip

Split-move

Holding the S key when you move a topology node and a selected edge makes the move a split-move. In a Split-move, the topological relationship of the node and any unselected edges is temporarily broken, then reestablished at the new location where you place the node. You must snap the node to an edge or another topology node to reestablish the topological relationship, otherwise the move will be cancelled.

Moving a shared endpoint node of an edge

1. Set the snapping environment to snap to Topology nodes. See Chapter 3, 'Creating new features'.

   You can also set snapping to edges of the feature to which you want to move the endpoint node.

2. Click the Topology Edit tool on the Topology toolbar.
3. Click the edge to which you want to move the endpoint node.
4. Hold the Ctrl key and click and drag the Selection Anchor to the place where you want to snap the edge's endpoint node.

   Holding the Ctrl key allows you to move the selection anchor.

5. Right-click and click Split Edge At Anchor.

   Splitting the edge creates a new node to which you can snap the endpoint node of the edge. ►

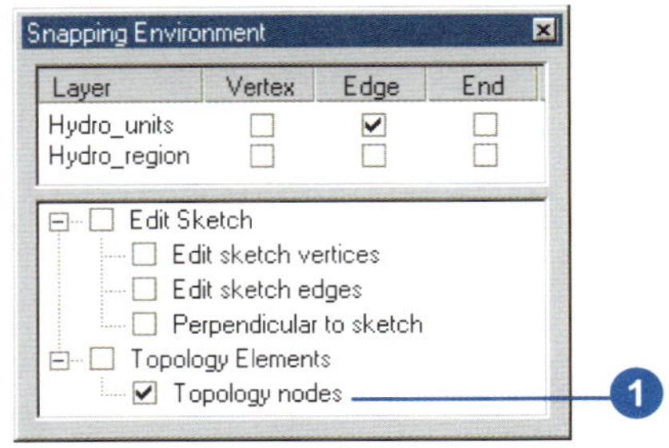

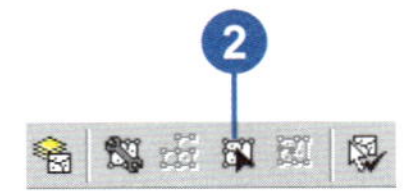

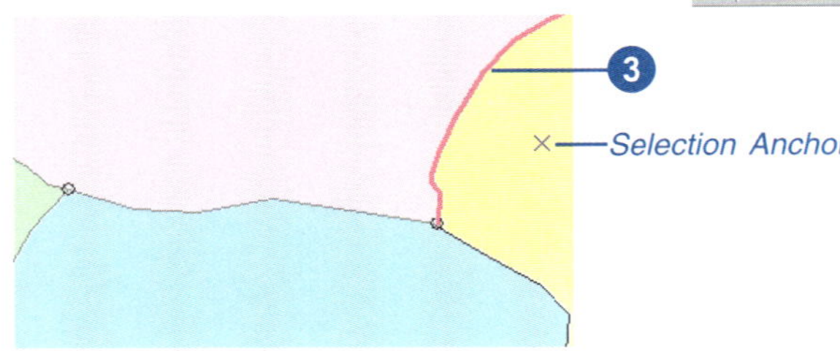

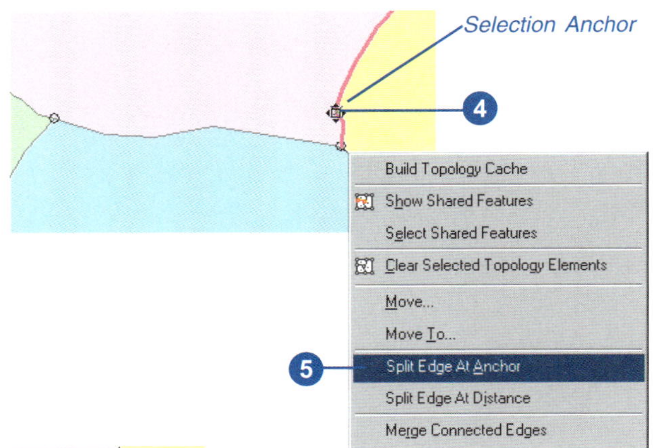

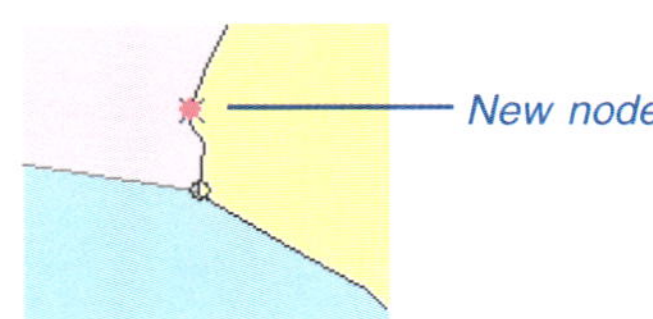

6. Click the edge that you want to move the end of.
7. Hold the N key and drag a rectangle around the endpoint node you want to move.

   The N key limits the Topology Edit tool selection to nodes.
8. Hold the S key and click and drag the node that you want to move to the node you created.

   Holding the S key changes the pointer to the Split-Move tool pointer.

   The endpoint node of the edge is moved to the new location and the topology is maintained. If you do not snap the node to the new node or to the edge, the split move will be cancelled.

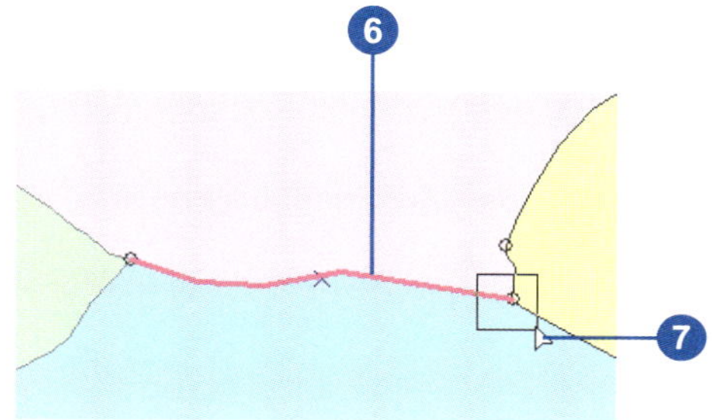

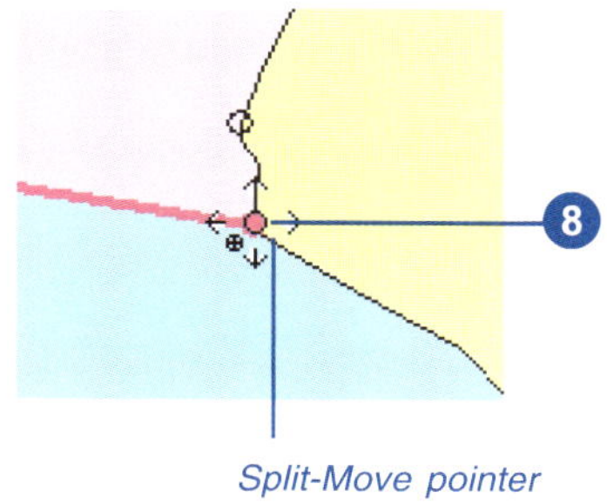

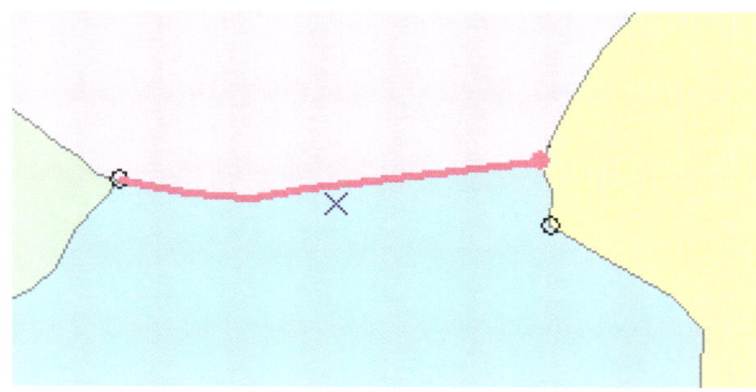

Rebuilding the topology cache

When you select a topology element using the Topology Edit tool, ArcMap creates a topology cache. The topology cache stores the topological relationships between edges and nodes of the features that fall within the current display extent. If you are editing with the map zoomed in to a very small area and you go back to a previous extent, some of the features in the new extent may not be in the topology cache. You can rebuild the topology cache to include these features. You can also rebuild the topology cache to remove temporary topology nodes that you created for snapping and editing.

1. Click the Topology Edit tool on the Topology toolbar.
2. Right-click on the map and click Build Topology Cache.

   The topological relationships between edges and nodes are rediscovered for all of the features in the current display extent.

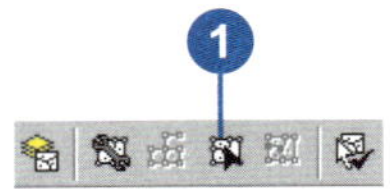

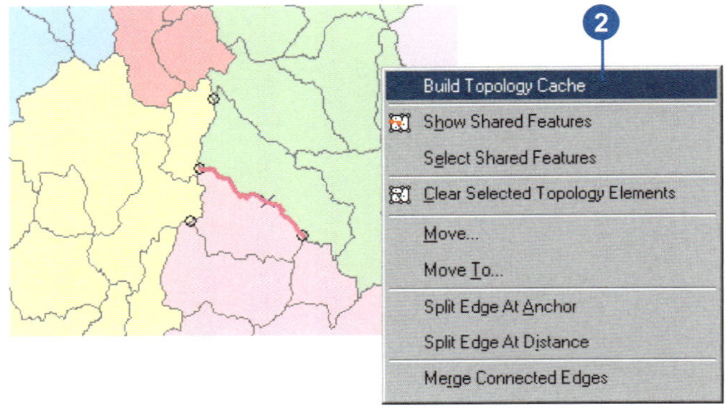

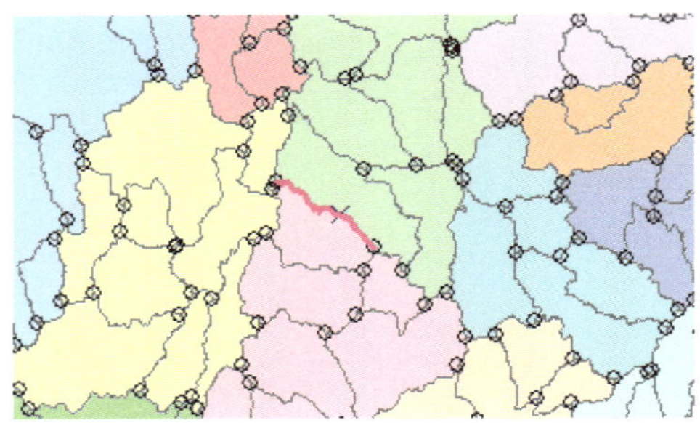

Clearing selected topology elements

When you are editing topological edges and nodes, you may sometimes want to unselect some elements. You can unselect a given element by holding the Shift key and clicking it, or unselect all selected edges and nodes by clearing all selected topology elements.

You can also click on the map with the Topology Edit tool away from edges and nodes to clear the selection.

Unselecting a single topology element

1. Click the Topology Edit tool on the Topology toolbar.
2. Hold the Shift key and click a selected edge or node to unselect it.

   The topology element is unselected.

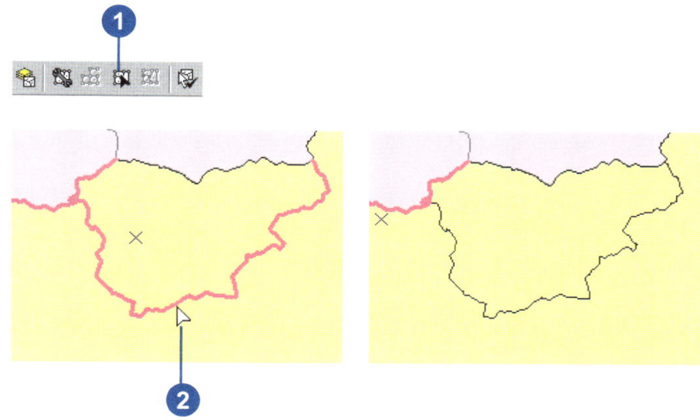

Unselecting all topology elements

1. Click the Topology Edit tool on the Topology toolbar.
2. Right-click and click Clear Selected Topology Elements.

   All of the topology elements are unselected.

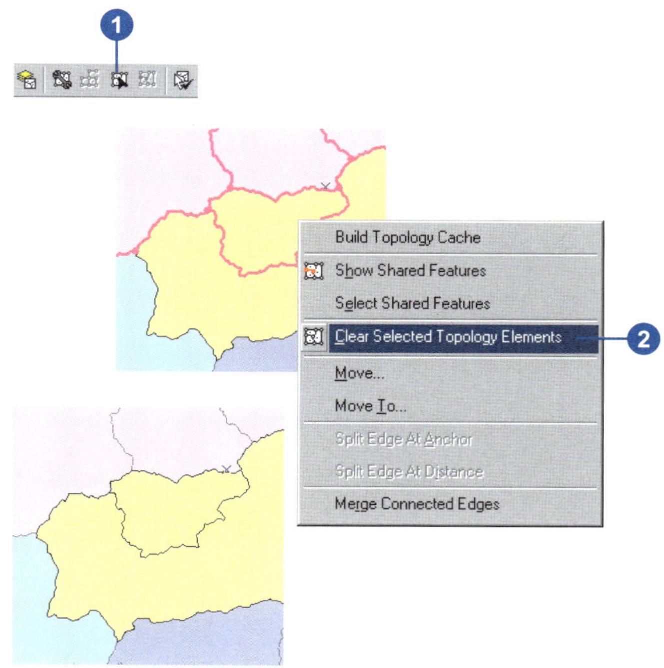

Finding out what features share topology elements

Topology elements may be shared by multiple features. It can be useful to know which features share a given node or edge. You can find out which features share a given topology element by selecting them, then choosing Show Shared Features.

You can also control whether or not the shared features will be affected by edits that you make to a given edge or node. By default, all features that share a topology element are updated when you edit that element with the Topology Edit tool. However, when you turn off a feature in the Shared Features dialog box, the feature will not be modified if you edit the topology element.

Showing shared features

1. Click the Topology Edit tool on the Topology toolbar.
2. Click an edge or node to select it.
3. Right-click and click Show Shared Features.

   You can also click the Show Shared Features button on the Topology toolbar.
4. Click the plus sign to show all of the features in a given feature class that are shared. A given topology element may be shared by feature in multiple feature classes, so more than one feature class may be listed.
5. Click a feature in the list to make it flash on the map.

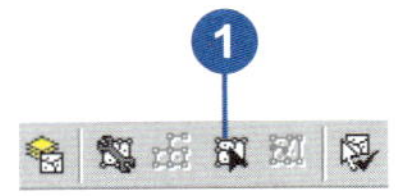

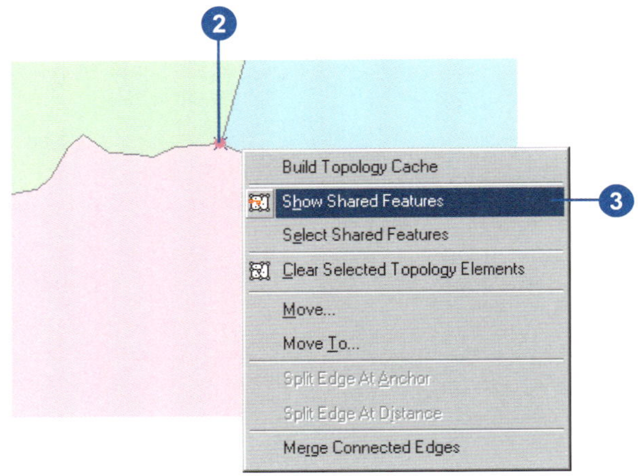

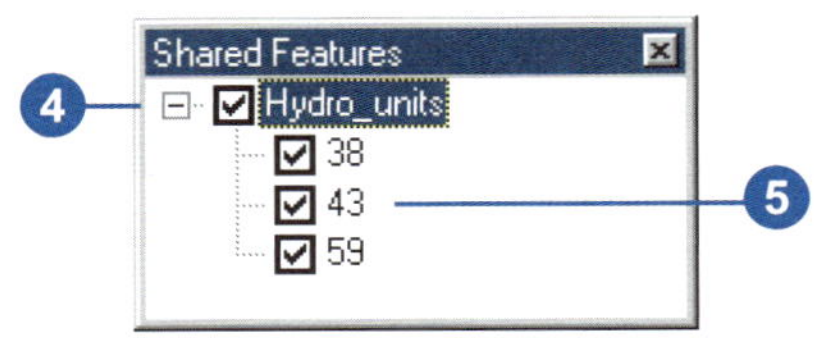

Temporarily turning off topology element sharing for a feature

1. Click the Topology Edit tool on the Topology toolbar.
2. Click an edge or node to select it.
3. Right-click and click Show Shared Features.

   You can also click the Show Shared Features button on the Topology toolbar.
4. Click the plus sign to show all of the features in a given feature class that are shared. A given topology element may be shared by features in multiple feature classes, so more than one feature class may be listed.
5. Uncheck a feature in the list to turn off topology element sharing. Edits that you make with the Topology Edit tool to the topology element will not update this feature.

   The unshared status of the feature is temporary. It only lasts while the topology element is selected.
6. Click the Close button to close the Shared Features dialog box.

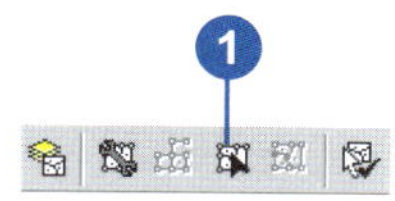

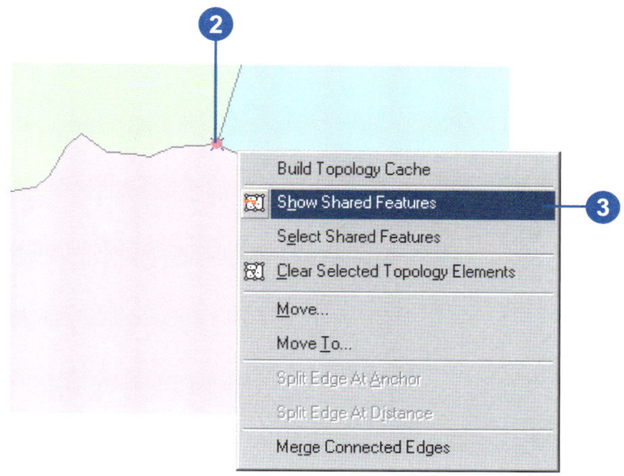

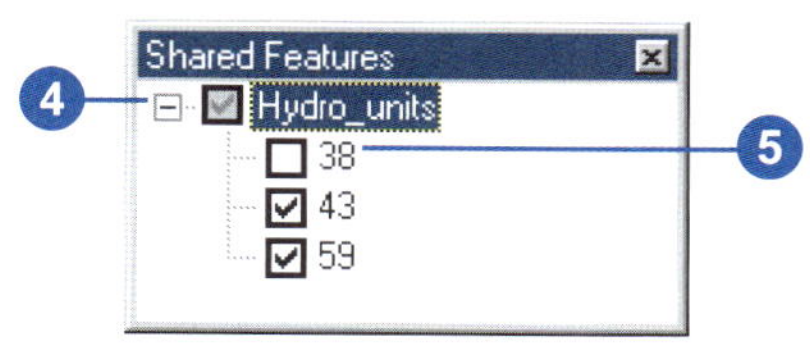

Selecting features that share a topology element

1. Click the Topology Edit tool on the Topology toolbar.
2. Click an edge or node to select it.
3. Right-click and click Select Shared Features.

   The features that share the topology element are selected.

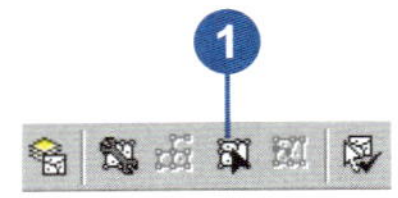

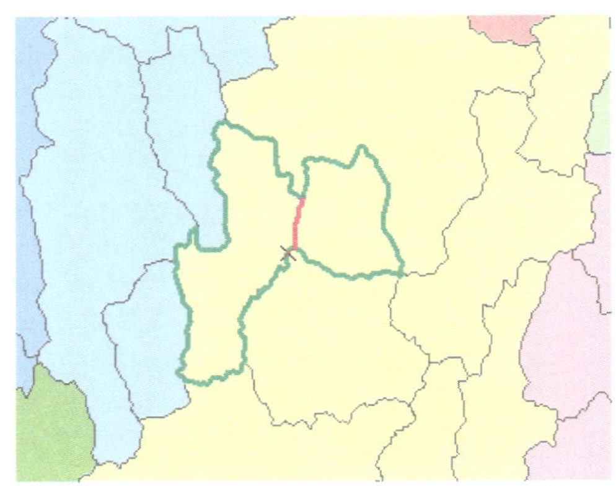

Merging connected edges within a feature

1. Click the Topology Edit tool on the Topology toolbar.
2. Click an edge of a feature that has been topologically split by adding nodes.
3. Right-click and click Merge Connected Edges.

   The selected edge is merged with the adjacent edge and the topology node is removed.

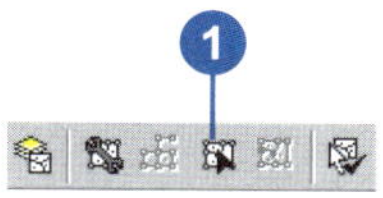

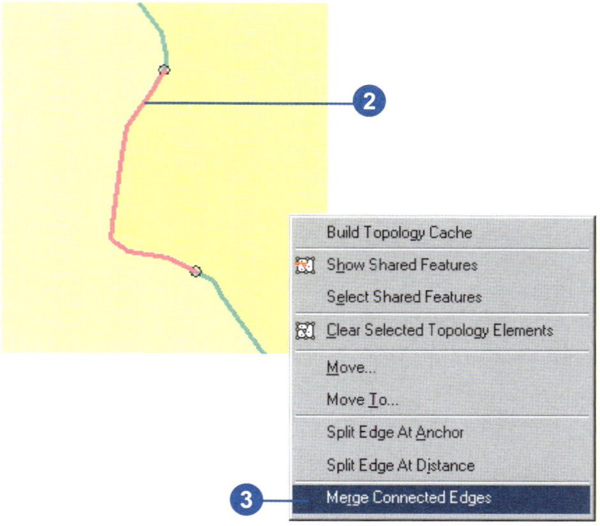

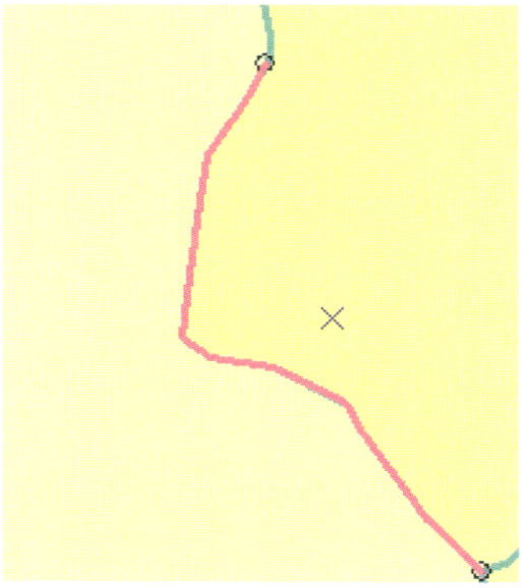

Using the edit sketch to make topology edits

In addition to editing topology elements with the Topology Edit tool, you can also modify and reshape a selected topology edge using an Edit Sketch.

The Modify Edge topology edit task takes the selected edge and makes an edit sketch from it. You can then use the standard editing tools to insert, delete, or move the vertices that make up the edge.

With the Reshape Edge topology edit task you can use the basic editing tools to create a new line to replace an existing edge line.

Modifying an edge

1. Click the Topology Edit tool on the Topology toolbar.
2. Click an edge to select it.
3. Click the Task dropdown arrow on the Editor toolbar and click Modify Edge. ►

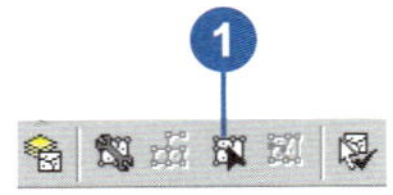

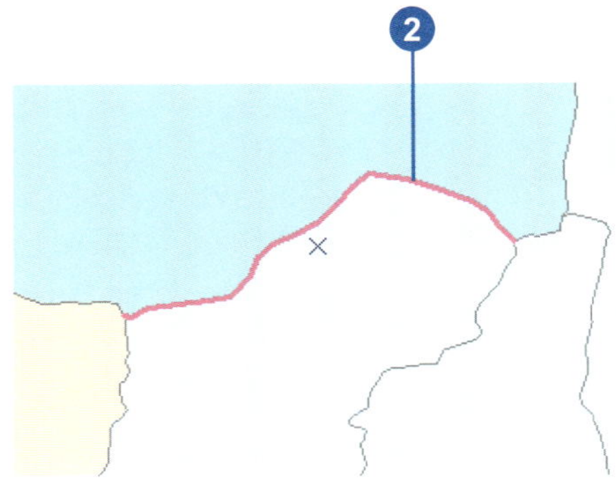

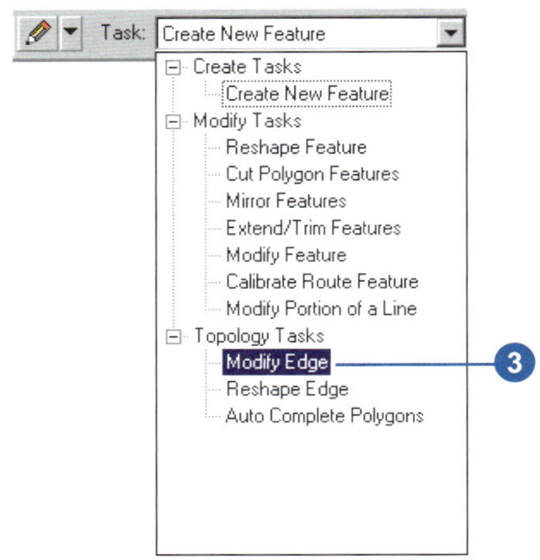

4. Optionally, right-click a segment of the edge that has no vertices and click Insert Vertex.

   A new vertex is inserted into the edge and into all features that share it.

5. Optionally, right-click a vertex and click Delete Vertex.

   The vertex is removed from the edge and from all features that share it.

6. Optionally, click a vertex and drag it to a new location. ▸

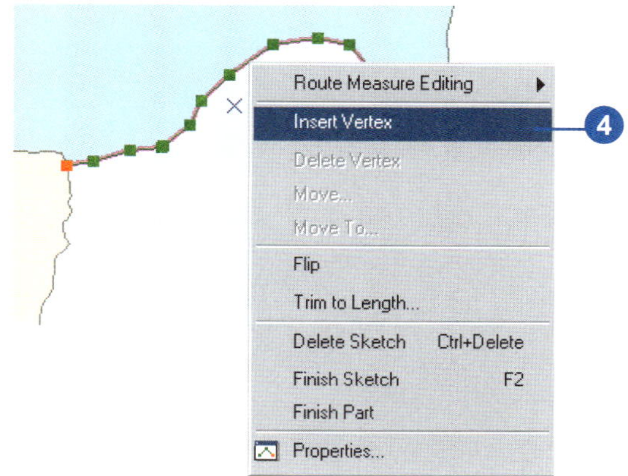

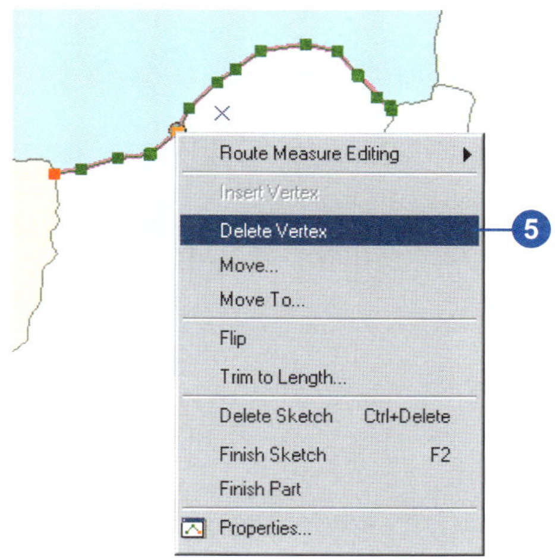

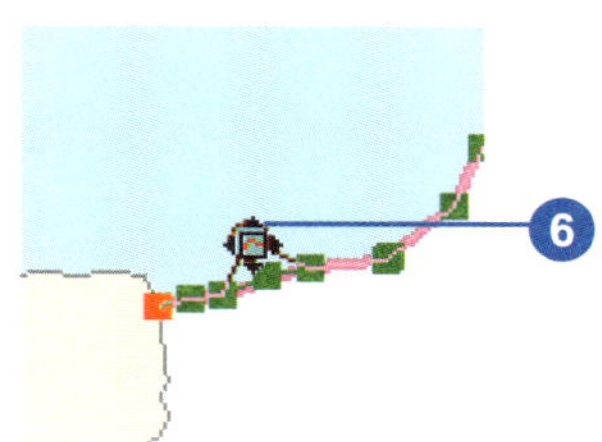

7. Optionally, right-click a vertex, then click Move.
8. Type an X and a Y distance and press Enter to move the vertex relative to its current position.
9. Optionally, right-click a vertex and click Move To.
10. Type the new coordinates for the vertex and press Enter. ▸

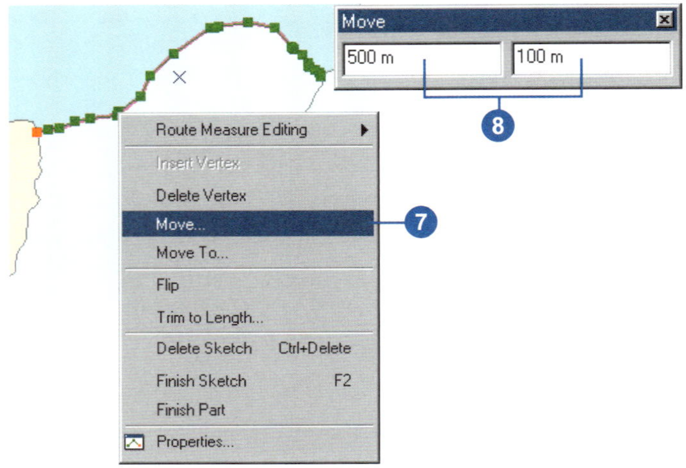

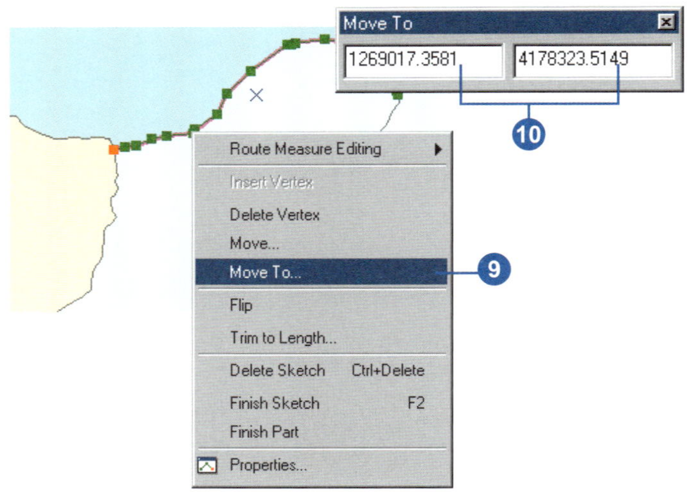

11. Right-click anywhere on the map and click Finish Sketch.

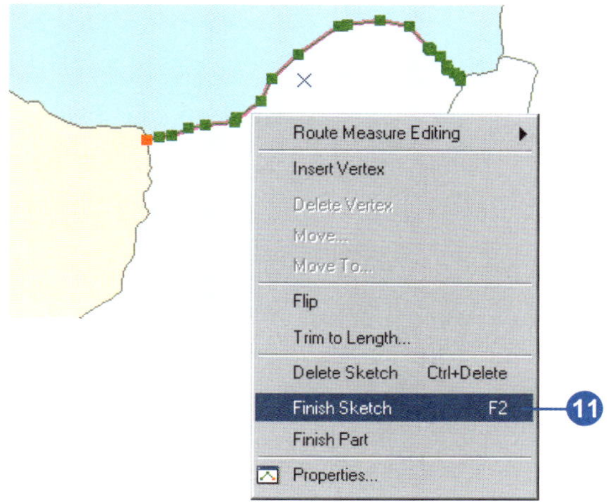

Reshaping an edge

1. Click the Topology Edit tool.
2. Click an edge to select it.
3. Click the Task dropdown arrow on the Editor toolbar and click Reshape Edge. ►

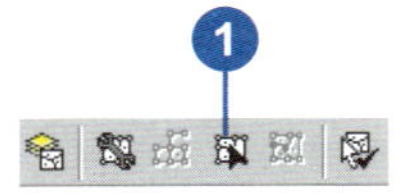

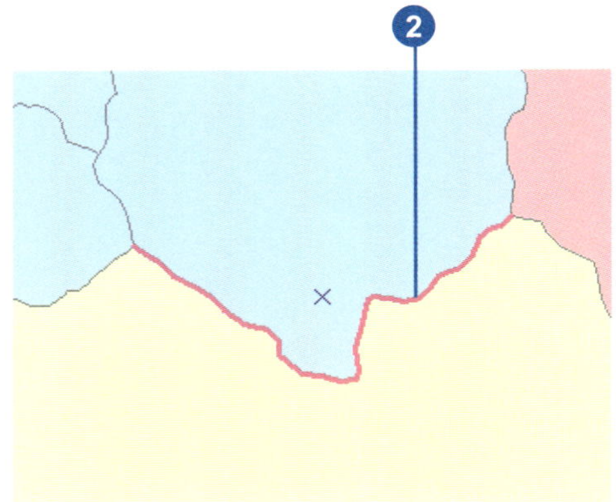

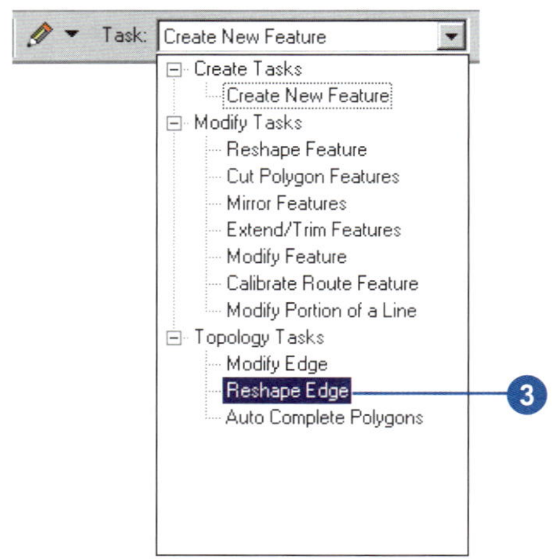

4. Click the Sketch Tool on the Editor toolbar.

5. Start an edit sketch.

   You can either snap the edit sketch to the selected edge or cross the edge to indicate where to start reshaping the edge.

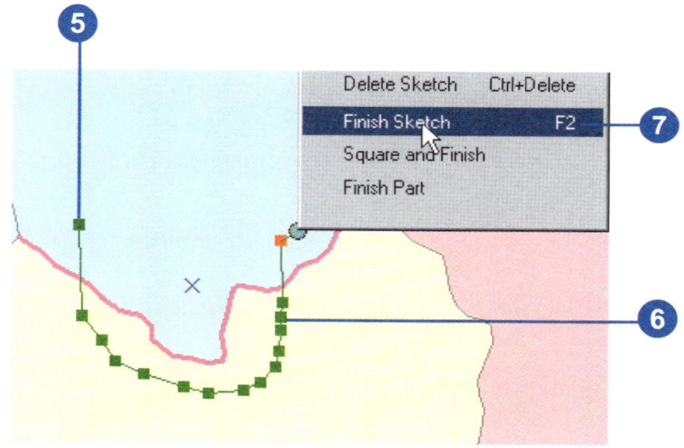

6. Use the Sketch tools to digitize a new shape for part of the selected edge.

   Once you've started the edit sketch, you can use any of the tools on the Sketch Tool Palette to create your edit sketch.

   You can either snap the edit sketch to the selected edge or cross the edge to indicate where to stop reshaping the edge.

7. Right-click anywhere on the map and click Finish Sketch.

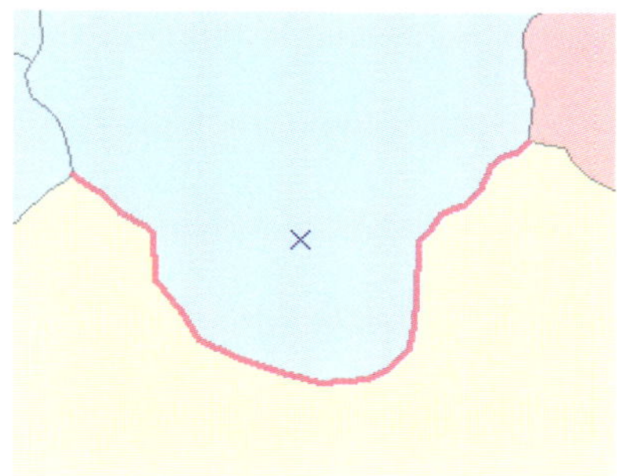

Stretching features when editing topology elements

Just as you can stretch a single feature geometry proportionately when moving a vertex, you can also stretch the geometry of features that share a topology element when moving a node or vertex on a topology edge.

1. Click Editor and click Options.
2. Click the General tab.
3. Check Stretch geometry proportionately when moving a vertex.
4. Click OK. ►

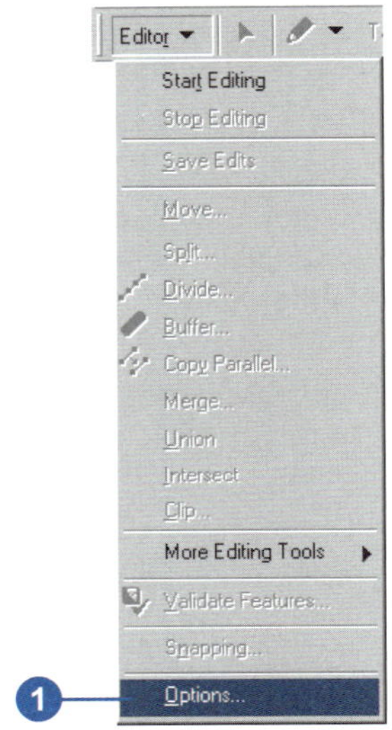

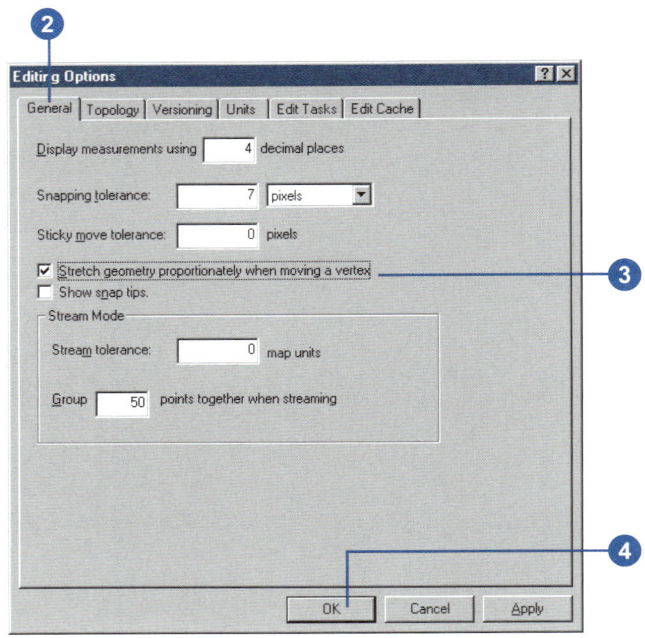

5. Click the Topology Edit tool on the Topology toolbar.
6. Click a topology node or double-click a topology edge and click a vertex.
7. Click and drag the node or vertex to a new location.

   The features that share the node or vertex stretch proportionately.

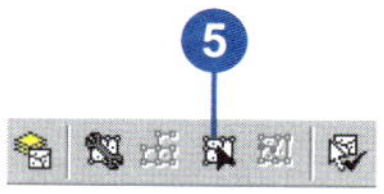

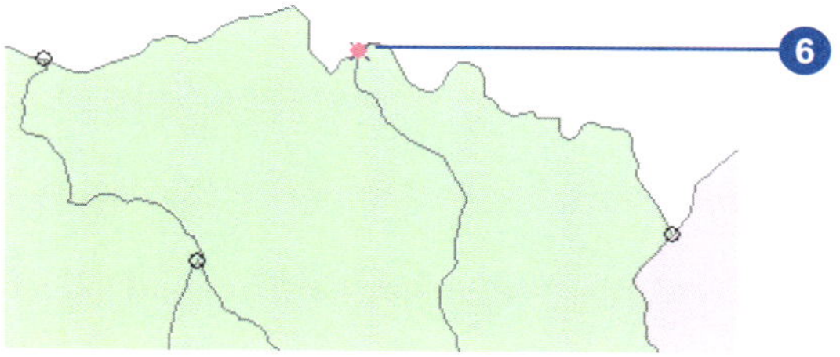

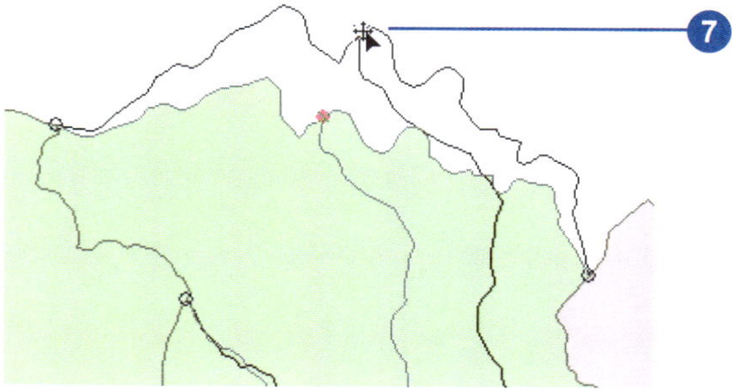

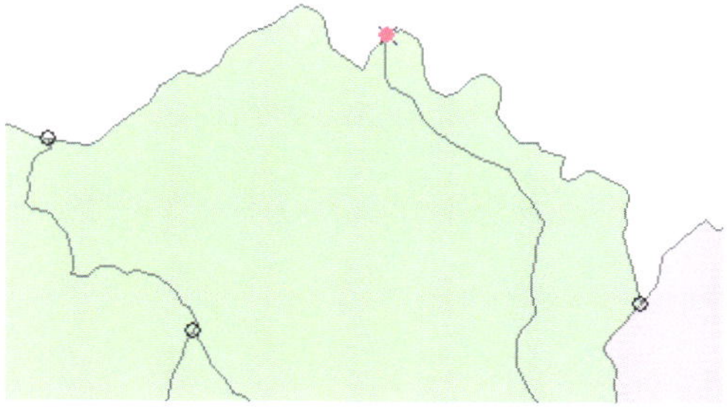

Snapping to topology nodes

When you are editing topology elements, it can be useful to snap to topology nodes. You can turn on snapping to topology nodes on the Snapping Environment dialog box.

1. Click Editor and click Snapping.
2. Check Topology nodes.

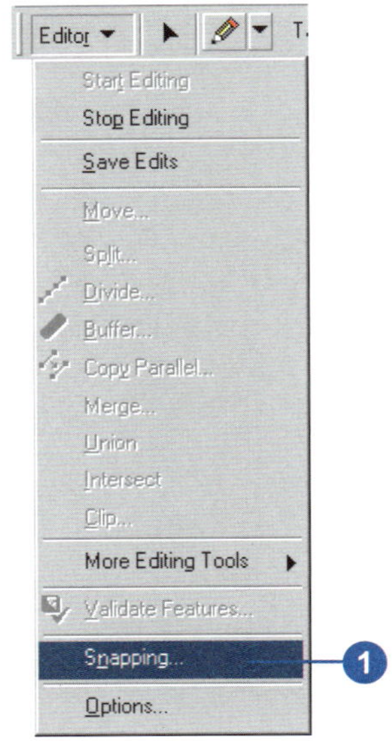

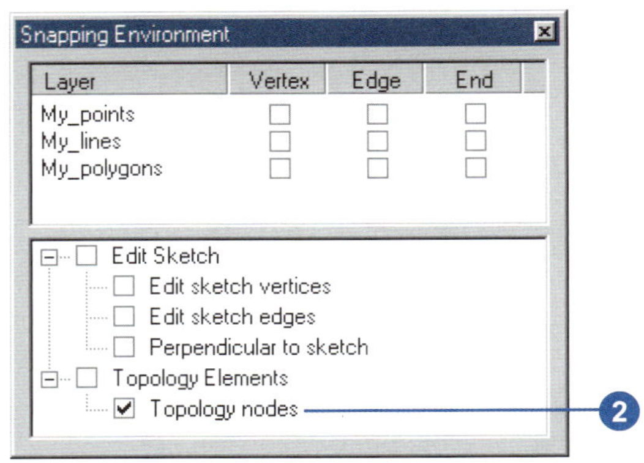

Changing the symbology for selected error features

You can change the way selected error features are drawn on the map. Error features are drawn with point, line, and area symbols of a given color by default. When you select errors, for example, when you are using the Fix Topology Error tool, the selected errors change color so that you can more easily identify the errors that you are fixing. You can change the symbology of selected topology errors to make them stand out better against a given map background.

1. Click Editor and click Options.
2. Click the Topology tab.
3. Click the Active Errors Symbology buttons to change the way error features look when they are selected.
4. Pick a new symbol for the selected error feature and click OK.
5. Click OK to close the Editing Options dialog box.

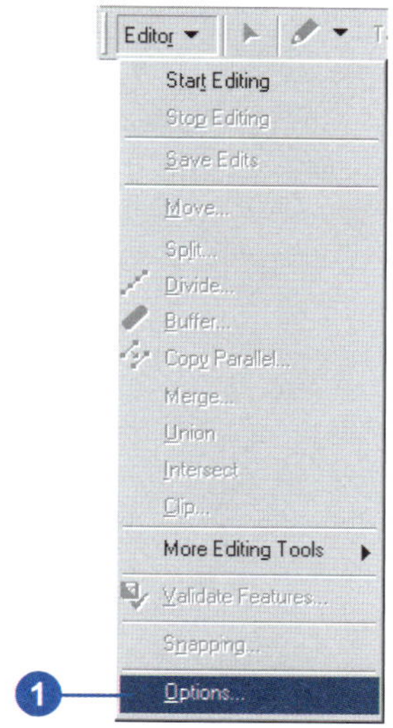

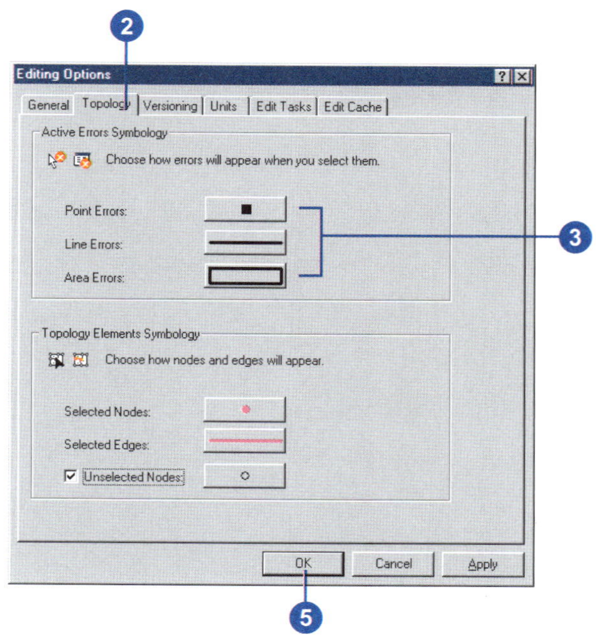

Changing the symbology for topology elements

You can change the way topology elements are drawn on the map. Topology nodes and edges are drawn with point and line symbols of a given color by default and unselected topology nodes are not drawn by default. You can change the symbol for selected topology nodes and edges and for unselected nodes. Turning on the symbol for unselected nodes can make it easier to identify nodes to snap to when you are editing topology elements.

1. Click Editor and click Options.
2. Click the Topology tab.
3. Click the Topology Elements Symbology buttons to change the way topology elements look on the map.
4. Pick a new symbol for the topology element and click OK.
5. Optionally, check Unselected Nodes to show the nodes in the topology cache that have not been selected.
6. Click OK.

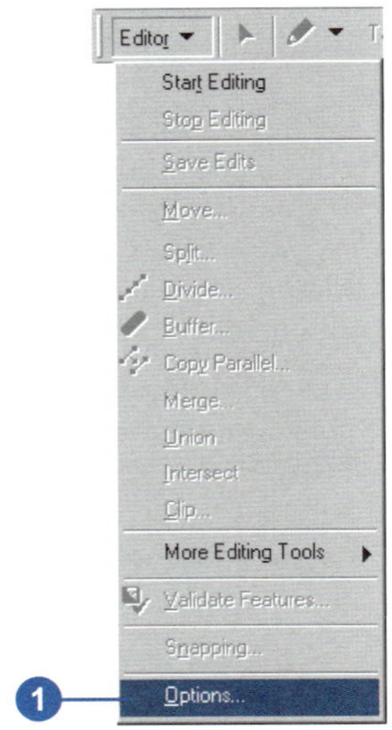

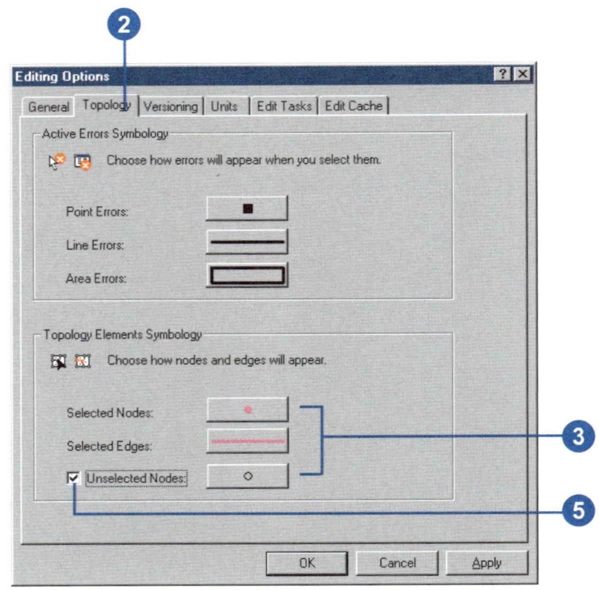

Changing the symbology for topology layers

You can change the way error features, exceptions, and dirty areas are drawn on the map. Point, line, and polygon errors are drawn with symbols of a single color by default. Exceptions and dirty areas are not drawn by default. Changing the symbology of error features and exceptions that relate to different topology rules can make it easier to understand what the problems are with your data. Drawing dirty areas can make it easier to see the areas that have been affected by edits and that have yet to be validated.

Changing topology error and exception symbology

1. Right-click the Topology layer in the ArcMap table of contents and click Properties.
2. Click the Symbology tab.
3. Check the error types that you want to see on the map.
4. Click the error type that you want to change the symbol for.
5. Optionally, click the button to draw all of the errors of this type with a single symbol.
6. Pick a new symbol for this type of error feature and click OK.
7. Optionally, click the button to draw the errors of this type with unique symbols.
8. Double-click the error symbol for the rule that you want to draw with a new symbol.
9. Pick a new symbol for this type of error feature and click OK.
10. Click OK to close the Layer Properties dialog box.

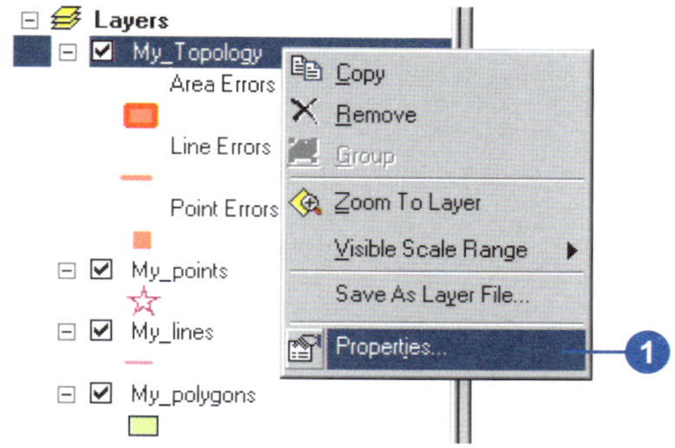

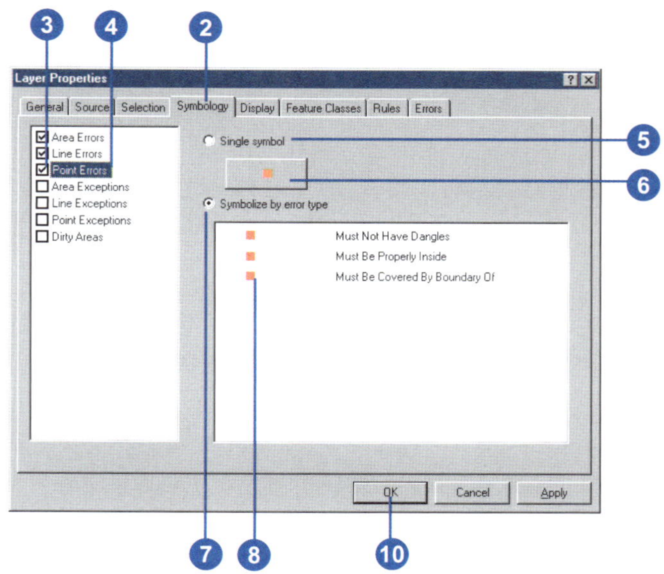

Showing dirty areas and changing their symbology

1. Right-click the Topology layer in the ArcMap table of contents and click Properties.
2. Click the Symbology tab.
3. Check Dirty Areas to draw dirty areas in the Topology layer.
4. Click Dirty Areas to set the symbology for dirty areas.
5. Click the button to change the dirty area symbol.
6. Pick a new symbol for dirty areas and click OK.
7. Click OK to close the Layer Properties dialog box.

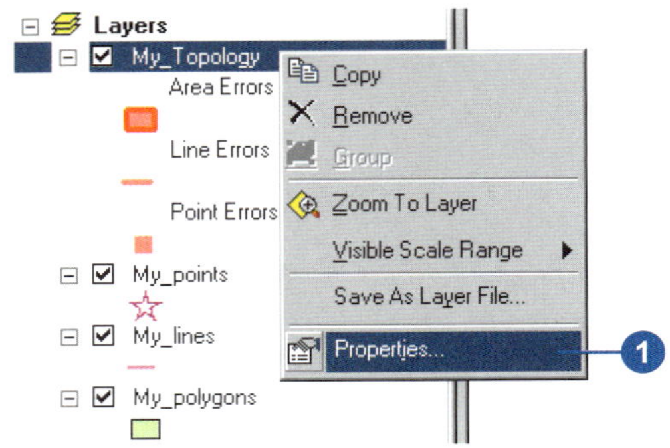

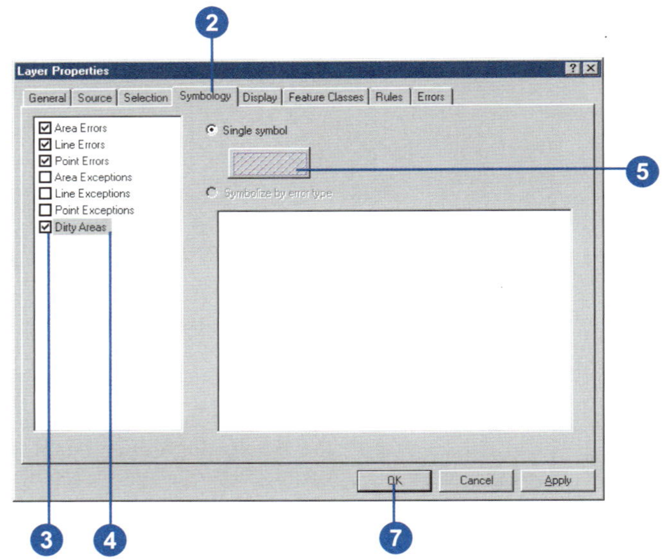

Validating edits to a topology

Once you've made edits to a feature that participates in a topology, the next step is to validate the topology. Validating the topology checks the features to identify any violations of the rules that have been defined for the topology.

You can validate the whole topology, validate the visible extent of your map, or drag a box around the area to validate.

Validating the whole topology

1. Click the Validate Entire Topology button on the Topology toolbar.

   Validating the entire extent may take a while for complex or large datasets or where there are many topology rules.

   You will be prompted to specify whether or not you wish to validate the whole topology.
2. Click Yes.

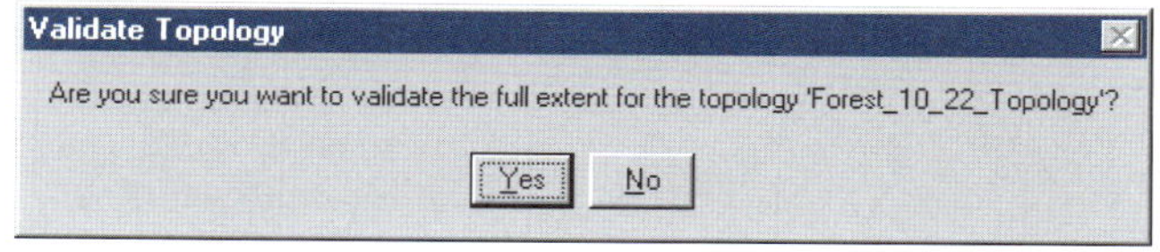

Validating topology in the visible extent of the map

1. Click the Validate Topology in Current Extent button on the Topology toolbar.

   The visible extent is validated. Areas that are not currently visible on the map are not validated.

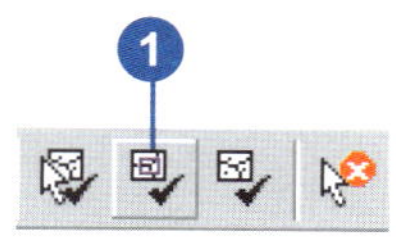

Validating topology in a selected area

1. Click the Validate Topology in Specified Area button on the Topology toolbar.
2. Drag a box around the area you want to validate.

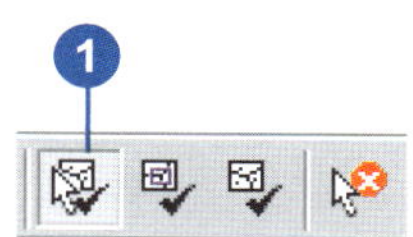

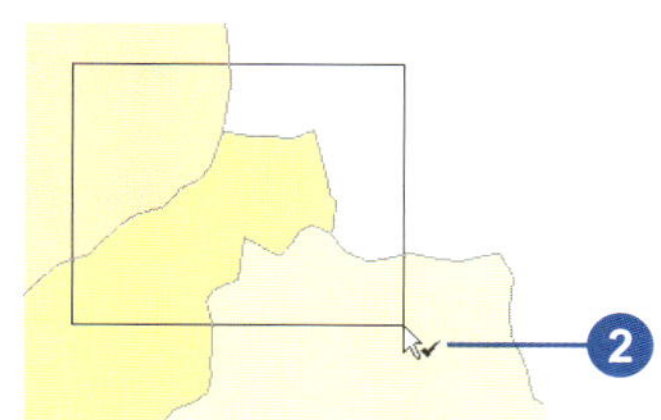

Summarizing topology errors

Once you've made edits to a feature that participates in a topology and validated your edits, you may see one or more topology errors. You can manage Topology errors with the Error Inspector.

The Error Inspector lets you view topology errors in a table that tells you the rule violated, the feature class or classes involved in the error, the geometry of the error, the feature ID of the features involved in the error, and whether or not the error has been marked as an exception. You can sort the errors by any of the fields in the table, so you can work with all of the errors of a given type. You can also limit the errors shown in the table to errors of a given type, errors that occur in the currently visible map extent, or errors that have been marked as exceptions.

In addition to letting you view and sort errors, the error inspector lets you select errors, pan or zoom to selected errors, and apply topology fixes of various types to errors.

Opening the Error Inspector

1. Click the Error Inspector button on the Topology toolbar.

   The Error Inspector can be docked to the ArcMap window or it can float free as a separate window.

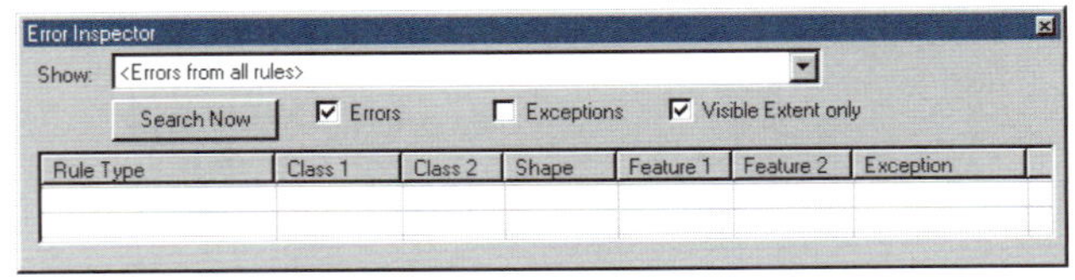

Finding all topology errors

1. On the Error Inspector, check Errors.
2. Click the Show dropdown arrow and click Errors from all rules.
3. Click Search Now.

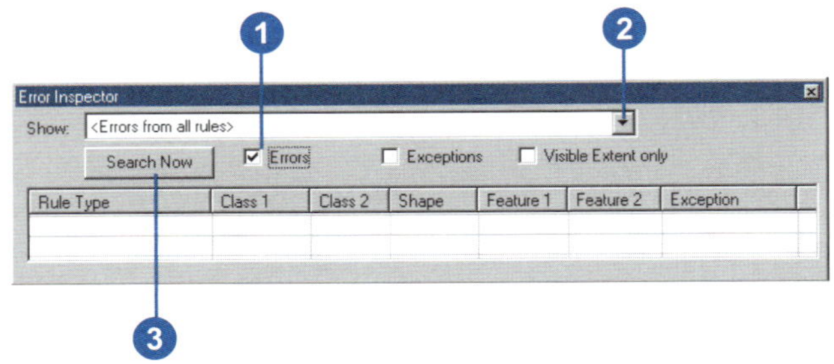

Finding only the errors in the visible extent

1. On the Error Inspector, check Visible Extent only.
2. Click Search Now.

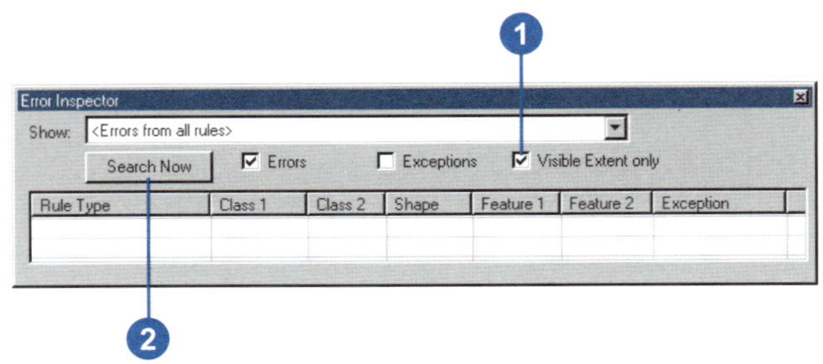

Finding errors for a particular topology rule

1. On the Error Inspector, click the Show dropdown arrow and click the rule that you want to search for violations of.
2. Click Search Now.

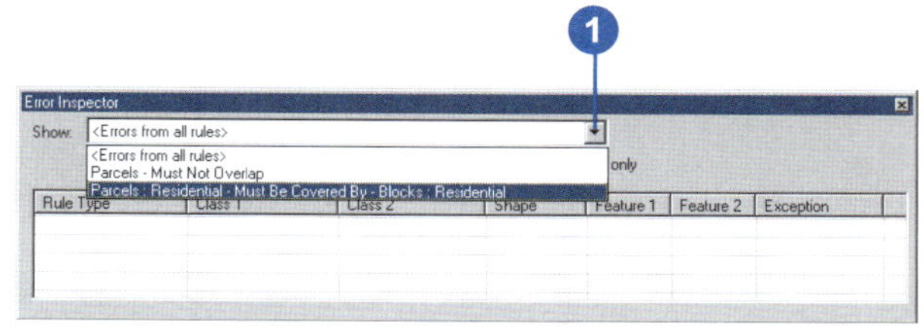

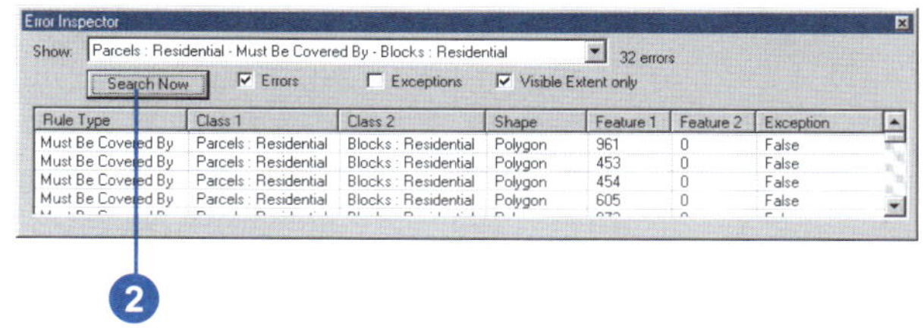

Finding exceptions

1. On the Error Inspector, check Exceptions.

   Exceptions are errors that have been marked as acceptable exceptions to the topology rule.
2. Uncheck Errors.
3. Click Search Now.

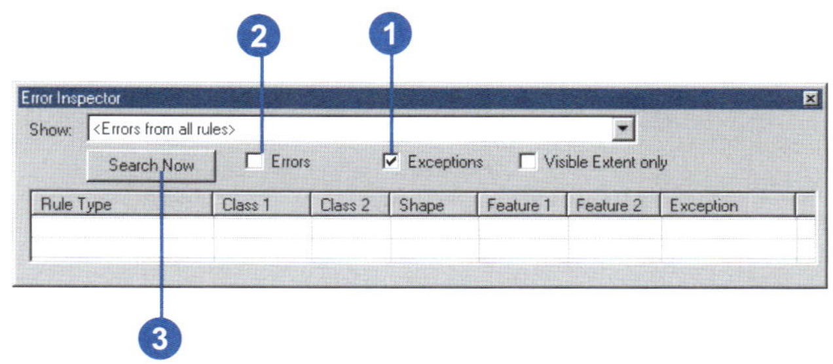

Correcting errors

You can apply a variety of predefined topology fixes to the errors you find. The sort of fix that will be appropriate for a given error depends on the error type and the geometry of the features involved. You can choose from different fixes or mark the error as an exception.

The topology fixes are available from the Fix Topology Error tool and by right-clicking a selected error in the Error Inspector. Depending on the type of error and the features involved, you can select, delete, merge, extend, trim, dissolve, subtract, or create features.

You can also pan and zoom to a selected error, show a description of the error, and mark an error as an exception.

Merging an error area into a polygon

1. Click Fix Topology Error on the Topology toolbar.
2. Click the error feature that you want to merge into one of the overlapping polygons.
3. Right-click and click Merge.
4. Click the feature that you want to merge the error feature into.
5. Click OK.

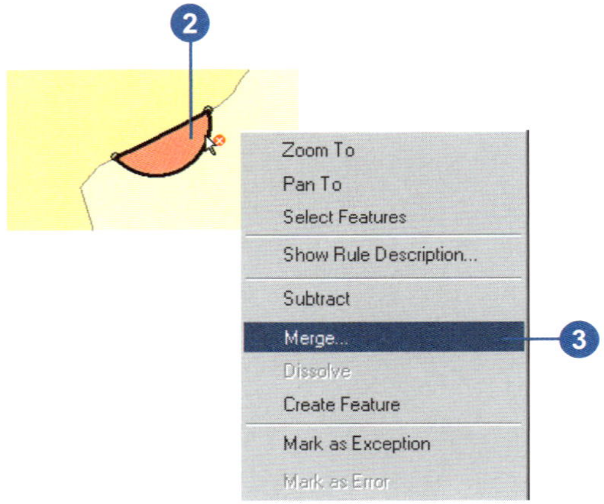

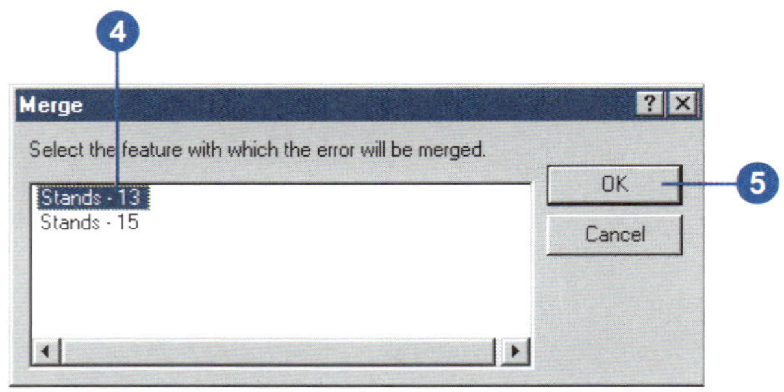

Merging an error area into a polygon from the Error Inspector

1. On the Error Inspector, click the error that you want to fix.
2. Right-click the error and click Merge.
3. Click the feature that you want to merge the error feature into.
4. Click OK.

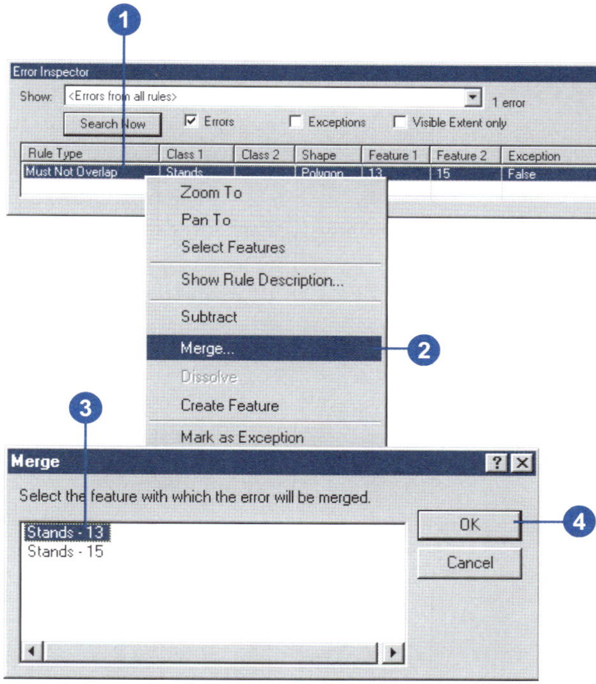

Finding the features that are affected by an error

1. On the Error Inspector, click the error.
2. Click the Feature 1 field in the error to see the first feature that is affected by it.
3. Click the Feature 2 field to see the second feature that is affected by it.

   The features flash on the map.

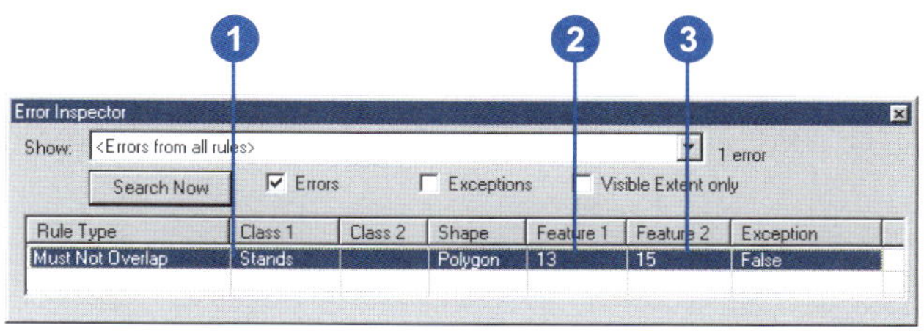

Marking an error as an exception

1. On the Error Inspector, click the error that you want to mark as an exception.
2. Right-click the error and click Mark as Exception.

   The error is marked as an exception. It is no longer symbolized as an error in the Topology layer on the map.

   You can use the Error Inspector to find exceptions as well as errors.

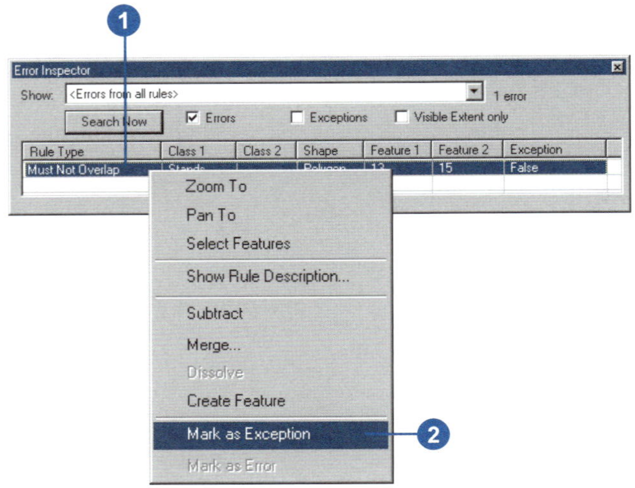

Getting a description of the rule that is violated for an error

1. On the Error Inspector, click the error that you want a description of.
2. Right-click the error and click Show Rule Description.

   A dialog box appears with a description of the error and some pictures of geometries that would and would not result in this error.

   The errors are marked in red.
3. Optionally, uncheck Show Errors to compare the feature geometries without the errors.
4. Click OK.

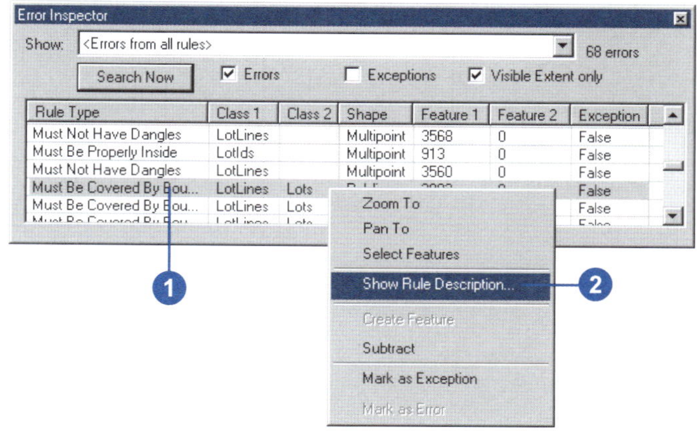

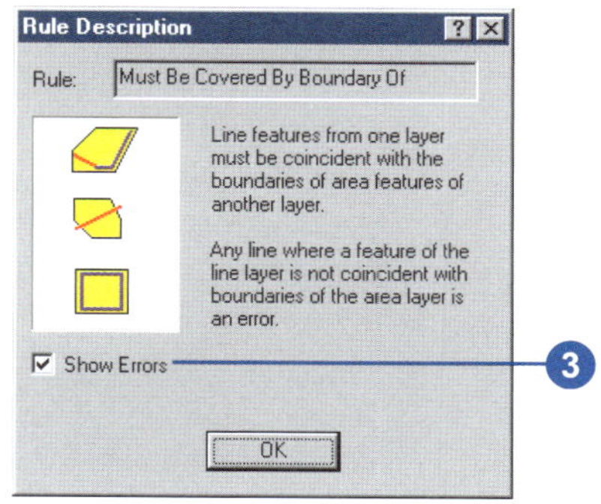

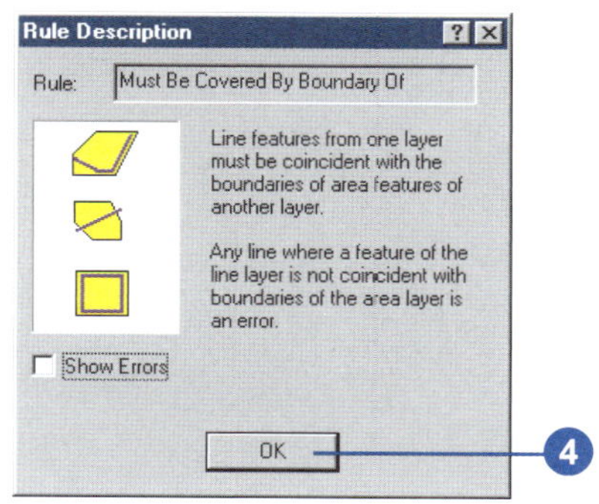

Creating new polygons from lines

Sometimes you need to create polygon features from line feature data. For example, you might have digitized the boundaries of a set of features into a line feature class, or you may have only been able to obtain line features from a data provider. Perhaps you have a detailed coastline feature class that you would like to use to update some existing, less detailed data. In ArcCatalog the Polygon Feature Class From Lines tool lets you create new polygon features from line and polygon features in one or more feature classes. You have the option of specifying a point feature class that will supply attributes for the new polygon features.

Creating a polygon feature class from lines

1. In ArcCatalog, navigate to the dataset in which you want to create a polygon feature class from an existing line feature class.
2. Right-click the dataset, point to New, and click Polygon Feature Class From Lines.
3. Type a name for the new polygon feature class.
4. Optionally, type a cluster tolerance.

   The default cluster tolerance is the minimum possible cluster tolerance.
5. Check the line feature classes that you want to be considered in creating the polygons.
6. Optionally, choose a point feature class in the dataset to provide attributes for the polygons.
7. Click OK.

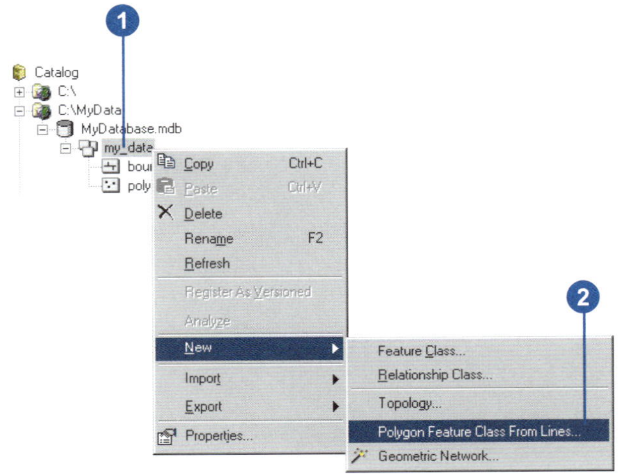

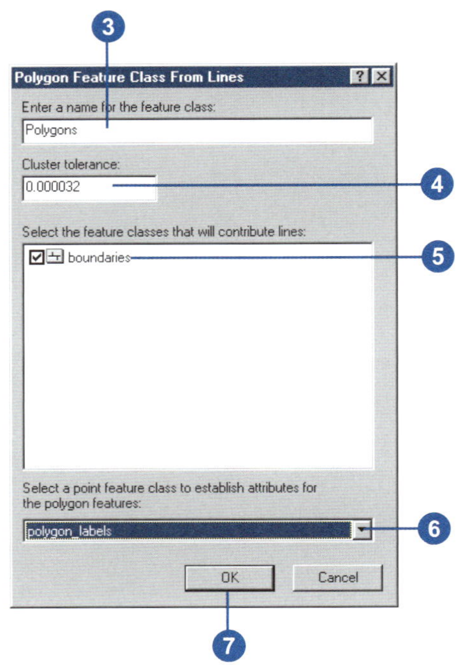

Creating new features from the geometry of existing features

Sometimes you need to create new features from the geometry of existing features. For example, you might need to create a new parcel feature from some parcel boundary lines or some parcel boundary lines from parcel features. In ArcMap you can select features and use their geometry to create new polygons or lines in the Editor Target Feature Class. You can create new features from existing features' geometry using the Construct Features tool.

You can create multiple line features by splitting longer features at the places where they intersect using the Planarize tool.

Constructing polygons from the geometry of other features

1. In ArcMap, click the Select Features tool.
2. Select the features whose geometry you want to use to construct new polygon features.
3. On the Editor toolbar, click the Task dropdown arrow and click Create New Feature.
4. On the Editor toolbar, click the Target dropdown arrow and click the polygon feature class that you want to create a new feature in.
5. On the Topology toolbar, click the Construct Features button.
6. Optionally, type a cluster tolerance.
7. Optionally, check the box to Consider existing features of the target layer in the current extent.

   This will use the boundaries of existing polygons as input geometry and will split such features where selected lines or polygons cross them.
8. Click OK.

   The new features are created in the target feature class.

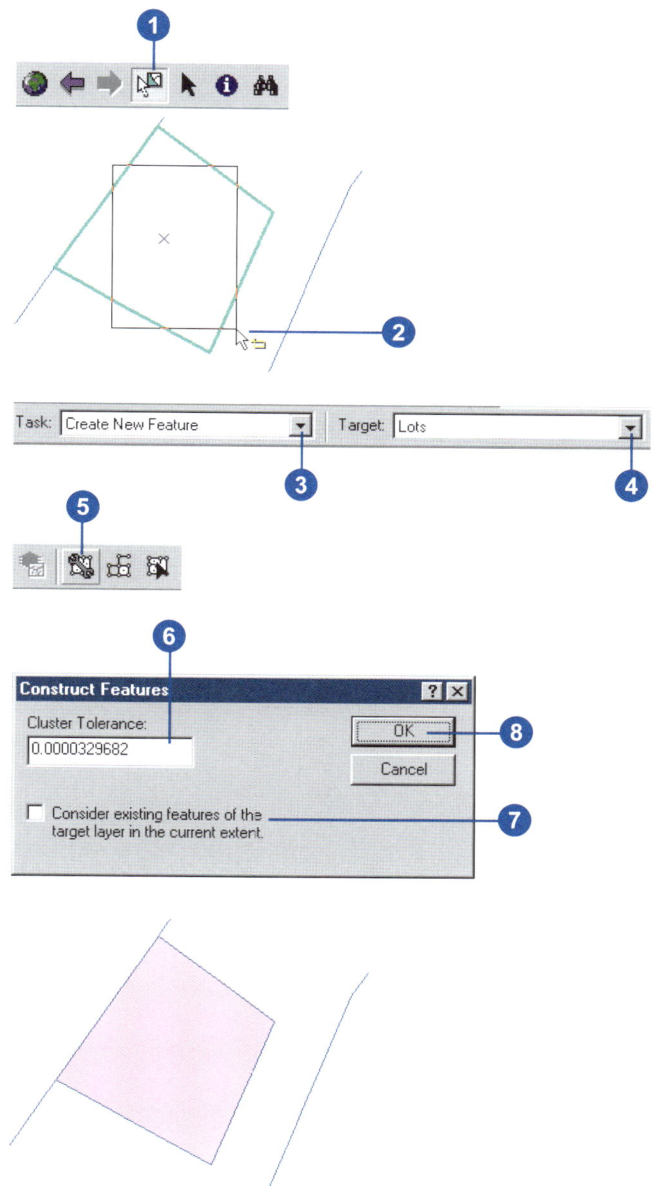

Constructing lines from the geometry of other features

1. In ArcMap, click the Select Features tool.
2. Select the features whose geometry you want to use to construct new line features.
3. On the Editor toolbar, click the Task dropdown arrow and click Create New Feature.
4. On the Editor toolbar, click the Target dropdown arrow and click the line feature class that you want to create a new feature in.
5. On the Topology toolbar, click the Construct Features button.
6. Optionally, type a cluster tolerance.
7. Optionally, check the box to Consider existing features of the target layer in the current extent.

   This will use existing lines as input geometry and will split such features where selected lines or polygons cross them.
8. Click OK.

   The new features are created in the target feature class.

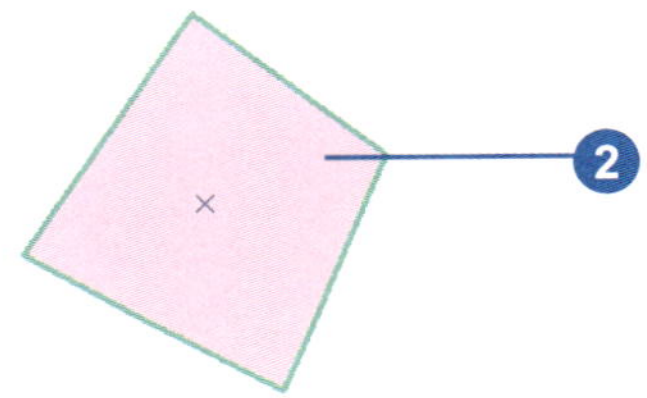

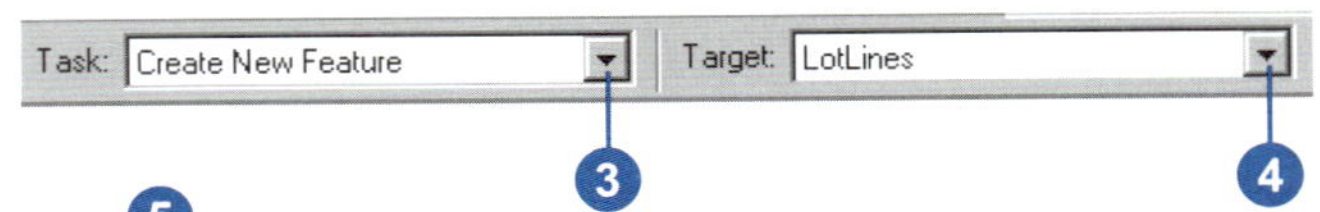

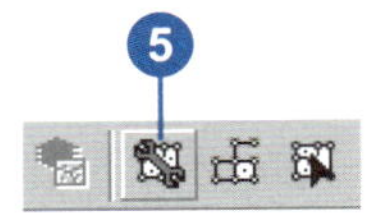

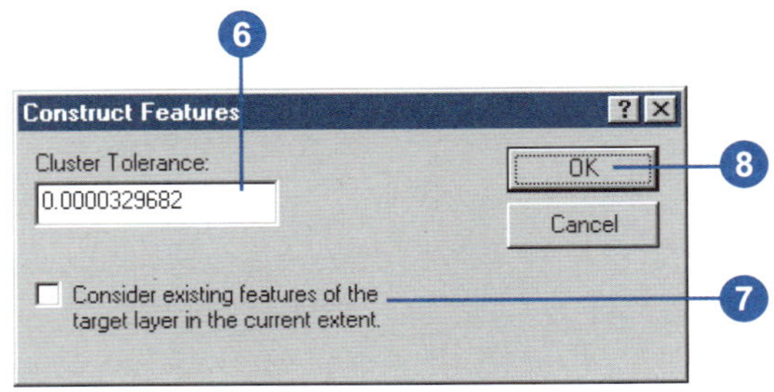

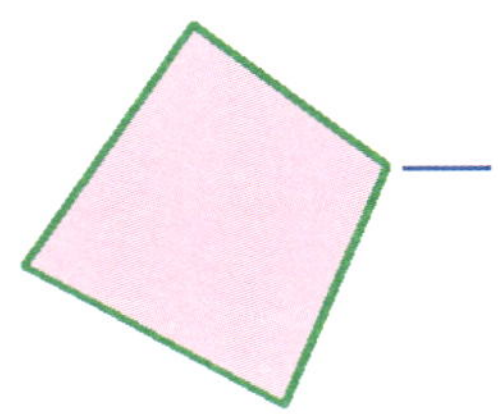

New line feature follows selected polygon feature boundary.

Using Planarize to split lines at intersections

1. In ArcMap, click the Select Features tool.
2. Select the line features that you want to split at intersections.
3. Click the Planarize button.
4. Optionally, type a cluster tolerance.
5. Click OK.

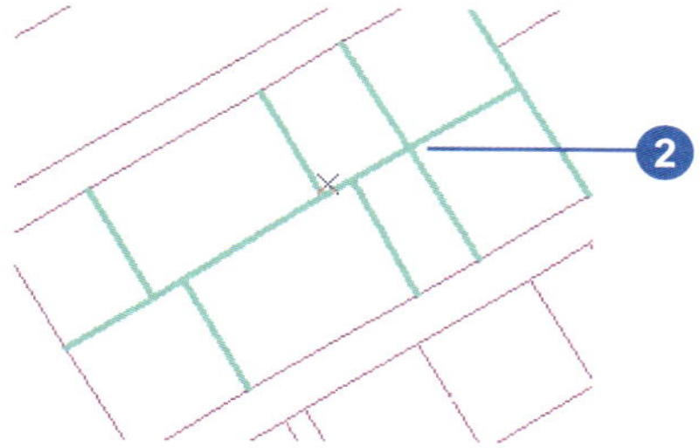

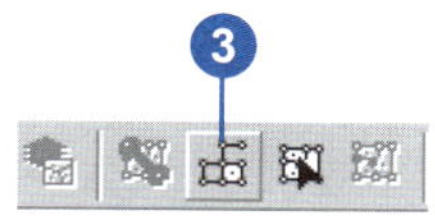

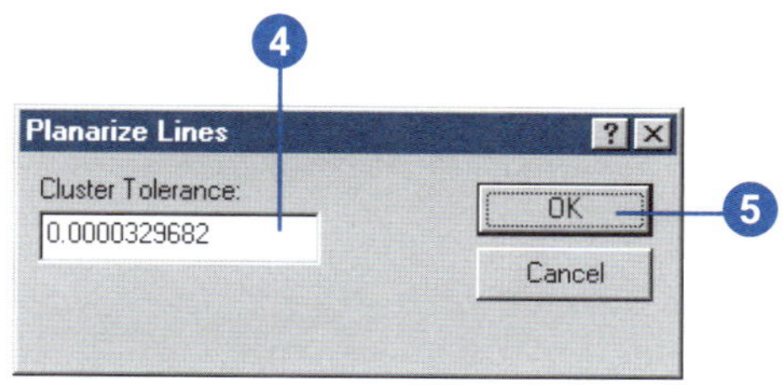

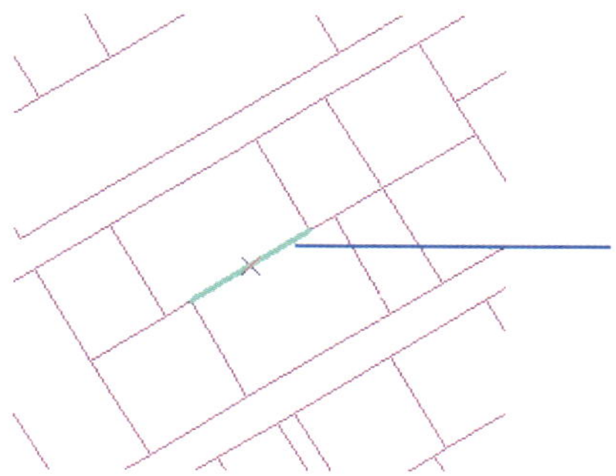

The lines are split into new features where they intersect.

Using a digitizer

5

IN THIS CHAPTER

Digitizing is the process of converting features on a paper map into digital format. To digitize a map, you use a *digitizing tablet* connected to your computer to trace over the features that interest you. The x,y coordinates of these features are automatically recorded and stored as spatial data.

Digitizing with a digitizing tablet offers another way, besides digitizing freehand, to create and edit spatial data. You can convert features from almost any paper map into digital features. You can use a *digitizer* in conjunction with the tools in ArcMap to create new features or edit existing features on a digital map.

You may want to digitize features into a new layer and add the layer to an existing map document or you may want to create a completely new set of layers for an area for which no digital data is available. You can also use a digitizer to update an existing layer on your digital map.

Chapter 3, 'Creating new features', introduced you to the Sketch tool and other useful editing tools in ArcMap and discussed how these are used to digitize features freehand. This chapter will teach you the fundamentals of editing features in ArcMap using a digitizer. You may want to read Chapter 3, 'Creating new features', first to get an understanding of editing before reading this chapter.

Setting up your digitizing tablet and preparing your paper map

Before you can start digitizing, you must set up your digitizing tablet and prepare your paper map. This can be done after you have installed the digitizer driver software.

Installing the driver software and configuring puck buttons

To use a digitizing tablet with ArcInfo, it must have WinTab™-compliant digitizer driver software. To find out if a WinTab-compliant driver is available for your digitizer, see the documentation that came with the tablet or contact the manufacturer.

After installing the driver software, use the WinTab manager setup program to configure the buttons on your *digitizer puck*. (You may have to turn on your digitizer and reboot your machine before you can use the setup program.) One puck button should be configured to perform a left mouse click to digitize point features and vertices; another button should be configured to perform a left double-click to finish digitizing line or polygon features.

With any development programming language, you can configure additional buttons to run specific ArcMap commands—such as the Zoom In or Sketch tools—normally accessed through toolbar buttons and menus. *Exploring ArcObjects* contains sample Visual Basic® for Applications (VBA) code that you can use to run a variety of ArcMap commands from the digitizer puck.

Preparing the map

After you have set up your digitizing tablet and configured the puck buttons, you can prepare your paper map for digitizing. Your map should ideally be reliable, up-to-date, flat, and not torn or folded. Paper expands or shrinks according to the weather. To minimize distortion in digitizing, experienced digitizers often copy paper maps to a more stable material such as Mylar®.

If you know what coordinate system (projection) your paper map is in, you should set the same projection for the layer you're digitizing into. If you are digitizing features into an existing feature layer, you must ensure that your paper map and digital layer share the same coordinate system. For more information on specifying a coordinate system in ArcMap, see *Using ArcMap*.

Establishing control points on your paper map

Before you can begin digitizing from your paper map, you must first establish *control points* that you will later use to register the map to the geographic space in ArcMap. If your map has a grid or a set of known ground points, you can use these as your control points. If not, you should choose between four and 10 distinctive locations such as road intersections and mark them on your map with a pencil. Give each location a unique number and write down its actual ground coordinates.

Once you've identified at least four well-placed control points, you can place your map on the tablet and attach it with masking tape. You don't have to align the map precisely on your tablet; ArcMap corrects any alignment problems when you register the map and displays such adjustments in the error report.

The error report includes two different error calculations: a point-by-point error and a root mean square (RMS) error. The point-by-point error represents the distance deviation between the transformation of each input control point and the corresponding point in map coordinates. The RMS error is an average of those deviations. ArcMap reports the point-by-point error in current map units. The RMS error is reported in both current map units and digitizer inches. If the RMS error is too high, you can reregister the appropriate control points. To maintain highly accurate data, the RMS error should be kept under 0.004 digitizer inches. For less accurate data, the value can be as high as 0.008 digitizer inches.

Registering your paper map

Before you can start digitizing, you must register your paper map into real-world coordinates. This allows you to digitize features directly in geographic space.

Registering your map involves recording the ground coordinates for the control points you identified while preparing your map. These are recorded using the Digitizer tab of the Editing Options dialog box. You must first use the digitizer puck to digitize the control points on the paper map; with the puck over each control point on the map, press the button you configured to perform a left mouse click. You must then type the actual ground coordinates for each control point.

When registering your map, you have the option of saving the ground coordinates you entered for later use—for example, if you want to reregister your map or register another map that uses the same control points. These ▸

See Also

For information on configuring puck buttons and establishing control points, see 'Setting up your digitizing tablet and preparing your paper map' in this chapter.

Registering your map for the first time

1. After adding a layer to your map, click the Editor menu and click Start Editing.
2. Click Editor and click Options.
3. Click the Digitizer tab.
4. With the digitizer puck, digitize the control points you established earlier on your paper map.

   A record appears in the X Digitizer and Y Digitizer columns for each control point you digitized.
5. Type the actual ground coordinates for each control point in the X Map and Y Map fields.

   An error in map units is displayed at each control point. An RMS error is displayed in map units and in digitizer inches.
6. Click OK to register the map and close the Editing Options dialog box.

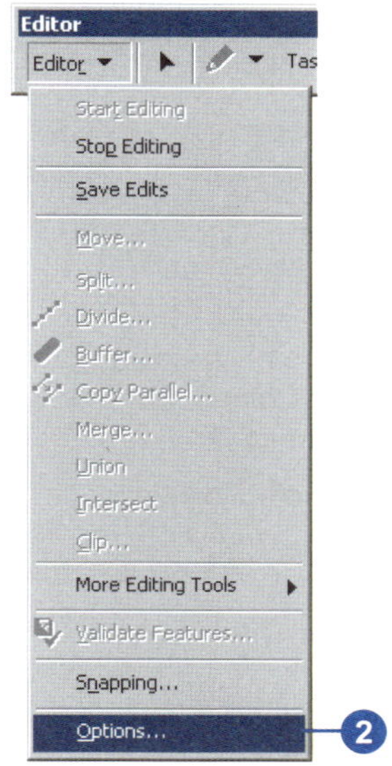

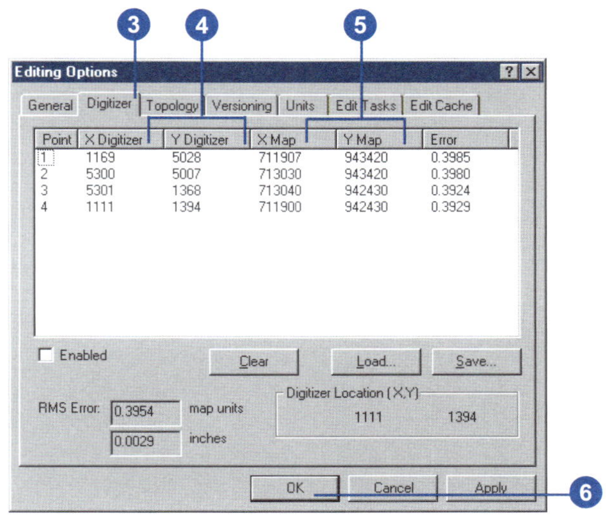

ground coordinates are stored in tic text files.

After you've entered the ground coordinates, ArcMap displays an error at each control point as well as an RMS error. If the RMS error is too high—greater than 0.004 digitizer inches for highly accurate data or greater than 0.008 digitizer inches for less accurate data—you can register the appropriate control points again. For more information on errors, see 'Setting up your digitizing tablet and preparing your paper map' in this chapter.

Tip

Missing Digitizer tab

If you installed ArcInfo before installing your digitizer, the Digitizer tab may be missing from the Editing Options dialog box. To add the tab, you must register the digitizer.dll file. Go to the DOS prompt, type "cd" followed by the path to the directory where you installed ArcInfo (%ARCHOME%\bin), and type "regsvr32 digitizer.dll". When you restart ArcMap, the Editing Options dialog box will have the Digitizer tab.

Saving new ground coordinates

1. Follow steps 1 through 5 for 'Registering your map for the first time' in this chapter.
2. Click Save.
3. Navigate to the directory in which you want to save the coordinates and type a filename.
4. Click Save.
5. Click OK.

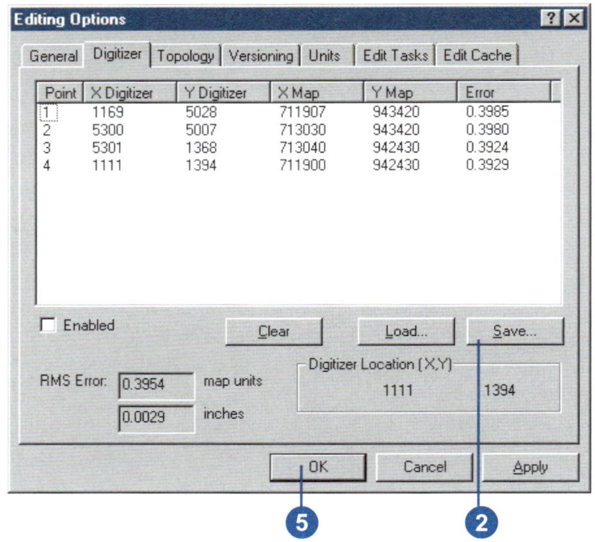

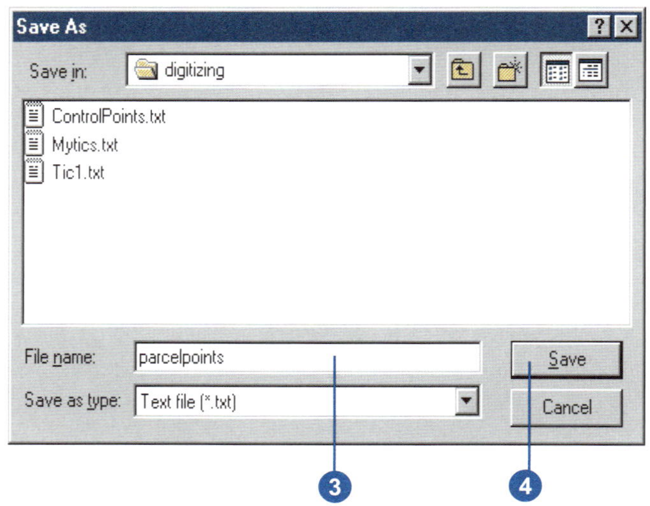

Registering your map using existing tic files or saved coordinates

1. After adding a layer to your map, click the Editor menu and click Start Editing.
2. Click the Editor menu and click Options.
3. Click the Digitizer tab.
4. Click Load.
5. Navigate to the file you want to use.
6. Click Open. ▸

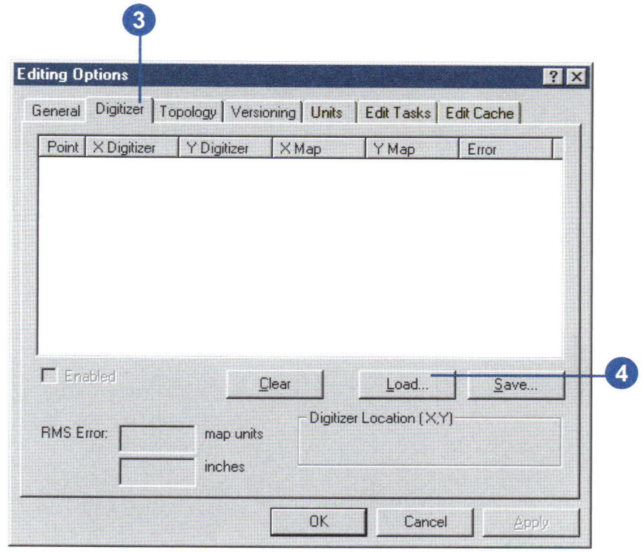

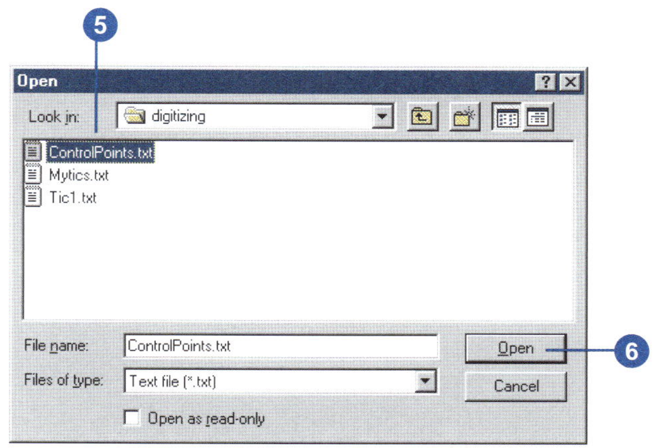

Tip

Removing records

If you want to remove all the ground coordinate records and start over, click Clear on the Digitizer tab. To remove an individual record, click the number in the Point column corresponding to the coordinates you want to remove and press the Delete key.

Tip

Adding records

If you want to add additional control points after entering a few, click below the last record with the mouse and digitize the new points with the digitizer puck.

Tip

Digitizer location

The Digitizer tab also displays the current x,y location of the digitizer puck on the tablet. The coordinates change as you move the puck along the tablet surface. This helps orient you to the location you're digitizing.

Tip

Digitizing accuracy

Always register your map at the start of each digitizing session, even if this means registering the same map more than once. Your paper map might shift between sessions; reregistering helps ensure that your digitizing is accurate.

The ground coordinates appear under the X Map and Y Map fields.

7. Click the first record and digitize the first control point with the digitizer puck.
8. Digitize each of the other control points.

   The digitized coordinates appear in the X Digitizer and Y Digitizer columns. An error is displayed for each control point, and an RMS error is displayed in map units and in digitizer inches.
9. Click OK to register the map.

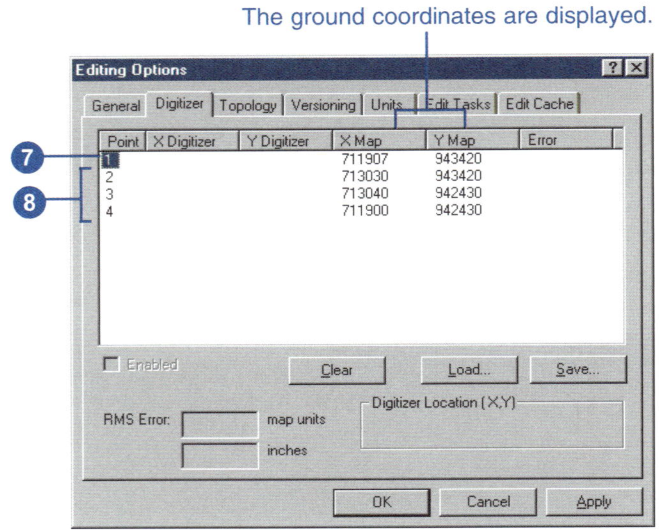

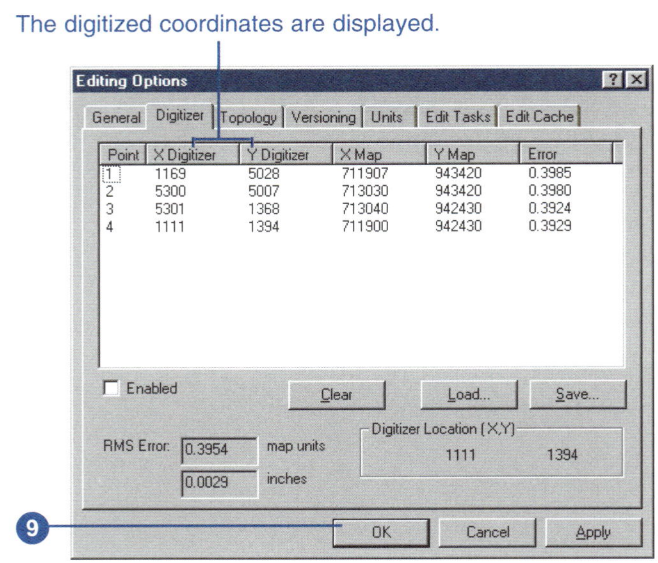

Creating features using a digitizer

It's easy to digitize features in ArcMap. You can digitize features into a new map layer or edit an existing layer.

Digitizing modes

Digitizing tablets generally operate in two modes: digitizing (absolute) mode and mouse (relative) mode.

In *digitizing mode*, the location of the tablet is mapped to a specific location on the screen. In other words, moving the digitizer puck on the tablet surface causes the screen pointer to move to precisely the same position. When you are in digitizing mode, you can only digitize features; you can't choose buttons, menu commands, or tools from the ArcMap user interface because the screen pointer is locked to the drawing area.

In *mouse mode*, the digitizer puck behaves just like a mouse; there is no correlation between the position of the screen pointer and the surface of the digitizing tablet, but you can choose interface elements with the pointer.

ArcMap lets you switch between digitizing and mouse modes using the Editing Options dialog box. This means you can use the digitizer puck to both digitize features and access user interface choices (as a substitute to the mouse) as you digitize.

Whether your digitizer is in mouse mode or digitizing mode, you can still use your mouse at any time to choose interface elements.

Two ways to digitize features on a paper map

You can digitize features on a paper map in two ways: using point mode digitizing or stream mode digitizing (streaming). You can switch back and forth between the two modes as you digitize by pressing F8.

Digitizing by point

When you start a digitizing session, the default is point mode. With *point mode digitizing*, you convert a feature on a paper map by digitizing a series of precise points, or vertices. ArcMap then connects the vertices to create a digital feature. You would use point mode when precise digitizing is required—for example, when digitizing a perfectly straight line.

Digitizing using stream mode

Stream mode digitizing (streaming) provides a quick and easy way to capture features on a paper map when you don't require as much precision—for example, to digitize rivers, streams, and contour lines. With stream mode, you create the first vertex of the feature and trace over the rest of the feature with the digitizer puck. When you're finished tracing, you use the puck to complete the feature.

As you stream, ArcMap automatically adds vertices at an interval you specify; this interval, expressed in current map units, is called the stream tolerance. You can change the stream tolerance at any time, even while you're in the process of digitizing a feature.

You can also digitize using stream mode when you create features "freehand" with the sketch construction tools. You can digitize in stream mode with the Sketch tool, for example, in the same way you do from a paper map. The only difference is that you use the mouse pointer to digitize freehand.

Adding topology to digitized features

Digitizing creates lines or points that have no topological relationships. ArcMap provides tools to improve such spagetti digitized data, for example, by splitting lines at intersections or creating polygons from lines. To learn more about topology, see Chapter 4, 'Editing topology'.

Digitizing features in point mode

Point mode digitizing works the same way with a digitizer as freehand digitizing with the Sketch tool; the only difference is that with the digitizer you're converting a feature from a paper map using a digitizer puck instead of a mouse.

Point mode digitizing involves converting point, line, and polygon features from a paper map by digitizing a series of precise points, or vertices. You digitize each vertex by pressing the puck button you configured to perform a left mouse click. To finish the feature, press the puck button you configured to perform a left double-click. ArcMap connects the vertices to create a digital feature.

Before you begin digitizing, you must set the digitizer to work in digitizing mode rather than in mouse mode; this constrains the screen pointer to the digitizing area. When the puck is in ▸

See Also

For information on configuring puck buttons and establishing control points, see 'Setting up your digitizing tablet and preparing your paper map' in this chapter.

1. Click Editor and click Options.
2. Click the Digitizer tab.
3. Check Enabled to use the puck in digitizing mode.
4. Click OK. ▸

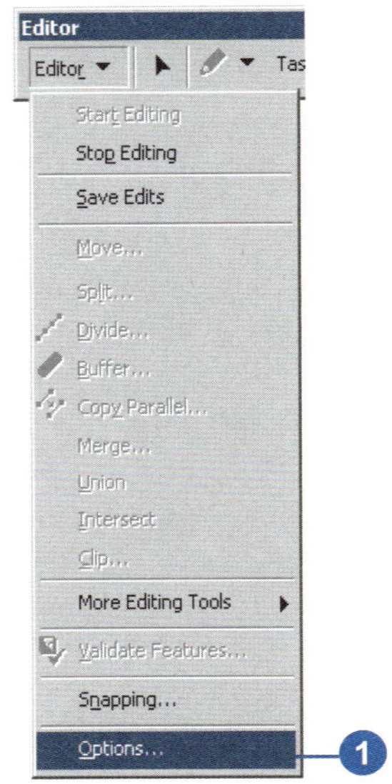

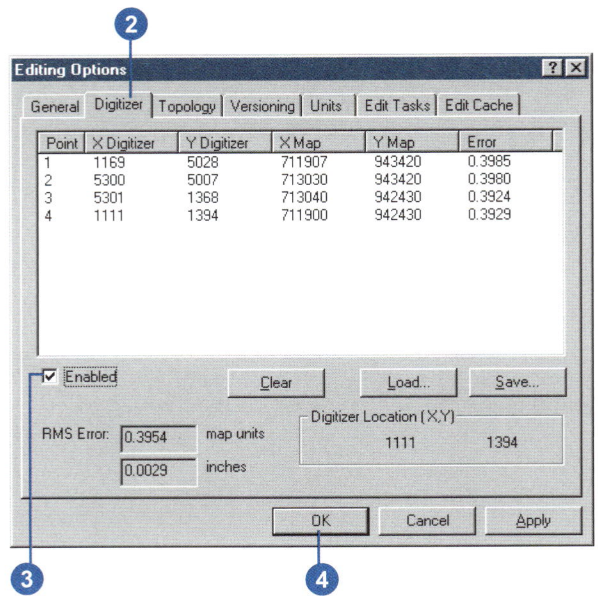

digitizing mode, you must use your mouse to choose items in the ArcMap interface—unless you have used VBA or another development programming language to configure additional puck buttons to run specific ArcMap commands.

Tip

Snapping

To help you digitize features in a precise location on an existing layer, you can use the snapping environment. For information on snapping, see Chapter 3, 'Creating new features'.

Tip

Deleting vertices

Click the Undo button on the ArcMap Standard toolbar to delete a vertex as you digitize.

See Also

For information on creating features by digitizing freehand with the sketch creation tools, see Chapter 3, 'Creating new features'.

See Also

For information on configuring puck buttons with programming code, see 'Setting up your digitizing tablet and preparing your paper map' in this chapter.

5. Click the tool palette dropdown arrow and click the Sketch tool.
6. With the digitizer puck, digitize the first vertex of the feature.
7. Trace the puck over the feature on the paper map, creating as many vertices as you need.
8. Finish the feature by pressing the appropriate puck button.

   The feature is created.

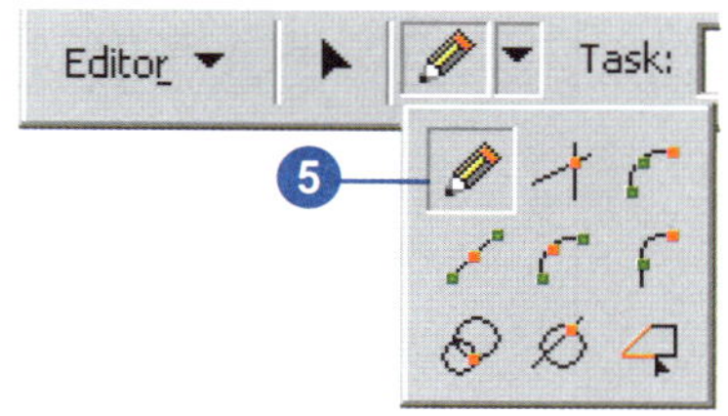

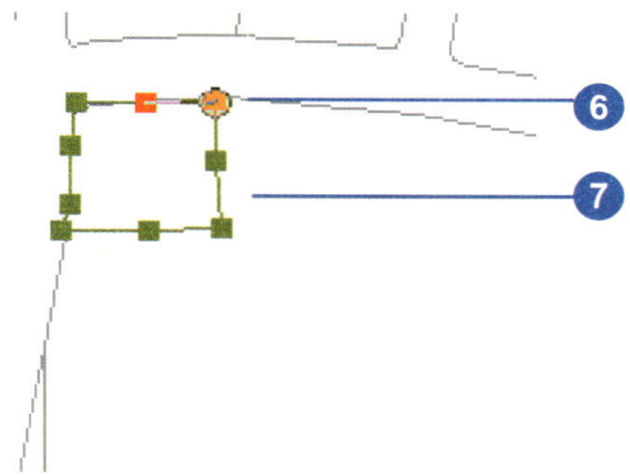

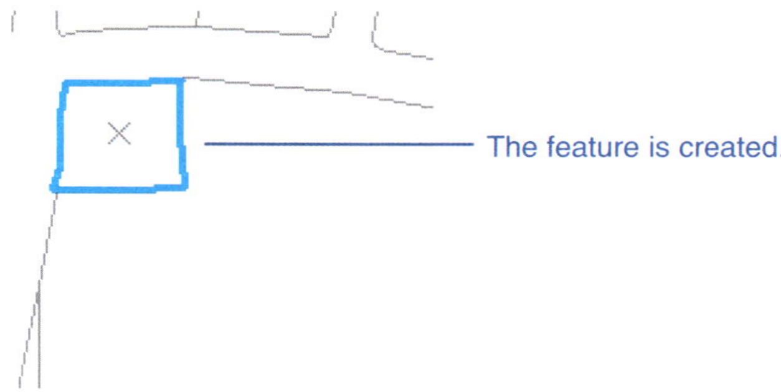

Digitizing features in stream mode

When you digitize line or polygon features from a paper map in stream mode (streaming), you create the first vertex of the feature by pressing the digitizer puck button you configured to perform a left mouse click. You then trace over the rest of the feature with the digitizer puck. When you're finished tracing, press the puck button you configured to perform a left double-click to complete the feature.

Before starting to digitize in stream mode, you must set the *stream tolerance*—the interval at which ArcMap adds vertices along the feature you're digitizing. Because the default stream tolerance is 0, you must enter a tolerance value before you start digitizing or the vertices will join together or overlap each other. You can change the stream tolerance any time in the digitizing process.

You must also specify the number of streaming vertices you want to group together. The number you set tells ArcMap how many vertices to delete when you click the Undo button. For example, if you set this ▸

Setting the stream tolerance

1. Click the Editor menu and click Start Editing.
2. Click the Editor menu and click Options.
3. Click the General tab.
4. Type the stream tolerance—in map units—in the Stream tolerance text box.
5. Click OK.

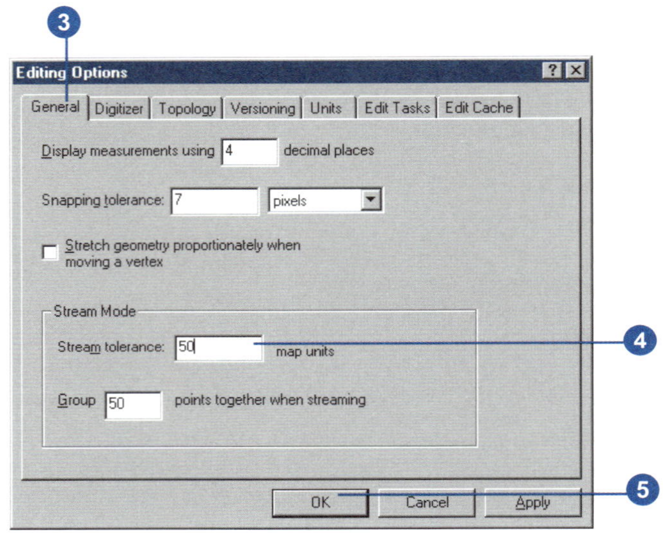

Setting the number of vertices to be grouped

1. Click the Editor menu and click Options.
2. Click the General tab.
3. Type the number of vertices you want to group together.
4. Click OK.

   Now when you click the Undo button while digitizing in stream mode, the number of vertices you specified are deleted.

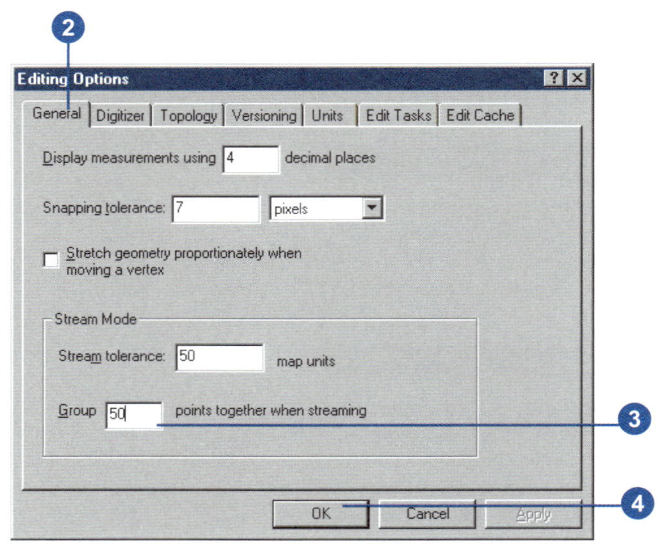

number to 20 and click the Undo button while you're digitizing a feature, ArcMap deletes the last 20 digitized vertices from your feature.

To begin digitizing in stream mode, you must choose Streaming from the sketch context menu. You can switch back to point mode at any time by pressing F8; press F8 again to switch to stream mode again.

Before streaming, remember to set the digitizer to work in digitizing mode rather than in mouse mode; this constrains the screen pointer to the digitizing area.

Tip

Snapping

To help you digitize features in a precise location on an existing layer, you can use the snapping environment. For information on snapping, see Chapter 3, 'Creating new features'.

Digitizing a feature in stream mode

1. Click the Editor menu and click Options.
2. Click the General tab.
3. Type the stream tolerance—in map units—in the Stream Tolerance text box.
4. Type the number of vertices you want to group together.
5. Click the Digitizer tab.
6. Check Enabled to use the puck in digitizing mode.
7. Click OK.
8. Click the tool palette dropdown arrow and click the Sketch tool. ▸

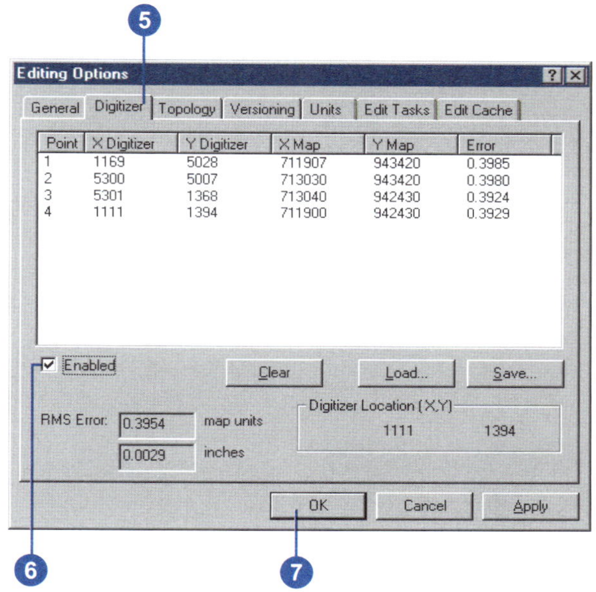

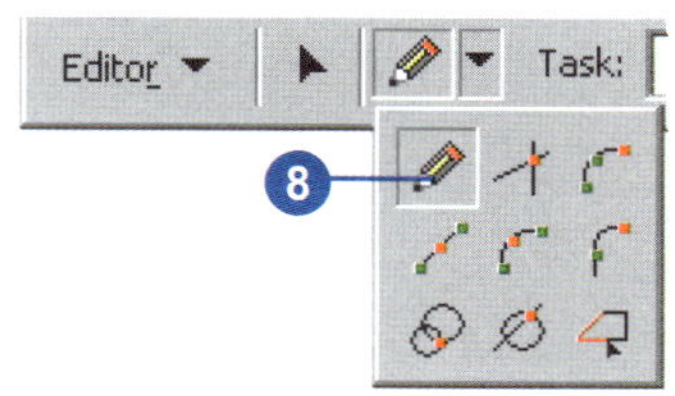

Tip

Choosing interface elements while streaming

When you're in the process of digitizing a feature in stream mode and want to interact with the ArcMap interface using your mouse—for example, to change the stream tolerance or undo an action—you must first switch back to point mode by pressing F8. After you have finished interacting with the interface, you can resume streaming by pressing F8 again.

Tip

Configuring a puck button for streaming

Instead of choosing Streaming from the context menu, you can configure one of your puck buttons using any development programming language, such as C++ or VBA, to activate stream mode digitizing. To learn more about configuring your puck buttons and customization in general, see Exploring ArcObjects.

9. With the mouse pointer, right-click anywhere on the map and click Streaming.
10. With the digitizer puck, digitize the first vertex of the line or polygon feature.
11. Trace the puck over the feature on the paper map.

    ArcMap creates vertices at the stream tolerance you specified.
12. Finish the feature by pressing the appropriate puck button.

    The feature is created.

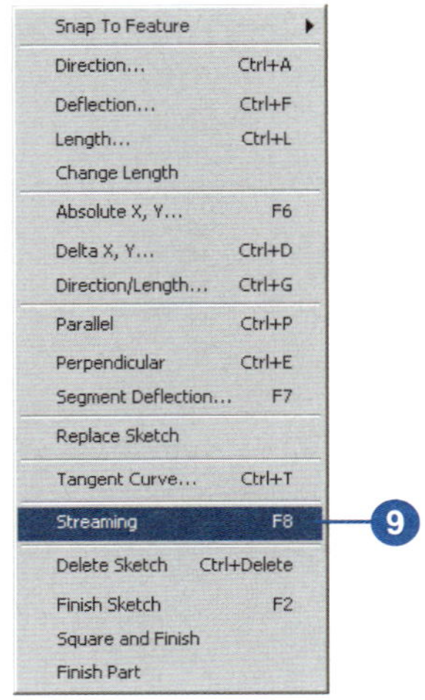

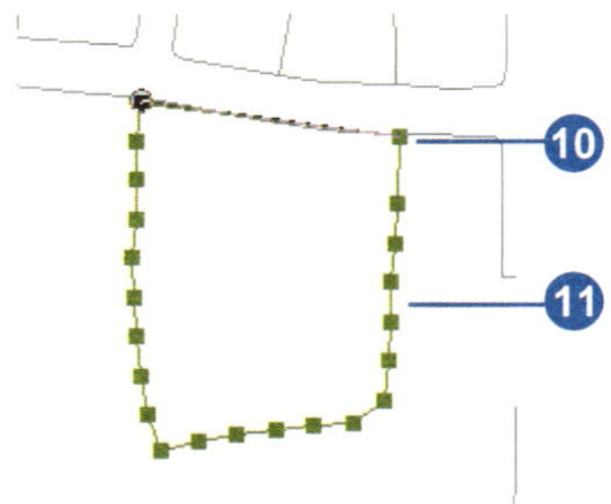

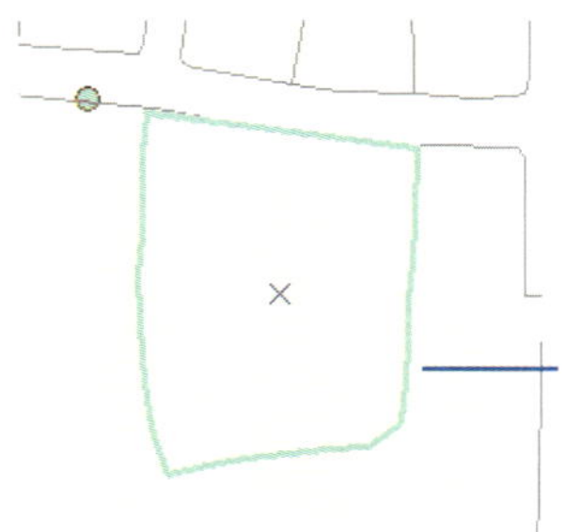

Creating features from other features

6

IN THIS CHAPTER

- **Copying a line at a specific interval**
- **Creating a buffer around a feature**
- **Creating a mirror image of a feature**
- **Merging features from the same layer into one feature**
- **Combining features from different layers into one feature**
- **Creating a feature from features with common areas**

In spatial data editing, many new features can be created using the shapes of other features. ArcMap has many tools you can use to create new features based on features already in your database.

For example, you can construct a line that is a parallel copy of an existing line to create a centerline on a street. You can create a buffer around a point, line, or polygon feature to show a specific area, such as a floodplain around a river. You can also create a new feature by combining or intersecting existing features or even create a mirror image of a feature or set of features.

In this chapter, you'll learn how easy it is to perform these tasks using various tools in ArcMap.

Copying a line at a specific interval

The Copy Parallel command copies a line parallel to an existing feature at a distance you specify. If you give a distance that is positive, the line is copied to the right side of the original feature. A negative distance value copies the line to the left.

You might use the Copy Parallel command to create a street centerline or to create a gas line that runs parallel to a road.

1. Click the Edit tool.
2. Click the line you want to copy.
3. Click the Target layer dropdown arrow and click the layer to which you want the new line to belong. ►

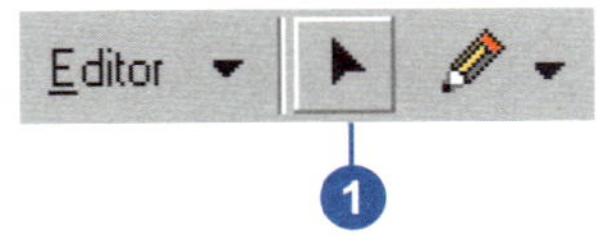

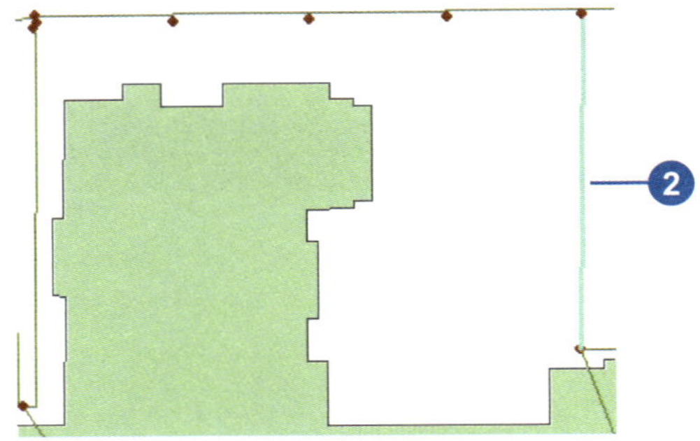

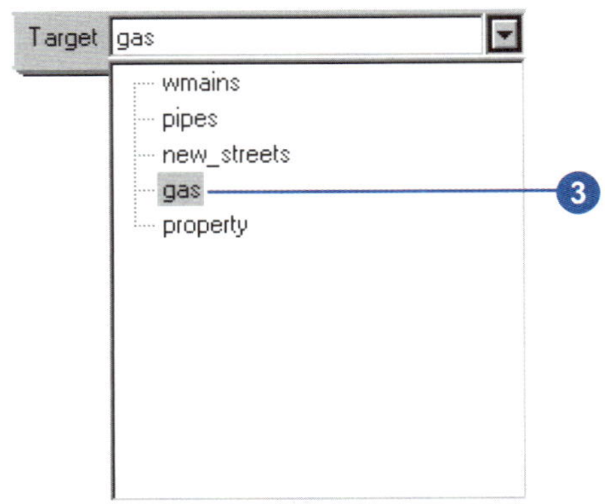

4. Click Editor and click Copy Parallel.
5. Specify the copy parameters including the distance—in map units—from the original feature where you want to copy the line, the corner style, and the behavior of intersecting loops.
6. Click OK or press Enter.

   A parallel copy of the line is created at the specified distance.

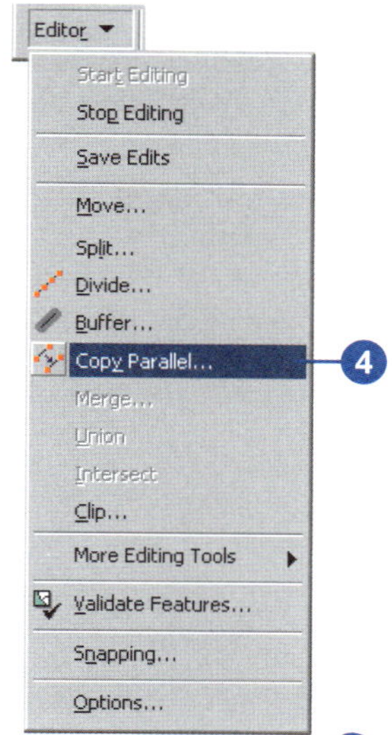

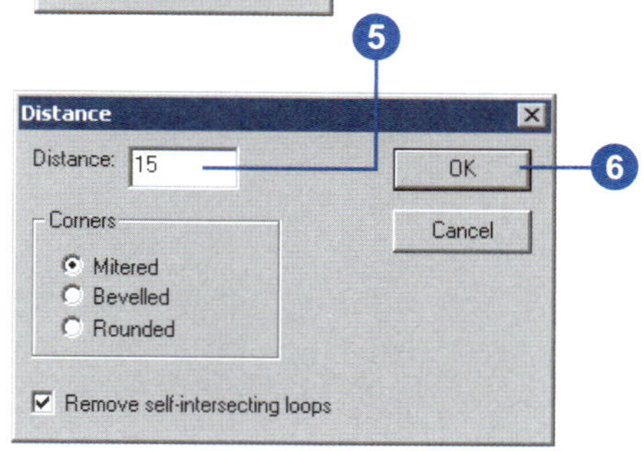

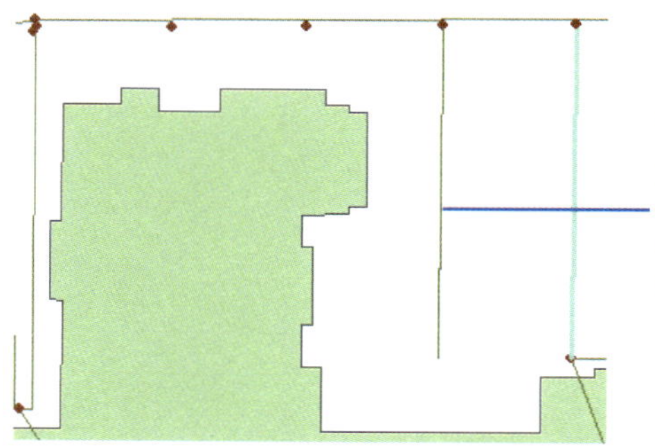

The line is copied parallel and to the left of the original feature at a distance of 15 map units.

Creating a buffer around a feature

You can create a buffer around a feature using the Buffer command. For instance, you might use Buffer to show the area around a well that's contaminated or to represent a floodplain around a river.

You can buffer more than one feature at a time, but a separate buffer will be created around each feature.

1. Click the Edit tool.
2. Click the feature or features around which you want to create a buffer.
3. Click the Target layer dropdown arrow and click the layer with the type of features you want the buffer to be. (This can only be a line or polygon layer.) ▸

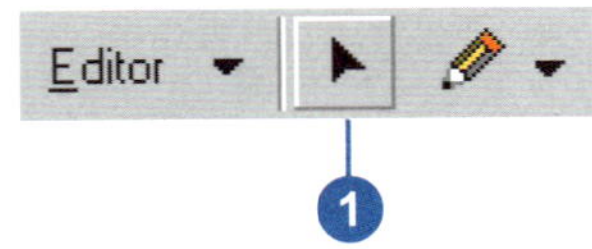

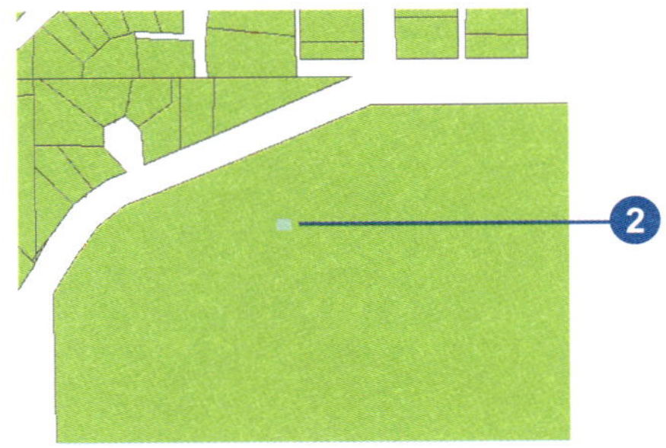

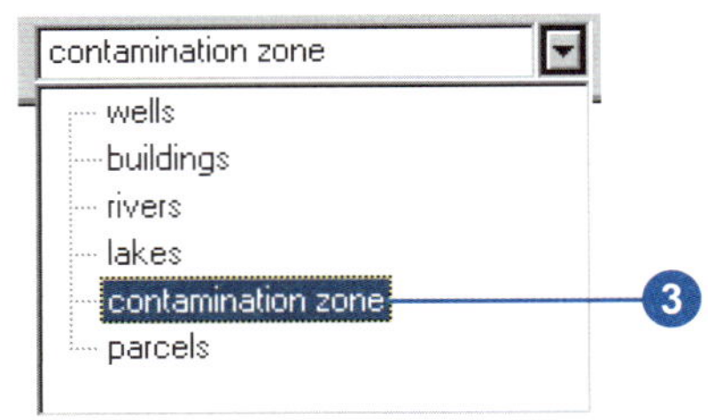

4. Click Editor and click Buffer.
5. Type the distance—in map units—from the feature around which you want to create the buffer and press Enter.

   A buffer is created at the specified distance.

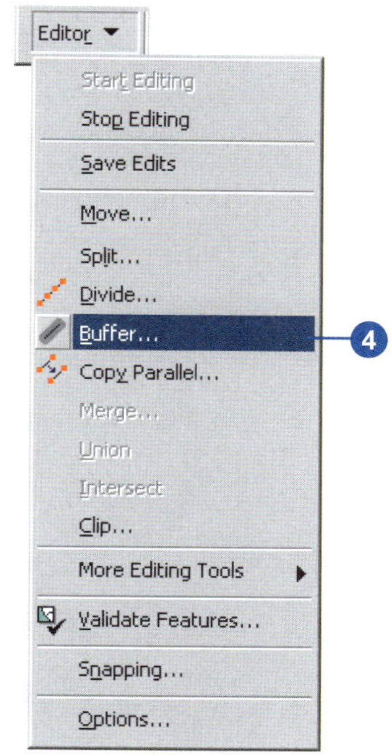

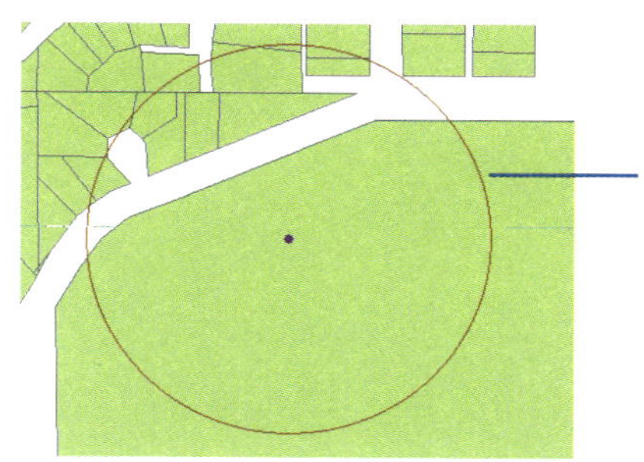

A buffer of 1,000 map units is created around the point.

Creating a mirror image of a feature

The Mirror task creates a mirror image of selected features on the other side of a straight line you create. You might use the Mirror task to create houses in a housing development where houses are mirror images of the ones on the opposite side of the street.

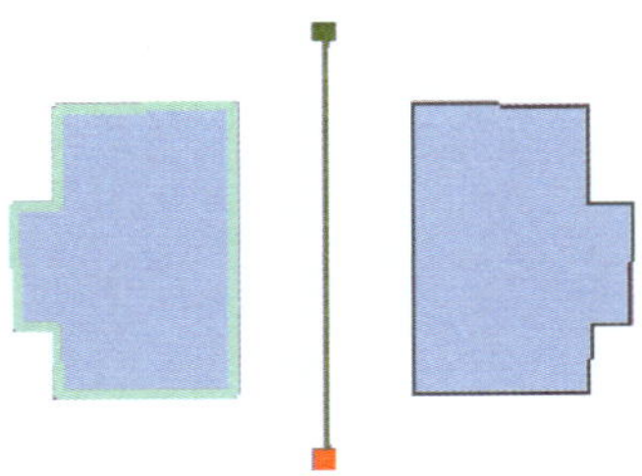

Also, as shown in the example, the Mirror task provides an easy way to add gas services to parcels that mirror the services on the other side of the street.

Tip

Other ways to construct a line

You can also use the Distance–Distance and Intersection tools to create the endpoints of the line. For more information, see Chapter 3, 'Creating new features'.

1. Click the Edit tool.
2. Click the feature or features that you want to mirror.
3. Click the Task dropdown arrow and click Mirror Features.
4. Click the tool palette dropdown arrow and click the Sketch tool.
5. Construct a line by clicking once on the start point and once on the endpoint. ▸

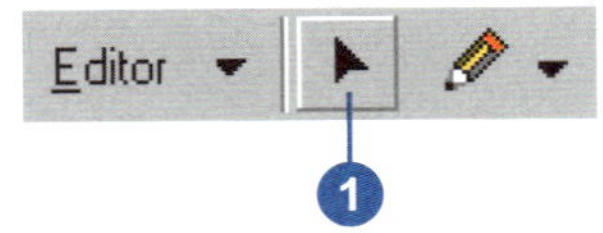

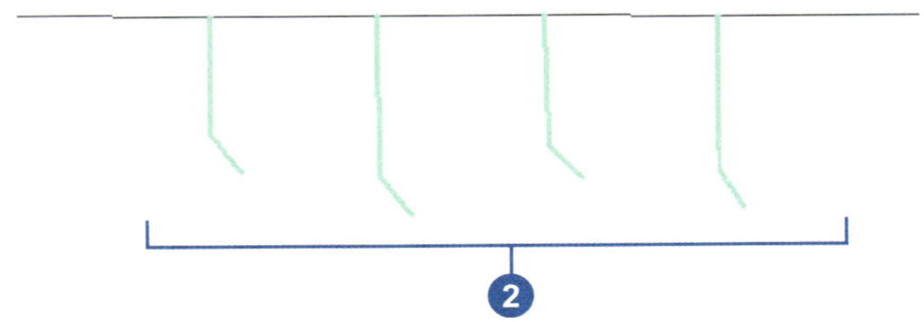

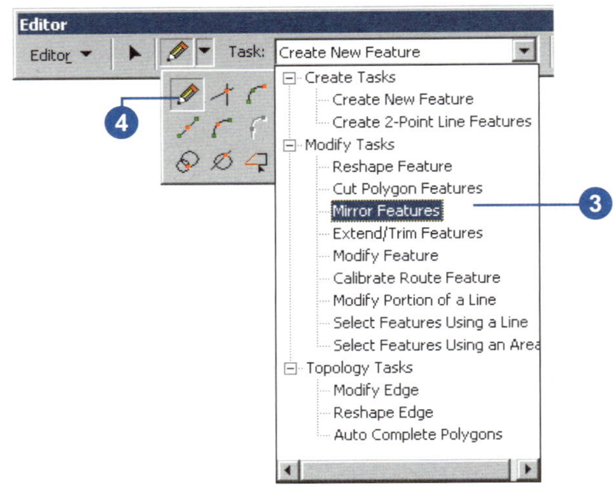

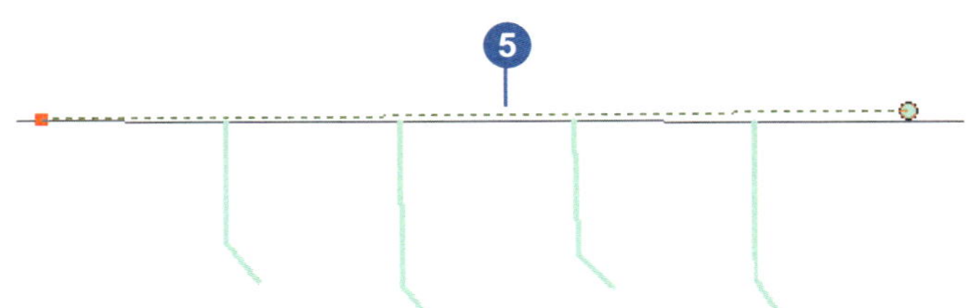

After you digitize the endpoint, a mirror image of the feature or features is created.

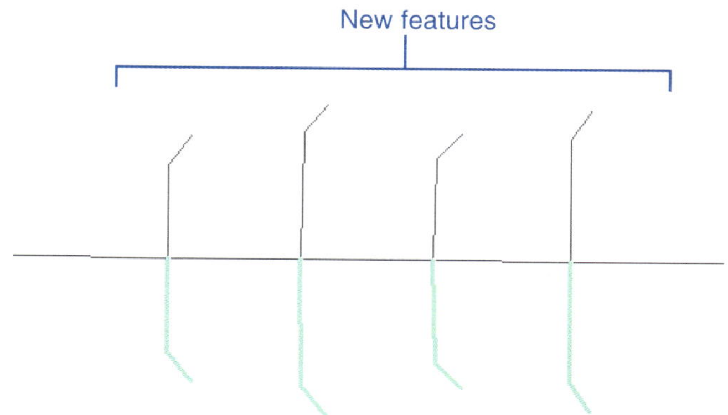

Merging features from the same layer into one feature

The Merge command combines features from the same layer into one feature. The features must be part of a line or polygon layer. You could use the Merge command to combine two parcels into one.

You might also want to merge nonadjacent features to create a multipart feature. For example, you could merge the individual islands that make up Hawaii to create a multipart polygon feature.

When you merge features in a geodatabase, the original features are removed and the new feature's attributes are copied from the feature that was selected first. If you merge shapefile features, the attributes of the feature with the lowest ID number (the oldest feature) are used.

1. Click the Edit tool.
2. Click the features that you want to merge.

   (The features must be from the same layer, either a line or polygon layer.)
3. Click the Target layer dropdown arrow and click the layer to which you want the new feature to belong. ►

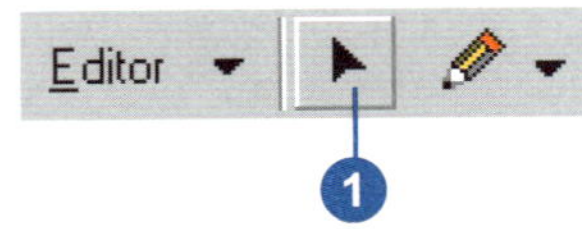

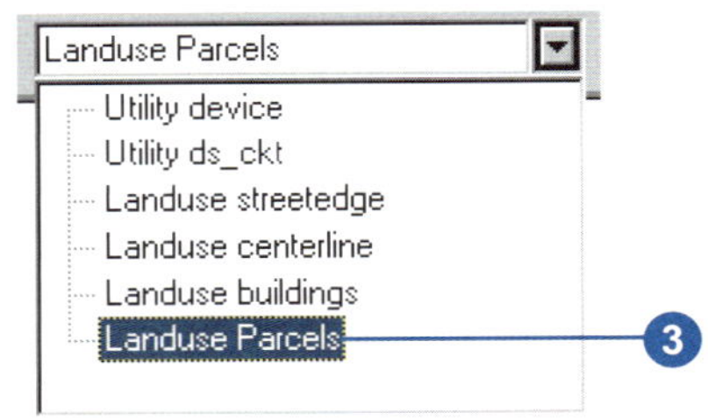

4. Click Editor and click Merge.
5. Click the feature that you want to merge the other feature or features into.

   The selected features are merged into one.
6. Click OK.

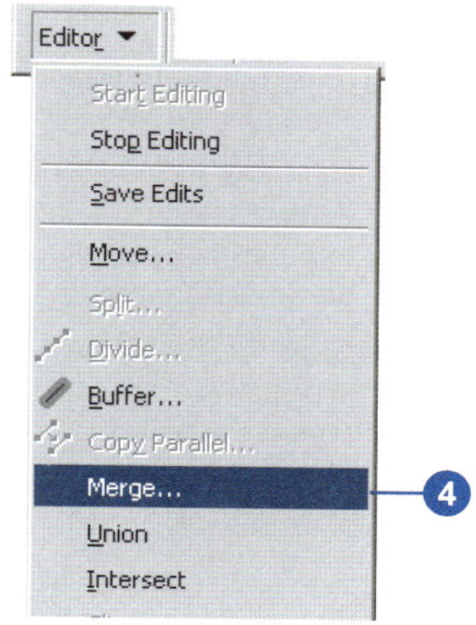

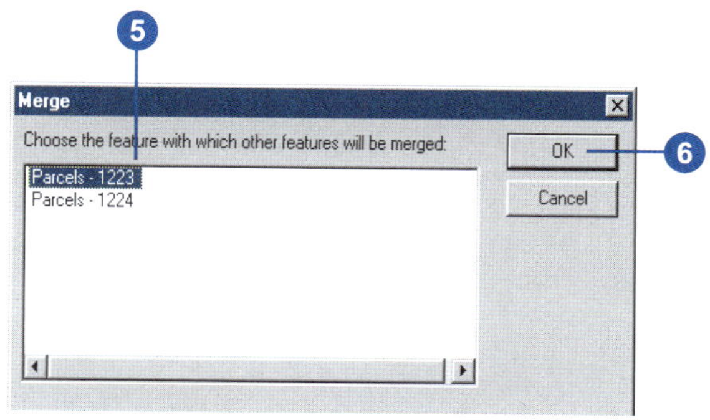

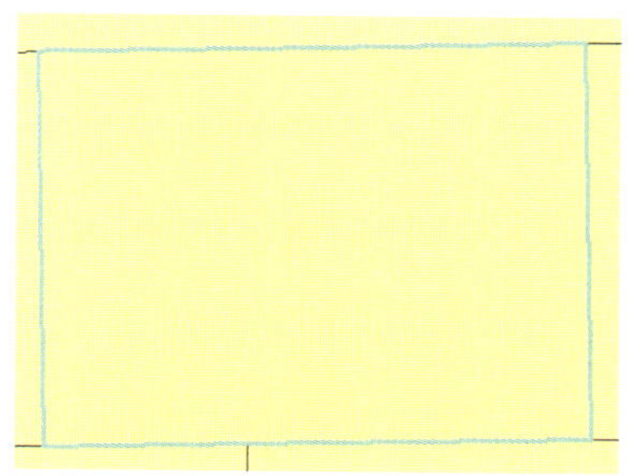

Parcels are merged into one.

Combining features from different layers into one feature

The Union command lets you combine features from different layers into one feature while maintaining the original features and attributes. You might use this command to create a sales territory from several ZIP Codes.

You can also create a multipart feature using the Union command by combining nonadjacent features from different layers. For example, suppose you want to create a sedimentary rock polygon in a new rock classification layer given selected clay and quartz polygons in an existing rock composite layer. You would use the Union command to combine the clay and quartz features to create a new, multipart sedimentary rock feature in the rock classification layer.

When you use the Union command, the features you combine must be from layers of the same type—line or polygon. The new feature is created in the current layer with no attribute values.

1. Click the Edit tool.
2. Click the features that you want to combine into one.

   (The features may be from different layers, although they must be the same layer type—line or polygon.)
3. Click the Target layer dropdown arrow and click the layer to which you want the new feature to belong. ►

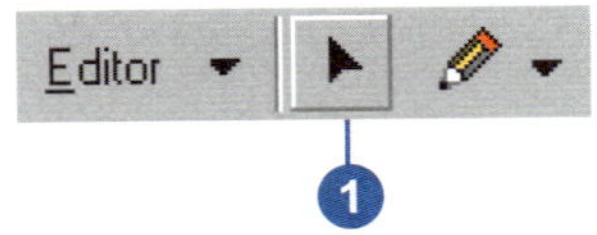

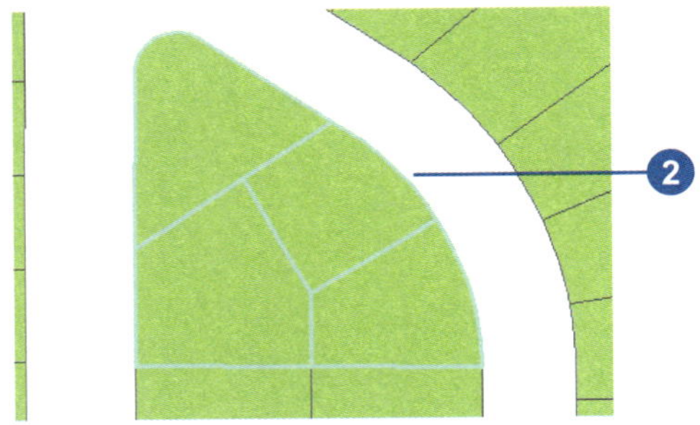

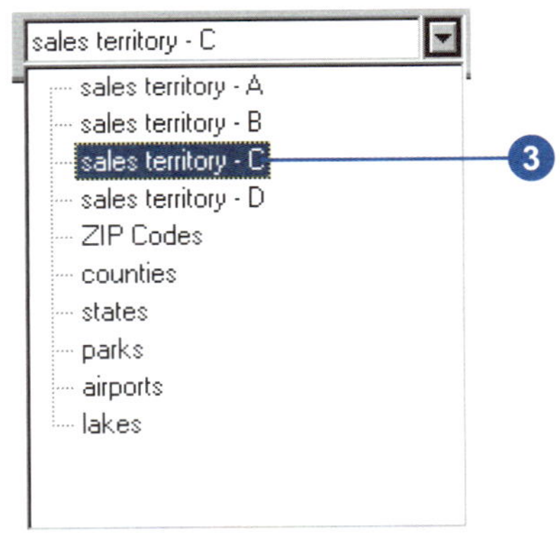

4. Click Editor and click Union.

   The selected features are combined into one.

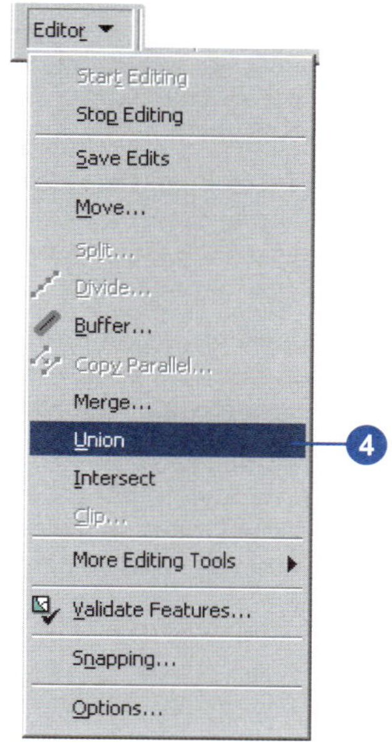

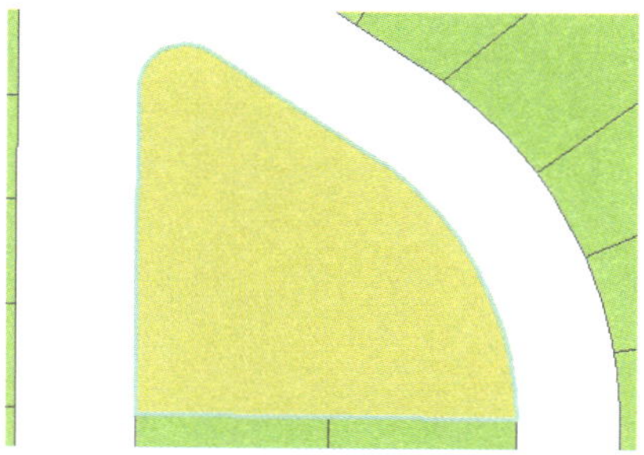

ZIP Codes are combined into one sales territory.

Creating a feature from features with common areas

The Intersect command creates a new feature from the area where features overlap. For instance, you might create a new sales territory out of overlapping trade areas.

You can find the intersection between features of different layers, but the layers must be of the same type—line or polygon. The original features are maintained, and the new feature is created in the current layer with no attribute values. You must enter attribute values for the new feature yourself.

1. Click the Edit tool.
2. Click the features from whose intersection you want to create a new feature.

   (The features may be from different layers, although they must be the same layer type—line or polygon.)
3. Click the Target layer dropdown arrow and click the layer to which you want the new feature to belong.

   (The layer must be of the same type as the selected features—line or polygon.) ►

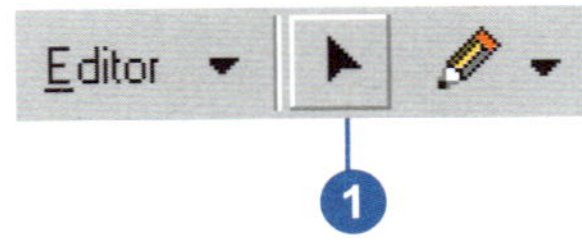

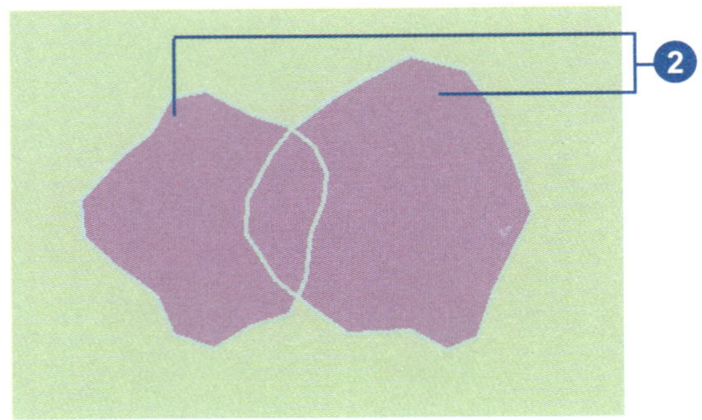

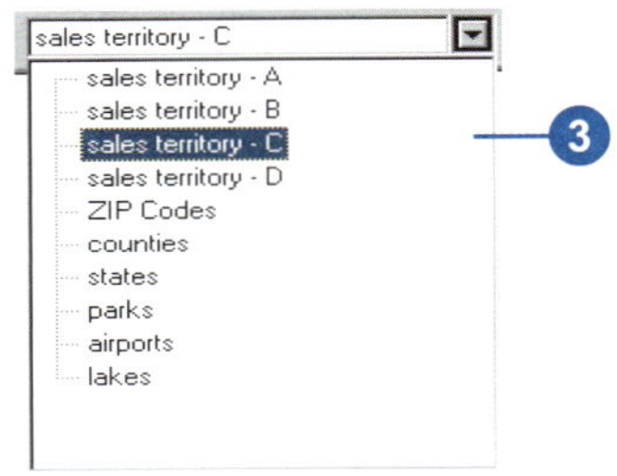

4. Click Editor and click Intersect.

   A new feature is created from the areas in common between all selected features.

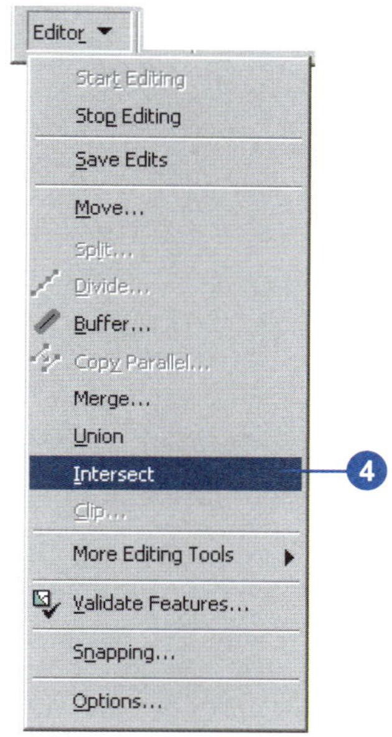

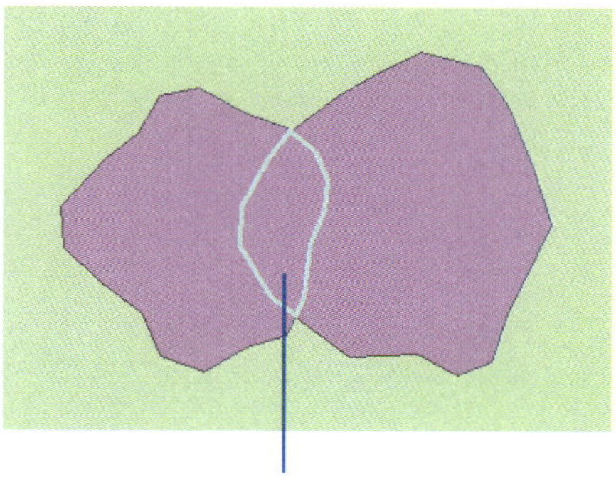

A single sales territory is created from the areas in common between two other sales territories.

Editing existing features

7

IN THIS CHAPTER

This chapter shows you how to modify features that already exist in your database. Suppose you need to change the shape of a parcel to accommodate a newly added cul-de-sac—you can use the Reshape Feature task to modify the parcel to the proper shape. Suppose the street you've digitized doesn't intersect with the correct cross street—you can use the Extend task to extend the line to the correct location. If you need to divide a parcel, you can use the Cut Polygon Feature task to cut the feature into two.

These are just a few examples of how easy it is to modify features while editing in ArcMap. The editing tools, commands, and tasks provide a variety of ways to make changes to existing features.

Many of the functions described in this chapter will react differently with data involved in a topology. If your project involves working with topological data, you should also read Chapter 4, 'Editing topology', to further understand the tools and functions related to topological rules and relationships.

Splitting a line or polygon

Using the editing tools, you can easily split line and polygon features.

To manually split one line into two, use the Split tool. The line is split at the location where you clicked with the mouse. The attributes of the original line are copied to each of the new lines. In the example shown, the Split tool is used to divide a street centerline into two features in anticipation of a new centerline being added between the parcels.

You can also split a line into two using the Split command on the Editor menu. Use the Split command when you know the distance at which you want to split the line, measured from either the first or last vertex. You can also use this command when you want to split a line at a certain percentage of the original length. You might use the ►

Tip

Using snapping to split a line

If you want to use the Split tool to split a line at a specific vertex, use the snapping environment to snap the pointer precisely to the vertex. For more information on snapping, see Chapter 3, 'Creating new features'.

Splitting a line manually

1. Click the Edit tool.
2. Click the line you want to split.
3. Click the Split tool.
4. Click the spot on the line where you want it to split.

   The line is split into two features.

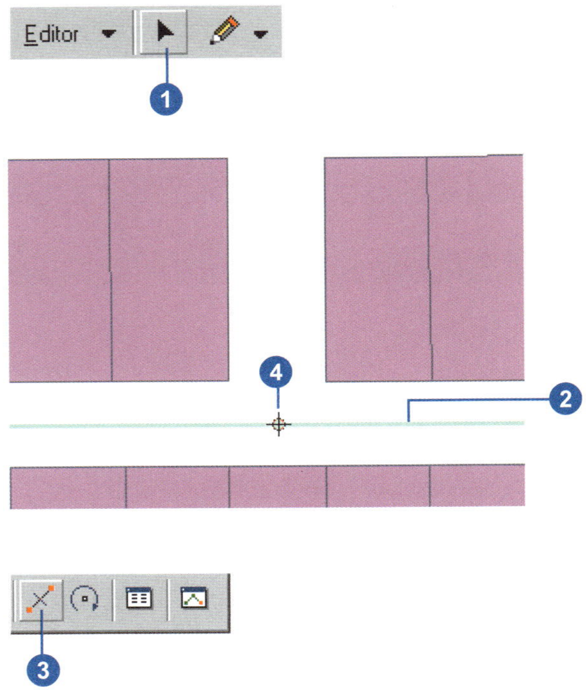

Split command to split a power line at a known distance along the line when you want to add an electrical pole that requires its own service.

The Split dialog box displays the length of the original feature in current map units to help you split it accurately. When you split the line using the Split command, the attributes of the original line are copied to each of the new lines.

To split one polygon into two, use the Cut Polygon Features task. The polygon is split according to a line sketch you create. The attributes of the original feature are copied to each of the new features.

Splitting a line at a specified distance or percentage

1. Click the Edit tool.
2. Click the line you want to split.
3. Click Editor and click Split. ▸

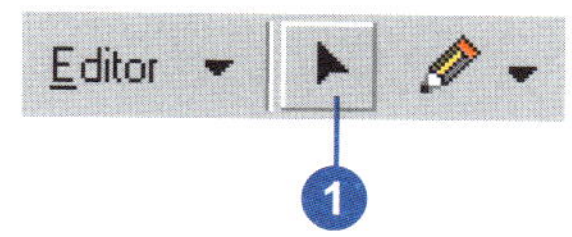

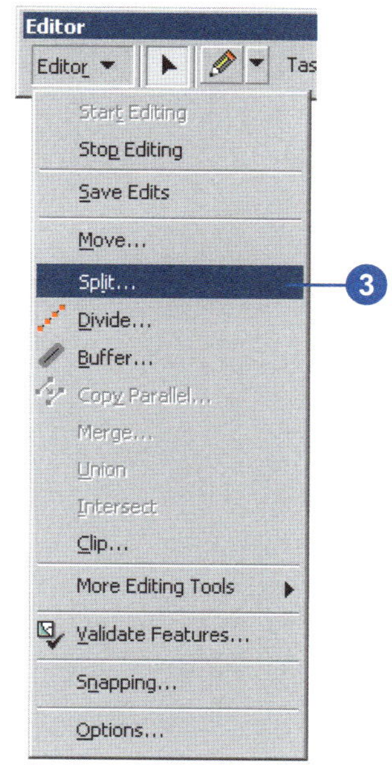

Tip

Controlling how attributes are handled

If you're working with geodatabase features, you can set up split policies that control the behavior of an object's attributes when it is split. For more information on split policies, see Building a Geodatabase.

4. Click the first Split option to split the feature at a certain distance.

   Click the second Split option to split the feature at a certain percentage of the whole.

5. Type a distance or percentage, as desired.

6. Select From Start Point of Line if you want to split the feature starting from the first vertex.

   Select From End Point of Line if you want to split the feature starting from the last vertex.

7. Click OK.

   The line is split into two features according to the parameters you specified.

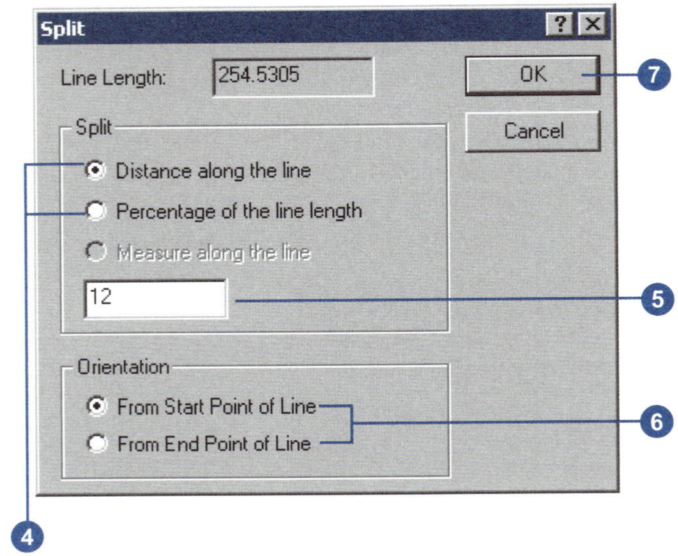

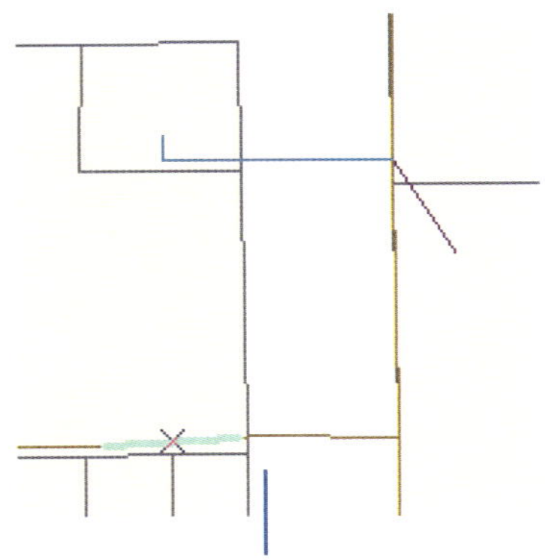

The line is split into two according to the distance and orientation you specified.

Tip

Cutting a polygon shape out of a polygon

You can use the Cut Polygon Features task to create a sketch that acts like a "cookie cutter," splitting the polygon in two. Simply create a line sketch that closes in on itself by double-clicking precisely on the first vertex of the sketch to finish it.

Tip

Other ways to construct a sketch

You can also use the Distance–Distance tool, the Arc tool, or the Intersection tool to create a sketch. For more information, see Chapter 3, 'Creating new features'.

Splitting a polygon

1. Click the Edit tool.
2. Click the polygon you want to split.
3. Click the Task dropdown arrow and click Cut Polygon Features.
4. Click the tool palette dropdown arrow and click the Sketch tool.
5. Construct a line or polygon sketch that cuts the original polygon as desired.
6. Right-click anywhere on the map and click Finish Sketch.

   The polygon is split into two features.

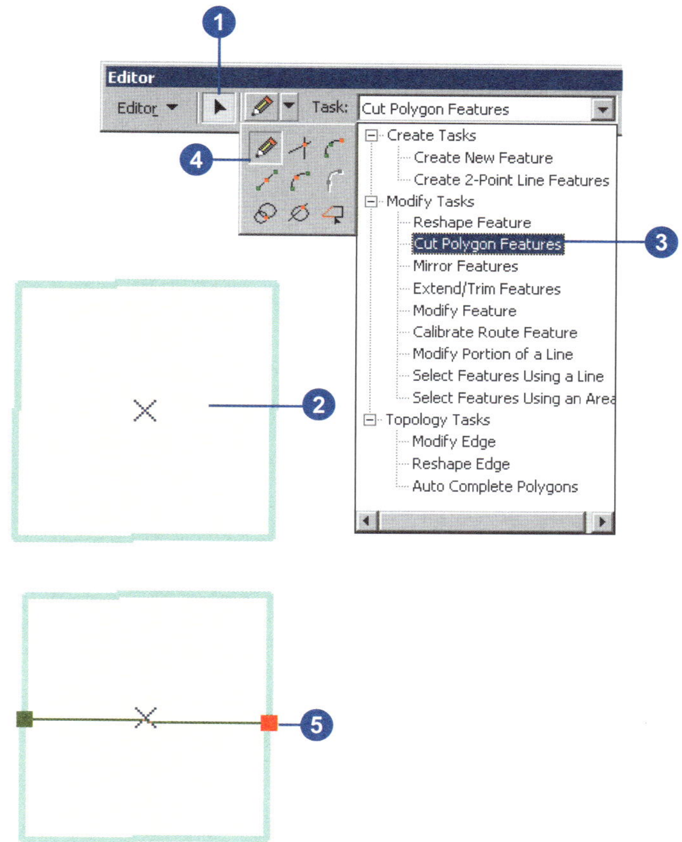

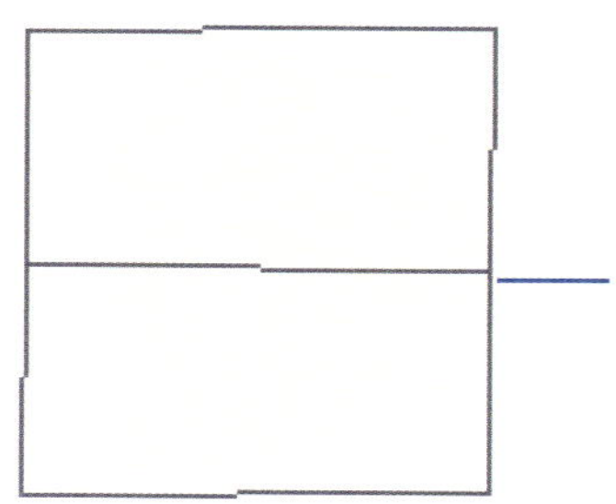

The sketch divides the polygon into two features.

Trimming a line

The Trim command on the Sketch context menu reduces the length of a line, trimming a distance you specify from the last vertex.

The Trim task in the Current Task dropdown list also trims lines, but instead of trimming them a given distance, the Trim task uses a *sketch* you draw. ►

Tip

Shortcut for modifying features

Instead of using the Modify Feature task to change a feature to its sketch, you can click the Edit tool and double-click the feature you want to modify.

Tip

Trimming from the first vertex of a line

You can trim a line from the first vertex instead of the last. See 'Flipping a line' in this chapter.

Tip

Shortcuts for finishing a sketch

When you're finished modifying a sketch, you can press F2 to finish it. Simply selecting another feature with the Edit tool will also finish the sketch.

Trimming a specific length from the last point

1. Click the Task dropdown arrow and click Modify Feature.
2. Click the Edit tool.
3. Click the line that you want to trim.

   The line appears as a sketch with vertices.
4. Right-click over any part of the line and click Trim.
5. Type the length to trim the line to and press Enter.

   If you type a positive value it trims the line to the specified length. If you type a negative value it removes that much from the length of the line, starting from the last vertex.

   The line is trimmed.
6. When finished modifying the line, right-click over any part of the sketch and click Finish Sketch.

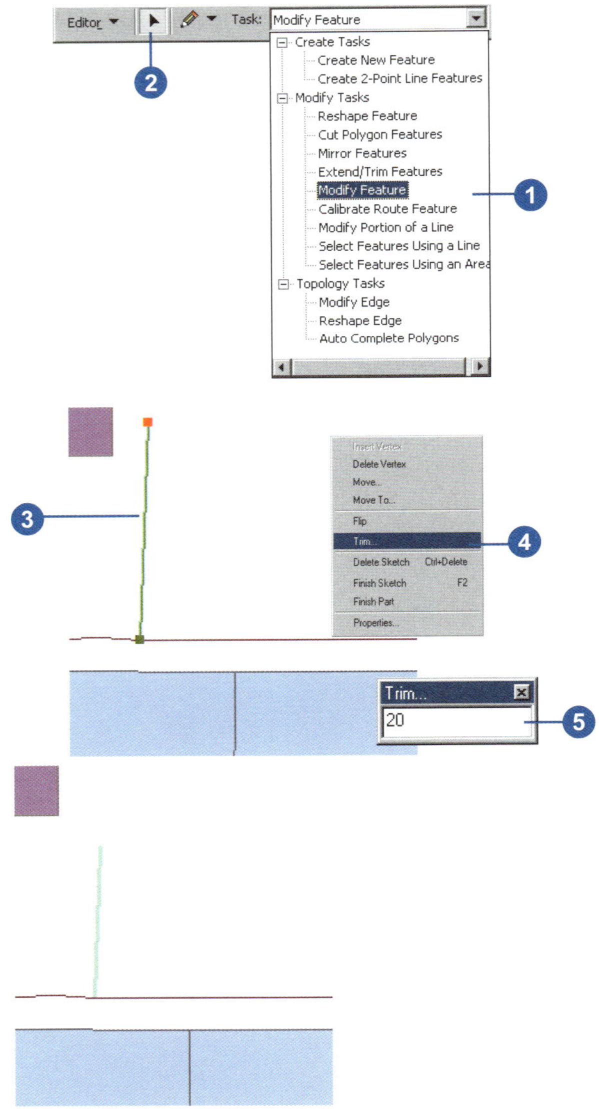

20 map units are trimmed from the original line.

This is useful if you don't know the exact distance you want to trim but have a physical boundary where the features should end or begin.

Suppose your database has some roads that should end at the coastline, but overshoot it instead. Using the Trim task, you can draw a line sketch on top of the coastline and the lines will be trimmed where you have drawn the sketch.

Portions of the lines that are on the right side of the sketch are trimmed. The right side of the sketch is based on the direction in which the sketch was drawn. Imagine riding a bicycle along the sketch in the direction in which the vertices were added. ▸

Tip

Other ways to construct a sketch

You can also use the Distance–Distance tool, the Arc tool, the Trace tool, or the Intersection tool to create a sketch. For more information, see Chapter 3, 'Creating new features'.

Tip

Shortcuts for finishing a sketch

You can double-click on the last vertex of a sketch to finish it. You can also press F2.

Trimming based on a line you draw

1. Click the Task dropdown arrow and click Extend/Trim Features.
2. Click the Edit tool.
3. Click the line or lines you want to trim.
4. Click the tool palette dropdown arrow and click the Sketch tool.
5. Construct a line that trims the selected line or lines as desired. The direction of the sketch line determines the part of the features to be removed. The portion of the selected features to the right of the sketch is trimmed.
6. Right-click anywhere on the map and click Finish Sketch. ▸

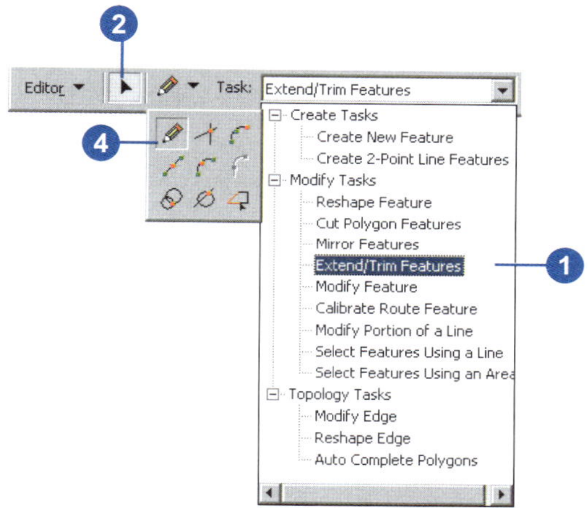

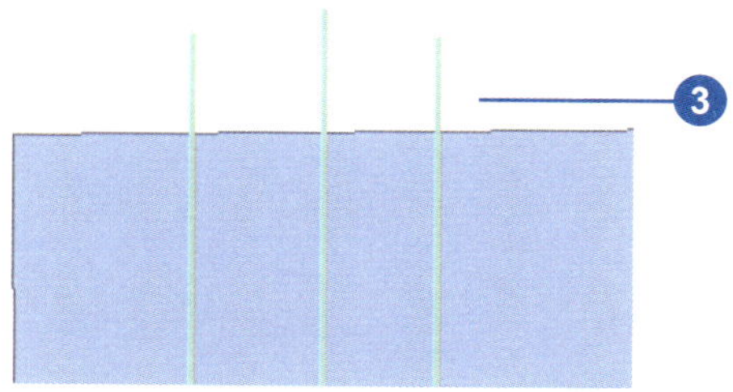

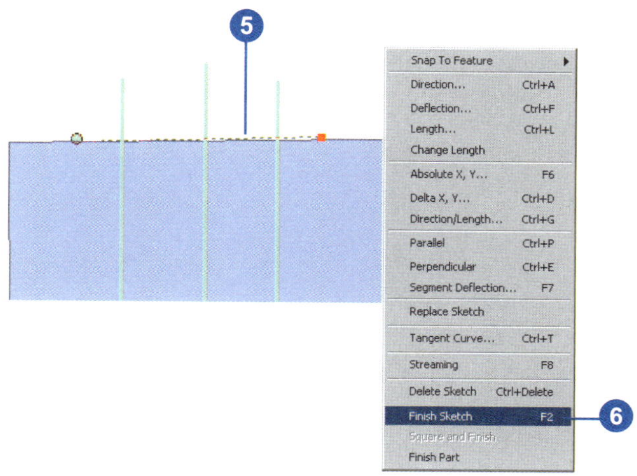

If you looked to your right, you would be looking at the right side of the sketch.

The lines are trimmed on the right side of the line you constructed.

The lines are trimmed where the sketch was drawn.

Extending a line

The Extend task is the opposite of the Trim task, extending selected lines to a line you construct. Consider the roads and coastline example shown in the Trim task. If your database has some roads that should end at the coastline, but instead stop short, you could use the Extend task. By drawing a sketch on top of the coastline, you can extend the roads to the sketch you drew.

Tip

Other ways to construct a sketch

You can also use the Distance–Distance tool, the Arc tool, the Trace tool, or the Intersection tool to create a sketch. For more information, see Chapter 3, 'Creating new features'.

1. Click the Task dropdown arrow and click Extend/Trim Features.
2. Click the Edit tool.
3. Click the line or lines you want to extend.
4. Click the tool palette dropdown arrow and click the Sketch tool.
5. Construct a line to which you want to extend the selected line or lines.
6. Right-click anywhere on the map and click Finish Sketch. ▸

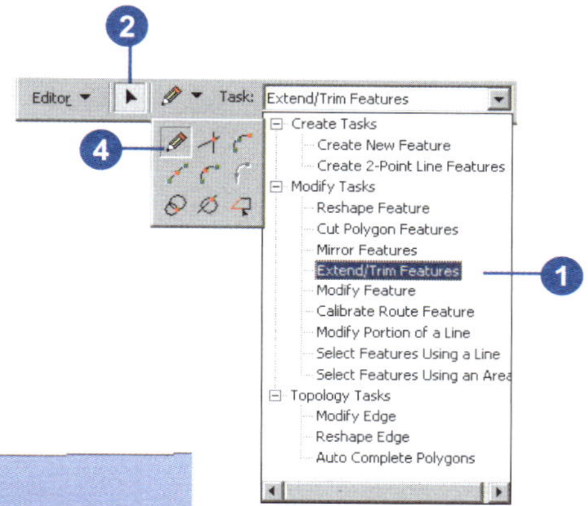

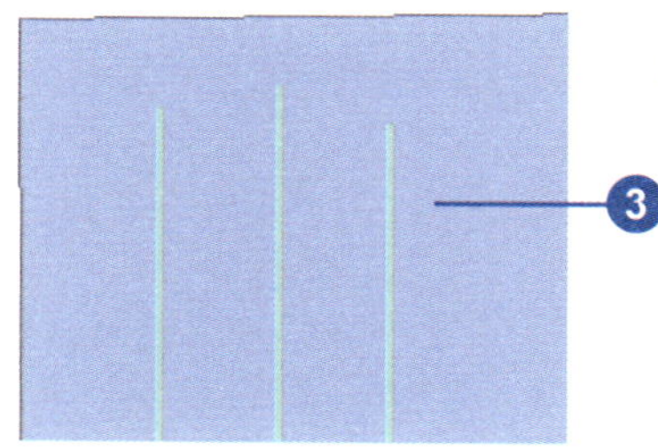

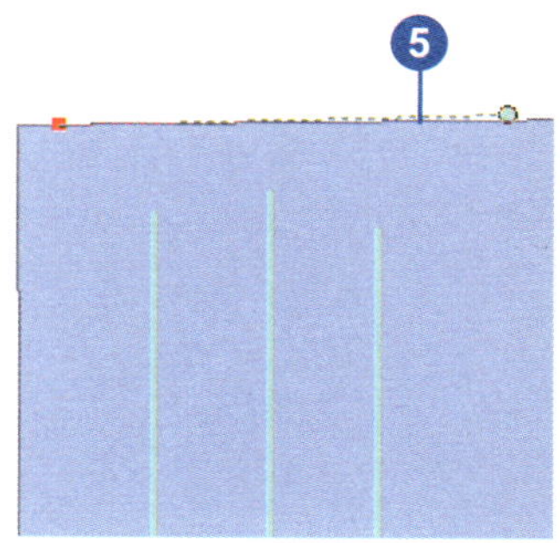

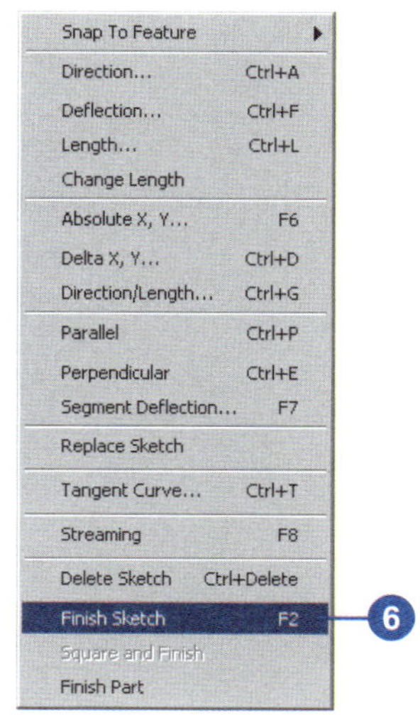

Tip

Other methods of extending or trimming a line

Beyond the basic Extend/Trim Features task, lines can be extended using the advanced editor Trim and Extend tools as well as specific topology correction methods when working with topologies. These additional functions are only available with ArcEditor and ArcInfo software packages.

The lines are extended to the line you constructed.

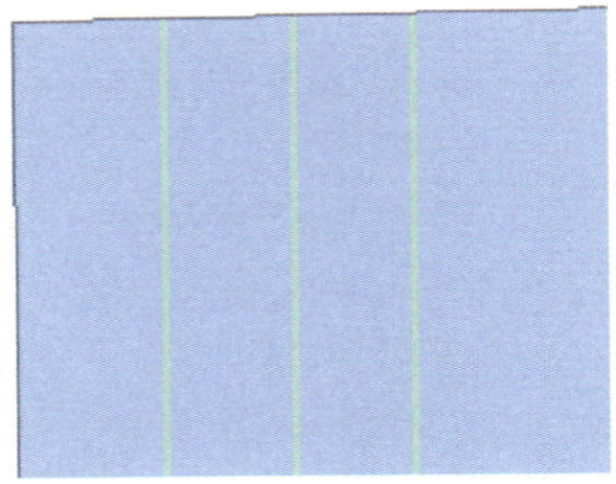

The lines are extended to where the sketch was drawn.

Flipping a line

When you modify a line by trimming or extending it, the line is automatically trimmed or extended from its last vertex.

However, if you prefer to trim or extend a line from the first vertex instead of the last, you can use the Flip command. The Flip command reverses the direction of a line so that the last vertex of the sketch becomes the first.

Tip

Working with topologies

For line feature classes or shapefiles that are part of a topology, the line direction is involved with the topological rules and the flip function will not be permitted. For more information on working with topologies, see Chapter 4, 'Editing topology', in this book.

1. Click the Task dropdown arrow and click Modify Feature.
2. Click the Edit tool.
3. Click the line whose direction you want to change.
4. Right-click over any part of the sketch and click Flip.

   The sketch becomes inverted (the first vertex becomes the last, marked in red).
5. When finished modifying the line, right-click over any part of the sketch and click Finish Sketch.

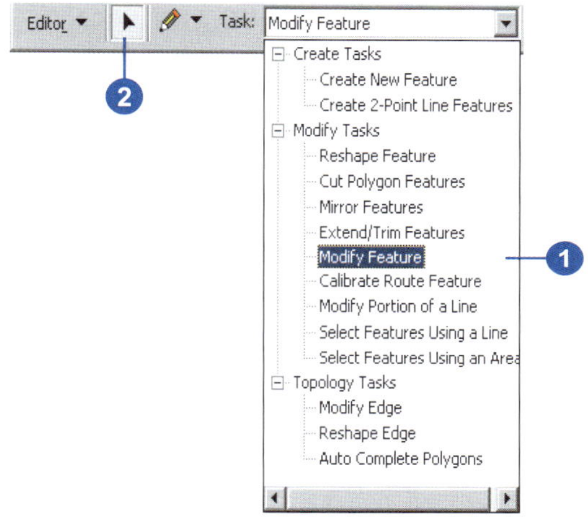

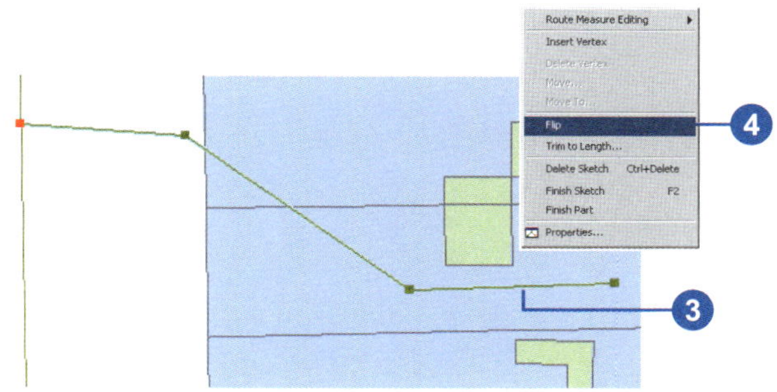

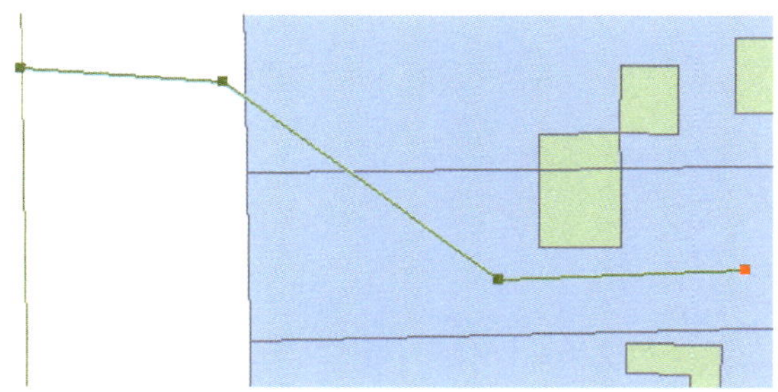

The first and last vertices of the line are reversed.

Placing points along a line

The Divide command creates points at a given interval along a line. For instance, you could use Divide to place utility poles along a primary.

You can create a specific number of points that are evenly spaced or you can create points at a distance interval you choose.

1. Click the Edit tool.
2. Click the line you want to divide.
3. Click the Target layer dropdown arrow and click the point layer containing the type of points you want to place along the line.
4. Click Editor and click Divide. ▸

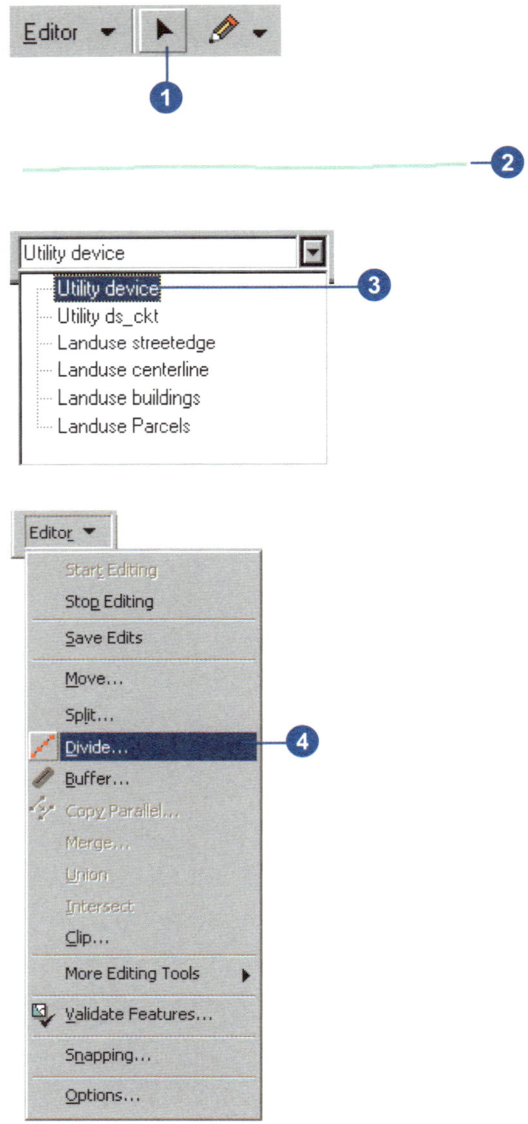

5. Click the first option and type a number to place a specific number of points evenly along the line.

   Or click the second option and type a number to place the points at a specific interval in map units.

   If the data has M values, you can click the third option and type a number to place the points at a specific interval of measure units

6. Click OK.

   The line is divided by points placed along the line as specified.

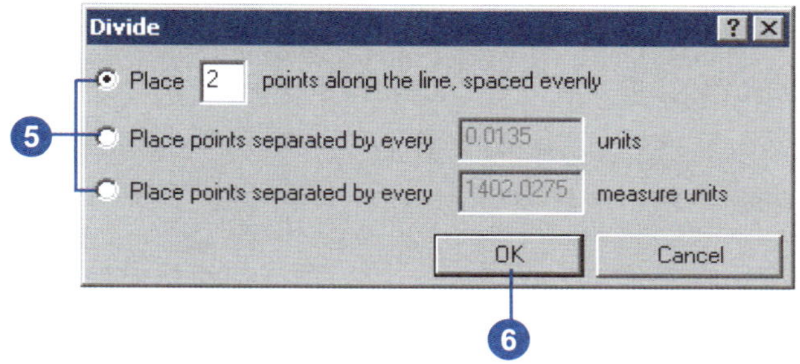

The line is divided by points.

Reshaping a line or polygon

The Reshape Feature task lets you reshape a line or polygon by constructing a sketch over the feature. The feature takes the shape of the sketch from the first place the sketch intersects the feature to the last.

When you reshape a polygon, if both endpoints of the sketch are within the polygon, the shape is added to the feature. ▸

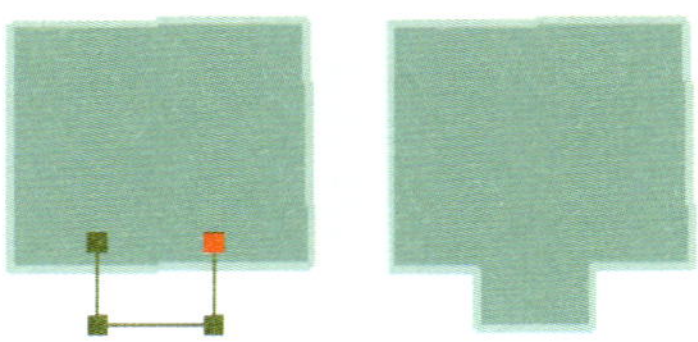

Tip

Other ways to construct a sketch

You can also use the Distance–Distance tool, the Arc tool, the Trace tool, or the Intersection tool to create a sketch. For more information, see Chapter 3, 'Creating new features'.

1. Click the Task dropdown arrow and click Reshape Feature.
2. Click the Edit tool.
3. Click the feature you want to reshape.
4. Click the tool palette dropdown arrow and click the Sketch tool.
5. Create a line according to the way you want the feature reshaped.
6. Right-click anywhere on the map and click Finish Sketch. ▸

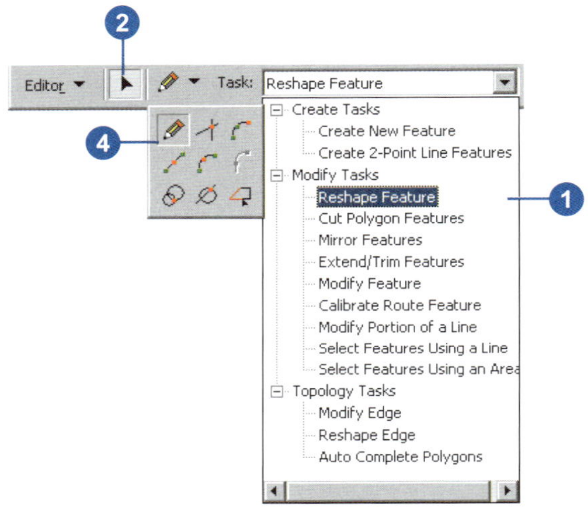

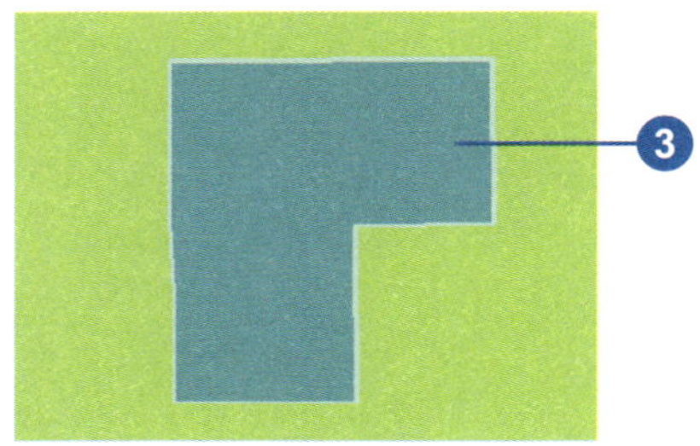

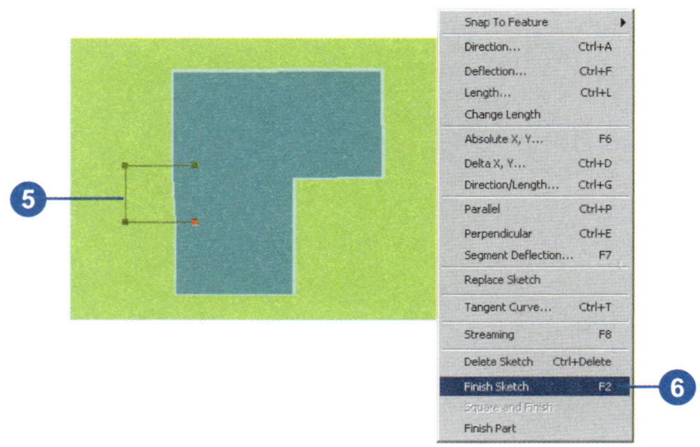

If the endpoints are outside the polygon, the feature is cut away.

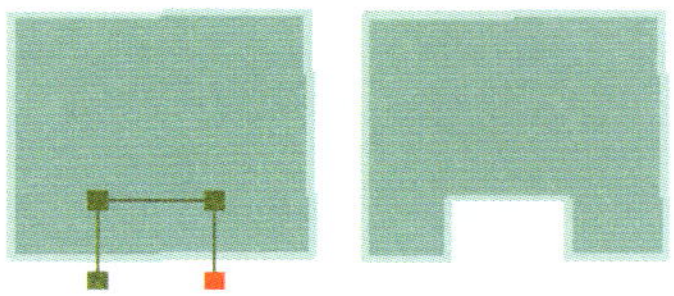

When you reshape a line, both endpoints of the sketch must be on the same side of the line. The line takes the shape of the sketch you draw.

The feature is reshaped.

The feature is reshaped according to the sketch you constructed.

Adding and deleting sketch vertices

You can easily add vertices to or delete vertices from a sketch using the Insert Vertex and Delete Vertex commands on the Sketch context menu. By adding or deleting vertices, you can reshape a feature when you obtain new or better geographic data.

Suppose you have an existing layer with curb lines and receive an aerial photo that shows that the lines in the layer are incorrectly shaped. Using the aerial ►

Tip

Adding vertices from the last vertex

You can add vertices to a feature beginning from the last vertex of the sketch. Click the Edit tool and double-click the feature to see its sketch. Then, click the Sketch tool to begin digitizing vertices.

Adding a vertex to a sketch

1. Click the Task dropdown arrow and click Modify Feature.
2. Click the Edit tool and click the line or polygon to which you want to add a vertex.
3. Move the pointer to where you want the vertex inserted and right-click.
4. Click Insert Vertex.

   A vertex is added to the sketch.
5. When finished modifying the line, right-click over any part of the sketch and click Finish Sketch.

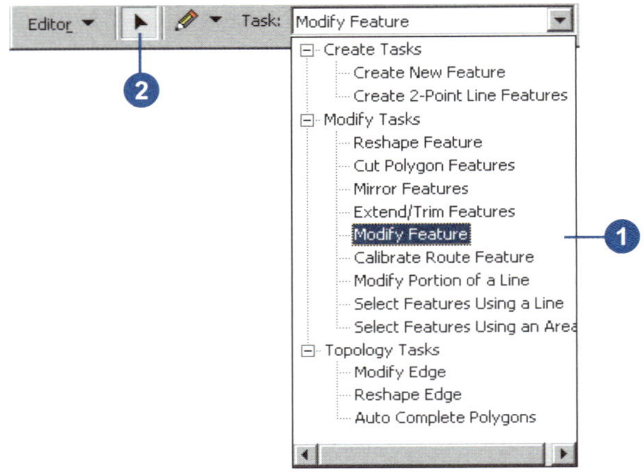

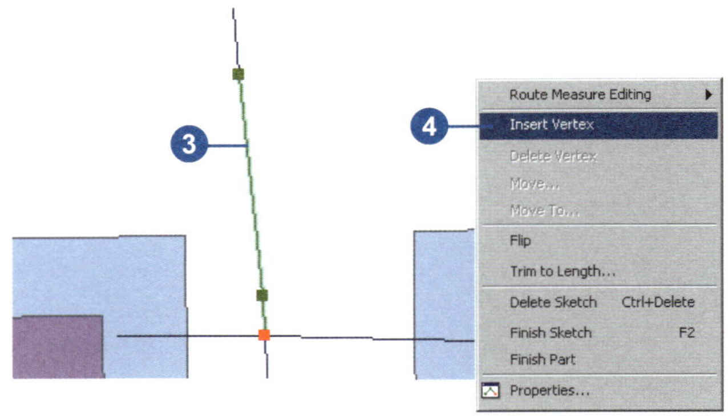

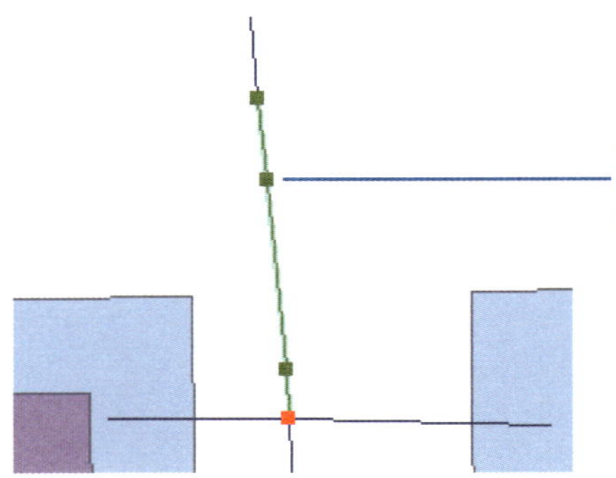

A new vertex is added to the sketch.

photo as a backdrop, you can add vertices to the curb lines as needed, then reshape the feature to match the photo by moving the vertices to new locations. You can also reshape the curb line features by deleting existing vertices from their sketches.

See Also

To learn how to move a vertex, see 'Moving a vertex in a sketch' in this chapter.

Deleting a vertex from a sketch

1. Click the Task dropdown arrow and click Modify Feature.
2. Click the Edit tool.
3. Click the line or polygon from which you want to delete a vertex.
4. Position the pointer over the vertex you want to delete.

   The pointer will change appearance to have four small arrows surrounding a circle.
5. Right-click and click Delete Vertex.

   The vertex is deleted from the sketch.
6. Right-click over any part of the sketch and click Finish Sketch.

   The feature is reshaped.

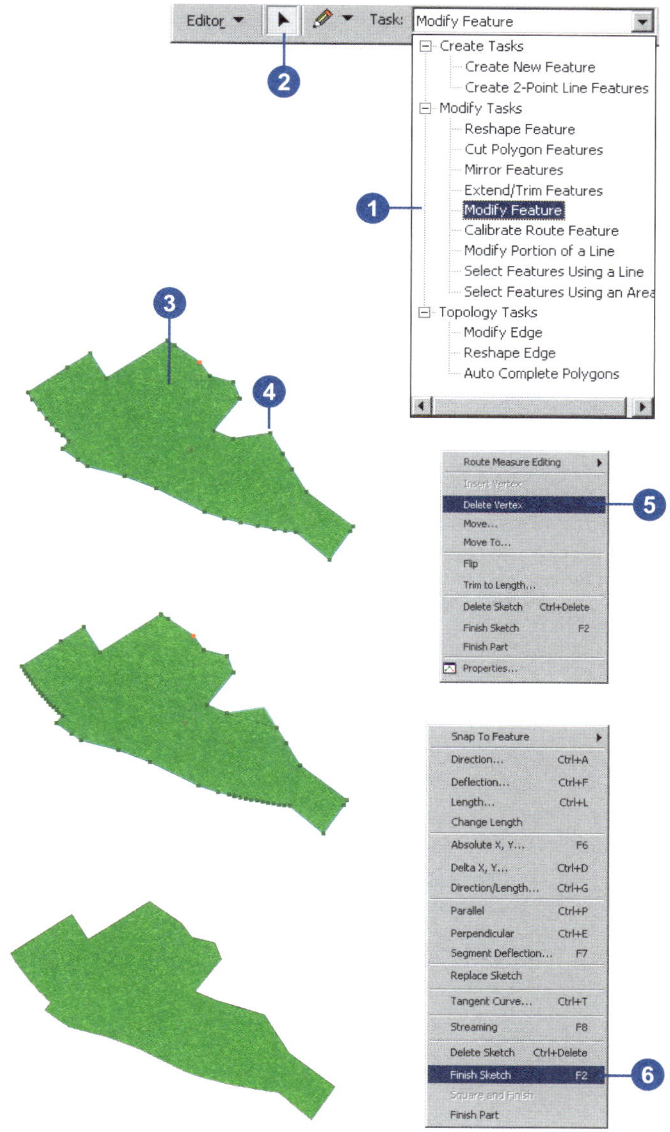

The vertex is deleted, and the feature is reshaped.

Moving a vertex in a sketch

Moving a vertex in a sketch offers another way to modify or reshape a feature.

ArcMap lets you move a vertex in several ways: by dragging it, by specifying new x,y coordinates, or by moving it relative to its current location.

You might choose to drag a vertex to a new location when you want to reshape a feature according to additional data you receive. For instance, you can drag a vertex to reshape a road feature in an existing layer in order to match it to the feature in a more accurate aerial photo.

You might move a vertex by specifying new x,y locations when you obtain additional data that provides the exact coordinate location at which the vertex should be. For example, suppose a parcel is resurveyed and a new global positioning system (GPS) point is obtained for the parcel corner. You can move the corner of the parcel to match the location found by the GPS by specifying the equivalent location in x,y coordinates. ►

Dragging a vertex

1. Click the Task dropdown arrow and click Modify Feature.
2. Click the Edit tool and click the line or polygon whose vertex you want to move.
3. Position the pointer over the vertex you want to move.

   The pointer will change appearance to have four small arrows surrounding a circle.
4. Click and drag the vertex to the desired location.
5. Right-click over any part of the sketch and click Finish Sketch.

   The feature is reshaped.

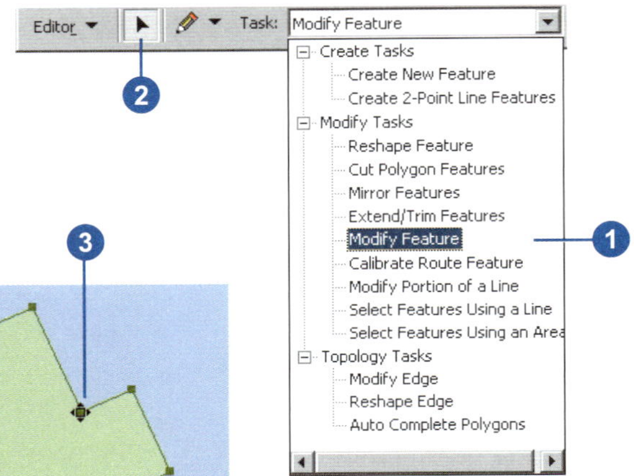

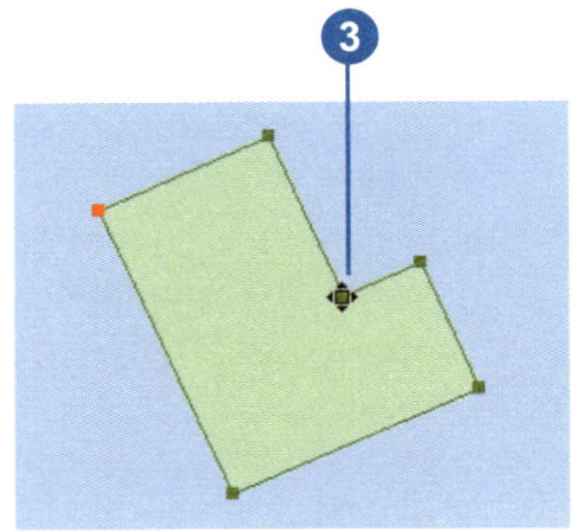

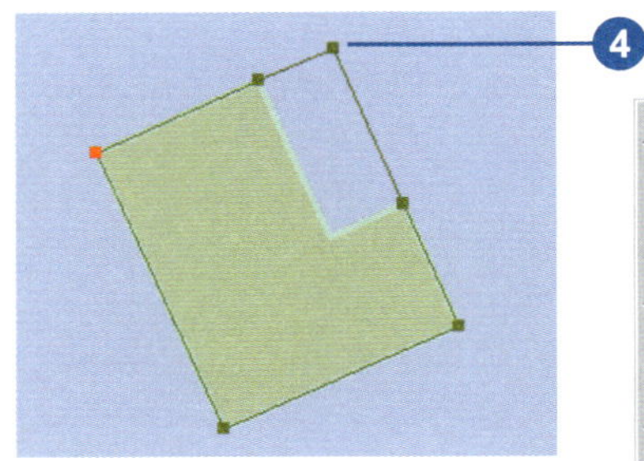

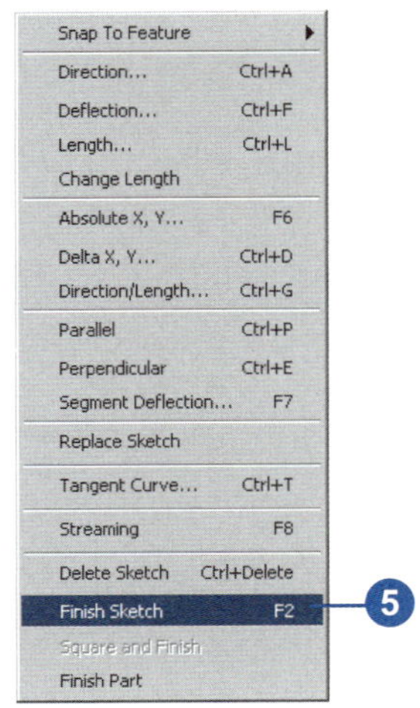

The vertex is moved, and the feature is reshaped.

The sketch context menu also provides a way to move a vertex relative to its current location. Suppose an electrical pole must be moved 15 feet east and 5 feet north of its current location due to a road widening. Before moving the pole, you must reshape its electrical line so that the pole can connect to the line in the new location; you can do this by moving the vertex of the electrical line on which the pole sits using relative (delta) x,y coordinates.

The original location of the vertex as the origin (0,0) is used, and the vertex is moved to the new location using the map unit coordinates you specify [(15,5) in this example]. After the vertex is moved and the electrical line is reshaped, you can snap the pole feature to the vertex in its new location.

Tip

Maintaining a feature's shape when moving a vertex

You can also move a vertex without changing the shape of the feature. For more information, see 'Stretching a feature's geometry proportionately' in this chapter.

Moving a vertex by specifying x,y coordinates

1. Click the Current Task dropdown arrow and click Modify Feature.
2. Click the Edit tool and click the line or polygon whose vertex you want to move.
3. Position the pointer over the vertex you want to move until the pointer changes.
4. Right-click and click Move To.
5. Type the x,y coordinates where you want to move the vertex.

   The vertex is moved. ▸

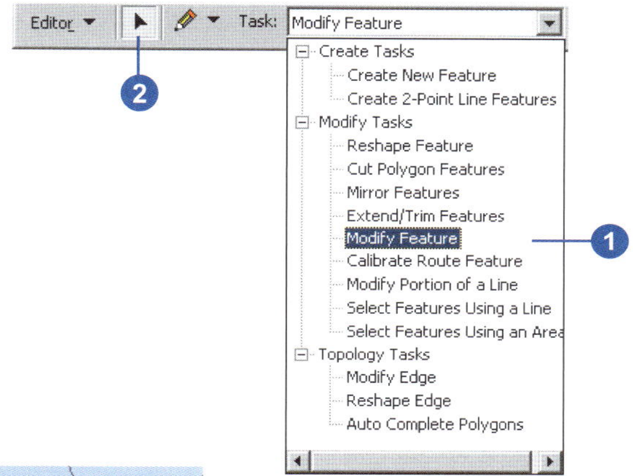

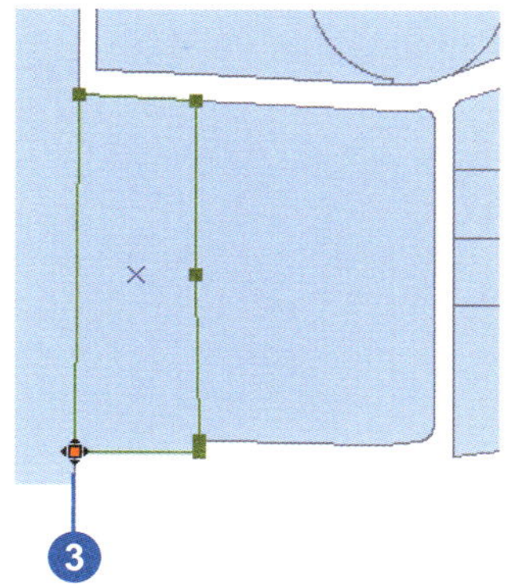

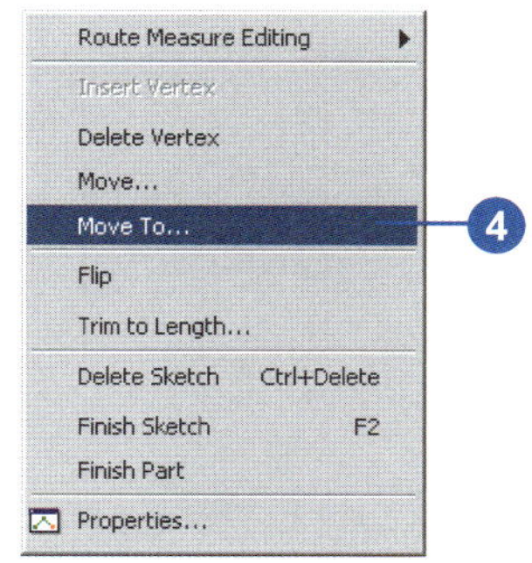

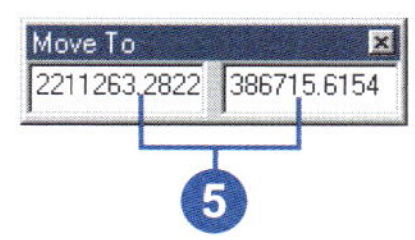

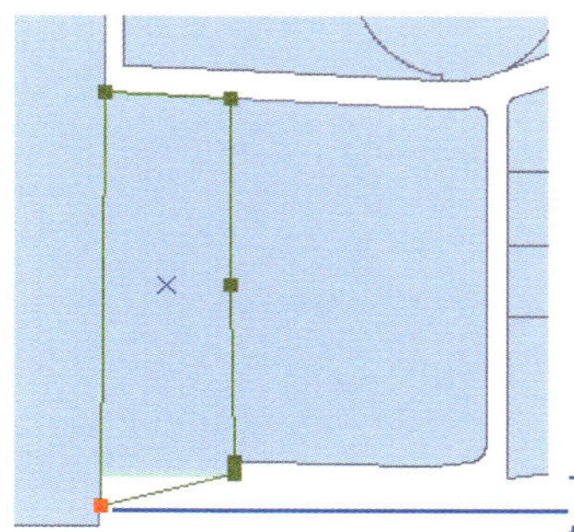

The vertex is moved according to the coordinates you specified.

Tip

Undoing a vertex move

If you move a vertex and don't want it to stay in the new location, click the Undo button on the ArcMap Standard toolbar. The vertex returns to its last position. Click the Redo button if you want to move the vertex back to the new location.

6. Right-click over any part of the sketch and click Finish Sketch.

   The feature is reshaped.

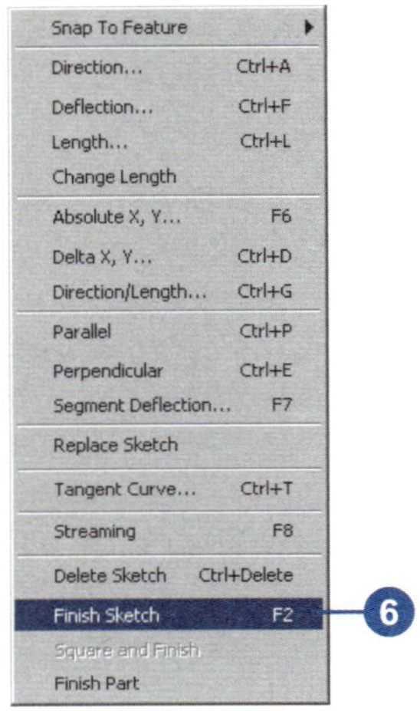

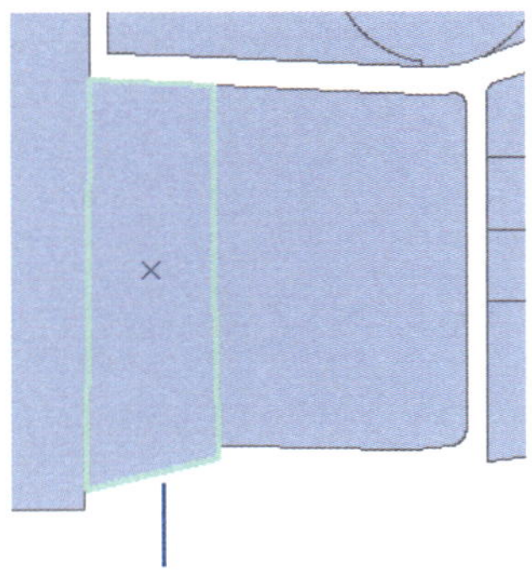

The feature is reshaped.

Moving a vertex relative to its current location

1. Click the Task dropdown arrow and click Modify Feature.
2. Click the Edit tool and click the line or polygon whose vertex you want to move.
3. Position the pointer over the vertex you want to move until the pointer changes.
4. Right-click and click Move.
5. Type the delta x,y coordinates where you want to move the vertex. ▸

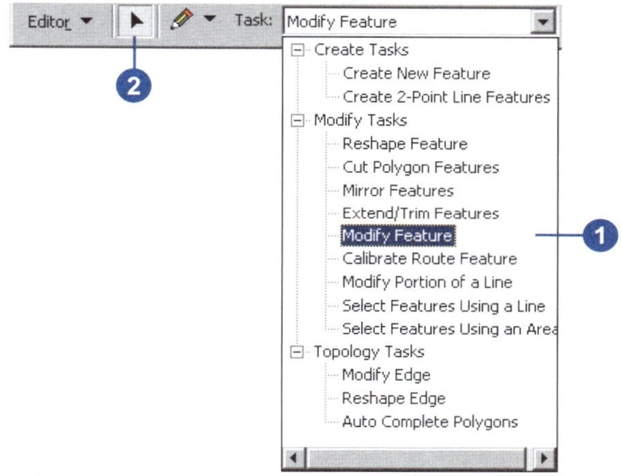

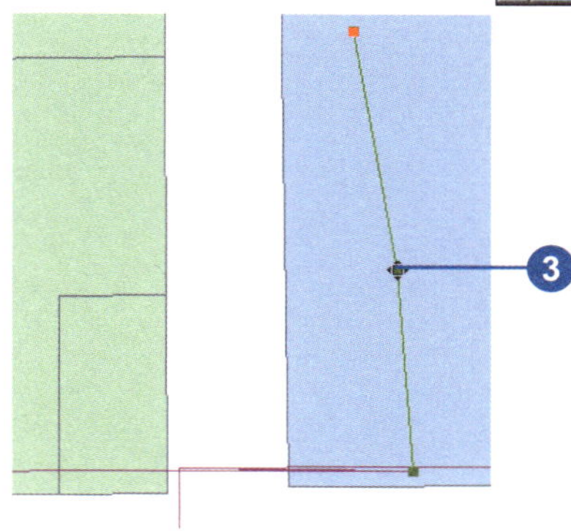

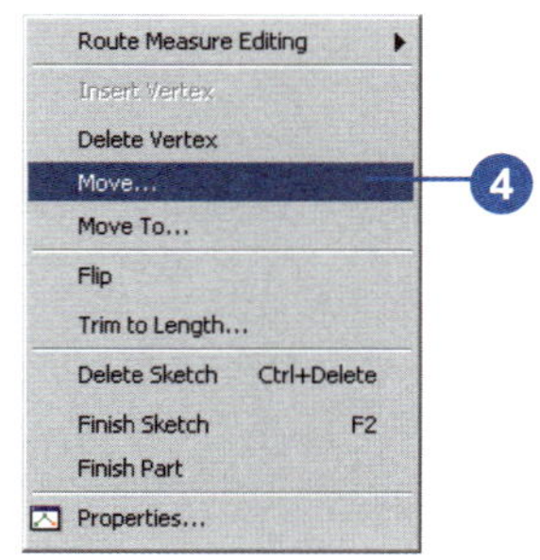

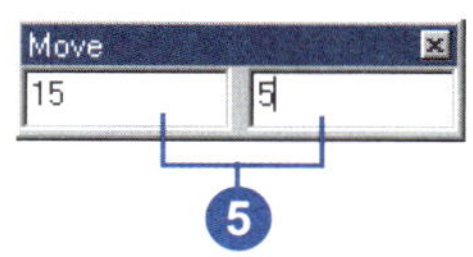

The vertex is moved.

6. Right-click over any part of the sketch and click Finish Sketch.

The feature is reshaped.

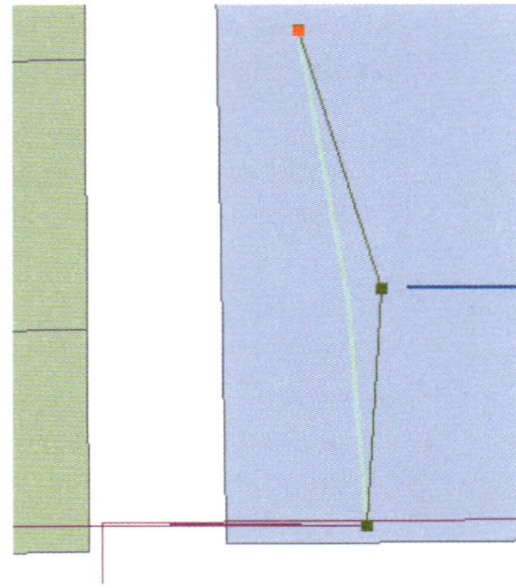

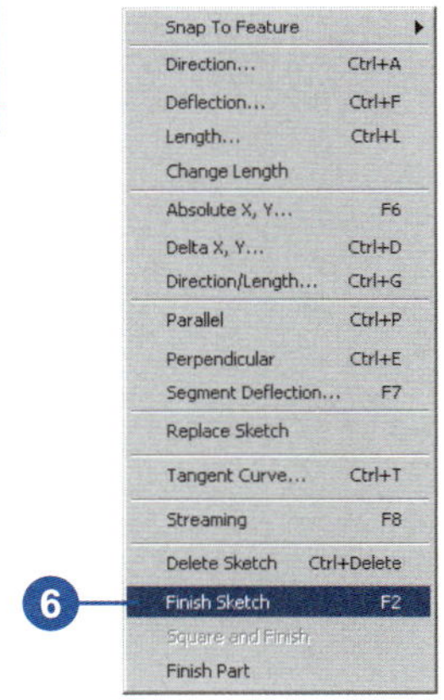

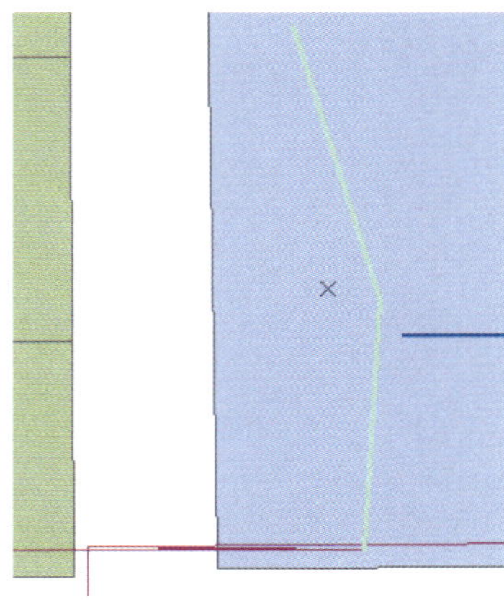

Changing the properties of a sketch

When creating a new feature or modifying an existing one, you can easily change the properties of the sketch shape using the Sketch Properties dialog box.

Using the Sketch Properties dialog box, you can remove parts from a multipart feature, insert and delete vertices, and alter the m- and z-values of vertices.

Suppose you are editing a layer that contains river features whose shapes contain too many vertices. You could use the Sketch Properties dialog box to select unwanted vertices and delete them.

Tip

How do I know which vertices I have selected?

As you select vertices in the dialog box, they change color on the map.

Deleting multiple vertices from a feature

1. Click the Edit tool and select the feature whose shape you want to modify.
2. Click the Task dropdown arrow and click Modify Feature to place the shape of the feature in the edit sketch.
3. Right-click the sketch and click Properties.
4. Select the vertices that you want to remove by holding down the Shift key and clicking vertices from the table. Use the Shift and Ctrl keys to select more than one vertex.
5. Click the Delete key or right-click over the selected vertices and click Delete.

   The selected vertices are deleted from the sketch.
6. Click Finish Sketch.

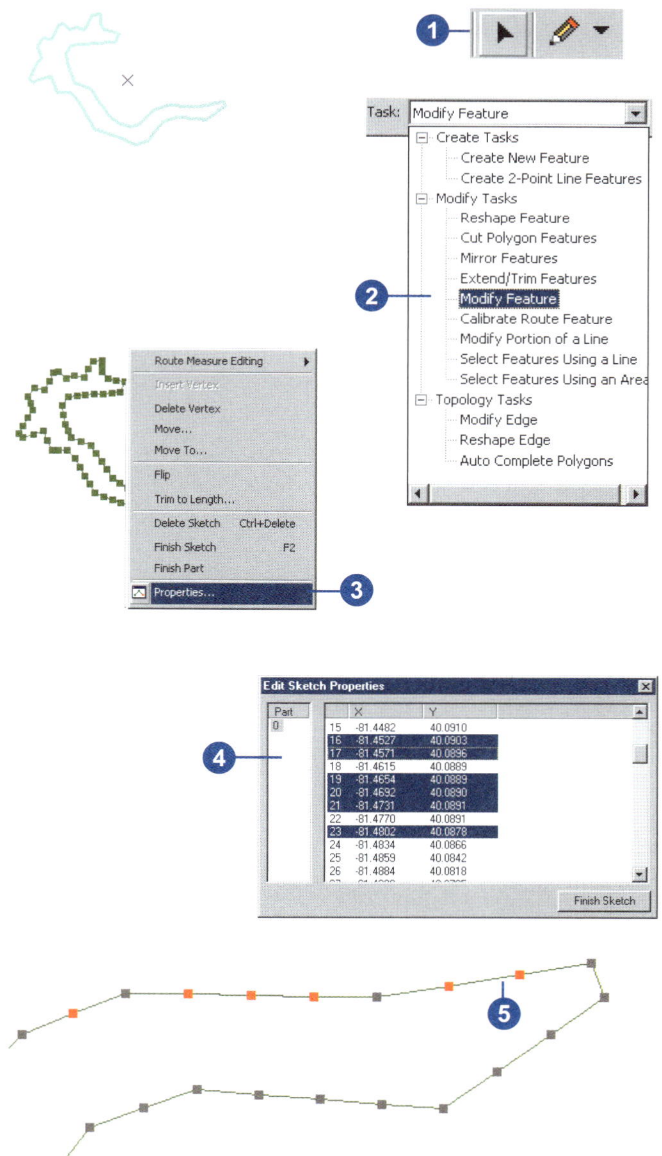

> **Tip**
>
> **Modifying the x,y coordinates**
>
> *If you don't want the added point to be exactly at the midpoint between two vertices, click the x or y column and type in a new coordinate for the point.*

> **Tip**
>
> **Insert vertices after a selected vertex**
>
> *You can insert vertices either before or after the vertex that you right-click on top of.*

Inserting a vertex at the midpoint of a segment

1. Right-click over a segment of the edit sketch and click Properties.
2. Select the vertex before which you wish to insert a new vertex.
3. Right-click the selected vertex and click Insert Before.

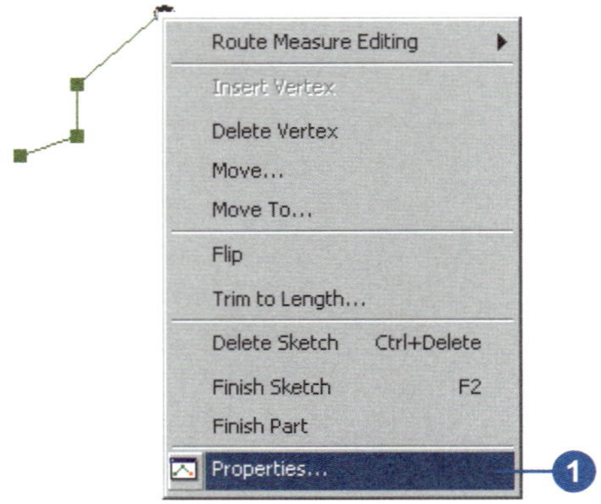

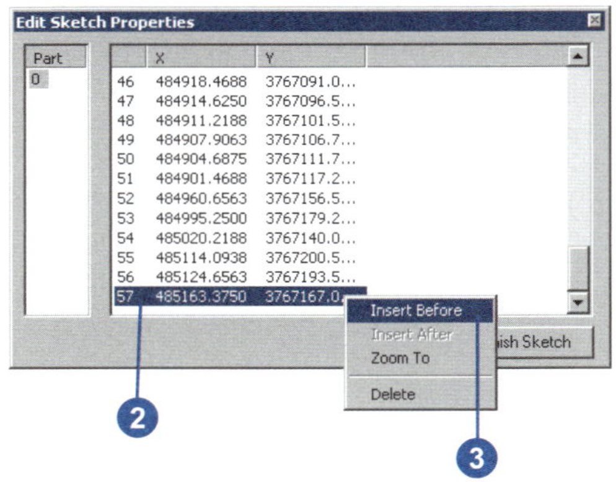

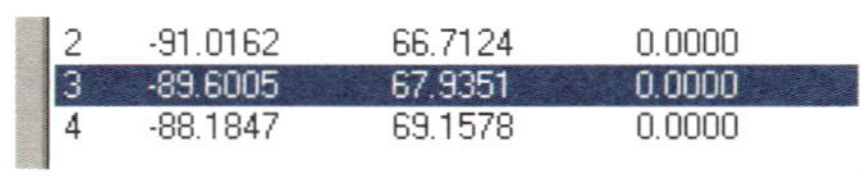

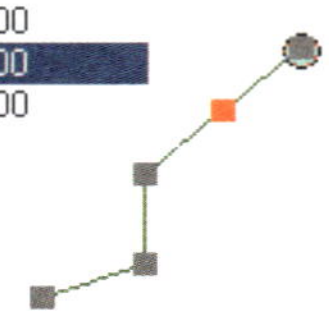

Vertex inserted at midpoint

How do I know which parts I have selected?

When you select a part from the Sketch Properties dialog box, the segments for that part will appear thicker.

Removing a part from a multipart feature

1. Click the Edit tool and select the feature you want to remove a part from.
2. Click the Task dropdown arrow and click Modify Feature to place the multipart shape in the edit sketch.
3. Right-click the sketch and click Properties.
4. Right-click the part that you want to remove and press the delete key or right-click and click Delete.
5. Click Finish Sketch.

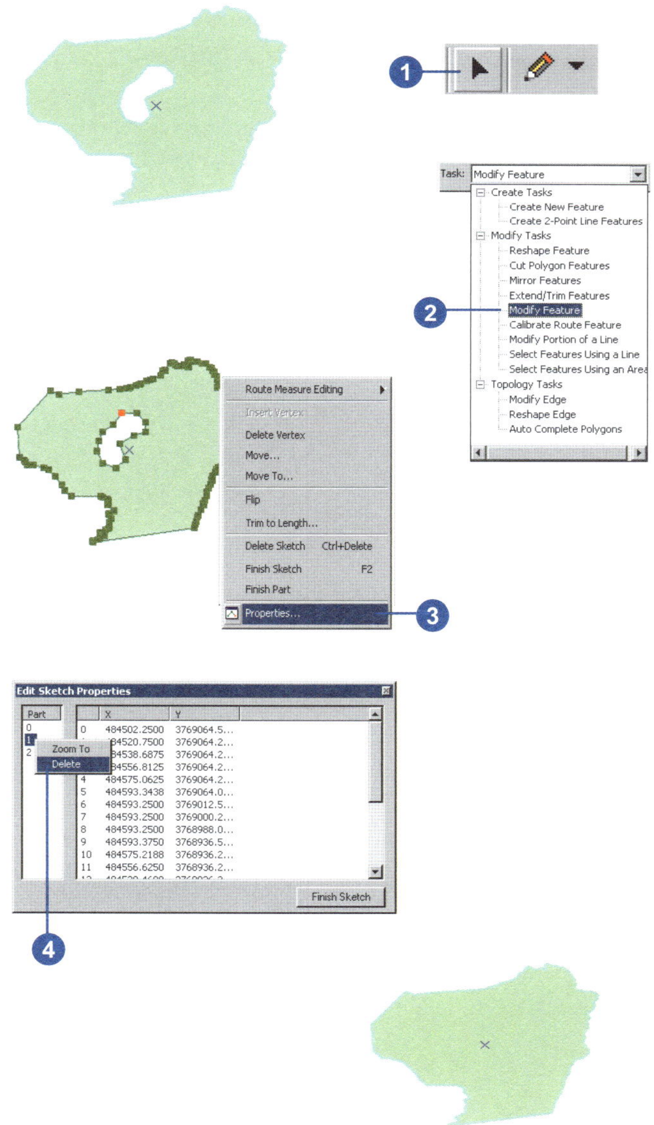

Tip

Using the Current Z control

When you add points to the edit sketch, you can control the z-value for each vertex using the Current Z tool.

To use the Current Z tool, you must first add it to a toolbar from the Commands tab of the Customize dialog box. The Current Z tool is listed in the Editor category.

Editing z- and m-values of a feature

1. Click the Edit tool and select the feature whose z- or m-values you wish to edit.
2. Click the Task dropdown arrow and click Modify Feature.
3. Right-click on top of the sketch and click Properties.
4. Select the vertex you wish to modify.
5. Click the z or m field in the table and type a new value.
6. Click Finish Sketch.

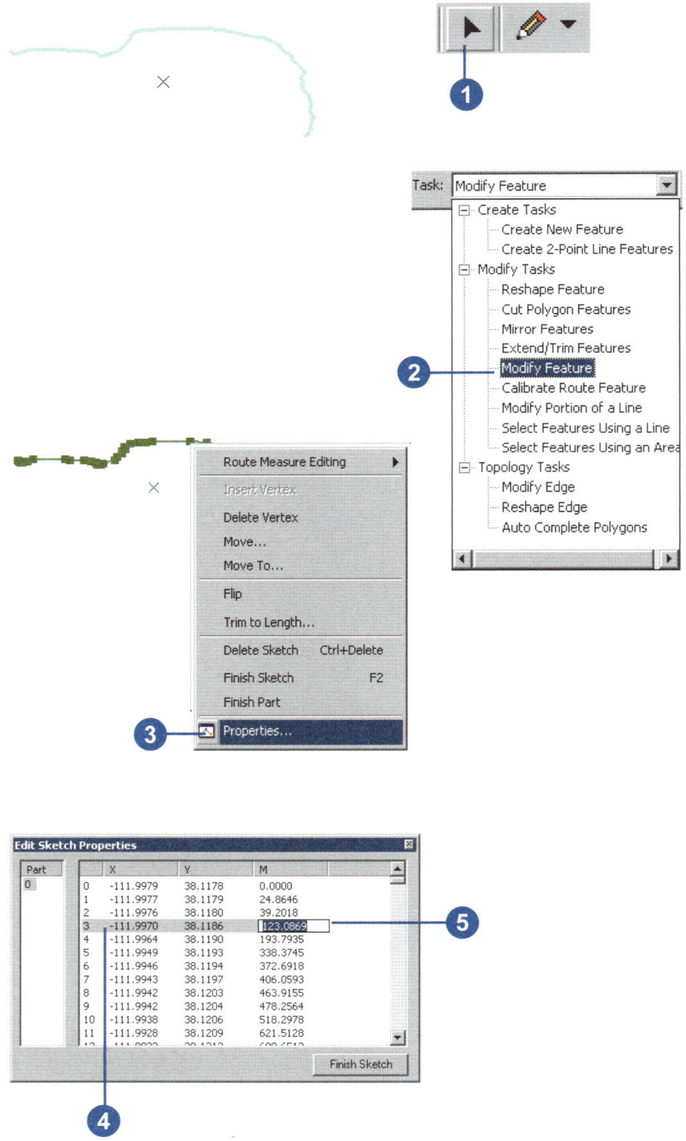

Scaling features

You can scale a feature—make the entire feature larger or smaller—using the Scale tool. The feature is scaled based on the location of the selection anchor—the small x located in the center of selected features.

You might use the Scale tool when working with data from a new source in which the scale is slightly different—for example, subdivision parcels from a surveyor. You can use the Scale tool to scale parcels so that they fit together properly.

To use the Scale tool, you must first add it to a toolbar from the Commands tab of the Customize dialog box. The Scale tool is available from the Editor category. For more information on adding a tool to a toolbar, see *Exploring ArcObjects* or *Using ArcMap*.

Tip

Moving the selection anchor

To move the selection anchor of a feature you want to scale, hold the scaling pointer over the anchor until the icon changes. Then, click and drag the anchor to a new location.

1. Click the Edit tool.
2. Click the feature you want to scale.
3. Click the Scale tool.
4. Move the selection anchor if necessary.
5. Click and drag the pointer over the feature to scale it as desired. ▸

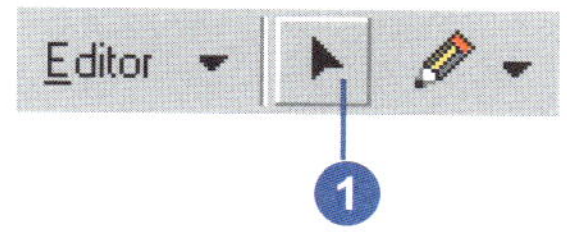

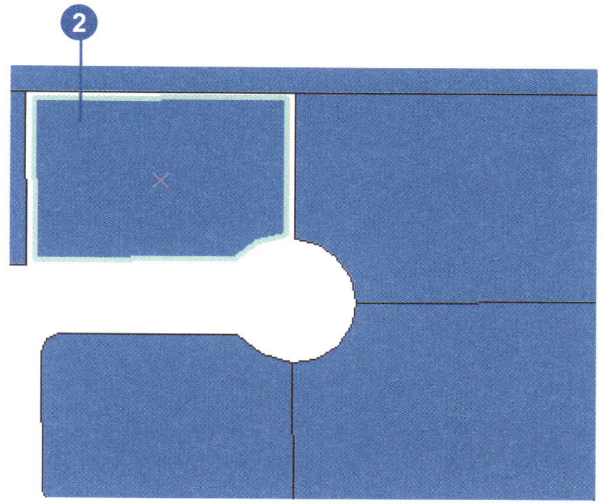

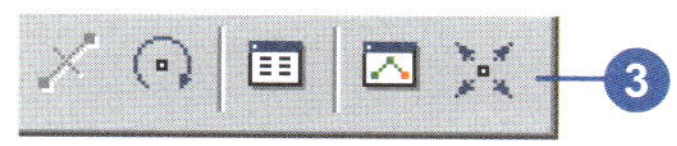

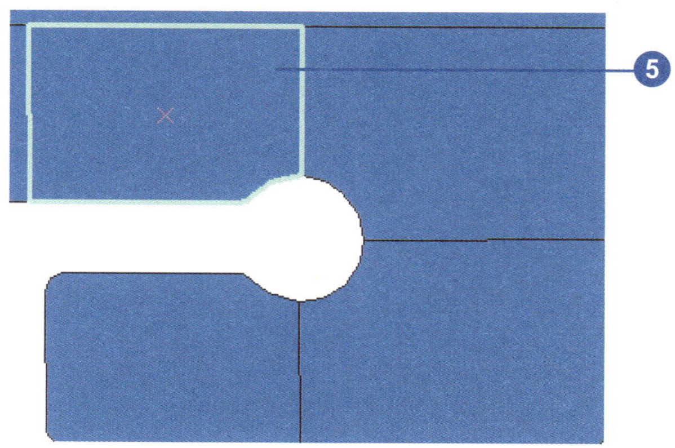

Tip

Scaling more than one feature

You can scale more than one feature at the same time. Simply select all the desired features and move the selection anchor to the desired location before using the Scale tool.

Tip

Undoing scaling

To return a feature to its original size after scaling it, click the Undo button on the ArcMap Standard toolbar.

Tip

Scale factor

You can scale features using a scale factor instead of dragging the mouse. Press the F key to set the scale factor.

Tip

Scaling with snapping

Press the S key to add an auxiliary selection anchor to the feature that you are scaling. The auxiliary selection anchor can be dragged anywhere on the feature and will snap to the features specified in the current snapping environment.

6. Release the mouse button when you're finished scaling the feature.

   The feature is scaled.

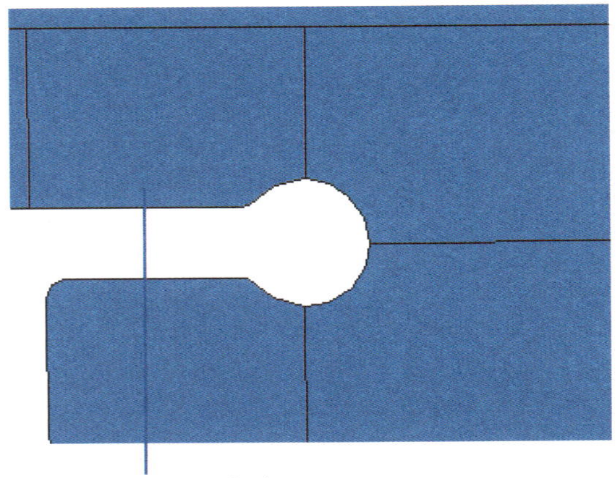

The feature is scaled.

Clipping features

You can easily clip features that touch or are within a buffered distance of selected features.

Suppose that you want to model the effect of a proposed road-widening project on the lots of a subdivision block. You can do this using the Clip command. Select the road centerline where the proposed widening is to occur, then click Clip from the Editor menu. Type the length measurement of the widening and click the option to Discard the area that intersects to clip the subdivision lots.

When using the Discard the area that intersects option, the Clip command will buffer the selected road feature and then clip all portions of editable features that are within the buffered region. Using the Preserve the area that intersects option, all features that touch the buffered feature will be deleted.

1. Select the feature you want to use to clip features.
2. Click Editor and click Clip.
3. Type a buffer value. You can leave the value as 0 if you are using a polygon feature to clip with.
4. Click the type of clip operation you wish to use.
5. Click OK to clip the feature.

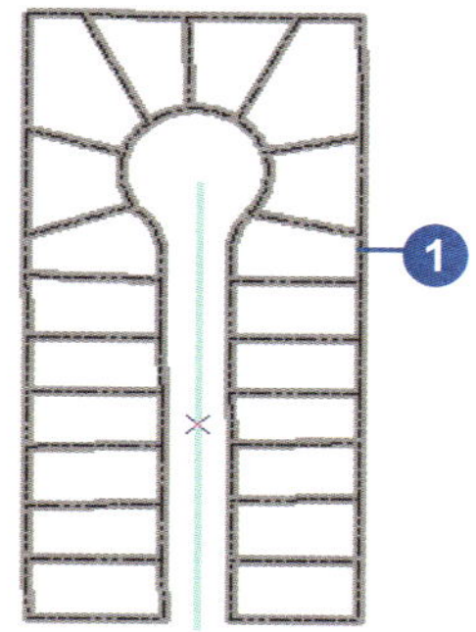

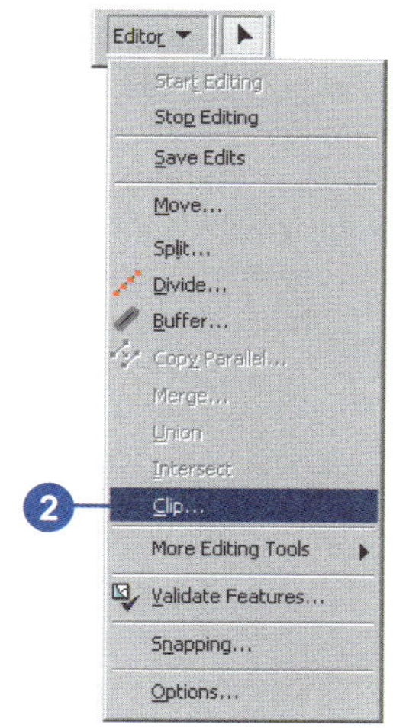

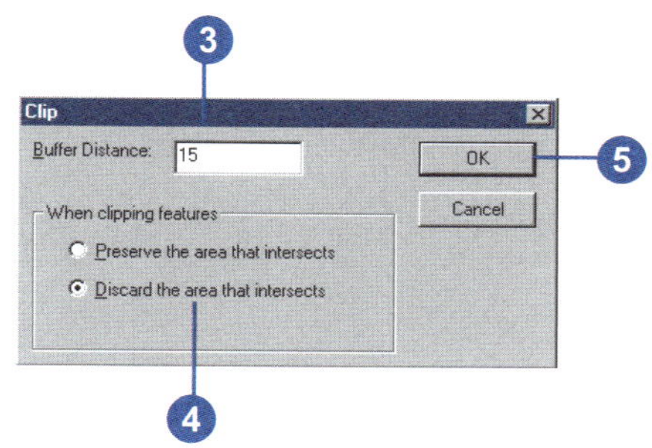

Stretching geometry proportionately

Sometimes you want to stretch a feature without changing its geometry (shape). Suppose you want to change the position of a feature in relation to other features by moving a vertex. For example, perhaps the data you have for an electric transmission system is not as accurate as you would like. However, you have other layers containing accurate surveyed points that coincide with some of the transmission towers, power generating plants, and substations. By moving the vertices of the transmission lines, you can adjust the positions of the lines to the known surveyed positions of the features in the more accurate layer. You can change the positions of these vertices without changing the general shape of the transmission lines by stretching the features proportionately.

When you stretch a feature proportionately, the proportions of the feature's segments are maintained, thereby maintaining the general shape of the feature. This is different from moving a vertex to reshape a feature.

The graphics below show the difference between moving a vertex to reshape a feature and moving a vertex while maintaining the shape of the feature. The three graphics on the top show how a feature is modified when its upper-right vertex is moved with proportionate stretching turned on. The three graphics on the bottom show how the same feature is reshaped when its upper-right vertex is moved with proportionate stretching turned off.

Proportionate stretching on

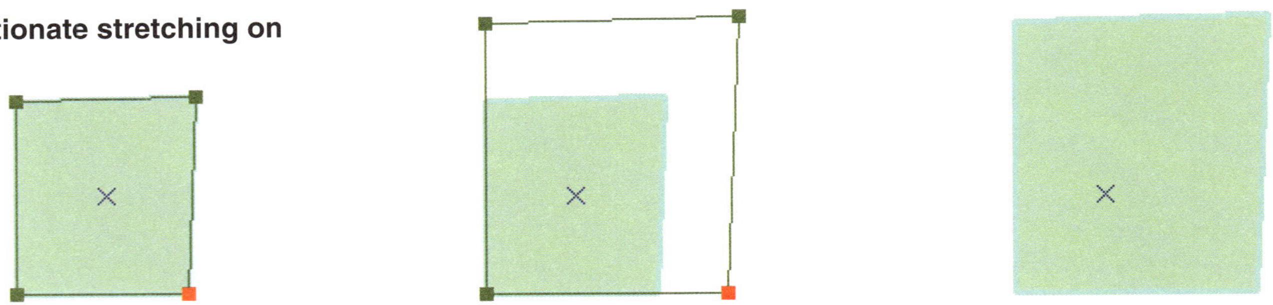

Proportionate stretching off

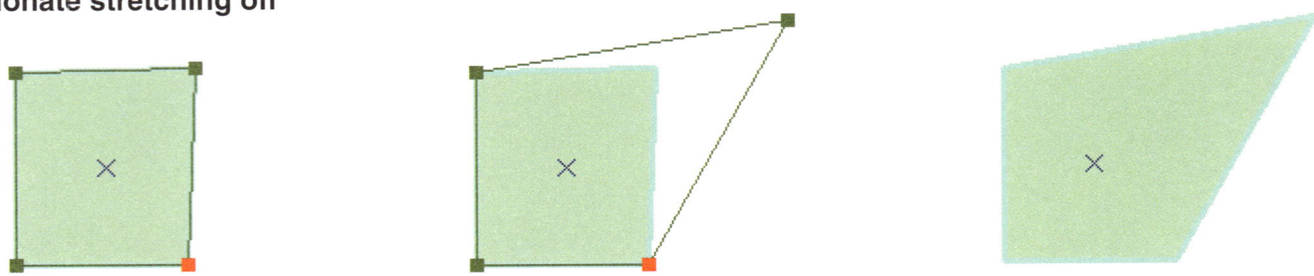

Stretching a feature's geometry proportionately

Within the Editing Options menu, you can choose to stretch the geometry of features proportionately when moving vertices. When you drag a vertex to a new location with this option turned on, the proportions of the feature's segments are maintained, thereby maintaining the general shape of the feature.

You might want to stretch features proportionately when merging data from different data sources—for example, utility lines from one source and subdivision parcels from another.

Suppose the data for the subdivision parcels is very accurate, but the data for the utility lines is not as accurate. While the shapes of the utility lines are generally correct, you want to change the position of one line relative to the parcels by moving a vertex. By stretching the utility line feature proportionately, you can make it fit accurately with the parcels without losing the general shape of the line. ▶

1. Click the Task dropdown arrow and click Modify Feature.
2. Click the Edit tool and click the feature you want to stretch.
3. Click Editor and click Options.
4. Click the General tab.
5. Check the check box to stretch the feature proportionately.

   Uncheck the check box if you want to reshape the feature without maintaining proportionate geometry.
6. Click OK. ▶

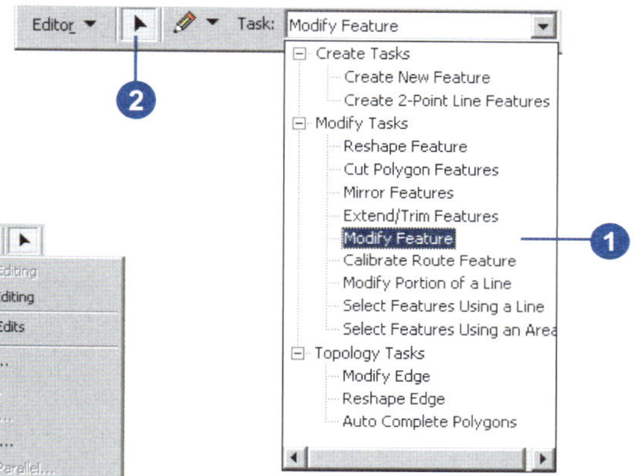

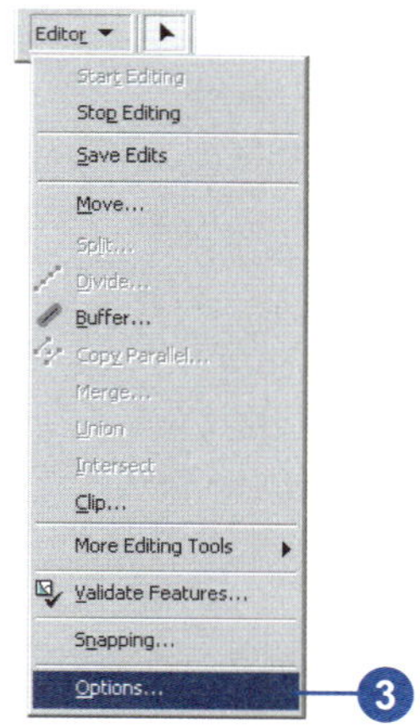

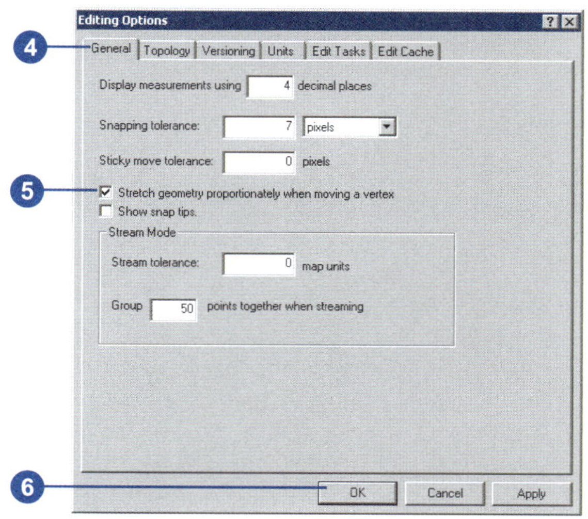

You can turn on proportionate stretching by checking a box on the General tab of the Editing Options dialog box. Uncheck the box if you simply want to reshape a feature without maintaining proportionate geometry.

See Also

To see how stretching a feature proportionately looks in comparison to stretching a feature to reshape it, see 'Stretching geometry proportionately' in this chapter.

7. Position the pointer over the vertex you want to move until the pointer changes.
8. Drag the vertex to the desired location.
9. Right-click over any part of the sketch and click Finish Sketch.

   The feature is stretched proportionately.

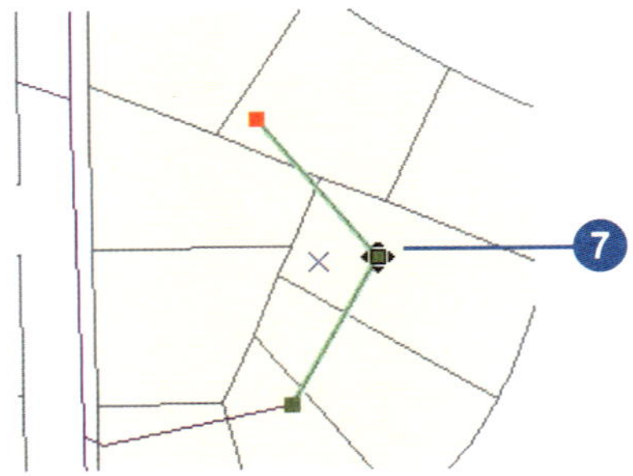

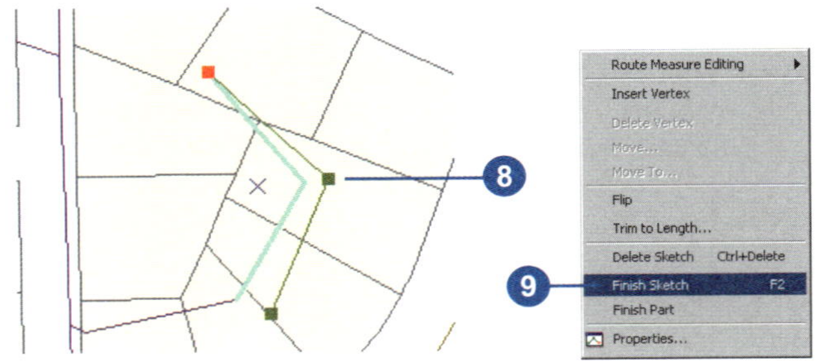

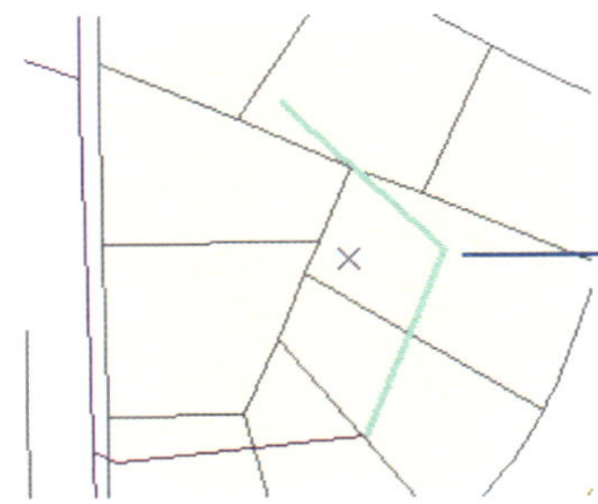

The feature is stretched proportionately.

Spatial adjustment

8

IN THIS CHAPTER

The Spatial Adjustment toolbar lets you transform, rubber sheet, and edgematch features in your map. It works within ArcMap Editor to provide a highly productive adjustment environment. Spatial adjustment supports a variety of adjustment methods and will adjust all editable data sources.

Spatial adjustment commands and tools are located on an additional editing toolbar called the Spatial Adjustment toolbar. These tools and commands allow you to define a spatial adjustment. Since spatial adjustment operates within an Edit session, you can leverage existing editing functionality, such as snapping, to enhance your adjustments.

Along with the ability to spatially adjust your data, the Spatial Adjustment toolbar also provides a way for you to transfer the attributes from one feature to another. This tool is called the Attribute Transfer tool and relies on matching common fields between two layers. Together, the adjustment and attribute transfer functions available in the Spatial Adjustment tool allow you to improve the quality of your data.

About spatial adjustments

The following section briefly describes the spatial adjustment methods and related concepts.

Transformations

Transformations convert data from one coordinate system to another. They are often used to convert data from digitizer or scanner units to real-world coordinates. Transformations can also be used to shift your data within a coordinate system, such as converting feet to meters.

The transformation functions are based on comparing the coordinates of source and destination points, also called control points, in special graphical elements called displacement links. You may create these links interactively, pointing at known source and destination locations, or by loading a link text file or control points file.

By default, ArcMap supports three types of transformations: affine, similarity, and projective.

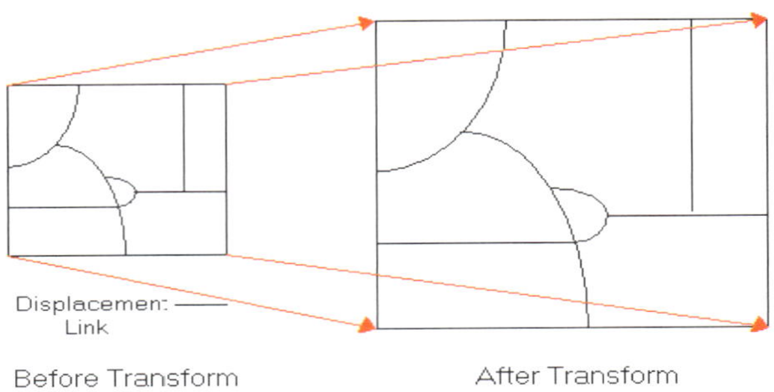

An *affine* transformation can differentially scale, skew, rotate, and translate the data. The graphic below illustrates the four possible changes:

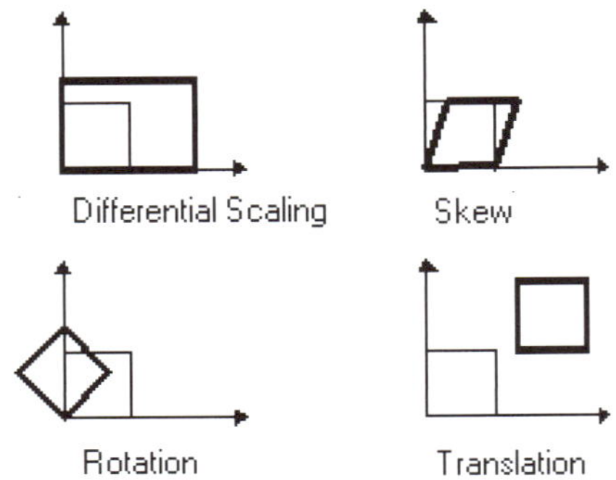

The affine transformation function is:

$$x' = Ax + By + C$$
$$y' = Dx + Ey + F$$

where x and y are coordinates of the input layer and x' and y' are the transformed coordinates. A, B, C, D, E, and F are determined by comparing the location of source and destination control points. They scale, skew, rotate, and translate the layer coordinates.

The affine transformation requires a minimum of three displacement links.

The *similarity* transformation scales, rotates, and translates the data. It will not independently scale the axes, nor will it introduce any skew. It maintains the aspect ratio of the features transformed.

The similarity transform function is:

$x' = Ax + By + Cy' = -Bx + Ay + F$

where:

$A = s \cdot \cos t$
$B = s \cdot \sin t$
C = translation in x direction
F = translation in y direction

and:

s = scale change (same in x and y directions)
t = rotation angle, measured counterclockwise from the x-axis

A similarity transformation requires a minimum of two displacement links.

The *projective* transformation is based on a more complex formula that requires a minimum of four displacement links:

$x' = (Ax + By + C) / (Gx + Hy + 1)$
$y' = (Dx + Ey + F) / (Gx + Hy + 1)$

This method is used to transform data captured directly from aerial photography. For more information, please refer to one of the photogrammetric texts listed in 'References' at the end of this section.

Understanding residual and root mean square (RMS)

The transformation parameters are a best fit between the source and destination control points. If you use the transformation parameters to transform the actual source control points, the transformed output locations won't match the true output control point locations. This is known as the residual error; it is a measure of the fit between the true locations and the transformed locations of the output control points. This error is generated for each displacement link.

An RMS error is calculated for each transformation performed. It indicates how good the derived transformation is. The following example illustrates the relative location of four destination control points and the transformed source control points:

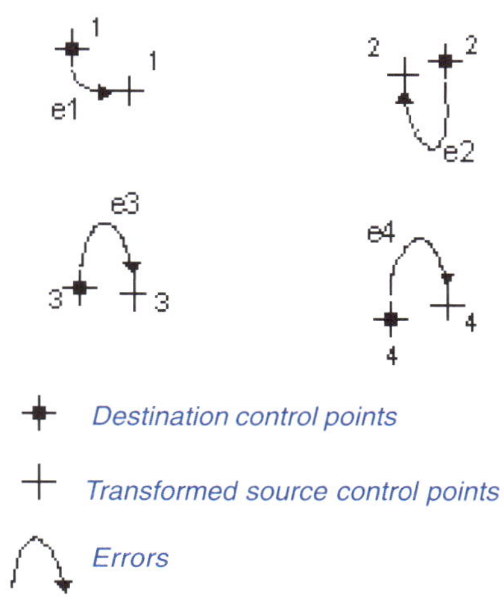

The RMS error measures the errors between the destination control points and the transformed locations of the source control points.

$$\text{RMS error} = \sqrt{\frac{e_1^2 + e_2^2 + e_3^2 + \cdots + e_n^2}{n}}$$

The transformation is derived using least squares, so more links can be given than are necessary.

Rubber sheeting

Geometric distortions commonly occur in source maps. They may be introduced by imperfect registration in map compilation, lack of geodetic control in source data, or a variety of other causes. Rubber sheeting corrects flaws through the geometric adjustment of coordinates.

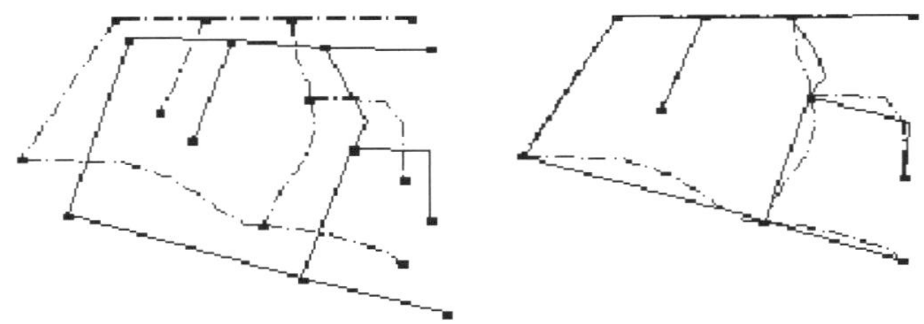

The source layer—drawn with solid lines—is adjusted to the more accurate target layer.

During rubber sheeting, the surface is literally stretched, moving features using a piecewise transformation that preserves straight lines. Similar to transformations, displacement links are used in rubber sheeting to determine where features are moved.

Conflation applications use rubber sheeting to align layers in preparation for transferring attributes.

Edgematching

The edgematching process aligns features along the edge of one layer to features of an adjoining layer. The layer with the least-accurate features is adjusted, and the other adjoining layer is used as the control.

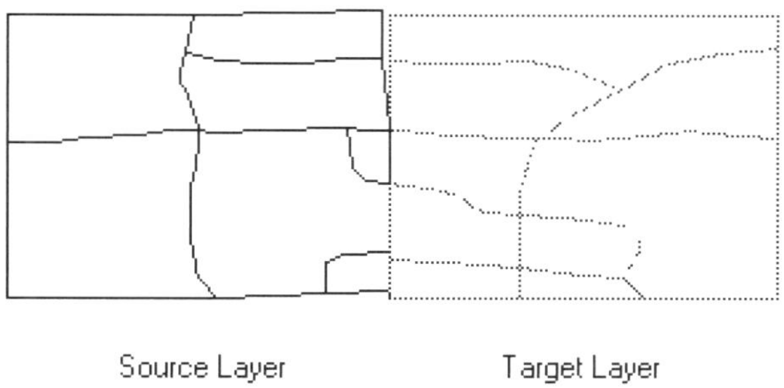

Attribute transfer

Attribute transfer is typically used to copy attributes from a less accurate layer to a more accurate one. For example, it can be used to transfer the names of hydrological features from a previously digitized and highly generalized 1:500,000-scale map to a more detailed 1:24,000-scale map.

In ArcMap, you can specify what attributes to transfer between layers, then interactively choose the source and target features.

References

Maling, D.H. *Coordinate Systems and Map Projections*. George Philip, 1973.

Maling, D.H. "Coordinate systems and map projections for GIS." *Geographical Information Systems: Principles and Applications*, Maguire, D.J., M.F. Goodchild, and D.W. Rhind (eds.) vol. 1, pp. 135–146. Longman Group UK Ltd., 1991.

Moffitt, F.H., and E.M. Mikhail. *Photogrammetry*. Third edition. Harper & Row, Inc., 1980.

Pettofrezzo, A.J. *Matrices and Transformations*. Dover Publications, Inc., 1966.

Slama, C.C., C. Theurer, and S.W. Henriksen (eds.). *Manual of Photogrammetry*. Fourth edition. Chapter XIV, pp. 729–731. ASPRS, 1980.

The Spatial Adjustment toolbar

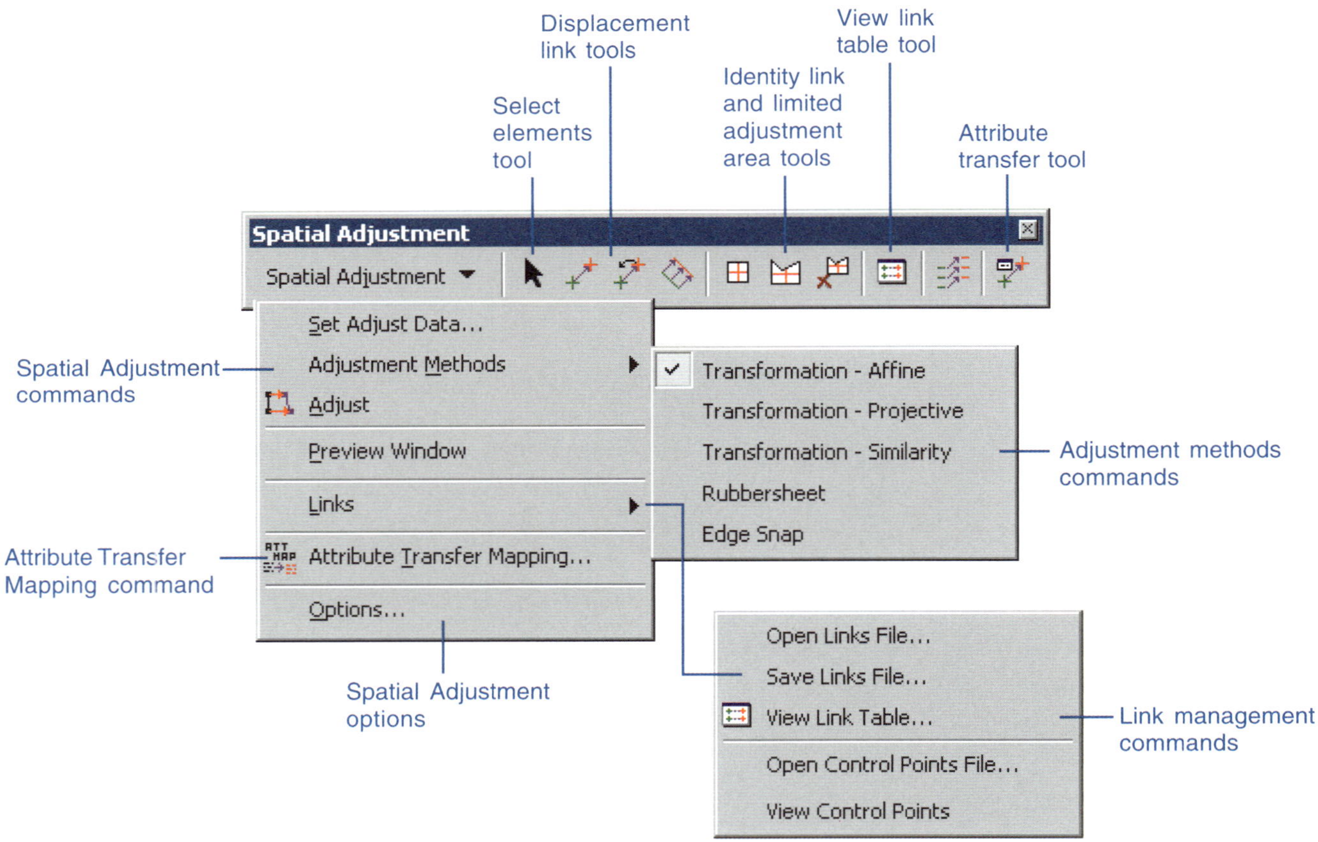

An overview of the spatial adjustment process

The following is a general overview of how to use ArcMap, the Editor toolbar, and the Spatial Adjustment toolbar to adjust your data. Each of the following steps is outlined in detail in this chapter or other chapters in this book.

1. Start ArcMap.
2. Create a new map or open an existing one.

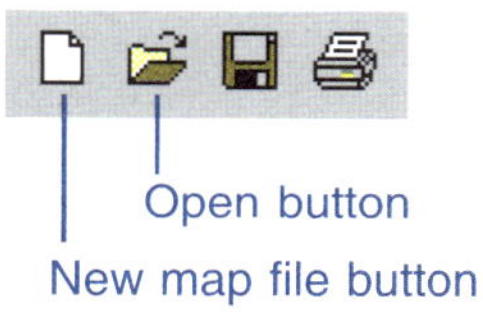

3. Add the data you want to edit to your map.

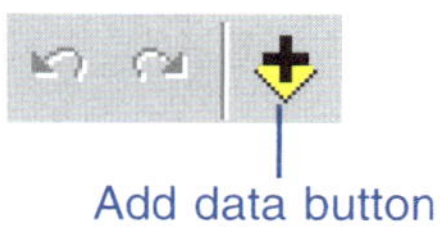

If there are no existing layers for the feature classes you want to edit, you can create them using ArcCatalog. For more information on creating a feature layer, see *Using ArcCatalog*.

4. Add the Editor toolbar to ArcMap.

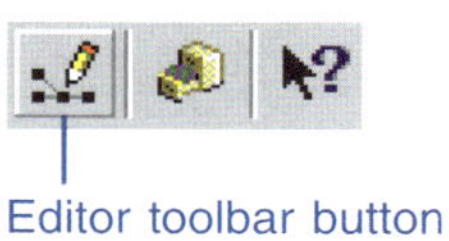

5. Add the Spatial Adjustment toolbar to ArcMap.

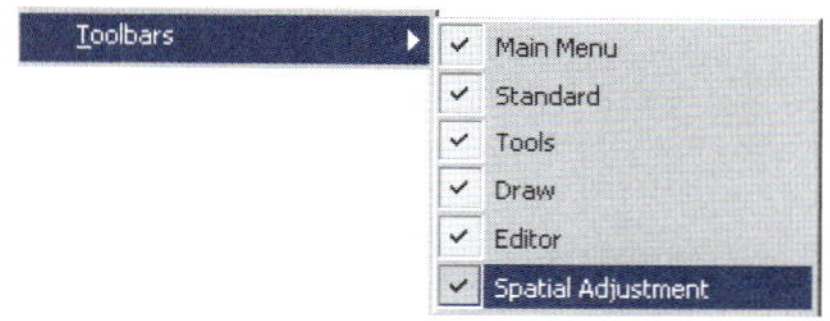

6. Choose Start Editing from the Editor menu.

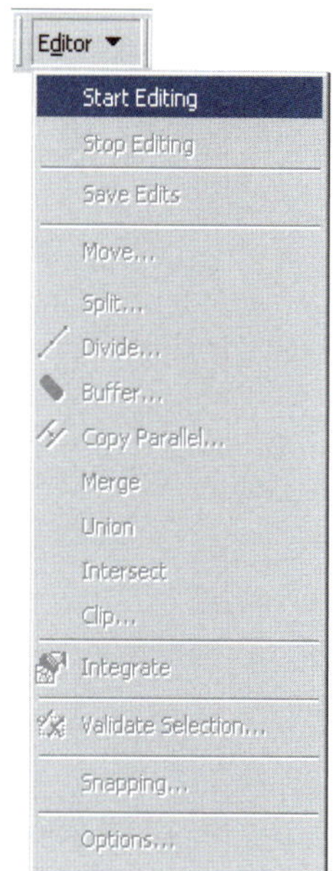

7. Click Spatial Adjustment, point to Adjustment Methods, and click one of the Transformation methods to choose a spatial adjustment method.

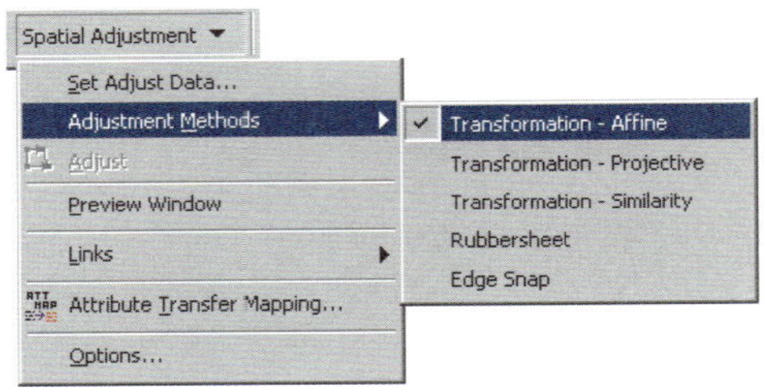

8. Click the Displacement Link tools to create displacement links.

9. Perform the adjustment.

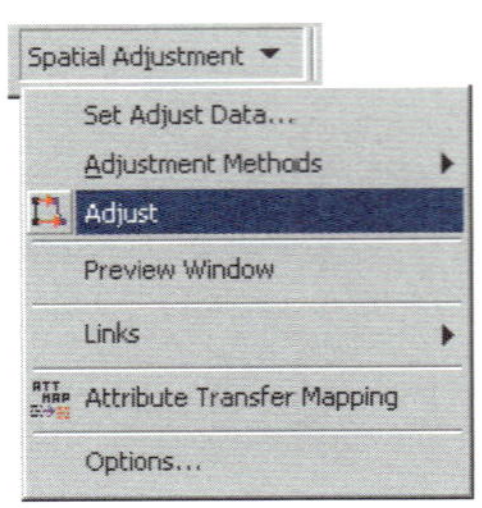

10. Choose Stop Editing from the Editor menu and click Yes when prompted to save your edits.

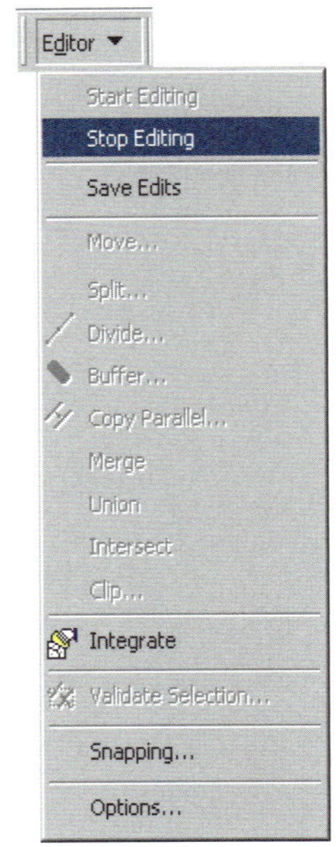

There is no need to save the map—all edits made to the database will automatically be reflected the next time you open the map.

Adding the Spatial Adjustment toolbar

Before adjusting geographic feature data within ArcMap, you must first add the Spatial Adjustment toolbar.

Tip

Adding the Spatial Adjustment toolbar using the Customize dialog box

Click the Tools menu and click Customize. In the Customize dialog box, click the Toolbars tab and check Spatial Adjustment.

1. Start ArcMap.
2. Click the View menu, point to Toolbars, and click Spatial Adjustment to display the Spatial Adjustment toolbar.
3. Click the toolbar's title bar and drag it to the top of the ArcMap application window.

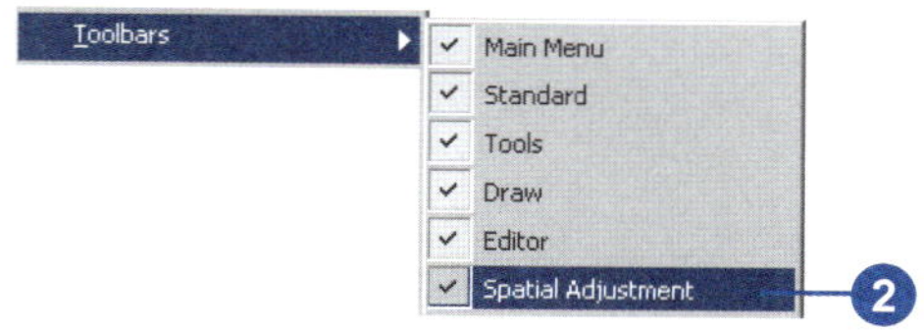

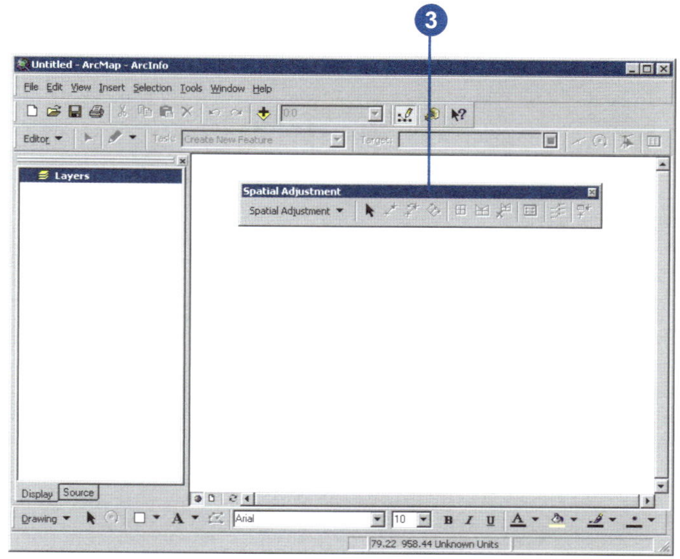

Choosing the input data for adjustment

The first step in the spatial adjustment process is to choose the input data for the adjustment. You have the option to adjust selected features or all the features in the layer. These settings are available in the Choose Input for Adjustment dialog box.

Tip

All selection methods are supported

The Spatial Adjustment tool will honor selections performed interactively or by an attribute query.

1. Click the Spatial Adjustment menu and click Set Adjust Data.

   The Choose Input For Adjustment dialog box appears.

2. Choose whether to adjust selected features in a layer or all features in a layer.

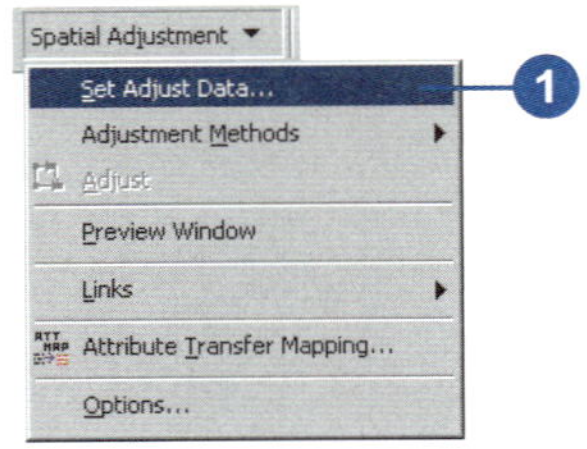

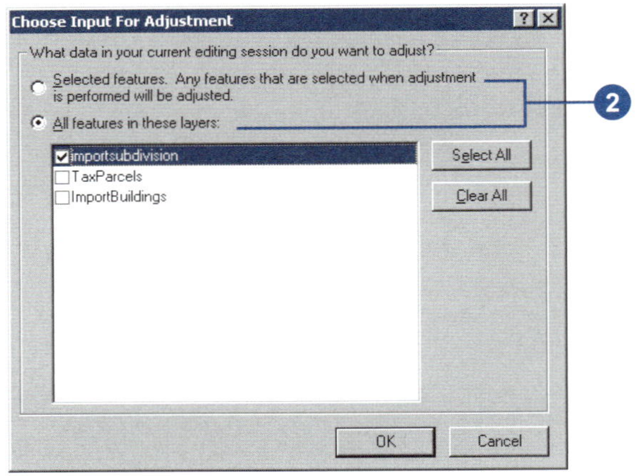

Choosing a transformation method

The Spatial Adjustment tool supports three types of transformation methods: affine, projective, and similarity. Choose a transformation method from the Adjustment Methods submenu.

See Also

For more information on the spatial adjustment methods, see the introductory material with this chapter.

1. Click the Spatial Adjustment menu, point to Adjustment Methods, and click a Transformation method.

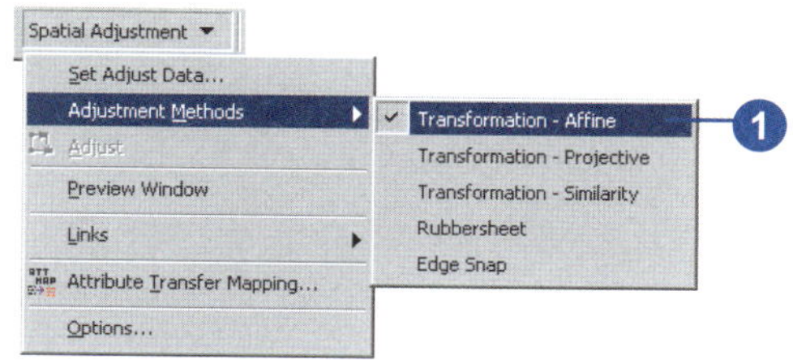

Choosing a rubber sheet method

The Spatial Adjustment tool supports two types of rubber sheet methods: Natural Neighbor and Linear. Choose Rubbersheet from the Adjustment Methods submenu. Once you have chosen Rubbersheet, you may specify a rubber sheet method in the Adjustment Properties dialog box. The Natural Neighbor method is the default.

1. Click the Spatial Adjustment menu, point to Adjustment Methods, and click Rubbersheet.
2. Click the Spatial Adjustment menu and click Options.

   The Adjustment Properties dialog box appears.
3. Click the Adjustment method dropdown arrow and choose Rubbersheet.
4. Click the Adjustment method Options button.

   The Rubbersheet properties dialog box will appear.
5. Click the Natural Neighbor or Linear method and click OK.
6. Click OK to dismiss the Adjustment Properties dialog box.

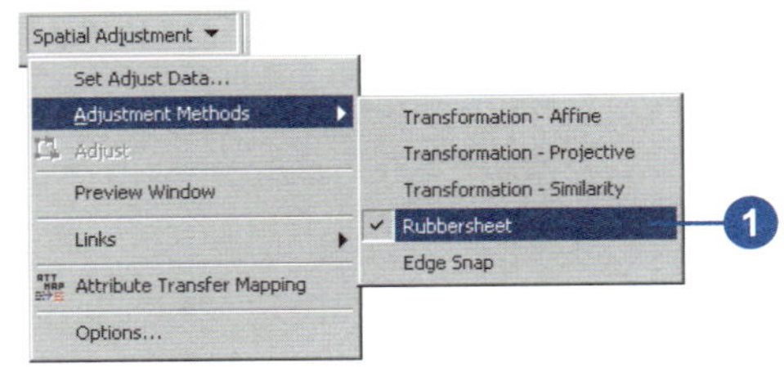

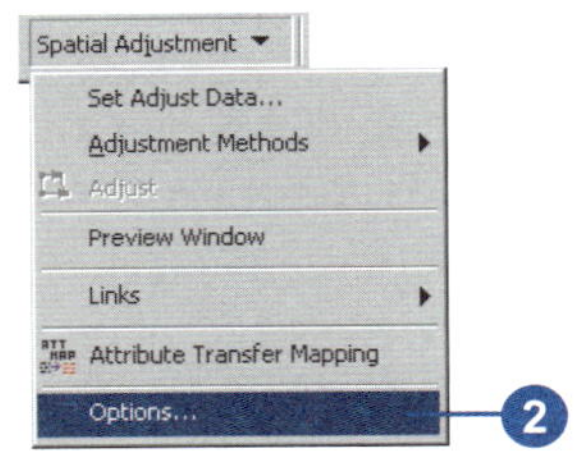

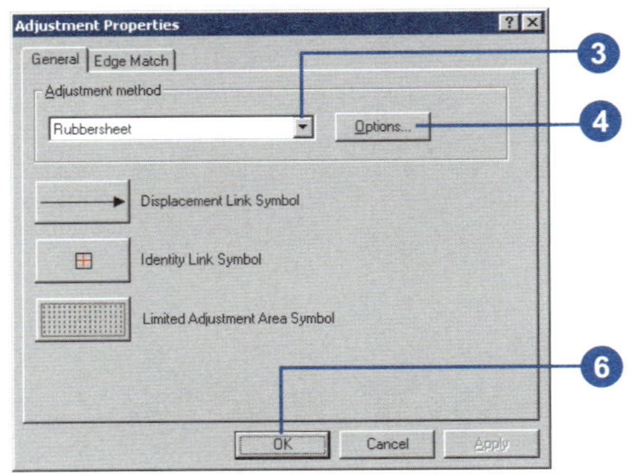

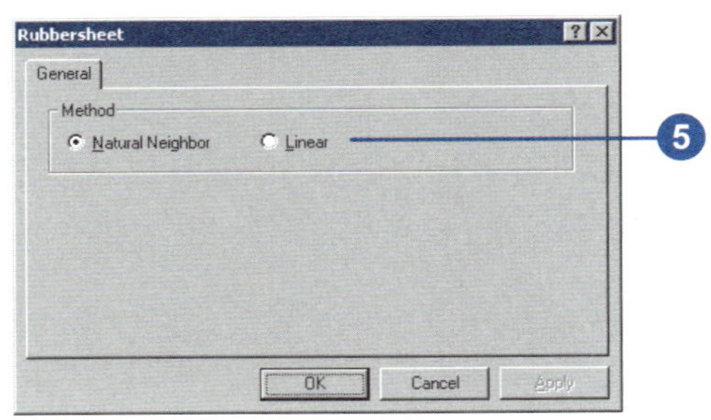

Choosing an edge snap method

The Spatial Adjustment tool supports two types of Edge Snap methods: Smooth and Line. Choose Edge Snap from the Adjustment methods submenu. Once you have chosen Edge Snap, you may specify an edge snap method in the Adjustment Properties dialog box.

When using the Smooth edge snap method, vertices at the link source point are moved to the destination point. The remaining vertices are also moved to give an overall smoothing effect.

When using the Line edge snap method, only the vertices at the link source point are moved to the destination point. The remaining vertices on the feature remain unchanged.

The Smooth method is the default.

1. Click the Spatial Adjustment menu, point to Adjustment Methods, and click Edge Snap.
2. Click the Spatial Adjustment menu and click Options.

   The Adjustment Properties dialog box appears.
3. Click the Adjustment method dropdown arrow and click Edge Snap.
4. Click the Adjustment method Options button.

   The Edge Snap dialog box will appear.
5. Click the Smooth or Line method.
6. Check the box if you want to adjust to the midpoint of the links, then click OK.
7. Click OK to close the Adjustment Properties dialog box.

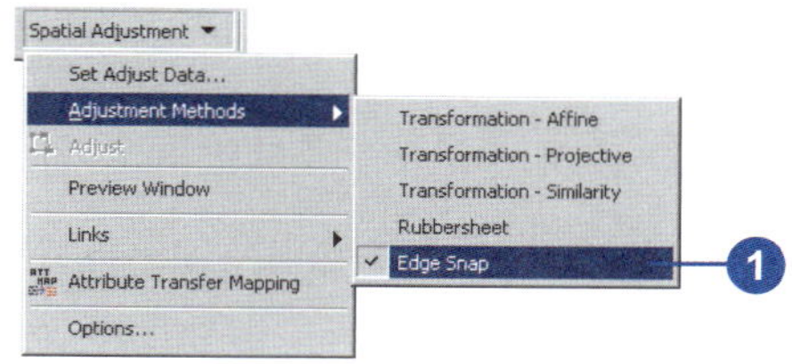

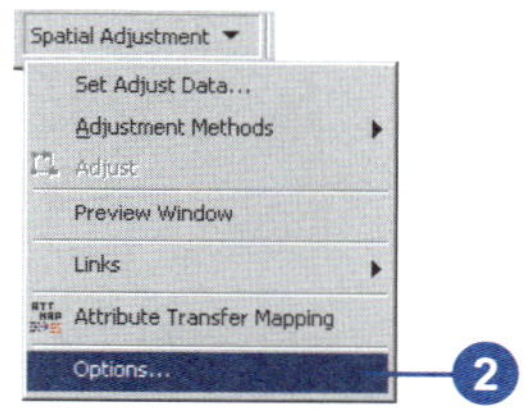

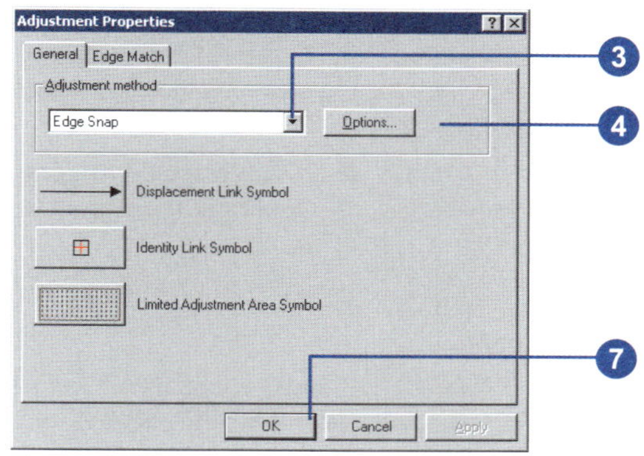

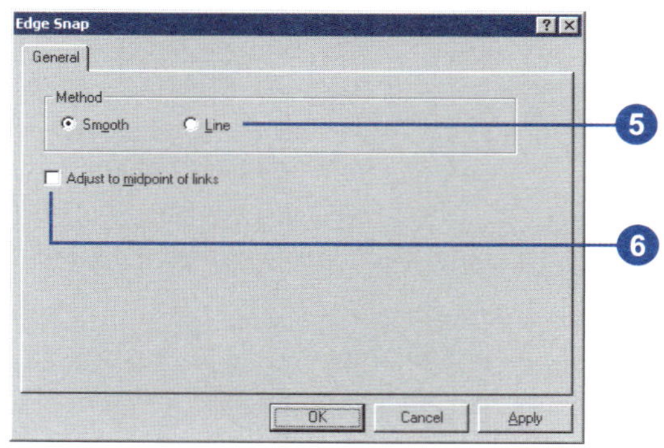

Setting the edge snap properties

The Edge Snap adjustment method requires more property settings than other methods. These property settings are located in the Edge Match tab of the Adjustment Properties dialog box.

Choose the source and target layers. The source layer's features will be edgematched to the target layer's features. If you choose to adjust to the midpoints of the links, features from both layers will be adjusted.

You have the option to specify one link per destination point and to prevent duplicate links. These settings can help you avoid creating unnecessary links. The Spatial Adjustment tool supports the ability to use attributes to enhance the edgematching process. Based on the Attribute Transfer Mapping dialog box, you have the option to match fields between the source and target layers and use common attributes to define the edgematch. This function can help ensure the accuracy of the edgematch.

1. Click the Spatial Adjustment menu and click Options.

   The Adjustment Properties dialog box appears.

2. Click the Edge Match tab.
3. Click the Source Layer dropdown arrow and choose a source layer.
4. Click the Target Layer dropdown arrow and choose a target layer.
5. Check Use Attributes if you want to use attributes to enhance the edgematch.
6. If you only want one link for each destination point, check the appropriate box.
7. If you want to prevent duplicate links, check the appropriate box.
8. Click the Attributes button if you chose to use attributes.

   The Edgematch Attributes dialog box appears.

9. Match the source and target layer fields.
10. Click Add.
11. Click OK when finished matching fields.
12. Click OK to close the Adjustment Properties dialog box.

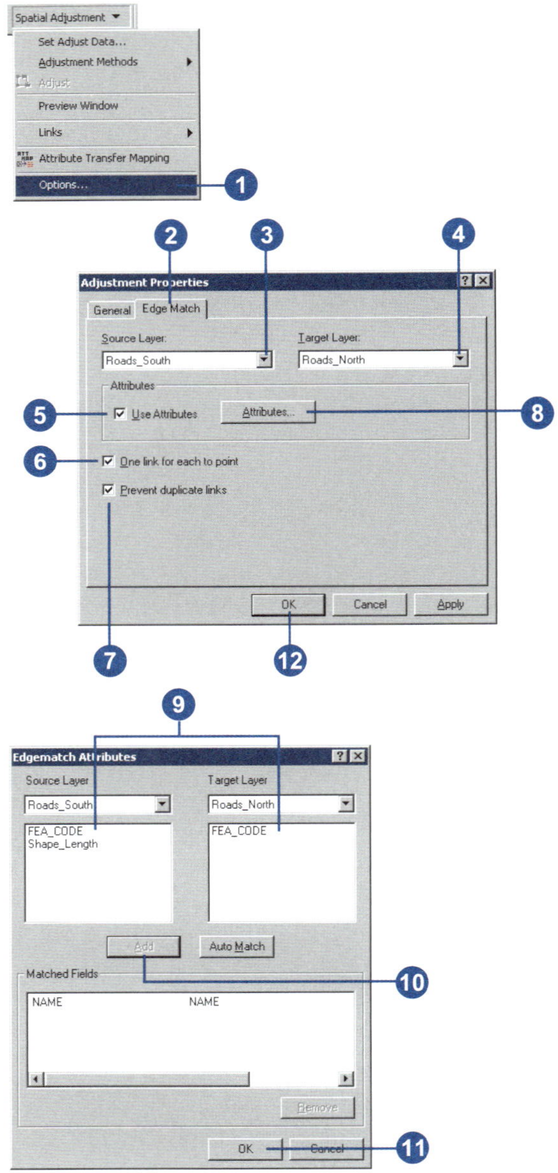

Creating displacement links

Before you adjust your data, you must create displacement links to define the source and destination coordinates for the adjustment. Links are represented as arrows with the arrowhead pointing toward the destination location. Links can be created manually or loaded by means of a link file.

Displacement links are represented as graphic elements in the map. You can change the symbol, size, and color of displacement links.

Tip

Use snapping to ensure accurate link placement

Use the Snapping Environment tool to set the snapping agents and features. Snapping will ensure that links are created at the vertices, edges, or endpoints of features.

1. Click the New Displacement link tool on the Spatial Adjustment toolbar.

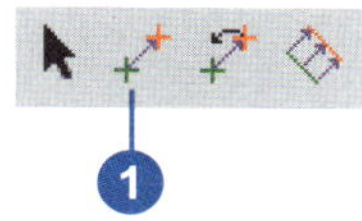

2. Position the cursor over the source location and click once to start adding a link.

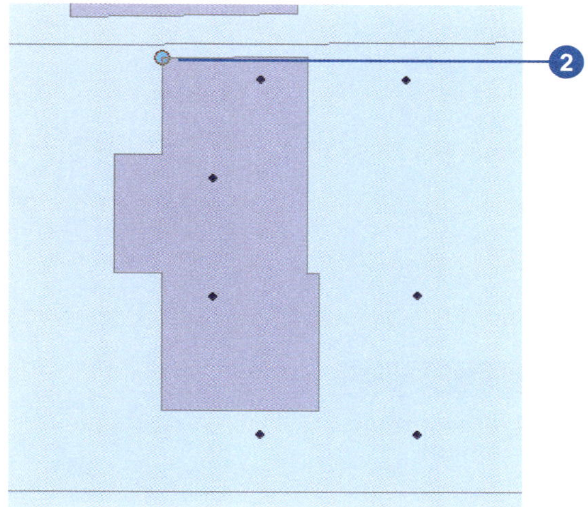

3. Position the cursor over the destination location and click once to finish adding the link.

   A displacement link now connects the source location to the destination location.

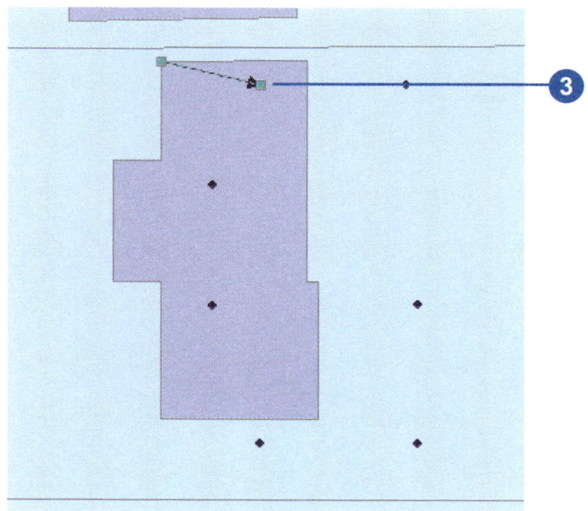

Creating multiple displacement links

You can create multiple displacement links using the Multi Displacement Links tool. The Multi Displacement Links tool is useful for areas that require many links, such as curve features. This tool can also help you save time by allowing you to create many links at once.

Tip

Snap to edges when using the Multi Displacement Links tool

It is best to snap to the edges of features when creating multiple links.

1. Click the Multi Displacement Links tool on the Spatial Adjustment toolbar.
2. Position the cursor over the source feature and click once.
3. Position the cursor over the target feature and click once.

   The Number of links dialog box will appear. This dialog box allows you to specify how many links to create. ▸

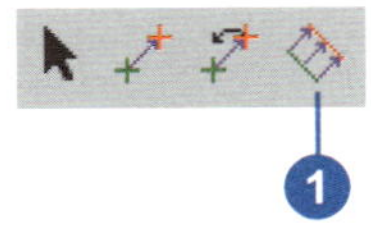

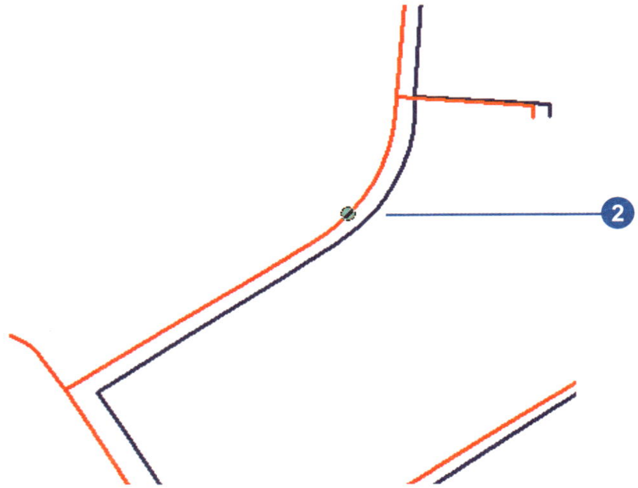

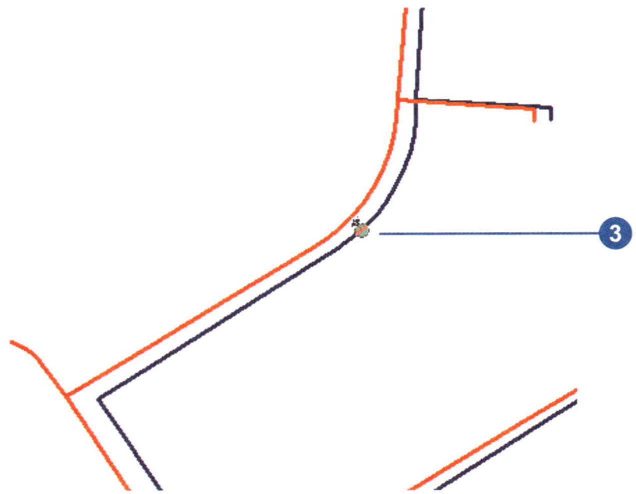

4. Enter a Number of Links value and press Enter. The default is 10.

   Based on the value you entered, the multiple links are created, and now connect the source feature to the target feature.

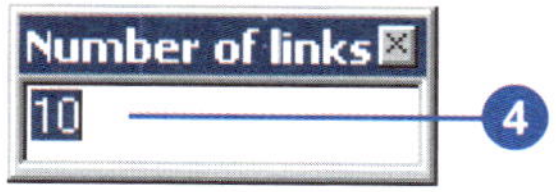

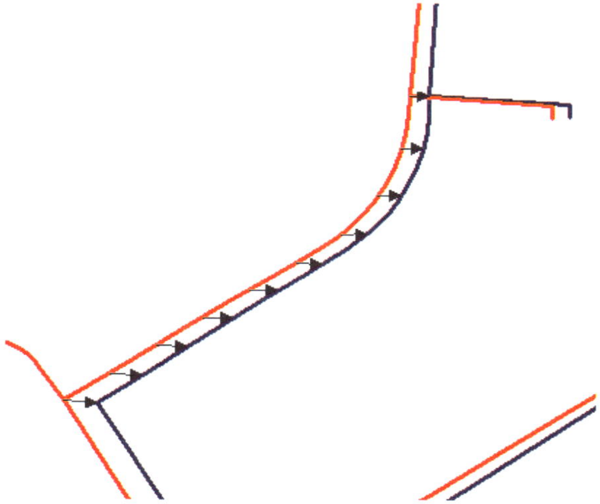

Creating identity links

Identity links can be used to hold features in place at specified locations. Identity links can serve as anchors because they prevent the movement of features during an adjustment.

Identity links are only available when using the Rubbersheet adjustment method. Like displacement links, identity links are represented as graphic elements in the map.

1. Click the New Identity Link tool on the Spatial Adjustment toolbar.
2. Position the cursor over the source location and click once.

   Add identity links to locations to prevent the movement of features during an adjustment.

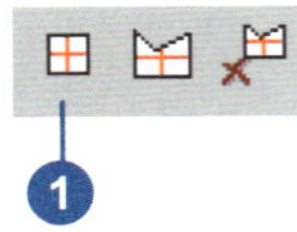

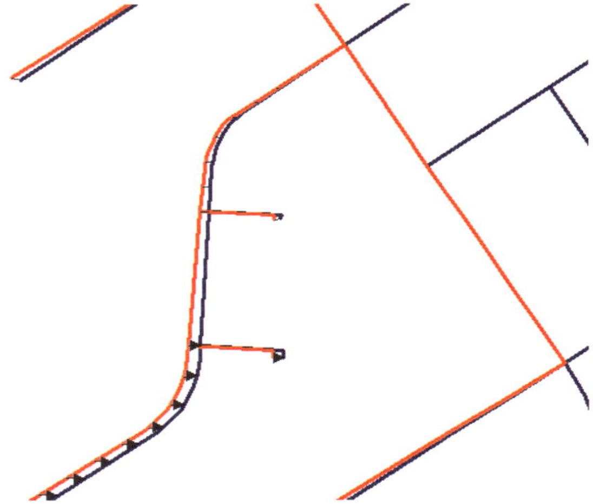

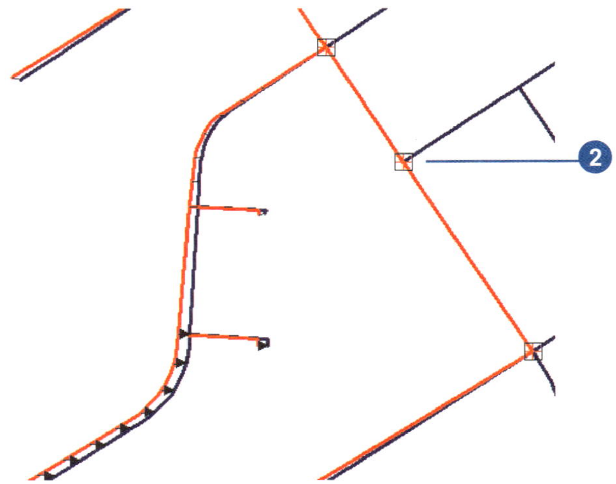

Using the Limited Adjustment Area tools

You can limit the scope of an adjustment area by using the Limited Adjustment Area tool. This tool is only available for the Rubbersheet adjustment method.

This tool allows you to draw a polygon shape around the features you wish to adjust. Any features that are outside this polygon area will not be affected by the adjustment, regardless of whether or not they are selected. The Limited Adjustment Area tool provides similar feature anchoring as identity links. However, in cases where you must add many identity links, using the Limited Adjustment Area may help you save time.

Creating a limited adjustment area

1. Click the New Limited Adjustment Area tool on the Spatial Adjustment toolbar.
2. Using the cursor on the map, draw a polygon around the area you wish to rubber sheet. Double-click to complete the polygon.

   Features outside of this polygon will not be affected during the adjustment.
3. Click the Clear Limited Adjustment Area tool on the Spatial Adjustment toolbar to remove the limited adjustment area polygon.

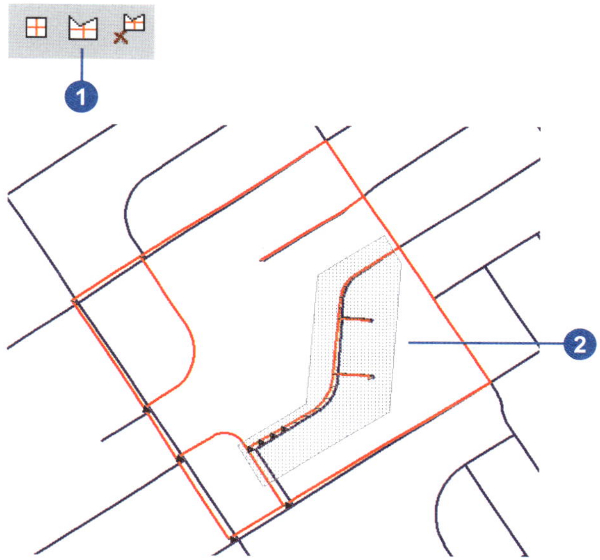

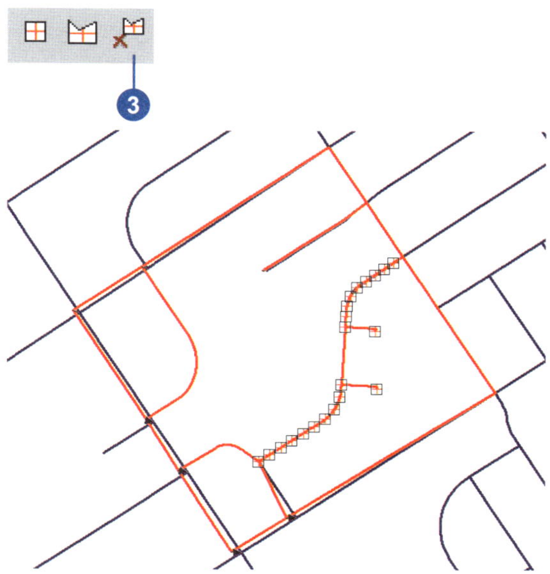

Using the Edge Match tool

The Edge Match tool allows you to create multiple displacement links that connect the edges of two adjacent layers. Once you have set the appropriate snapping agents and tolerance, use the tool to drag a box around the features you wish to edgematch. This will create links between the closest source and target features by default.

You can also use additional properties to enhance the link creation process, such as restricting one link per destination point and preventing duplicate links.

These properties, combined with the proper snapping settings, can help ensure an accurate edgematch.

Tip

Use attributes to refine the edgematch

The Use Attributes option located in the Edge Match properties dialog box can assist in the creation of links by ensuring they connect to features that share common attribute values.

1. Click the Edge Match tool on the Spatial Adjustment toolbar.

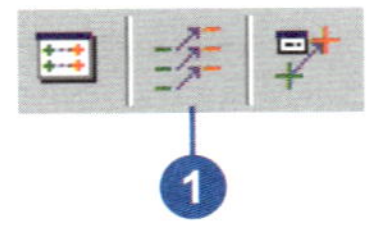

2. Drag a box around the features you want to edgematch.

   Links will now connect the edges of the source layer to the edges of the target layer.

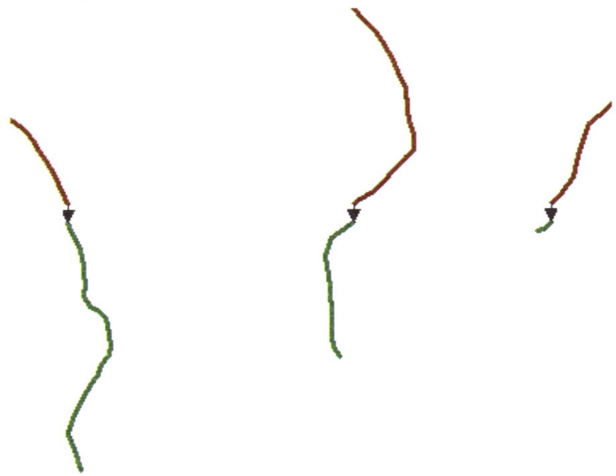

Modifying the link and limited adjustment area symbols

You can customize the symbology of the displacement links, identity links, and limited adjustment area. These settings are in the Adjustment Properties dialog box. When you click the Displacement Link Symbol, Identity Link Symbol, or Limited Adjustment Area Symbol buttons, the Symbol Selector dialog box appears. Use this window to choose a new style, size, and color for the links and limited adjustment area. The Symbol Selector will present options to modify the symbols based on the graphical element's geometry type.

1. Click the Spatial Adjustment menu and click Options.

   The Adjustment Properties dialog box appears.

2. Click the Displacement Link Symbol, Identity Link Symbol, or Limited Adjustment Area Symbol button.

   The Symbol Selector dialog box appears.

3. Choose a different symbol, change the symbol size, and/or specify a new symbol color, then click OK.

4. Click OK to close the Adjustment Properties dialog box.

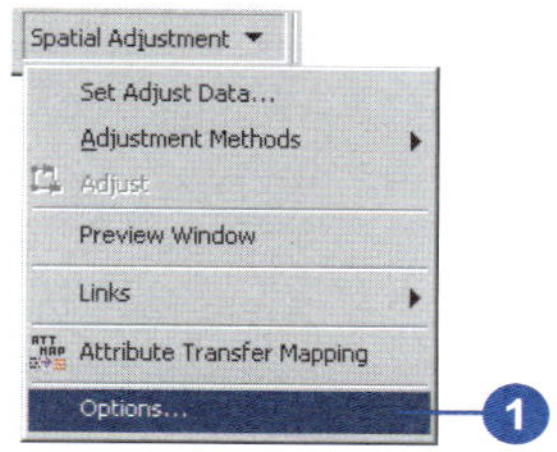

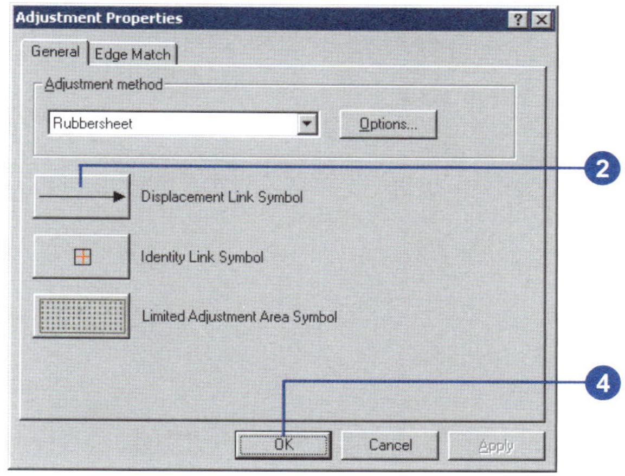

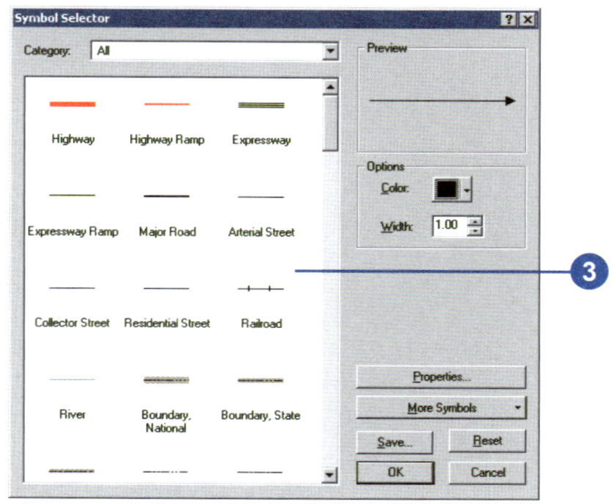

Selecting links

Displacement and identity links are selected with the Select Elements tool located on the Spatial Adjustment toolbar.

Links must be selected prior to modifying or deleting a link.

Tip

Use the Select Elements tool on the ArcMap standard toolbar

You can also select links by using the Select Elements tool.

Selecting a link

1. Click the Select Elements tool on the Spatial Adjustment toolbar.
2. Position the cursor over the link you want to select and click once.

   Selection grips will appear at the endpoints of the link.

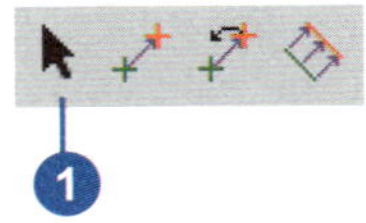

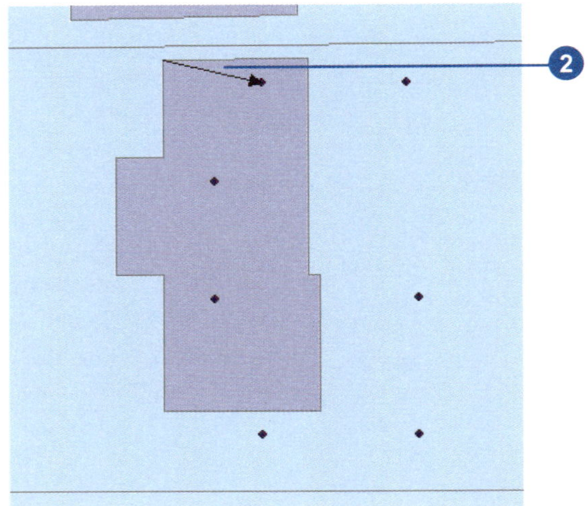

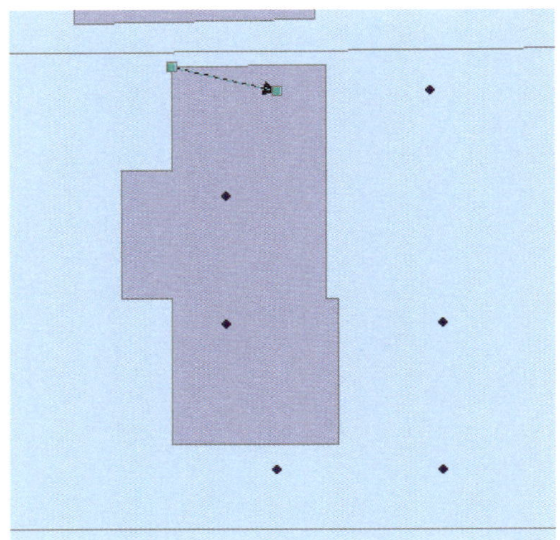

Tip

Use the Shift key to select more than one link at a time

You can select multiple links by holding down the Shift key while clicking links with the Select Elements tool.

Tip

Use the Select All Elements command

You can select all the links in the map by using the Select All Elements command. Click the Edit menu and click Select All Elements.

Selecting multiple links

1. Click the Select Elements tool on the Spatial Adjustment toolbar.
2. Drag a box around the links you want to select.

   Selection grips will appear at the endpoints of the links.

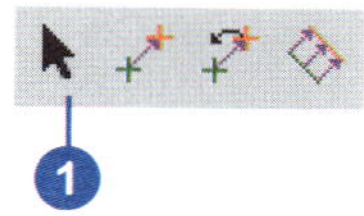

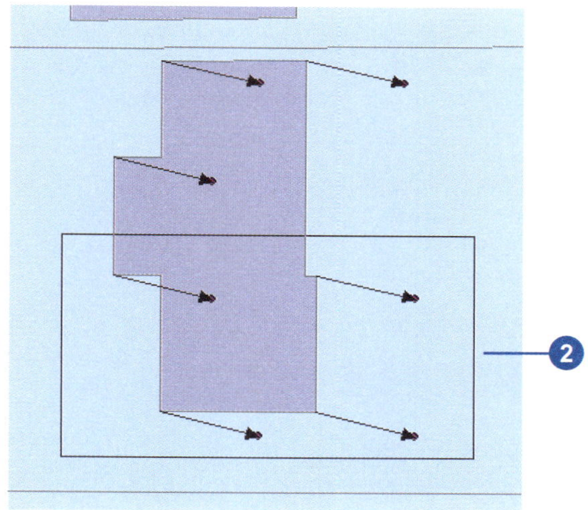

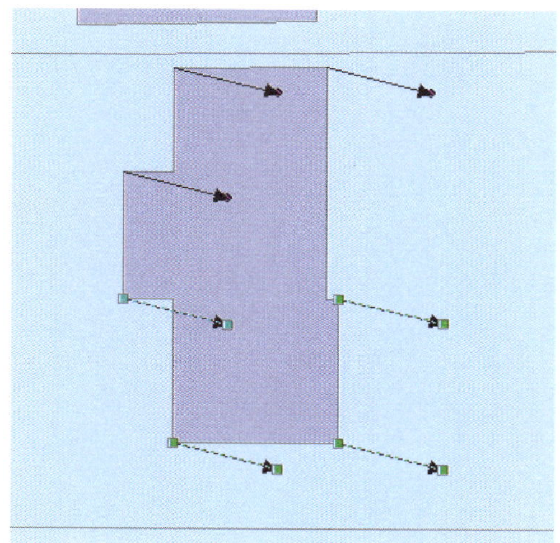

Modifying displacement links

Displacement links can be modified by using the Modify Link tool. They must be selected prior to modifying them. Links can be modified both inside or outside of an edit session. You can modify links at their source or destination location or move the entire link to a new location.

Tip

Use the Select Elements tool to modify identity links

You can modify identity links with the Select Elements tool. Simply click an identity link and drag it to a new location.

1. Click the Select Elements tool on the Spatial Adjustment toolbar.

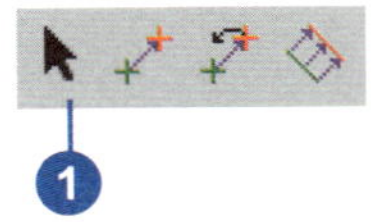

2. Position the cursor over the link you want to modify and click once.

   Selection grips will appear at the endpoints of the link. ►

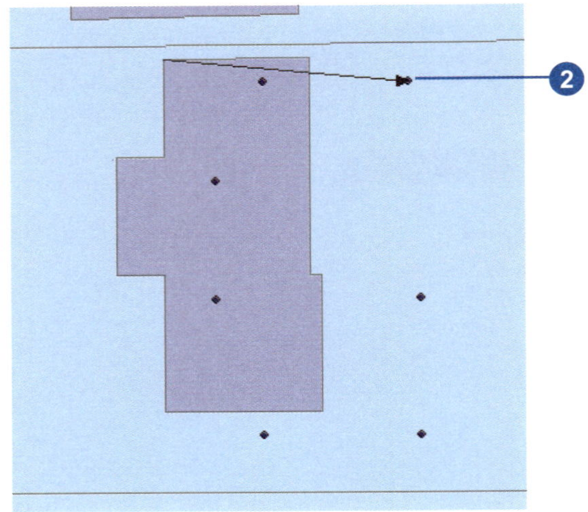

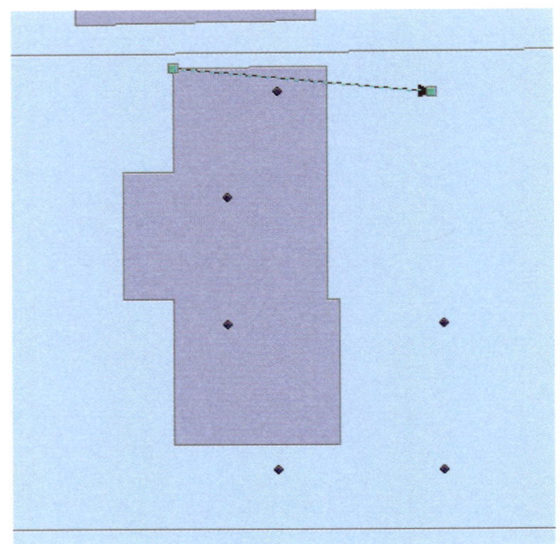

3. Click the Modify link tool on the Spatial Adjustment toolbar.
4. Position the cursor over the source or destination point of the link.

   The link cursor changes to an arrow cursor.
5. Move the endpoint of the link to the desired location.

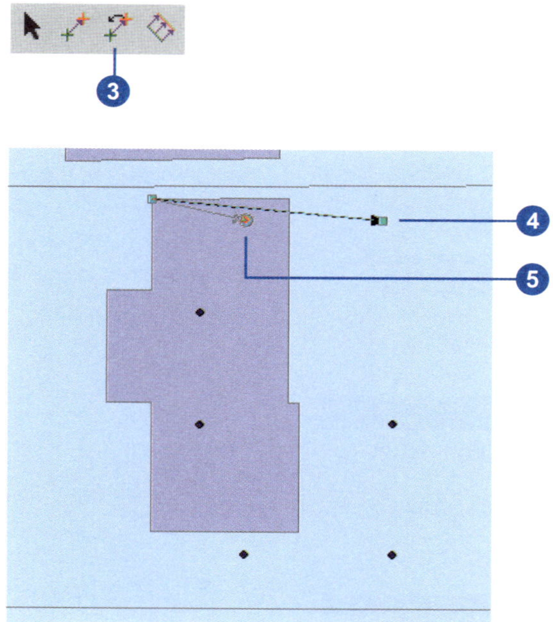

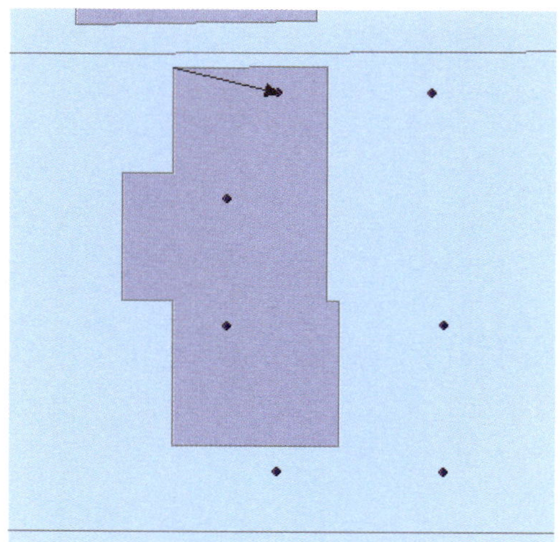

Deleting displacement links

Displacement links can be deleted by using the Delete command or pressing the Delete key. Links must first be selected in order to delete them. Links can be deleted both inside or outside of an edit session.

Tip

Use the Delete command to remove links

Click the Edit menu and click Delete to remove links.

Tip

Delete links with the Link Table

You can also delete links with the Link Table. Right-click a link record to open the context menu and click Delete link(s) to open the link table; see 'Viewing the Link Table' later in this chapter.

Deleting a link

1. Click the Select Elements tool on the Spatial Adjustment toolbar.
2. Position the cursor over the link you want to delete and click once.

   Selection grips will appear at the endpoints of the link.
3. Press the Delete key.

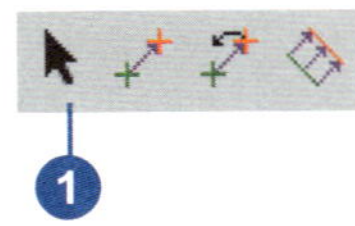

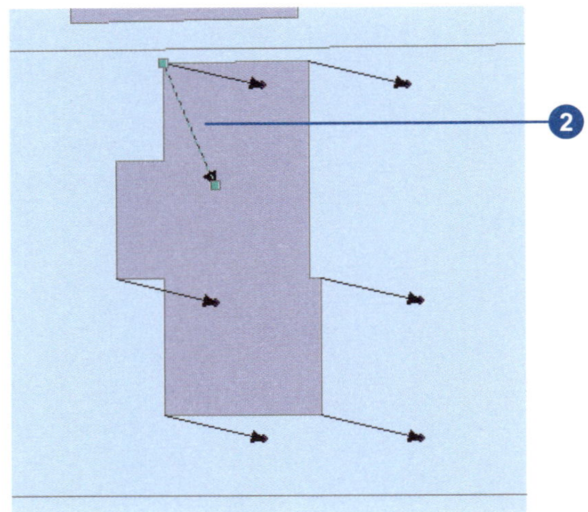

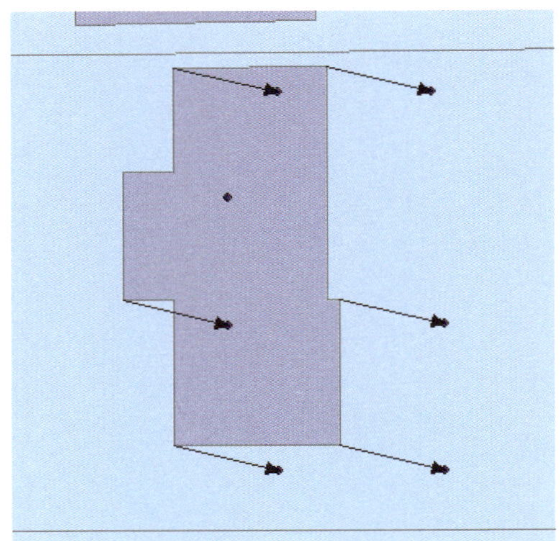

Deleting multiple links

1. Click the Select Elements tool on the Spatial Adjustment toolbar.
2. Drag a box around the links you want to delete.

   You can also select multiple links by holding down the Shift key while selecting links.
3. Press the Delete key.

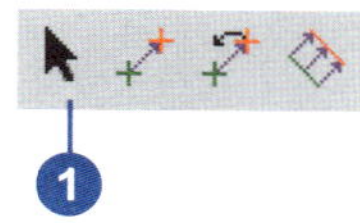

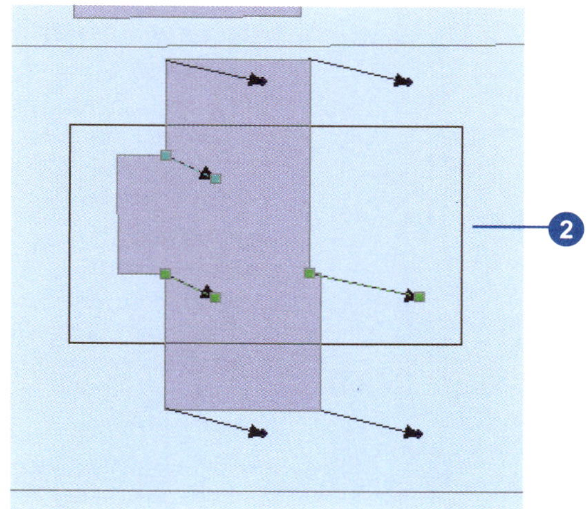

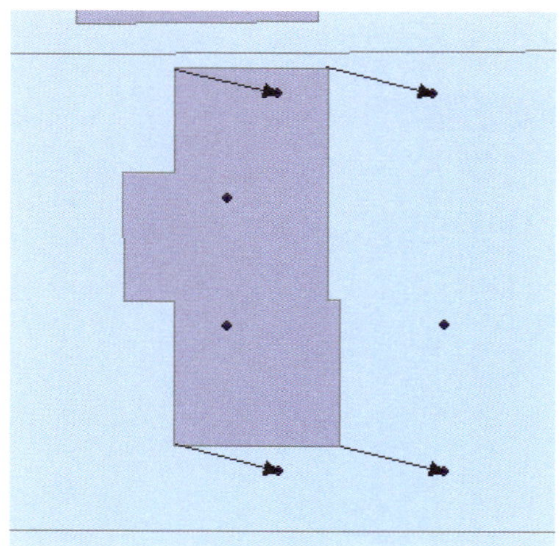

Viewing the Link Table

The Link Table displays displacement links in a tabular format. This table displays the source and destination coordinates of the links, the link IDs, and the residual error of the adjustment (residual errors will only be displayed for Transformation adjustments).

You can select links by clicking a row in the table. Coordinate values can be edited for the selected links. You can select multiple links by holding down Shift while selecting rows. When a link is selected, right-click to open the Link Table context menu. This menu supports commands to flash links, pan and zoom to links, and delete links.

Tip

Open the Link Table from the toolbar

You can also open the Link Table by clicking the View Link Table button (shown below), located on the Spatial Adjustment toolbar.

1. Click the Spatial Adjustment menu, point to Links, and click View Link Table.

   The Link Table dialog box appears.

2. Click a row in the link table to highlight a link.
3. With a link record highlighted, you can edit the coordinates of the link or delete the link by clicking Delete Link.
4. Right-click the highlighted link to access the Link Table context menu. You can pan to the link, zoom to the link, select the link, and delete the link with the commands offered in this menu.
5. When finished working with the Link Table, click Close to close the window.

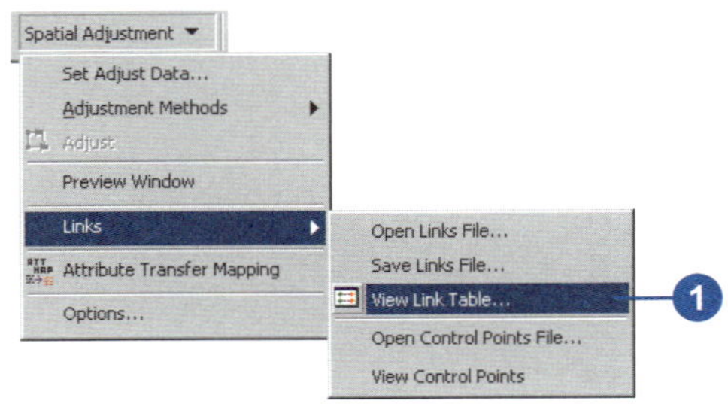

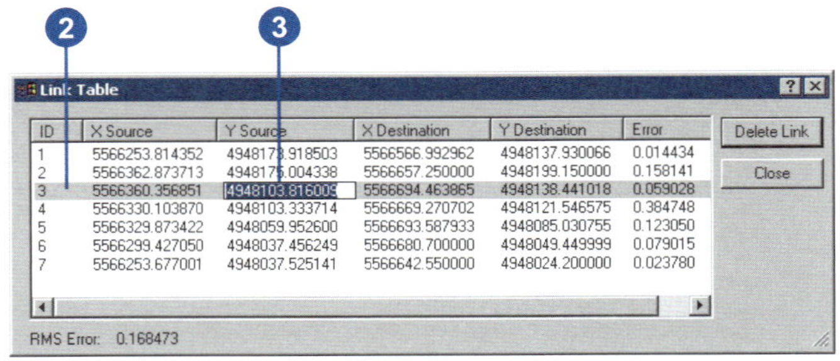

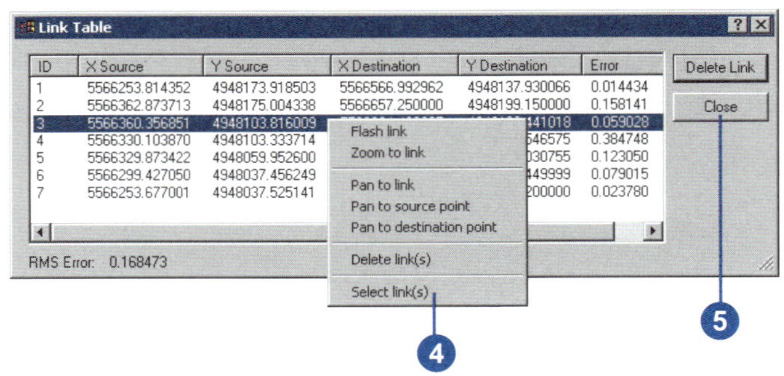

Opening a link file

Link files are text files that contain source and destination coordinates that define an adjustment. The Spatial Adjustment tool supports tab-delimited link files that contain either four or five columns. A four-column link file consists of two pairs of source and destination coordinate values. A five-column link file consists of an ID column (string or numeric) that precedes two pairs of source and destination coordinate values.

When you open a link file, the Spatial Adjustment tool automatically creates displacement links in the map based on the source and destination coordinate values in the file. Link files can help you save time by automating the link creation process.

1. Click the Spatial Adjustment menu, point to Links, and click Open Links File.

   The Open dialog box appears.

2. Click the Look in dropdown arrow and navigate to the folder where the link file resides.
3. Double-click the link file to load it.

   Displacement links will be automatically created in the map.

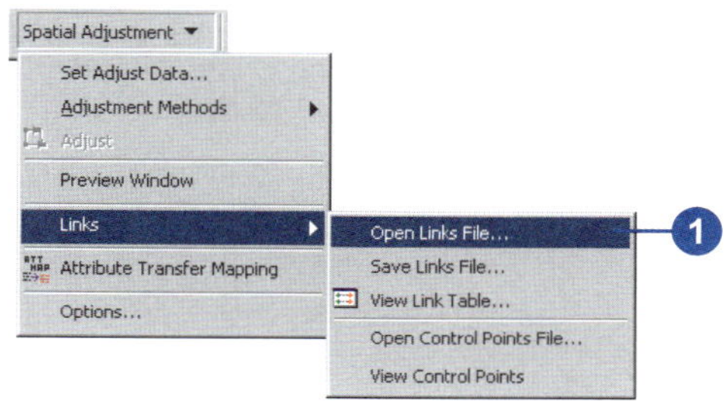

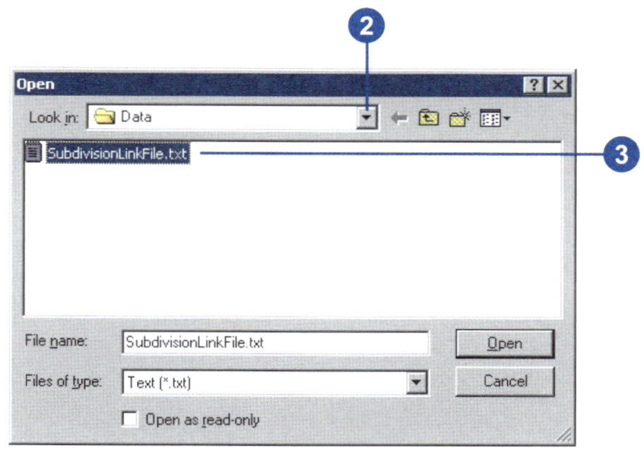

Saving a link file

You can create a link file from the existing displacement links in the map using the Save Links File command. This command opens a Save dialog box and allows you to navigate to the folder of your choice and name the new link file. You also have the option to save link IDs. The link file is saved as a tab-delimited text file.

1. Click the Spatial Adjustment menu, point to Links, and click Save Links File.

   The Save Links dialog box appears.
2. Click the Save in dropdown arrow and navigate to the folder where you want to save the link file.
3. Enter a name for the link file.
4. Click Save.
5. Choose whether or not to save link IDs.

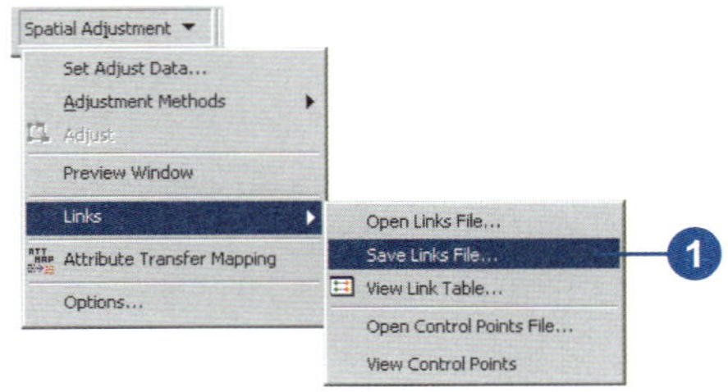

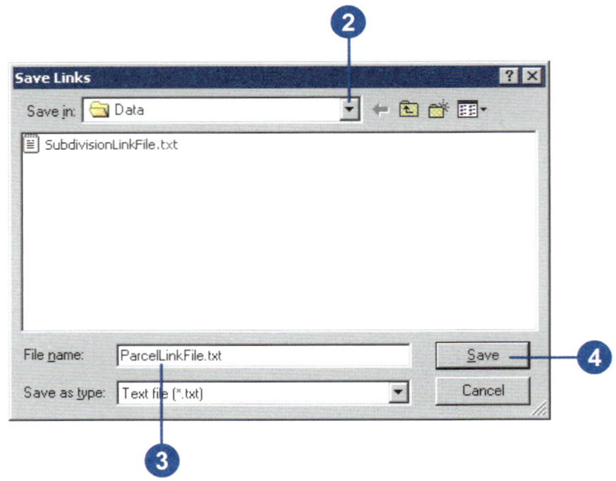

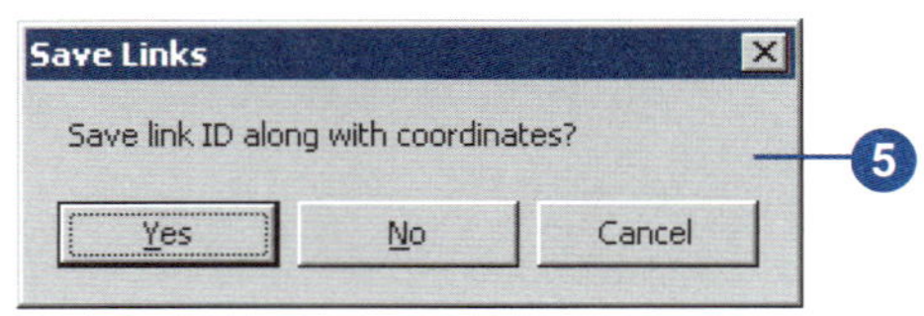

Opening a Control Point file

Control Point files are text files that contain destination coordinates that define part of an adjustment. The Spatial Adjustment tool supports tab-delimited Control Point files that contain either two or three columns. A two-column Control Points file consists of a pair of destination coordinate values. A three-column Control Points file consists of an ID column (string or numeric) that precedes a pair of destination coordinate values. Control points may represent known locations of features from GPS or ground survey and are displayed in the Control Points Window.

1. Click the Spatial Adjustment menu, point to Links, and click Open Control Points File.

   The Open dialog box appears.

2. Click the Look in dropdown arrow and navigate to the folder where the control point file resides.

3. Double-click the control point file to load it.

   The Control Points Window appears.

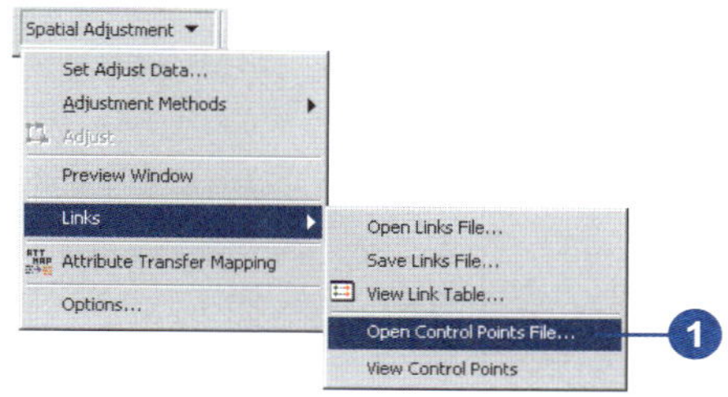

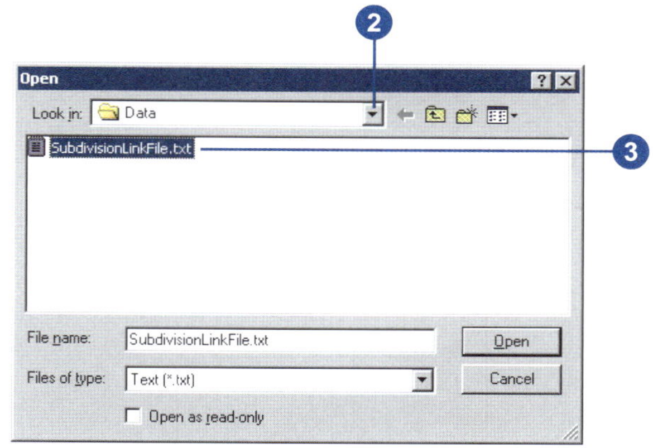

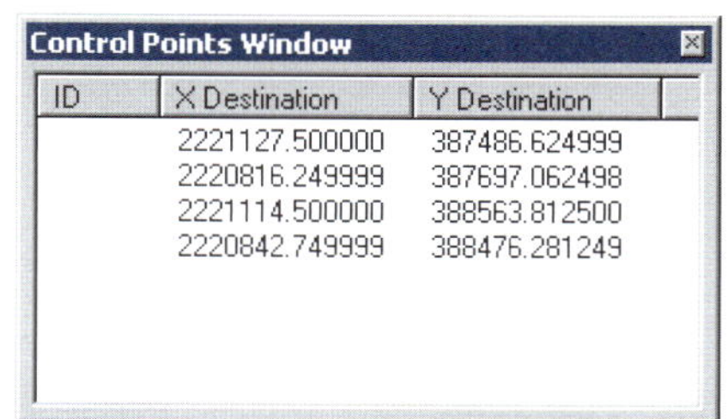

Creating displacement links from control points

Opening a control point file does not automatically create displacement links as in the case of link files. You must manually create the links from the control points. This requires that you open the Control Points Window with the View Control Points command, then double-click each row to create a destination link in the map. Once the link is created at the destination location, you must then finish adding the link to the source location. This will connect the source feature to the target feature. Repeat this process until all the control point rows are removed from the Control Points Window.

1. Click the Spatial Adjustment menu, point to Links, and click View Control Points.

   The Control Points Window appears.

2. Double-click a row in the Control Points Window.

   This will create a link that is snapped to a destination location. ►

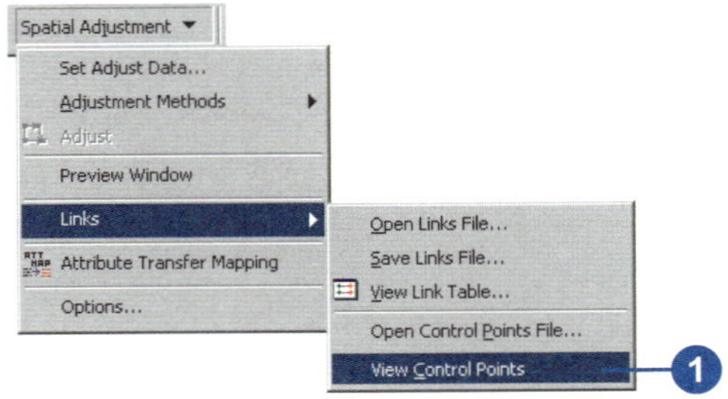

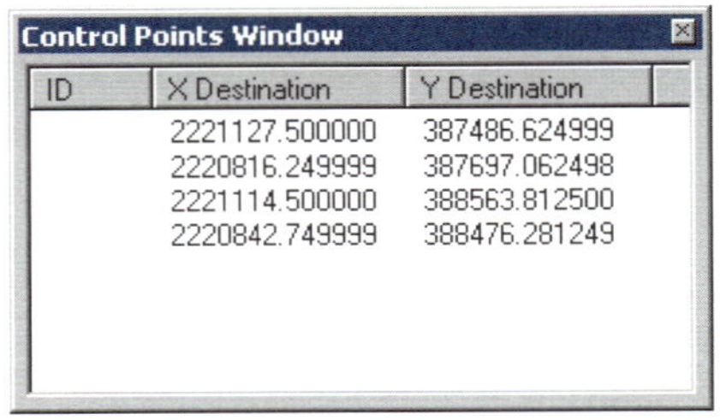

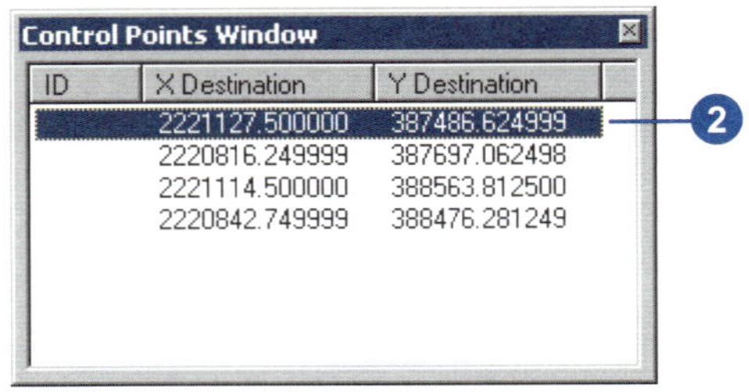

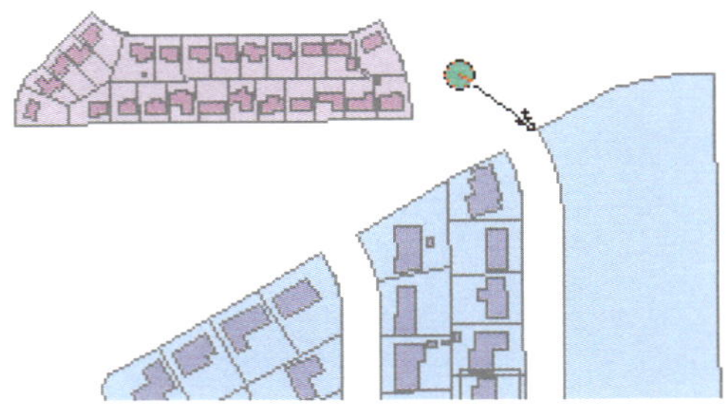

3. Snap the link to a source location.

   The row is now removed from the Control Points Window.

4. Repeat steps 2 and 3 until all the rows in the Control Points Window are removed and converted into displacement links.

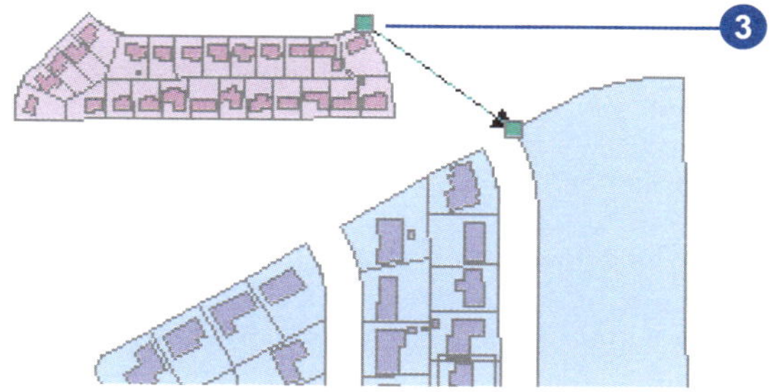

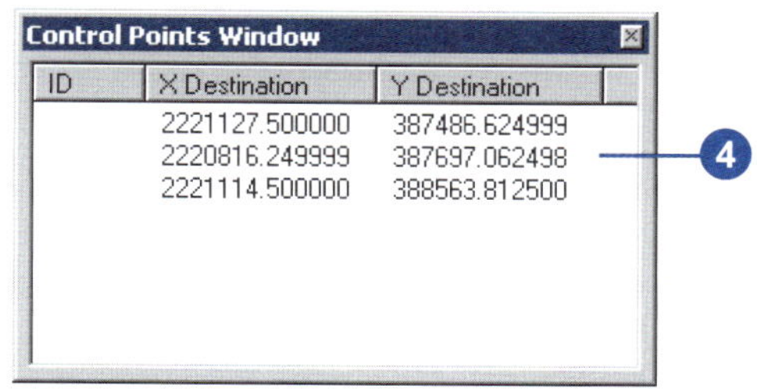

Previewing the adjustment

You can preview an adjustment using the Adjustment Preview Window. This window allows you to view the results of an adjustment prior to performing the adjustment in the map. You can use standard ArcMap Zoom and Pan commands in this window to closely examine how the adjustment will affect your features. This allows you an opportunity to go back in the map and make modifications before you adjust your data. Additionally, the Adjustment Preview Window supports its own display commands from a context menu, such as zooming to the dataframe's extent, and tracking the data frame's extent.

The Adjustment Preview Window can help you save time and unnecessary edits by giving you a glimpse of how the adjustment will turn out.

1. Click the Spatial Adjustment menu and click Preview Window.

   The Adjustment Preview Window appears.

2. Examine the adjustment more closely by using the Zoom and Pan commands on the standard map display toolbar.

3. You can access additional commands by right-clicking inside the Adjustment Preview Window to open the context menu.

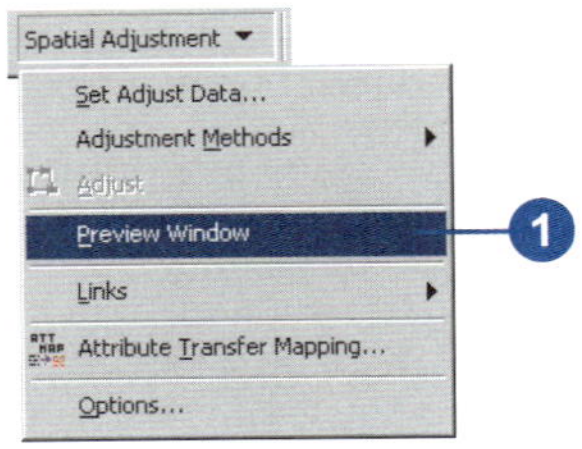

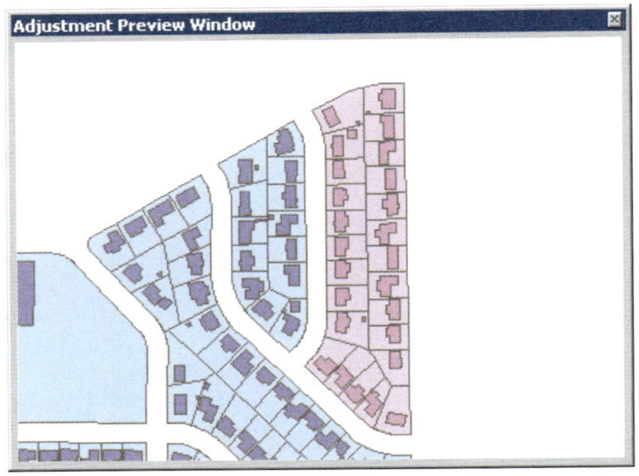

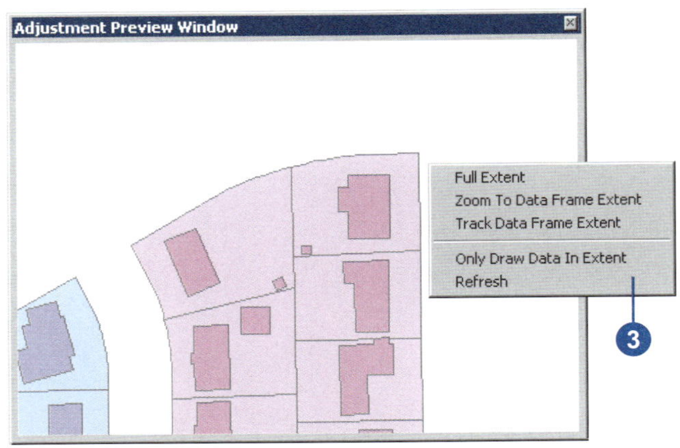

Performing the adjustment

Once you have chosen which data to adjust, selected an adjustment method, set the adjustment properties, and created links and limited adjustment areas, you can adjust the data. Clicking the Adjust command will execute the spatial adjustment.

Tip

Use the Undo command to undo an adjustment

All adjustments can be undone by clicking the Undo button:

1. Click the Spatial Adjustment menu and click Adjust.

   The data is now adjusted.

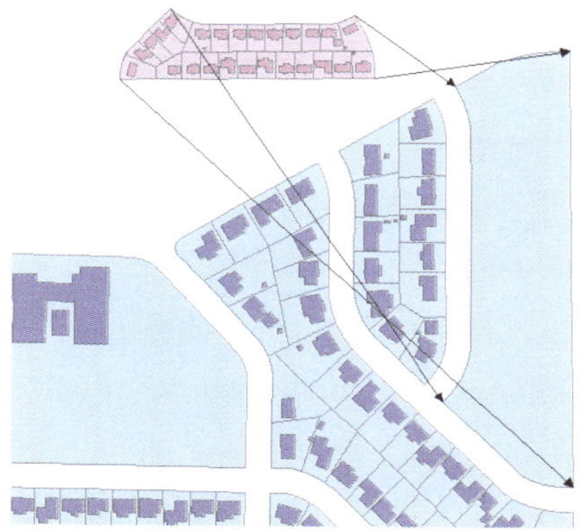

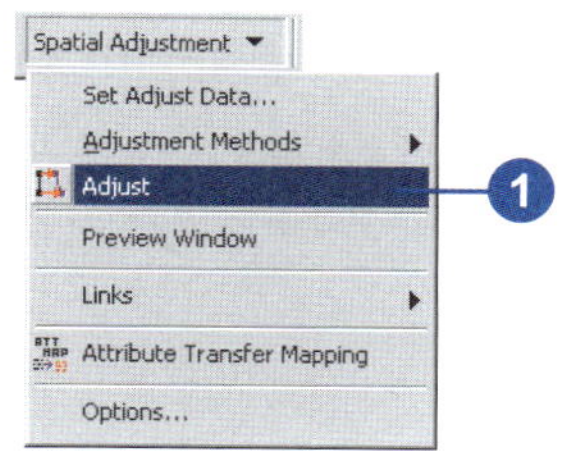

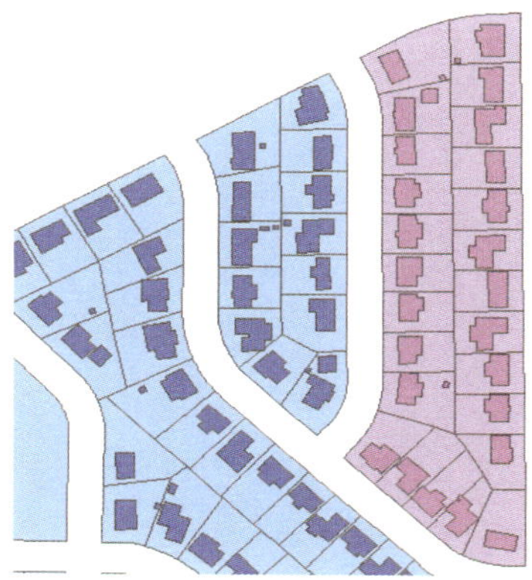

Attribute Transfer Mapping

The Attribute Transfer Mapping functions support the interactive transfer of attributes between features. The Attribute Transfer Mapping dialog box allows you to set the source and target layers and specify which fields to use as criteria for the attribute transfer. Once you have identified the common fields in the source and target layers, they are matched. These matched fields define which attributes are transferred when using the Attribute Transfer tool. You also have the option to transfer the geometry of the feature by checking the Transfer Geometry check box.

Tip

Using the Auto Match command to match multiple fields at once

You can use the Auto Match command to match multiple fields at once based on common field names.

1. Click the Spatial Adjustment menu and click Attribute Transfer Mapping.

   The Attribute Transfer Mapping dialog box appears.

2. Click the Source Layer dropdown arrow and choose a layer.
3. Click the Target Layer dropdown arrow and choose a layer.
4. Click a field in the Source Layer's field list box. ▸

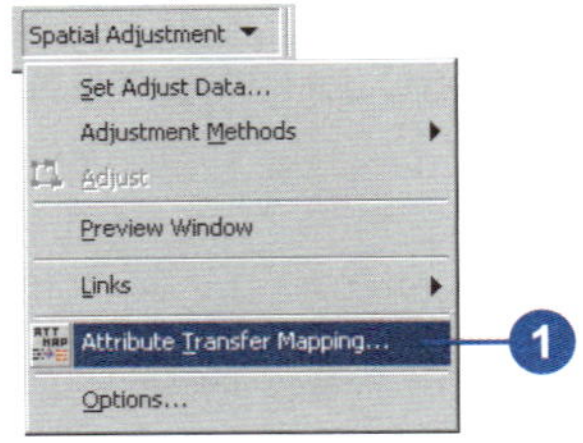

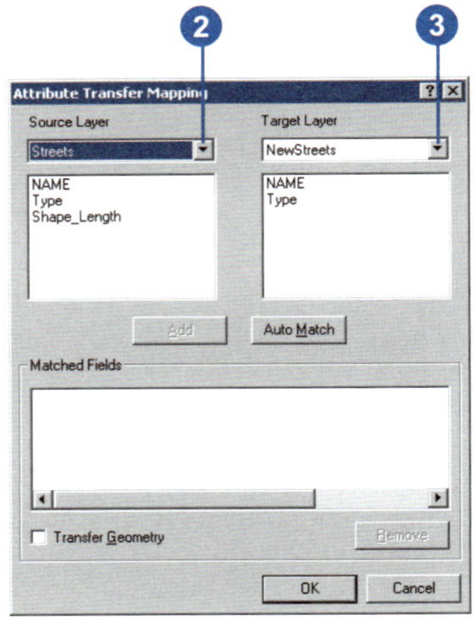

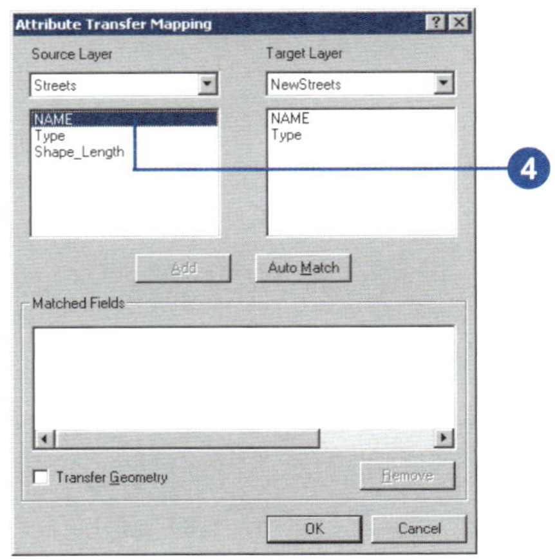

5. Click a corresponding field in the Target Layer's field list box.

   Both fields will be highlighted.

6. Click Add.

   Repeat the process for all other fields that are to be used as criteria for the attribute transfer.

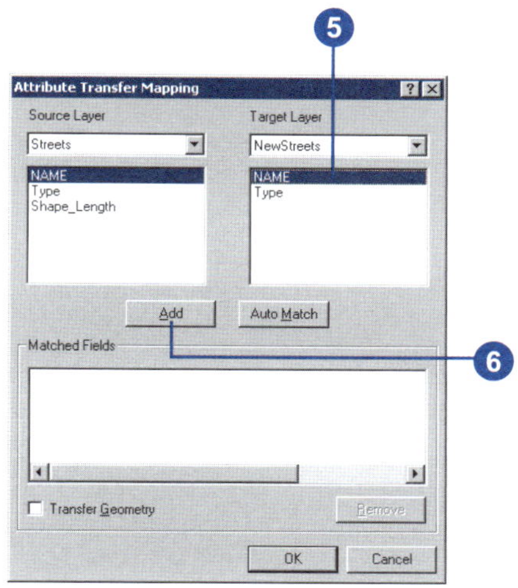

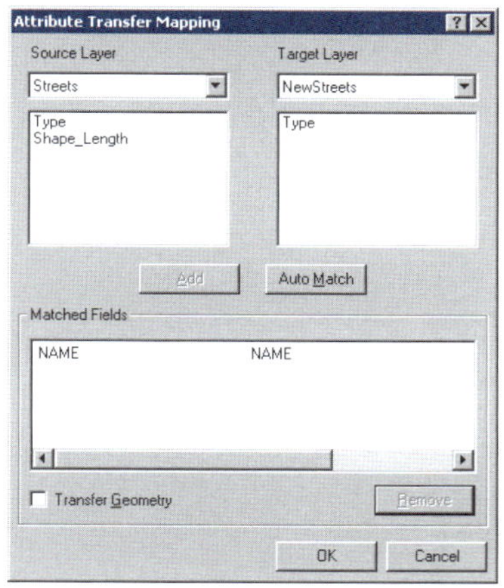

Using the Attribute Transfer tool

The Attribute Transfer tool allows you to transfer the attributes of a feature to another feature. This transfer is performed by matching the fields that are specified in the Attribute Transfer Mapping dialog box.

Use the Attribute Transfer tool to select a source feature followed by a target feature. When the transfer is complete, verify the target feature's attributes using the Identify tool.

Tip

Transfer attributes to multiple features

You can transfer attributes to multiple features by holding down the Shift key while selecting the target features.

See Also

For information on using the Attribute Transfer Mapping dialog box, see 'Attribute Transfer Mapping' in this chapter.

1. Click the Attribute Transfer tool on the Spatial Adjustment toolbar.

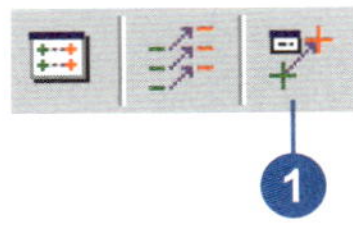

2. Position the cursor over the source feature and click once.

   This is the feature that contains the desired attribute data.

3. Position the cursor over the target feature and click once to transfer the attribute data of the source feature.

   The target feature is now updated with the source feature's attribute data.

Editing attributes

9

IN THIS CHAPTER

- **Viewing attributes**
- **Adding and modifying attributes**
- **Copying and pasting attributes**

Creating and editing features in a GIS usually entails creating or editing some *attributes* of the features, in addition to creating their shape.

ArcMap makes it easy to view and update the attributes of features in your database. You can edit feature attributes in two ways: using the Attributes dialog box or using a feature layer's attribute table. This chapter focuses on editing attributes using the Attributes dialog box. With the Attributes dialog box, you can view the attributes of selected features on your map; add, delete, or modify an attribute for a single feature or multiple features at the same time; and copy and paste individual attributes or all the attributes of a feature.

You can perform similar functions using a feature layer's attribute table. However, with tables you can also do computations—such as adding and sorting records—with attribute values. To learn how to edit attributes in an attribute table—including performing computations with attribute values—see *Using ArcMap*.

In the next chapter you will learn how to take advantage of the tight integration of ArcMap with Geodatabases to make editing attributes quicker and more accurate.

Viewing attributes

The Attributes dialog box lets you view the attributes of features you've selected in your map. The left side of the dialog box lists the features you've selected. Features are listed by their primary display field and grouped by layer name. The number of features selected is displayed at the bottom of the dialog box.

The right side of the Attributes dialog box is called the property inspector. The property inspector contains two columns: the attribute properties of the layer you're viewing, such as Type or Owner, and the values of those attribute properties.

Tip

Finding the feature on the map

You can find a selected feature on the map by either highlighting or zooming to it. To highlight the feature, click the primary field and the feature will flash on the map. Right-click the field and click Zoom To in the context menu to get a close-up view of the feature. Click the Back button on the Tools toolbar to return to the previous map extent.

1. Click the Edit tool.
2. Select the features whose attributes you want to view.
3. Click the Attributes button.
4. Click the layer name that contains the features whose attributes you want to view.

   The layer's attribute properties appear on the right side of the dialog box. ►

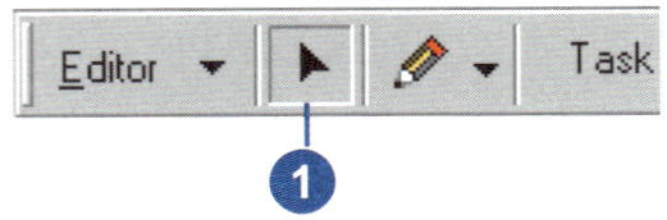

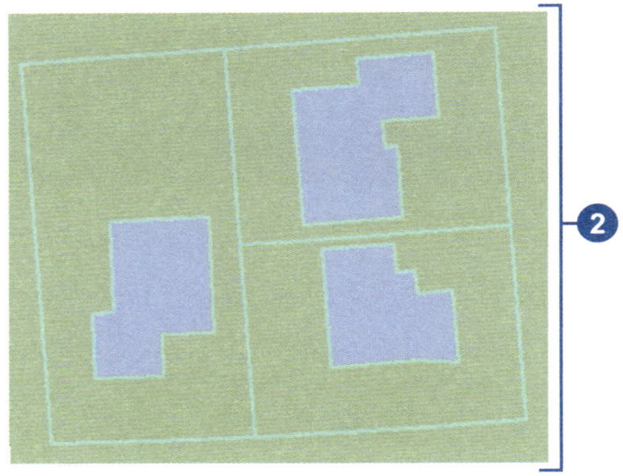

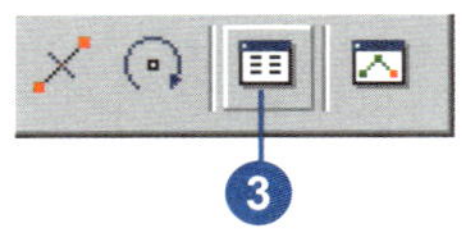

The layer's attribute properties appear.

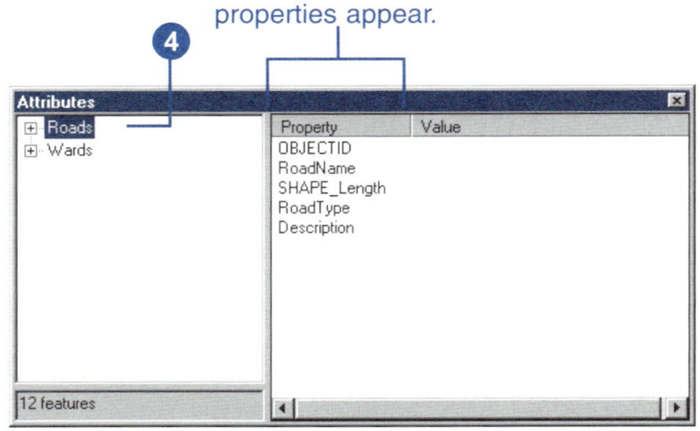

Tip

Unselect a feature

You can remove features from the selection without having to click on the map. To remove a feature from the selection, right-click on the feature and click Unselect from the context menu.

Tip

Delete a selected feature

If you want to delete a feature without losing your selection, simply right-click on the feature and click Delete from the context menu.

Tip

Viewing attributes in the Identify Results window

To view the attributes of a feature quickly, click the Identify Features button on the Tools toolbar, then click the feature whose attributes you want to view. View the feature's attributes in the Identify Results window.

Tip

Changing the primary display field

You can change the primary display field for a layer on the Fields tab of the layer's Properties dialog box. To open the dialog box, right-click the layer name in the table of contents.

5. Double-click a layer name to see the primary display fields, representing the selected features in the layer.

   Double-click again to hide the primary display fields.

6. Click a primary display field to see the corresponding feature's attribute values.

   The corresponding feature flashes on the map.

7. Click the Close button to close the dialog box.

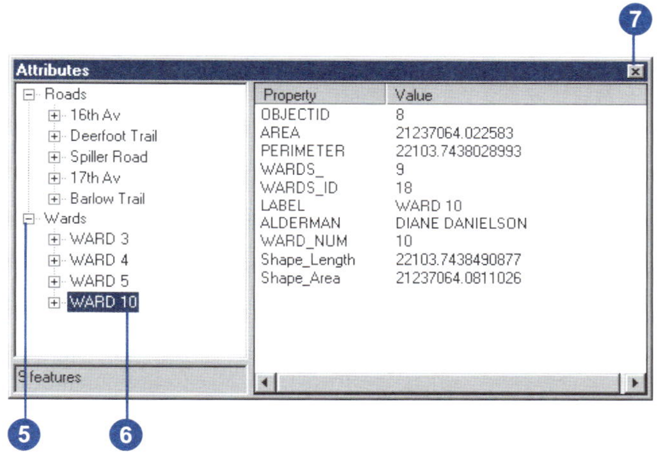

Adding and modifying attributes

The easiest way to make changes to the attributes of a selected feature is by using the Attributes dialog box.

You can add or modify attributes of selected features as needed. For example, you might want to update the attribute values—such as its name and maintenance information—for a park feature you created.

To add or modify an attribute value for a single feature, click the primary display field for the feature on the left side of the dialog box and make your changes in the Value column on the right. ▸

Tip

Saving your edits

Click the Editor menu and click Save Edits.

Tip

Attribute domains

You can use attribute domains to create a list of valid values for a feature in a geodatabase. You can also use the Validate command to ensure attribute quality. For more information, see Building a Geodatabase.

Adding an attribute value to a single feature

1. Click the primary display field of the feature to which you want to add an attribute value.
2. Click in the Value column where you want to add the attribute value.
3. Type the attribute value and press Enter.

   The attribute value is added to the feature.

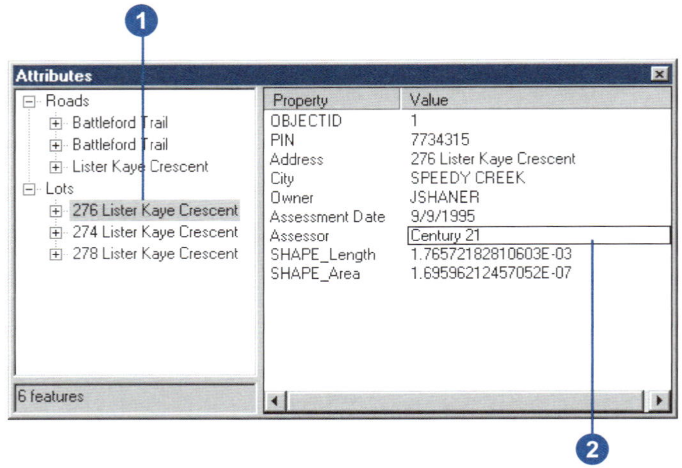

Adding an attribute value to all selected features in a layer

1. Click the layer to which you want to add an attribute value.
2. Click in the Value column where you want to add the attribute value.
3. Type the attribute value and press Enter.

   The attribute value is added to all selected features in the layer.

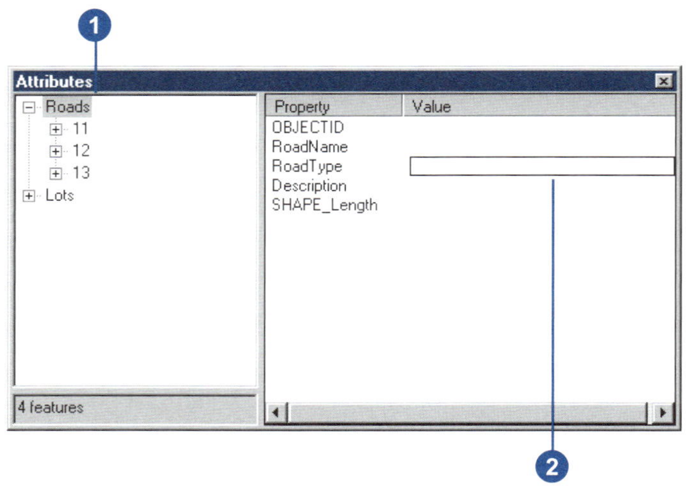

You can also add or modify an attribute value for all selected features in a layer at the same time. Simply click the layer name on the left and make your changes in the Value column on the right.

Tip

Deleting attributes

To delete an attribute value, right-click over the value and click Delete. You can also press the Delete key to delete an attribute value.

Tip

Undoing your edits

To undo any edit to feature attributes, click the Undo button on the ArcMap Standard toolbar.

Tip

Performing calculations

When editing attributes, you might need to perform calculations using the field calculator in the feature layer's attribute table dialog box. For more information, see Using ArcMap.

Tip

Adding attribute properties

You can add an attribute property for a feature by working with its attribute table in ArcCatalog. For more information, see Using ArcCatalog.

Modifying an attribute value for a feature

1. Click the primary display field of the feature for which you want to modify an attribute value.
2. Click the value you want to modify.
3. Type a new attribute value and press Enter.

   The attribute is modified for the feature.

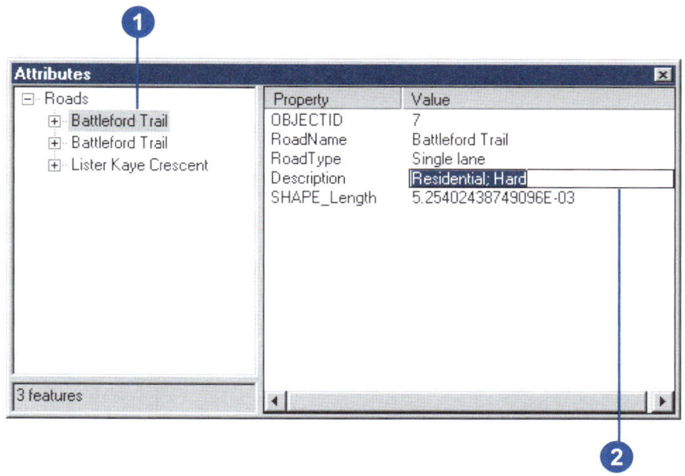

Modifying an attribute value for all selected features in a layer

1. Click the layer for which you want to modify an attribute value.
2. Click in the Value column next to the attribute property you want to modify for all selected features in the layer.
3. Type a new attribute value and press Enter.

   The attribute is modified for all selected features in the layer.

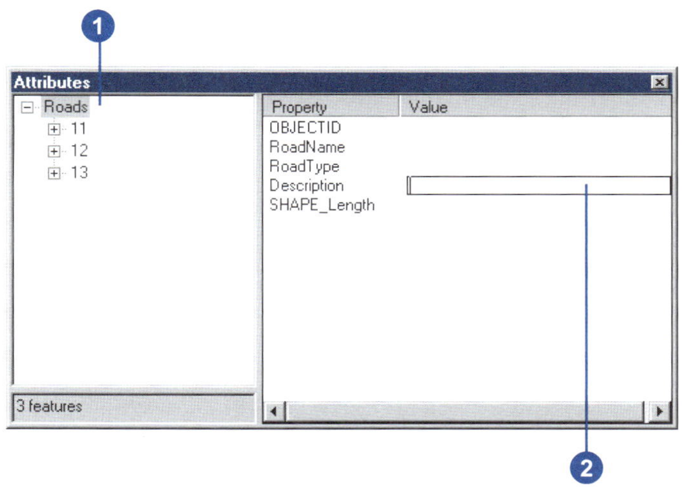

Copying and pasting attributes

Copying and pasting is an easy way to edit the attributes of features on your map. You can copy individual attribute values or all the attribute values of a feature. Attribute values can be pasted to a single feature or to all selected features in a layer.

Tip

Copying and pasting individual attribute values to an entire layer

To copy an attribute value to a layer, click the value you want to copy, right-click, and click Copy. Then, click the layer name and right-click in the Value column next to the appropriate property. Click Paste and the attribute value is copied to every selected feature in the layer.

Tip

Cutting and pasting attributes

Cutting and pasting attributes is similar to copying and pasting them. Right-click and click Cut from the context menu to remove the attribute value from its current location in the Attributes dialog box, then click Paste to paste it elsewhere.

Copying and pasting individual attribute values from feature to feature

1. Click the attribute value you want to copy.
2. Right-click the value you want to copy and click Copy.
3. Click the primary display field of the feature to which you want to paste the value.
4. Click where you want to paste the value.
5. Right-click where you want to paste the value and click Paste.

   The attribute value is pasted to the feature.

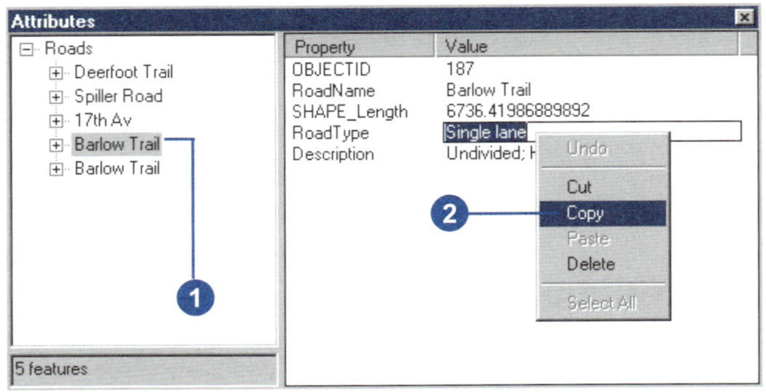

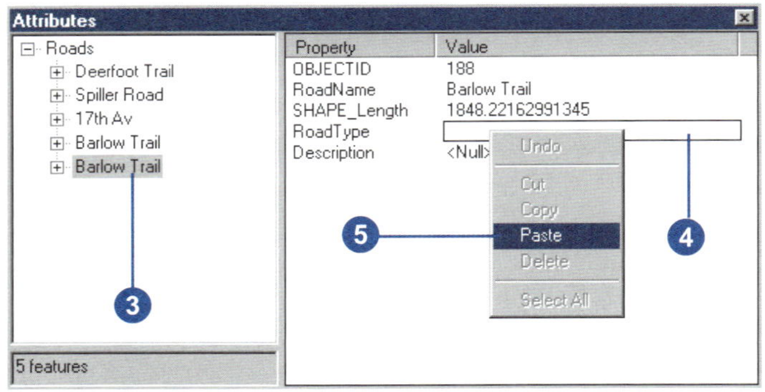

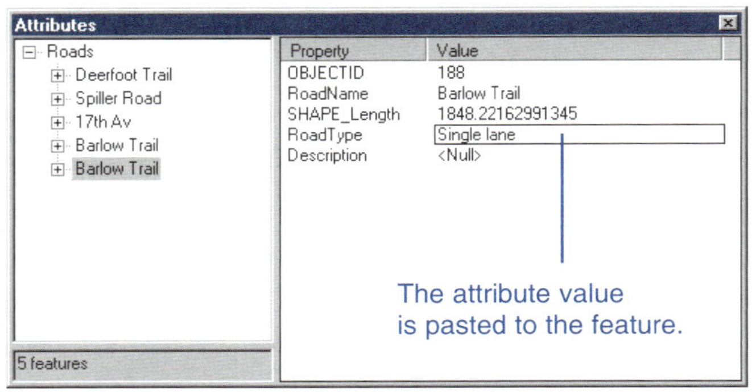

Tip

Copying and pasting all attribute values from one feature to an entire layer

You can copy all attribute values from one feature to all selected features in a layer. Right-click the primary display field of the feature whose attribute values you want to copy and click Copy. Right-click the layer name to which you want to paste the attribute values. Click Paste and the attribute values are copied to every selected feature in the layer.

Copying and pasting all attribute values from feature to feature

1. Right-click the primary display field of the feature whose attribute values you want to copy and click Copy.
2. Right-click the primary display field of the feature to which you want to paste the attribute values and click Paste.

   The attribute values are pasted to the feature.

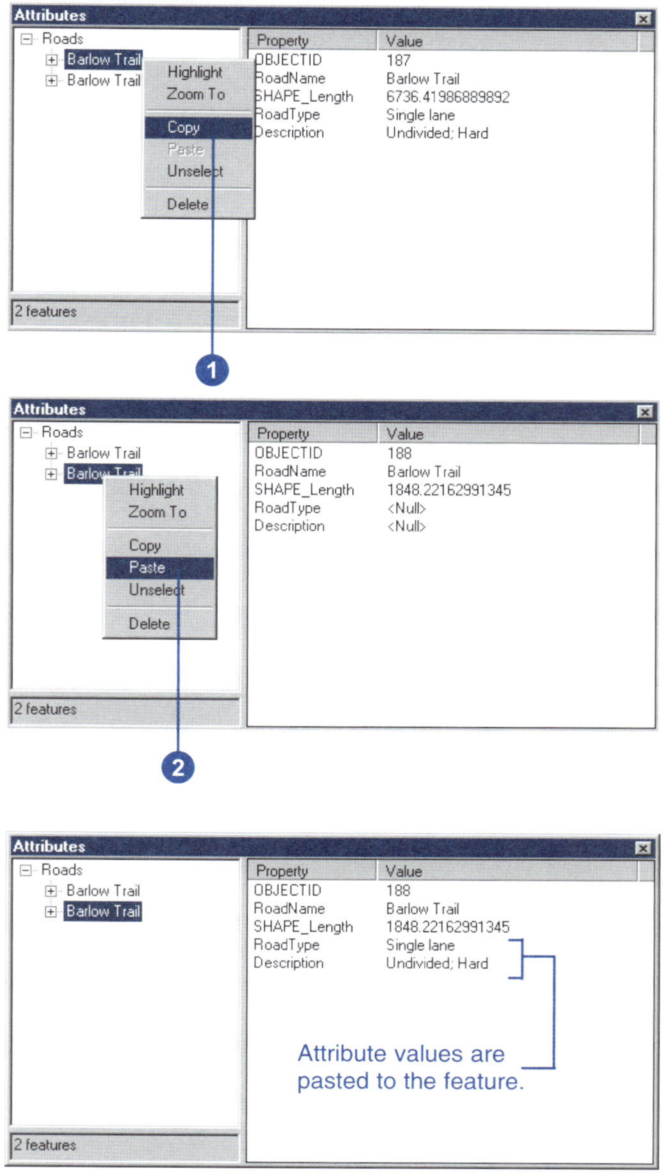

Editing geodatabase attributes 10

IN THIS CHAPTER

- **Editing a geodatabase with ArcMap**
- **Editing features with subtypes and default values**
- **Editing attribute domains**
- **Validating features**

Some features in a geodatabase are designed with subtypes, default values, and attribute domains. These can make it quicker and easier to edit feature attributes and can help prevent data entry errors.

Subtypes are logically distinct categories of a given type of feature that may have different attributes, network roles, or topology rules. When you create a new feature in a feature class with subtypes, you can choose which subtype of feature to create. For example, when creating a new building footprint, you might get the choice of residential, high-density residential, mobile home, commercial, industrial, school, and public administration building subtypes.

Because subtypes represent logical groups within a class of feature, they may typically have different attributes. In addition to a code identifying the subtypes, each subtype might have codes identifying how the data was collected. Residential buildings might get a default source description of 'digitized from plats', while mobile homes might get a default source description of 'digitized from ortho photographs'. These might be two out of a list of five permissible source descriptions.

Attributes in a geodatabase can have rules that specify that their values must fall within a particular range or be one of a list of permissible values. Building features might all have an occupancy figure for emergency planning. Residential buildings might have a valid occupancy range from 0–15, while commercial buildings might have a valid range of 0–500.

After you enter attributes for features that have domains specified in the geodatabase, you can validate your edits to check that the attributes fall in the permissible domain.

Editing a geodatabase with ArcMap

ArcMap editing capabilities are tightly integrated with the various aspects of the geodatabase, such as *geometric networks* and validation rules. While each component of the geodatabase can act independently, the true power of the geodatabase becomes evident when you bring all of these things together.

In this chapter, you will learn how editing in ArcMap takes advantage of the aspects of a geodatabase that help you maintain a valid *database*.

Validation rules

The geodatabase supports several broad types of validation rules: attribute validation rules, network *connectivity rules,* and relationship rules. It is important to understand that these validation rules can be broken; in certain cases, a geodatabase permits invalid *objects* to be stored in the database.

For example, if you have an attribute rule stating that the valid pressure range for a water distribution main in your water network is between 50 and 75 psi, the geodatabase won't prevent you from storing a value outside that range. However, a distribution main with a water pressure outside of this range will be an invalid object in the geodatabase. ArcMap has many editing tools that help you identify invalid features so that you can correct them.

The exceptions are *edge–edge connectivity rules*, *edge–junction connectivity rules*, and coded value attribute rules. In these cases, ArcMap takes a more active role when editing features with these rules associated with them. You will learn how editing in ArcMap behaves in these contexts later in this chapter.

The general approach to the issue of validating features is that the validation process should not result in valid features being flagged as invalid (false negatives); it is, however, allowable to have features that are invalid being reported as valid (false positives). If the geodatabase did not enforce any validation, every feature would effectively be valid. When performing validation on a particular feature, the validation occurs in five steps:

1. Validate the subtype.
2. Validate the attribute rules.
3. Validate the network connectivity rules—if network feature.
4. Perform custom validation—using optional class extension.
5. Validate the relationship rules.

This strategy means the least expensive validation is performed first. The validation process stops once a feature is found to be invalid. So, for example, if a feature fails the validity test for check number 1, then checks 2, 3, 4, and 5 are never executed.

When checking connectivity and relationship rules, all associated rules must be valid. With network connectivity rules, if you specify one rule, you must specify them all. Thus, if a type of connectivity exists that doesn't have an associated connectivity rule, the network feature is deemed invalid.

In addition to these rules, topology rules can also be established. To learn more about topology, see Chapter 4, 'Editing topology'.

To learn more about attribute validation rules, see *Building a Geodatabase*. To learn more about connectivity rules, see Chapter 12, 'Editing geometric networks' in this book, or *Building a Geodatabase*.

Editing features with subtypes and default values

Feature classes and individual subtypes of feature classes in a geodatabase can have default values.

Default values help streamline the attribute editing process and help you maintain realistic values for the attributes of the features in your database. If most of the building features that you create in a feature class are residential, you might have a default value of residential for the building type attribute. Then when you create a new building it will automatically be given the residential attribute value. In the few cases where the building is another type, you can change the attribute.

If your feature class contains subtypes, then when you change the subtype of a feature, the feature takes on the default field values of the new subtype.

Creating new features of a subtype

1. Click the Task dropdown arrow and click Create New Feature.
2. Click the Target layer dropdown arrow and click the layer with the type of features you want to create.
3. Click the Tool Palette dropdown arrow and click the Sketch tool.
4. Click the map to digitize the feature's vertices. ▸

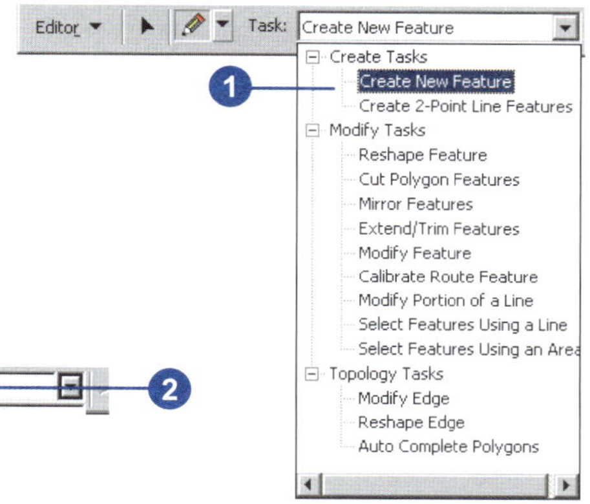

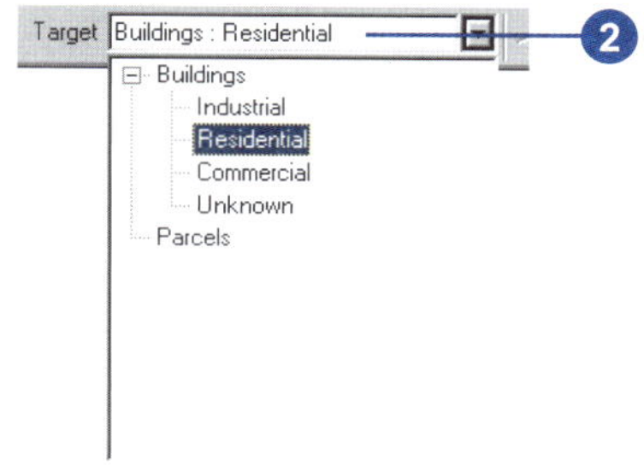

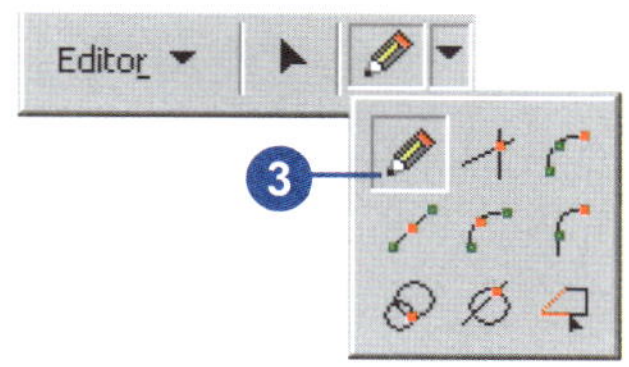

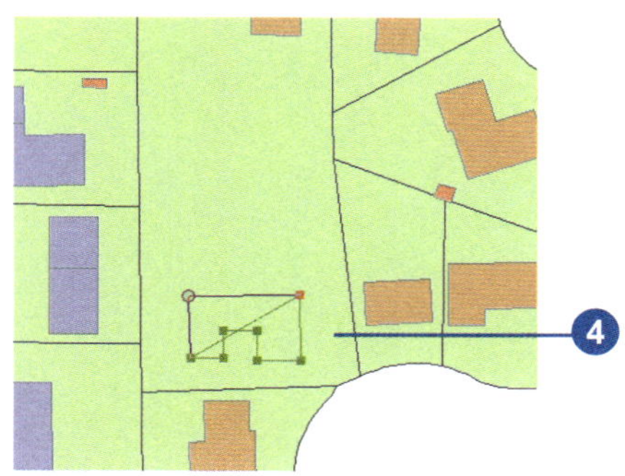

5. Double-click the last vertex to finish the feature.
6. Click the Attributes button.

   The Attributes dialog box appears. Notice that some of the fields already have values. These are the default values that were specified when this feature class was created.
7. Click the fields whose values you want to modify and type the new values.
8. Click the Close button to close the Attributes dialog box.

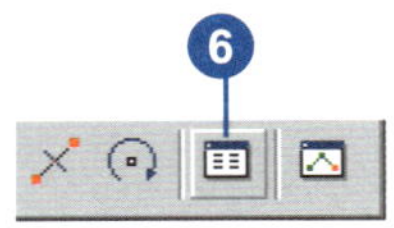

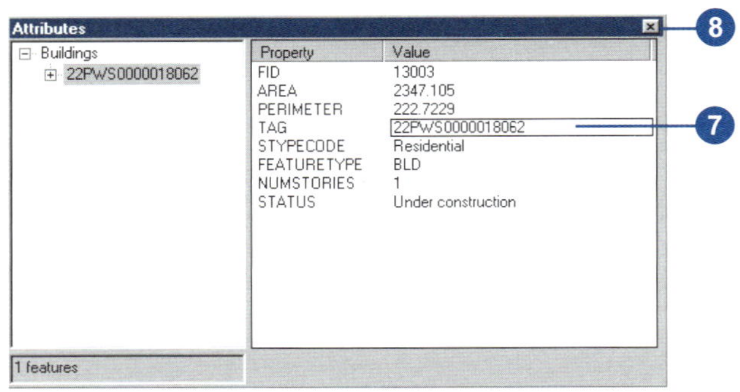

See Also

To learn more about subtypes and attribute domains, see Building a Geodatabase.

Changing a feature's subtype

1. Click the Edit tool.
2. Click the feature whose subtype you want to change.
3. Click the Attributes button.
4. Click the value of the subtype field.

   A dropdown list with all the available subtypes appears.
5. Click the subtype you want. ►

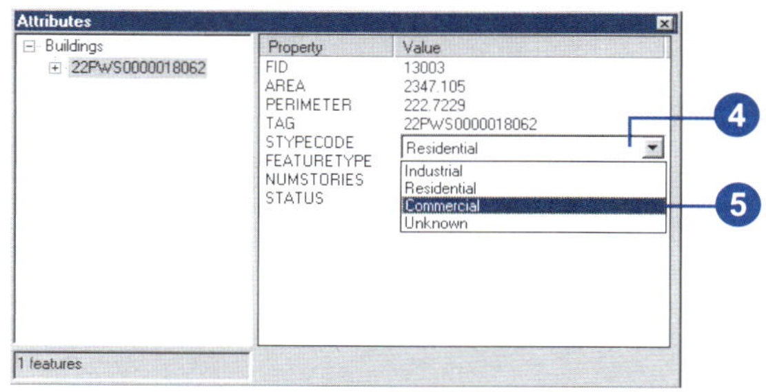

Tip

Default values

When you change a feature's subtype, the fields will take on the default values for the new subtype. If a field does not have a default value associated with it for the new subtype, its value remains unchanged.

The feature's symbology changes to match the new subtype. The fields with default values assume the default values for the new subtype.

The building is now marked with the symbology of the feature's new subtype.

Editing attribute domains

Feature classes and subtypes in a geodatabase can have *attribute domains*. These are rules that control the permissible values for a feature's attributes. These help maintain the data quality and consistency of the attributes of the features in your database.

There are two types of attribute domains: range domains and coded value domains. In a water network database you may have a feature class that stores water transmission mains. Transmission mains could have a range domain that specifies that the pressure value must be within the valid pressure range of between 40 and 100 psi.

If there are only three diameters of transmission main in your water system, you could have a coded value domain that specifies that transmission mains can have a diameter of 10, 24, or 30 inches. Coded value domains can speed attribute editing because when you edit them, ArcMap gives you a dropdown list of the permissible values that you can choose from.

Modifying coded value fields

1. Click the Edit tool.
2. Click the feature whose attributes you want to edit.
3. Click the Attributes button.
4. Click the value of the coded value field you want to modify.

   A dropdown list of all the coded value descriptions in the domain appears.
5. Click the value you want for the field.

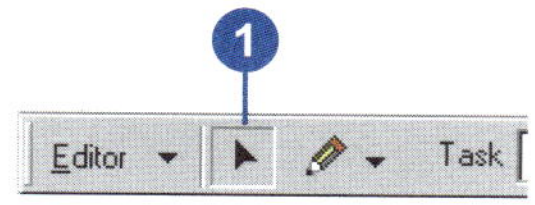

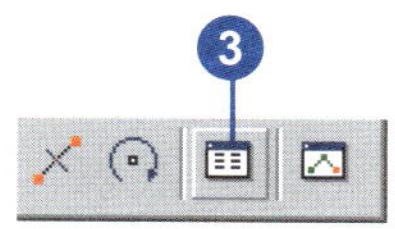

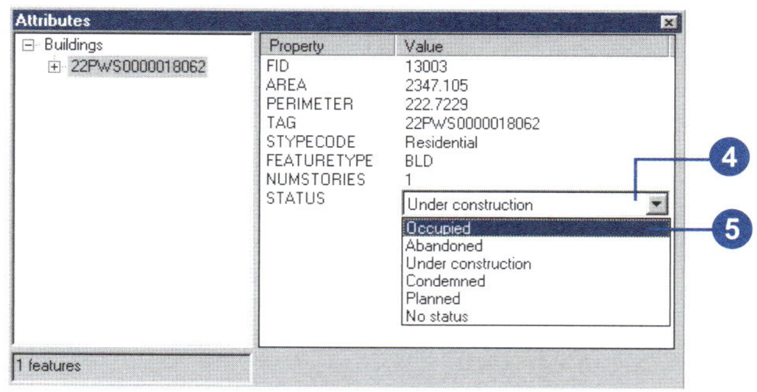

Validating features

When you edit features that have coded value or range domains, you should validate the features with the database to check that the attributes have appropriate values.

Validating features also validates any geometric network connectivity rules or relationship rules that may be defined for the feature class. For more information on relationship and connectivity rules, see Chapter 12, 'Editing geometric networks', and *Building a Geodatabase*.

Tip

Validating topology

Validating features and attributes is not the same as validating topology. For more information, see Chapter 4, 'Editing topology'.

Validating features

1. Click the Edit tool.
2. Click the features that you want to validate.
3. Click Editor and click Validate Features. ►

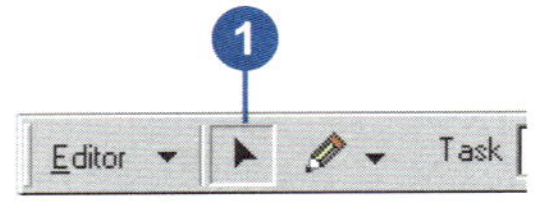

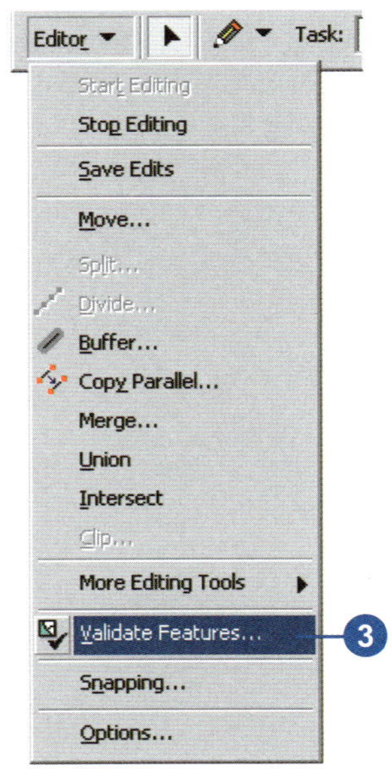

If your selection contains any invalid features, a message box appears with the number of invalid features. Only invalid features remain selected.

4. Click OK.
5. Click one of the invalid features.
6. Repeat step 3.
7. A message box appears telling you why the feature is invalid.
8. Click OK.
9. Click the Attributes button to view the attributes of the invalid feature.
10. Click the values that are invalid and change them.
11. Close the Attributes dialog box.
12. Repeat steps 5 through 11 for all of the invalid features.
13. Repeat steps 2 and 3. You should see a message box informing you that all the features are valid.
14. Click OK.

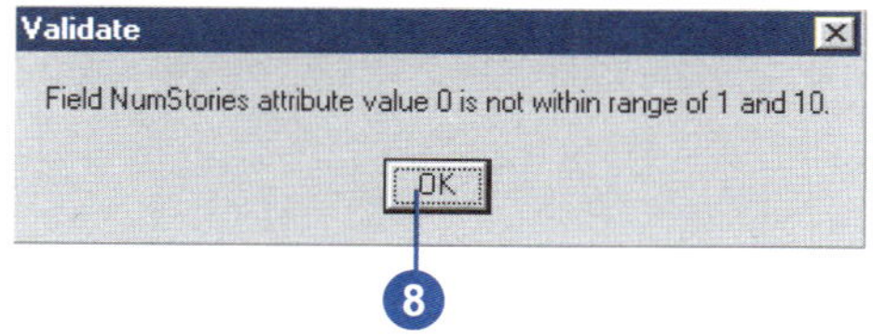

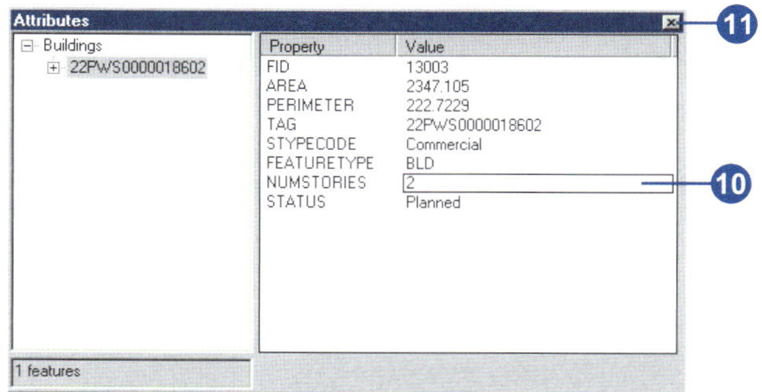

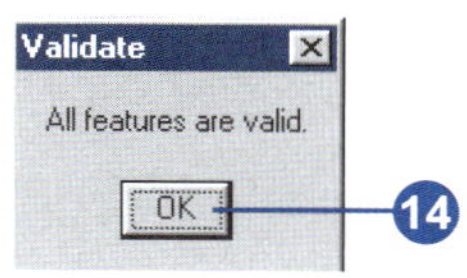

Editing related objects

11

IN THIS CHAPTER

- **Understanding relationships and related objects**
- **Editing relationships and related objects**

Some feature classes and tables in a geodatabase are designed to have relationships to other feature classes or tables. When you create, modify, or delete such geodatabase objects with ArcMap, messages can be automatically sent to the geodatabase to create, modify, or delete the related objects. These types of built-in relationships are stored in *relationship classes* in the geodatabase.

If there are relationship classes between feature classes and tables in your geodatabase, you can use ArcMap editing tools to take advantage of the relationships. Using ArcMap editing tools, you can find all of the objects related to a particular object and edit them. For example, you can select a parcel and find the owner of that parcel, then edit some of the *attributes* of that owner without ever having to add the *table* that stores the owners to your ArcMap session.

You can also use ArcMap editing tools to establish a new relationship between objects or to break existing relationships between objects. For example, if a parcel changes ownership, you can delete the relationship between the parcel and its original owner, then establish a new relationship to its new owner.

After you edit related geodatabase features or tables that have relationship rules you can validate your edits to check that the related objects still conform to the geodatabase relationship rules.

To learn more about relationship classes, see *Building a Geodatabase*.

Understanding relationships and related objects

Relationship classes allow you to maintain associations between objects in your geodatabase. These relationships can be simple and passive or they can be composite. Composite relationships imply parent/child relationships, or composition, and, therefore, have behavior, which is triggered through changes to objects on one side of the relationship to objects on the other side.

Relationships in a relationship class can be stored using primary and foreign keys in the object classes on either side of the relationship class. Alternatively, in the case of many-to-many relationship classes (M–N) and attributed relationship classes, the relationships are rows stored in a separate table.

You can use the Attributes dialog box or the table dialog box to find all objects related to any selected object. Once you have navigated to the related object, you can edit its attributes. You can also use the editing tool in ArcMap to break the relationship between any two objects or create new relationships between objects. When you edit objects and relationships in this way, all referential integrity is maintained.

Creating and deleting relationships

You can use the Attributes dialog box to create and delete relationships between two objects. If the relationship is managed by primary and foreign keys, the foreign key in the destination object is populated with the value of the primary key from the origin object. If a relationship between two objects is deleted, then the value for the foreign key in the destination object is replaced with a null.

If the relationship class is M–N or is attributed, then the relationships are stored in a separate table in the database. When a new relationship is created between two objects in this type of relationship class, a new row is added to that table. This new row is populated with the values from the primary keys in the origin and destination objects. If a relationship between two objects is

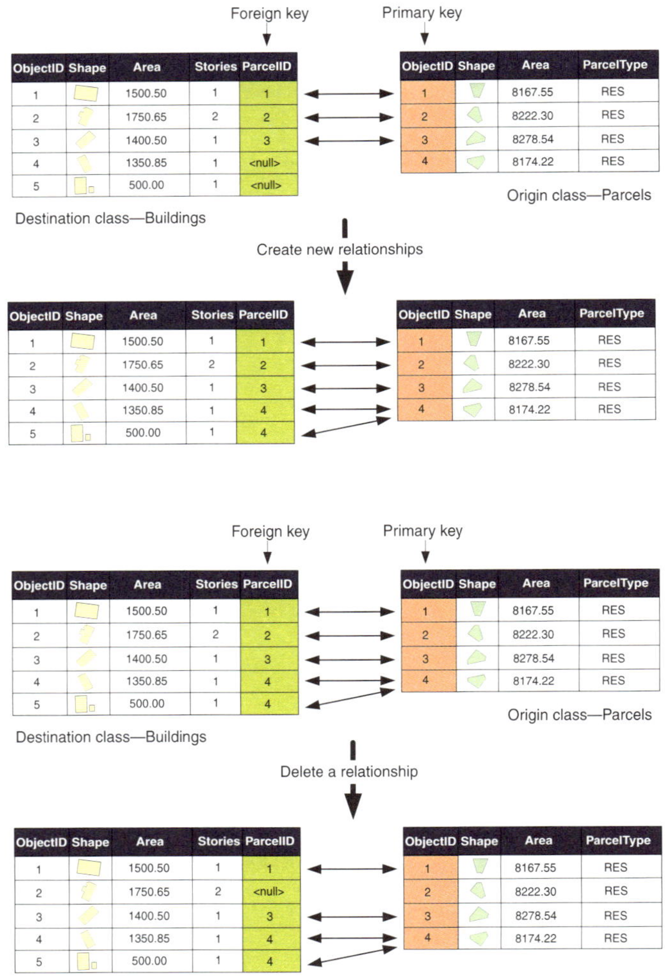

If the relationships in a relationship class are managed by primary and foreign keys (nonattributed 1–1 or 1–M relationships), creating and deleting relationships populate and null the foreign key in the destination class objects.

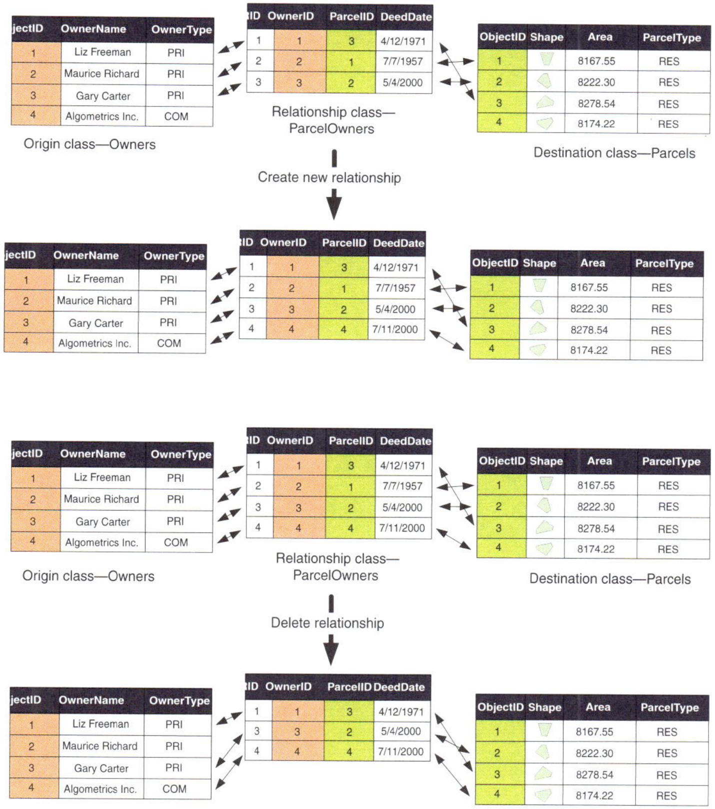

If the relationship class is M–N or is attributed, the relationships are stored as rows in the relationship class's table. Creating and deleting relationships adds and removes rows in the relationship class's table.

deleted, then the row corresponding to that relationship is deleted from the relationship table.

Deleting objects with relationships

When an object that participates in relationships with other objects is deleted from the database, all of its relationships are also deleted. If the relationships are maintained using primary and foreign keys, and the object deleted is the origin object, then the foreign key in the destination object is made null. If the object deleted is the destination object, then the origin object is not affected.

If relationships are maintained as rows in a relationship table (M–N relationships or attributed relationships) and either an origin or destination object and its relationships are deleted, then the rows corresponding to those relationships are also deleted from the relationship's table.

Creating new related objects

In ArcMap, you can select an object, then use the Attributes dialog box to create a new nonspatial object in a related class. When this new object is created, all of its attributes are populated with their appropriate default values (see *Building a Geodatabase*), and a relationship is created back to the object it was created from. You can only create nonspatial objects in this way; you cannot create new features.

If the relationships are maintained using primary and foreign keys, then the foreign key in the destination object is populated with the primary key of the origin object, regardless of whether the origin or destination object is created using the Attributes dialog box. If the relationships are maintained as rows in a relationship table (M–N relationships, attributed relationships), then a new row is added to the relationship class's table.

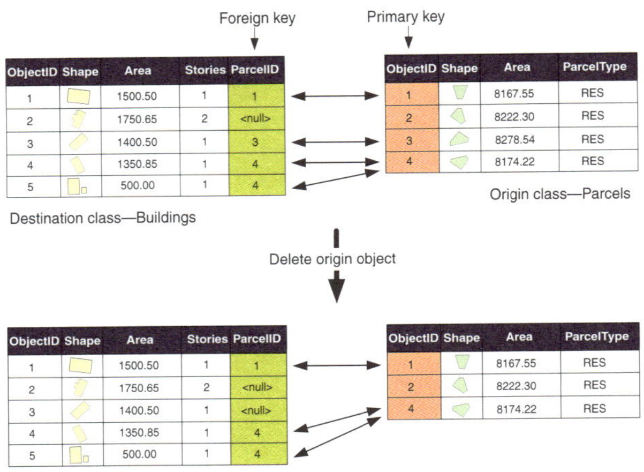

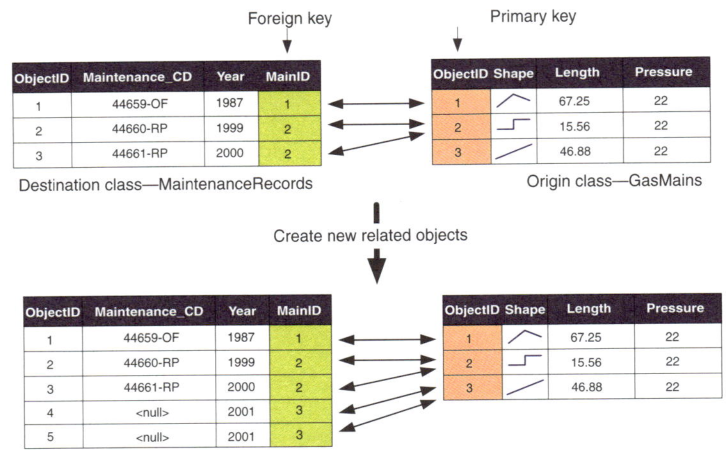

When you use the Attributes dialog box to create new related objects, a relationship is created back to the object from which it was created. If the relationships are maintained using primary and foreign keys, then the foreign key in the destination object is populated with the primary key of the origin object.

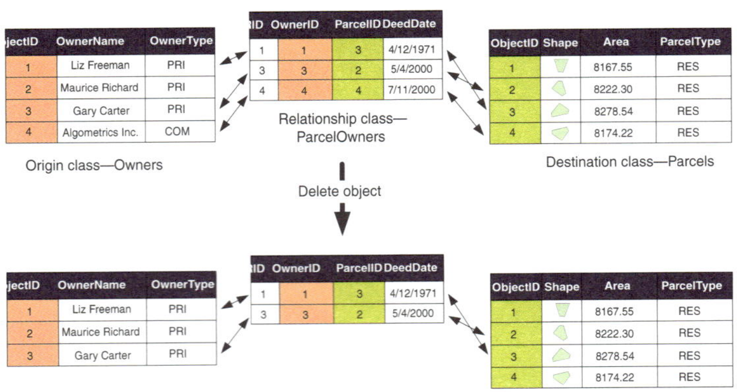

When an object that participates in relationships with other objects is deleted from the database, all of its relationships are also deleted.

Editing composite relationships

Composite relationships have some specialized behavior. When editing the objects that participate in a composite relationship, this behavior carries over to the editing process. Edits made to the origin object in a composite relationship often directly affect its related destination objects. This behavior is partially dependent on relationship class messaging.

By default, composite relationship classes have forward messaging (see *Building a Geodatabase*)—that is, when the origin object in a composite relationship is edited, it sends messages to its related destination objects. The related objects will respond to that messaging in a standard way: if the destination objects are nonspatial objects, then they will not change. However, if the destination objects are features when the origin object is moved, then the destination objects will also move the same distance. If the origin object is rotated, then the destination objects will also be rotated by the same angle.

Similar to simple relationships, composite relationships also maintain referential integrity when objects are deleted, but they do this in a different way. When the origin object in a composite relationship is deleted, all of the objects related to it through that composite relationship are also deleted. This cascade deletion will happen whether messaging is set to forward, back, both, or none.

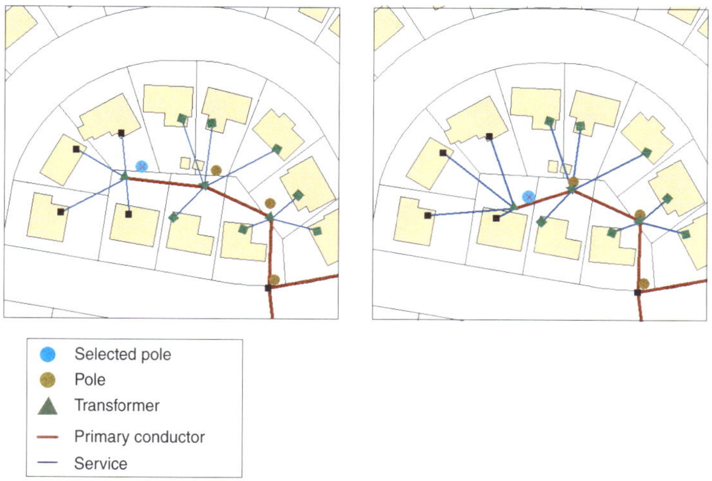

When an origin object in a composite relationship is moved and messaging is set to forward or both, if the related objects are features, they will move the same distance to follow the feature. In this example, the selected pole is the origin object and the transformer is the destination object.

Foreign key

ObjectID	Shape	TranType	PoleID
1	▲	Overhead1	1
2	▲	Overhead1	1
3	▲	Overhead2	2
4	▲	Overhead3	3
5	▲	Overhead3	3

Destination class—Transformers

Primary key

ObjectID	Shape	Height	PoleType
1	●	15	WOOD
2	●	15	WOOD
3	●	20	STEEL

Origin class—Poles

Delete origin object

ObjectID	Shape	TranType	PoleID
1	▲	Overhead1	1
2	▲	Overhead1	1
3	▲	Overhead2	2

ObjectID	Shape	Height	PoleType
1	●	15	WOOD
2	●	15	WOOD

When an origin object in a composite relationship is deleted, all destination objects related to it through a composite relationship are also deleted.

When a destination object is deleted, the relationship between it and the origin object is deleted; the origin object itself is not deleted or modified.

Splitting features that participate in relationships

Splitting a single geodatabase feature into two separate features is actually a delete and create operation—that is, the original feature is deleted and two new features are created. This has implications when the feature being split has relationships with other objects in the database.

With simple relationships, when an origin feature is split, the relationships between the original feature and its related destination objects are deleted. When the new features are created from the split operation, new relationships are created between the new feature with the larger portion of the original feature's geometry and the destination objects that were related to the original feature.

In the case of a composite relationship, the behavior is different. When an origin feature in a composite relationship is split, any objects related to it through that composite relationship are deleted before the two resulting new features are created from the split.

When splitting a destination feature in either a simple or composite relationship, the relationships between the original feature and the related origin objects are deleted, and the new relationships are created between the origin objects and both new features that result from the split.

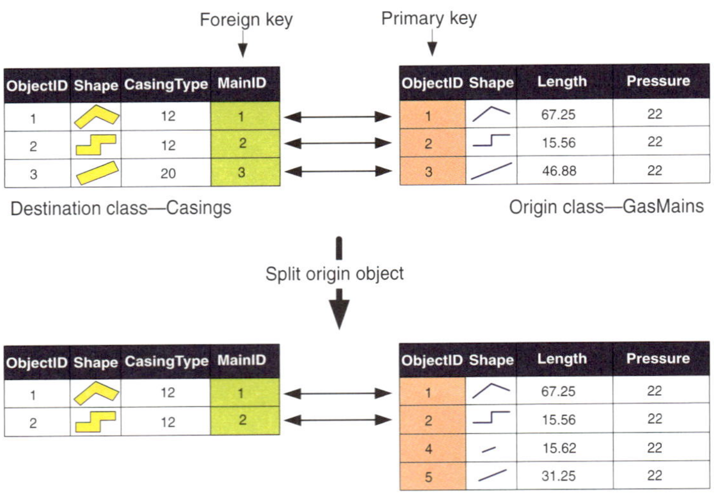

When an origin object in a composite relationship is split, its destination objects are deleted.

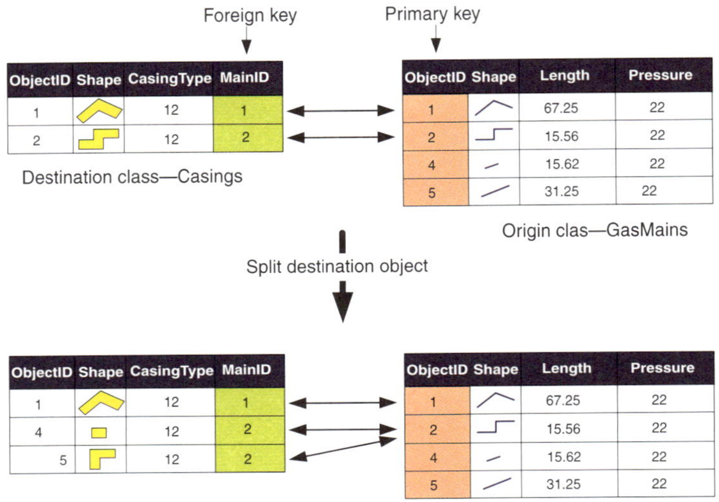

When splitting a destination feature in either a simple or composite relationship, the relationships between the original feature and the related origin objects are deleted, and the new relationships are created between the origin objects and both new features that result from the split.

The behavior of splitting objects with relationships described here is the default behavior. You can override this behavior at the class level by writing a class extension that implements the IFeatureClassEdit interface. The IFeatureClassEdit interface has a property called CustomSplitPolicyForRelationship that allows you to specify how relationships are handled when features are split. To learn more about class extensions and how to implement them, see *Exploring ArcObjects*.

Editing relationships and related objects

The tasks presented here are all examples of editing relationships between water laterals and hydrants in a water network. The rules of this relationship class state that a hydrant lateral must have a hydrant related to it and that hydrants cannot be related to other lateral types.

Tip

Relationship rules and topology

It is important to distinguish relationships and relationship rules from topology and its associated rules. To learn more about topology, see Chapter 4, 'Editing topology'.

Editing a related object

1. Click the Edit tool.
2. Click the hydrant lateral whose related hydrant you want to modify.
3. Click the Attributes button.
4. Double-click the lateral in the left panel of the Attributes dialog box.
5. Double-click the relationship path label.

   The related hydrant objects are listed below the path label.
6. Click the ID number of the related hydrant whose attributes you want to modify.
7. Modify the attributes of the hydrant object by clicking the value and typing a value or clicking the value and choosing the new value from the list.

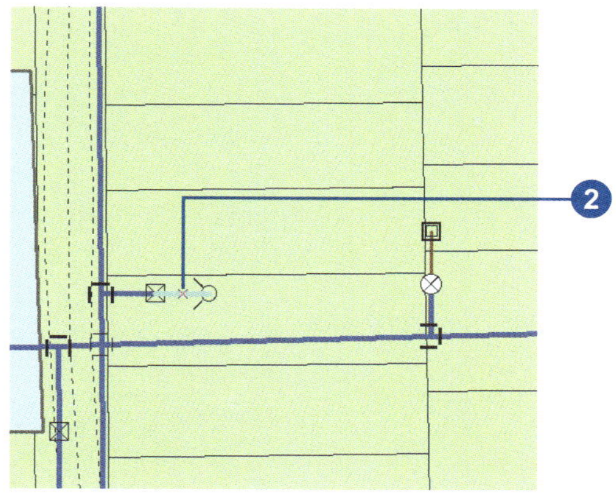

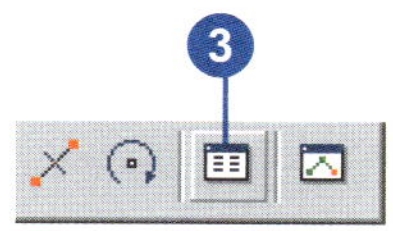

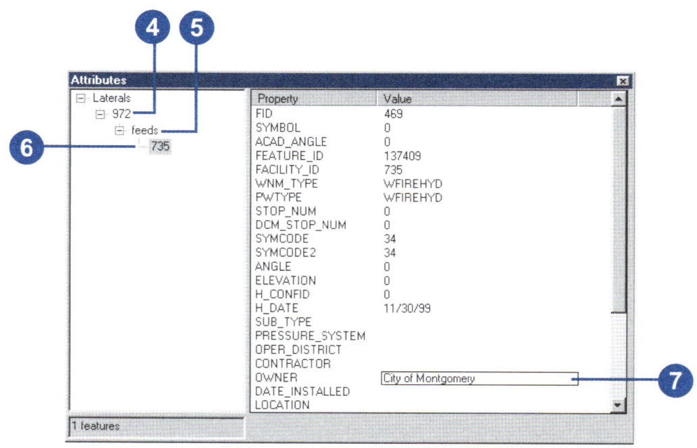

Selecting a related object

1. Click the Edit tool.
2. Click the hydrant lateral whose related hydrant you want to modify.
3. Click the Attributes button.
4. Double-click the lateral in the left panel of the Attributes dialog box.
5. Double-click the relationship path label.

   The related hydrant objects are listed below the path label.
6. Right-click the related hydrant you want to add to the map's selection and click Select. ▸

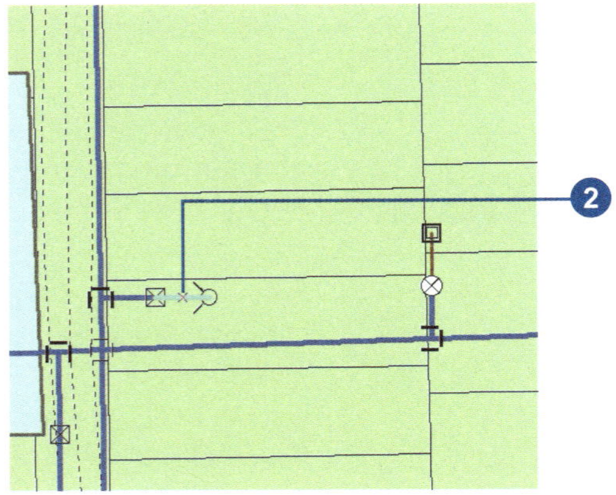

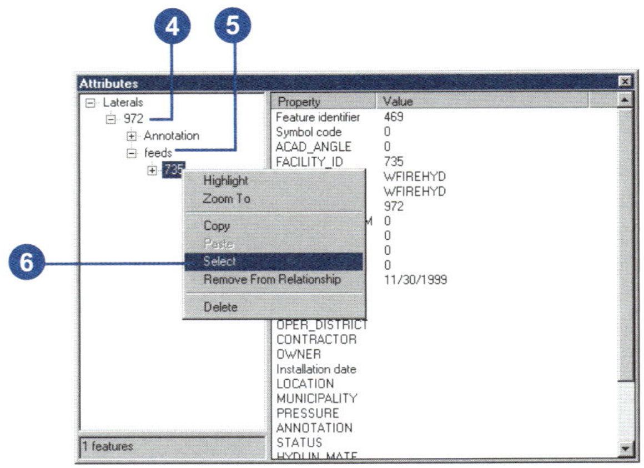

The hydrant is added to the selection.

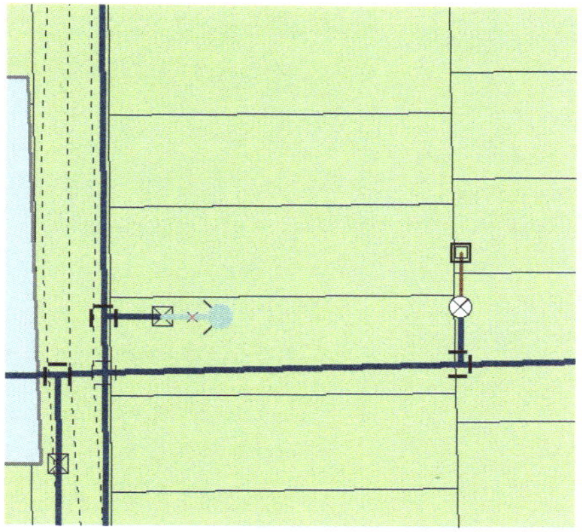

The hydrant is added to the selection.

ArcInfo and ArcEditor

Tip

Relating objects

Before adding a relationship between two objects, you must first create a relationship class between the feature classes or tables containing the objects you want to relate. To learn more about how to create relationship classes, see Building a Geodatabase.

Creating a new relationship between features

1. Click the Edit tool.
2. Click the features between which you want to create relationships.
3. Click the Attributes button.
4. Double-click one of the features in the left panel.
5. Right-click the relationship path label and click Add Selected. ▸

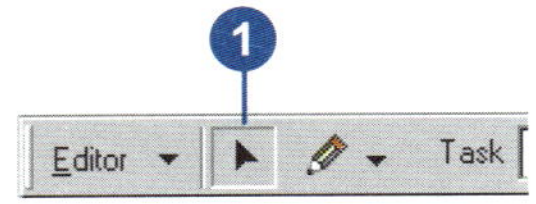

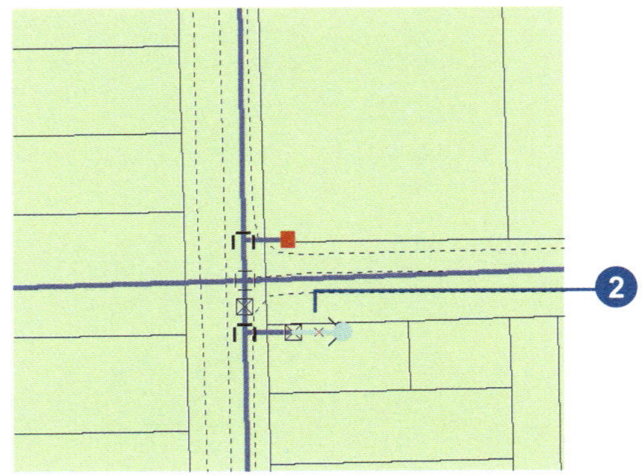

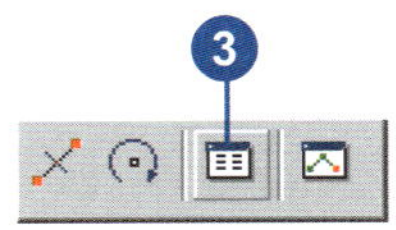

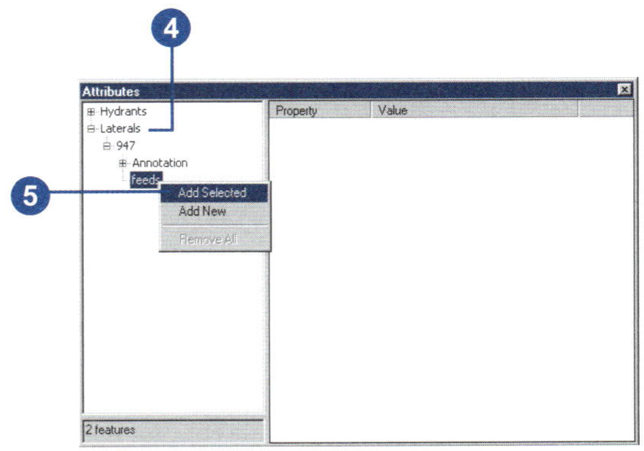

The selected object or objects are now added to the list of selected objects under the relationship class path label.

Relationships to the selected objects are added.

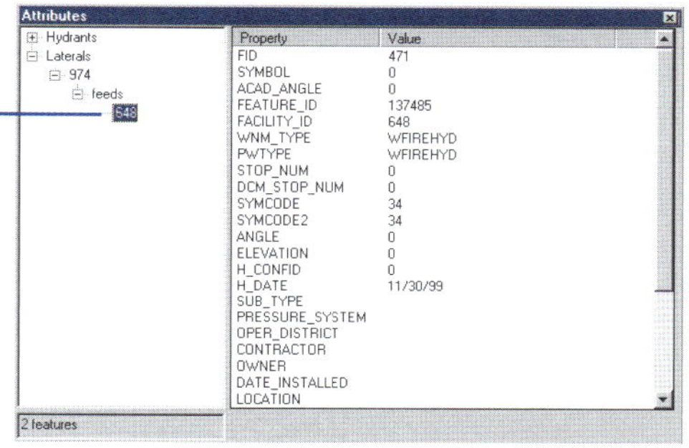

Creating a new relationship between a feature and a nonspatial object

1. Click the Edit tool.
2. Click the feature to which you want to create a relationship.
3. In the table of contents, right-click the table that contains the objects that you are relating to, then click Open. ▸

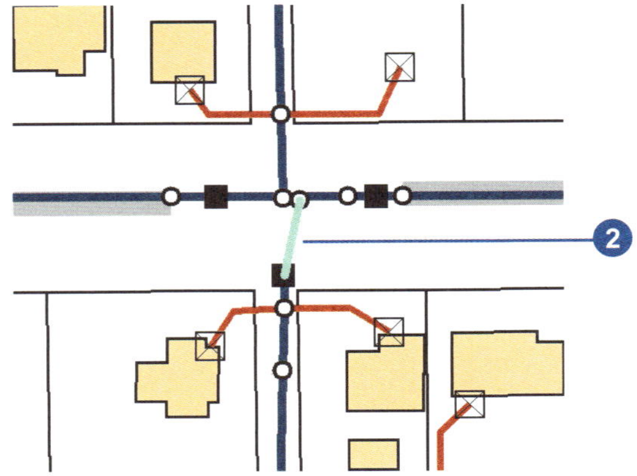

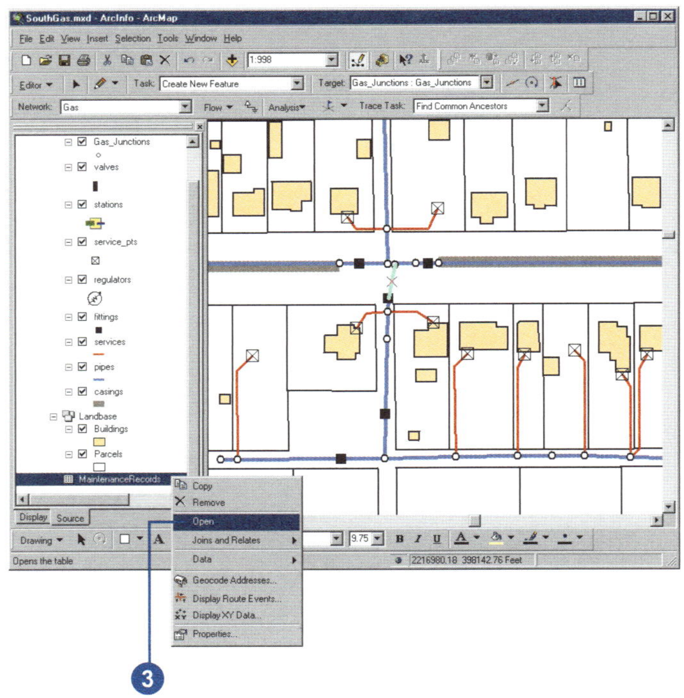

4. Click the object in the table with which you want to create a relationship to the selected feature.
5. Click the Attributes button.
6. Double-click the feature in the left panel.
7. Right-click the relationship path label and click Add Selected.

   The selected object or objects are now added to the list of selected objects under the relationship class path label.

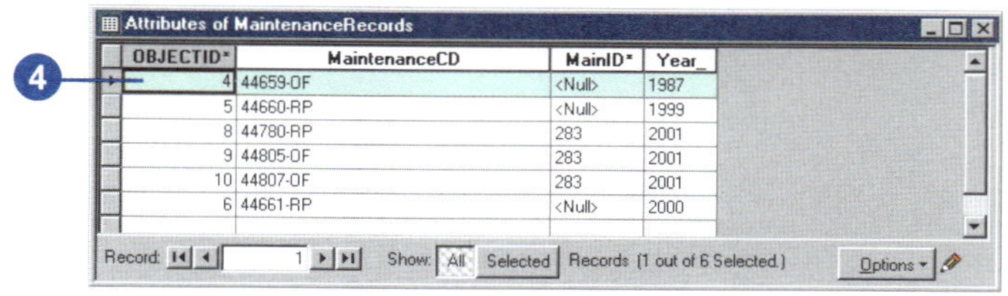

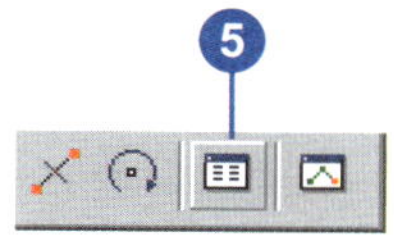

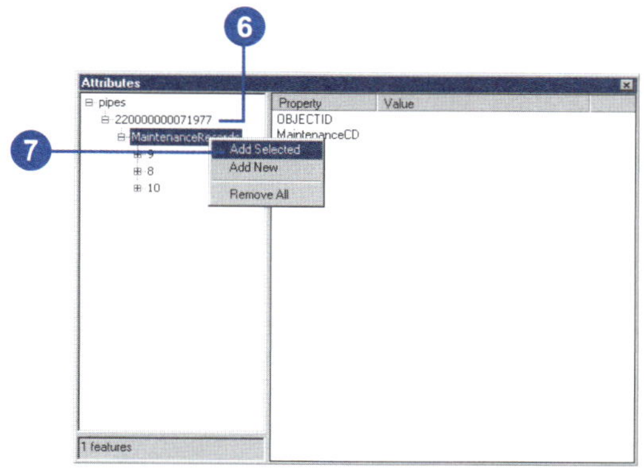

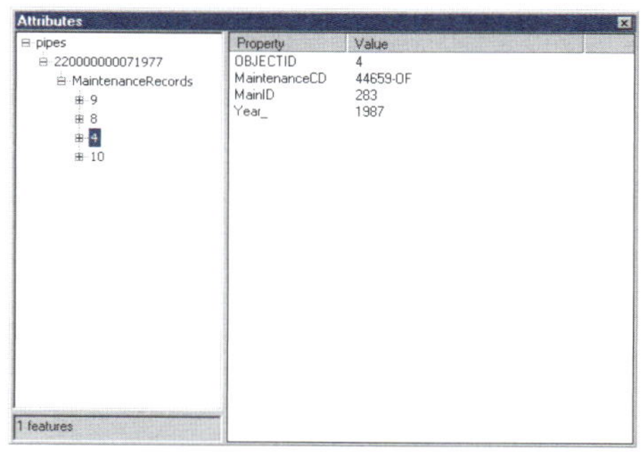

The object is listed under the relationship path label.

Deleting a relationship

1. Click the Edit tool.
2. Click the feature from which you want to delete a relationship.
3. Click the Attributes button.
4. Double-click the feature in the left panel.
5. Double-click the relationship path label to see a list of related objects.
6. Right-click the object from which you want to delete the relationship and click Remove From Relationship. ►

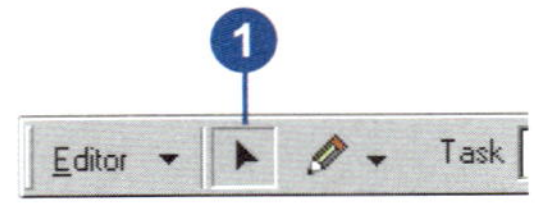

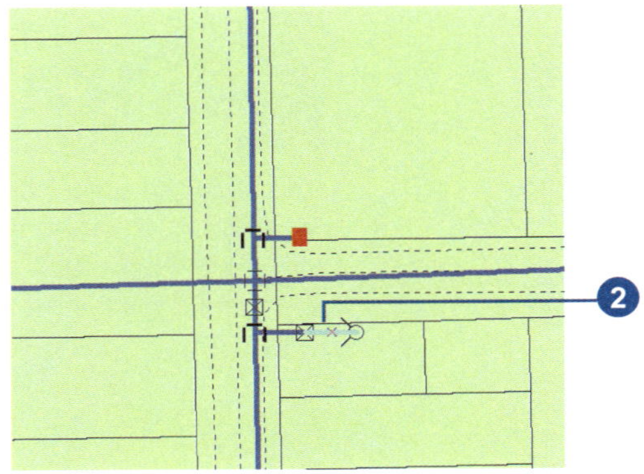

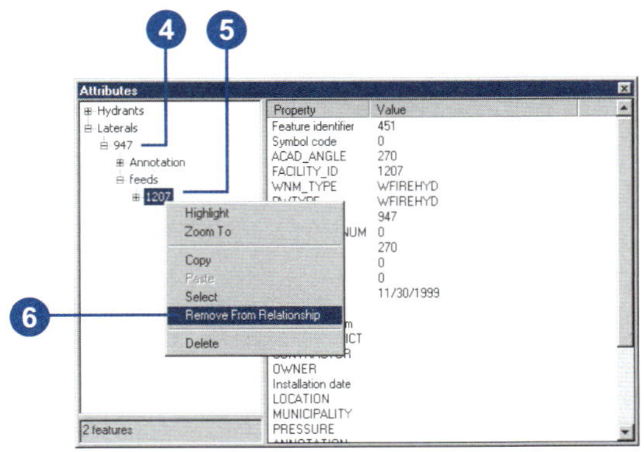

The object is no longer listed under the relationship path label.

7. Click the Close button to close the Attributes dialog box.

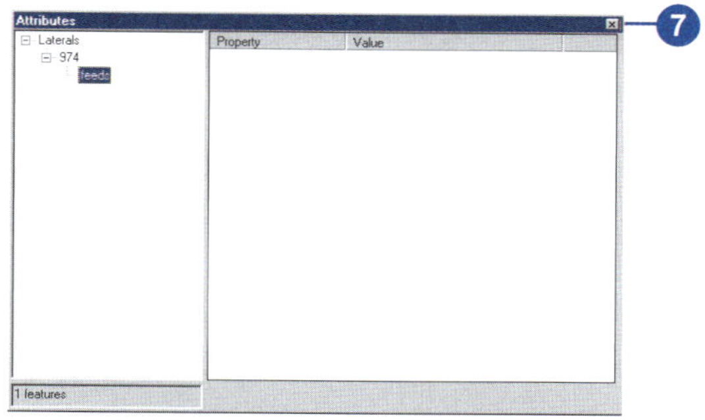

The object is no longer listed under the relationship path label.

Tip

Creating new related features

You cannot use the Add New command in the Attributes dialog box to create new related features. See 'Creating new related features', later in this chapter, to learn how to create new related features.

Creating new related nonspatial objects

1. Click the Edit tool.
2. Click the feature for which you want to create a new related object.
3. Click the Attributes button.
4. Double-click the feature in the left panel. ▸

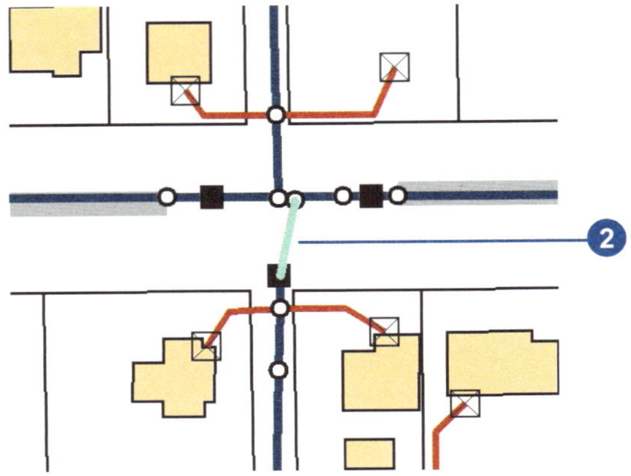

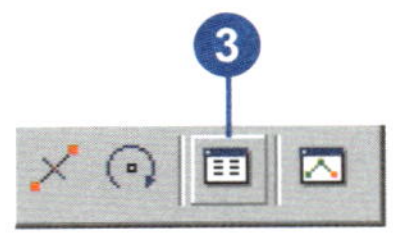

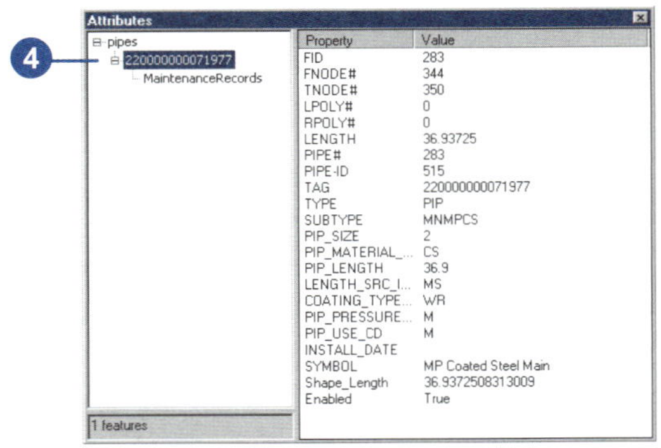

5. Right-click the relationship path label and click Add New.

   A new object is created and related to the selected feature.

6. Click the new object in the left panel to see its attributes.

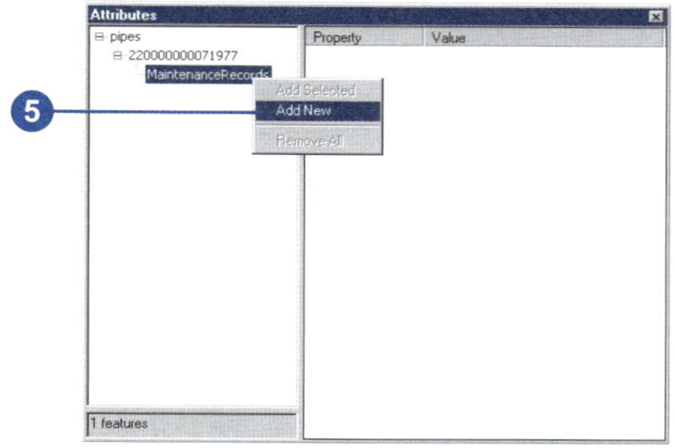

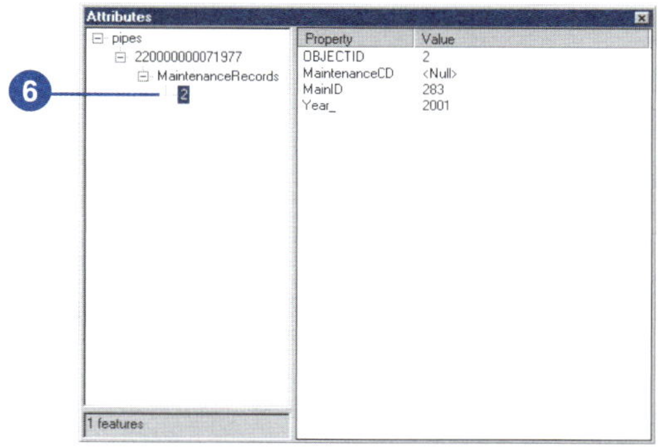

A new object is created in the related class and a relationship is created between it and the selected feature.

See Also

To learn more about ArcMap sketch tools and how to create new features, see Chapter 2, 'Editing basics', and Chapter 3, 'Creating new features'.

Creating new related features

1. Use ArcMap sketch tools to create the new feature.
2. Click the Edit tool.
3. Hold down the Shift key and click the feature for which you want to create a relationship to the new feature.

   Both the new feature and the feature you are relating it to should be selected. ►

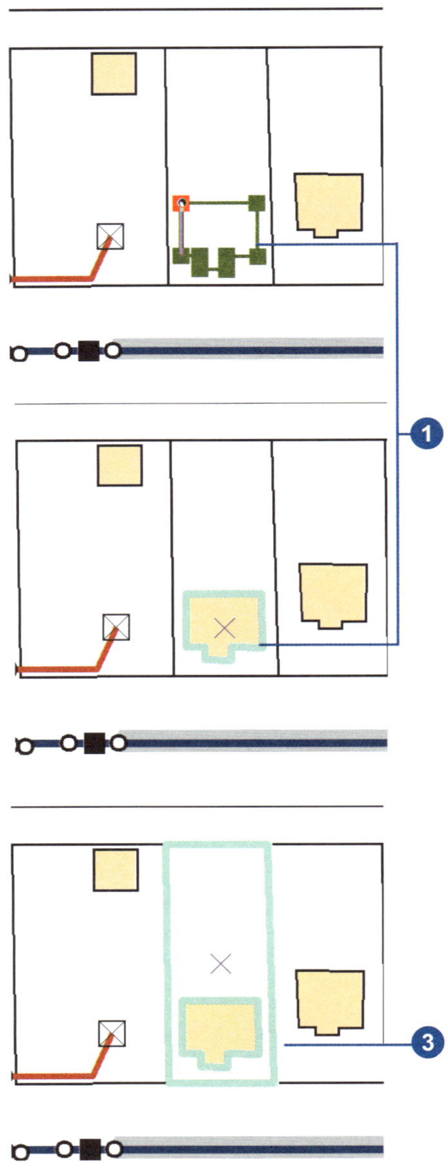

4. Click the Attributes button.
5. Double-click the feature in the left panel.
6. Right-click the relationship path label and click Add Selected.

   The selected object or objects are now added to the list of selected objects under the relationship class path label.

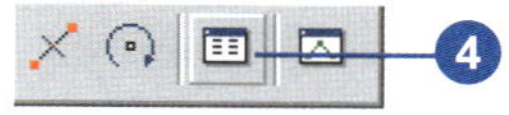

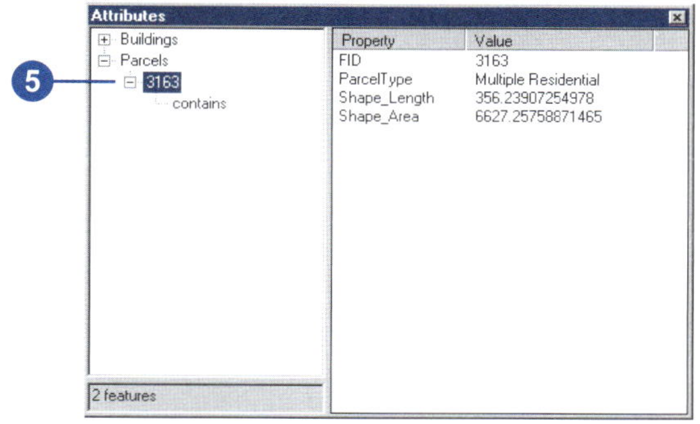

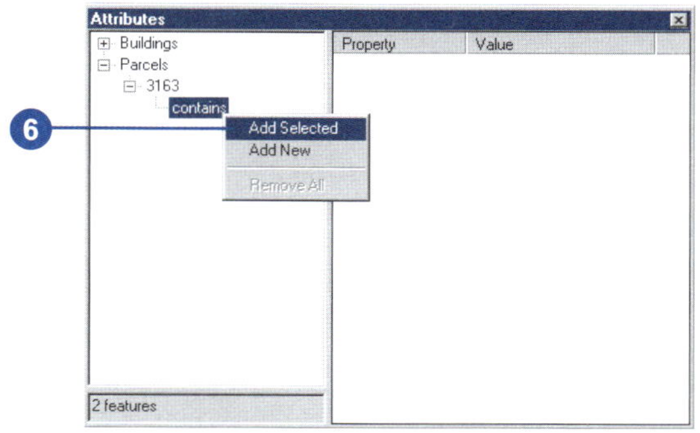

A relationship to the selected object is added.

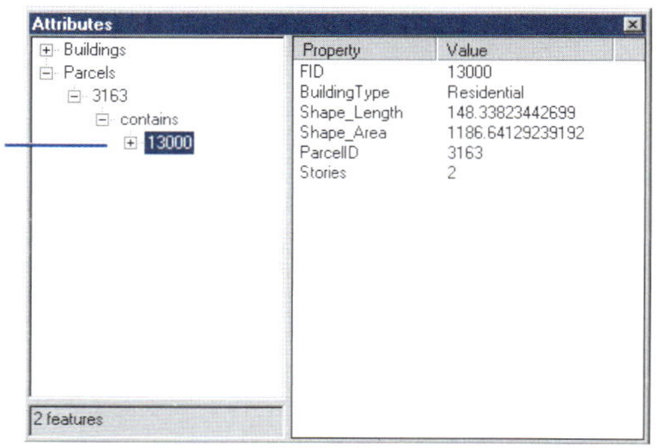

Tip

Deleting related features

You can also delete related features using the Delete command on the Relationship context menu in the Attributes dialog box.

Deleting related objects

1. Click the Edit tool.
2. Click the feature whose related object you want to delete.
3. Click the Attributes button.
4. Double-click the feature in the left panel.
5. Double-click the relationship path label to see a list of related objects.
6. Right-click the object you want to delete and click Delete.

   The object is deleted and no longer listed under the relationship path label.

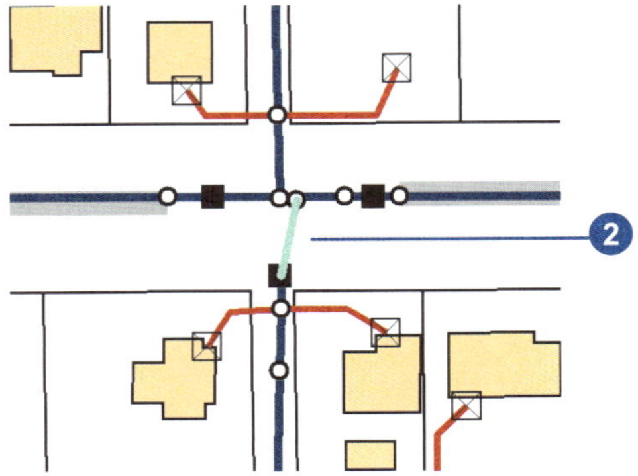

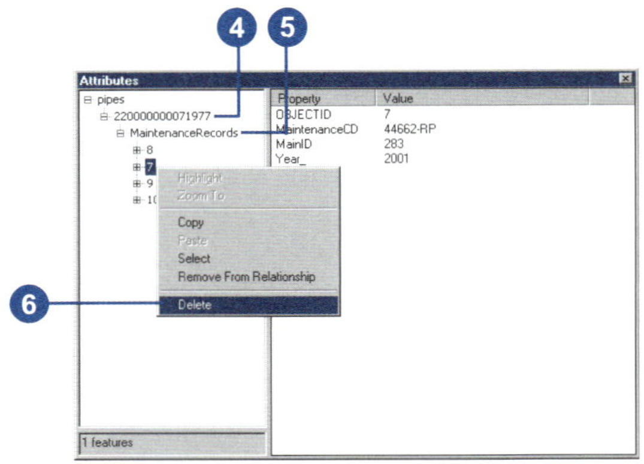

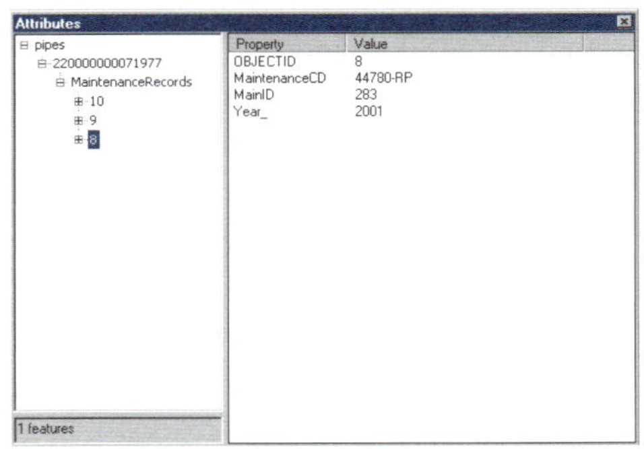

The object is no longer listed under the relationship path label.

Tip

The Attributes dialog box

The Attributes dialog box behaves the same way with composite relationships as it does with creating or deleting new features or relationships.

Editing features with composite relationships

1. Click the Edit tool.
2. Click the origin feature in the composite relationship you want to edit.
3. Click and drag the feature to a new location.

   The related features move the same x,y distance as the origin feature you moved. ▸

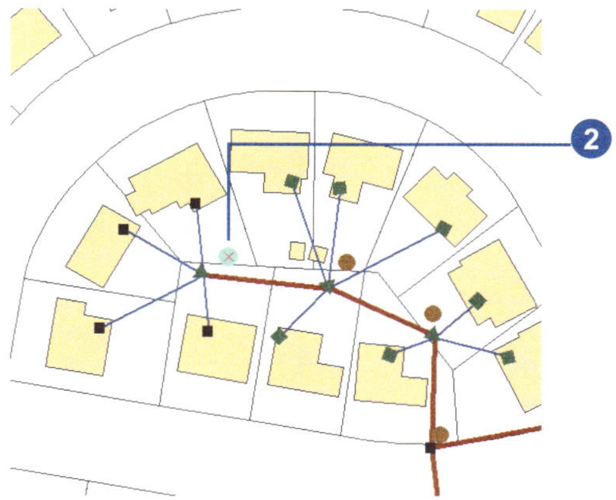

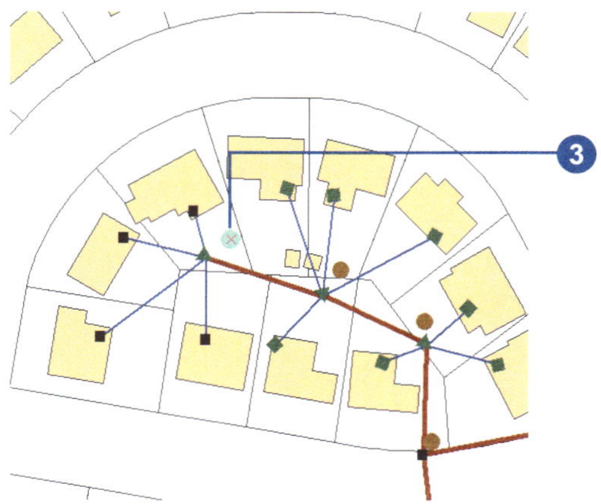

The related features move the same distance as the feature you moved.

4. Click the Rotate tool.
5. Click anywhere on the map and drag the pointer to rotate the feature to the desired location.

   The related features rotate with the feature.
6. Click the Edit tool and click a destination feature in a composite relationship.
7. Click and drag the feature to a new location.

   The origin feature in the relationship doesn't move. ►

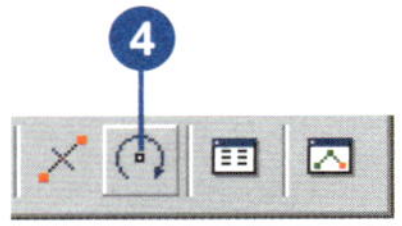

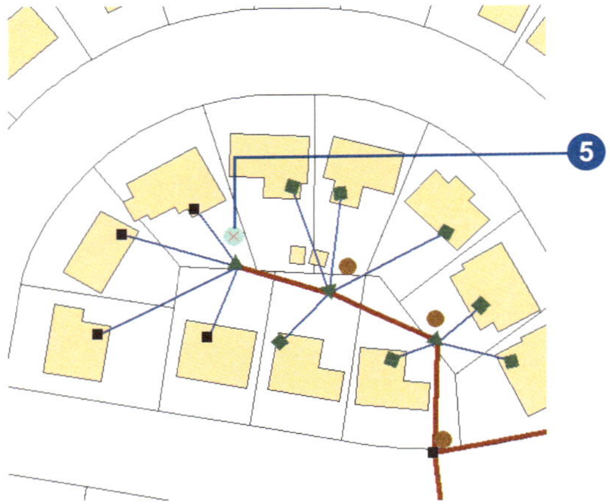

The related features rotate with the features.

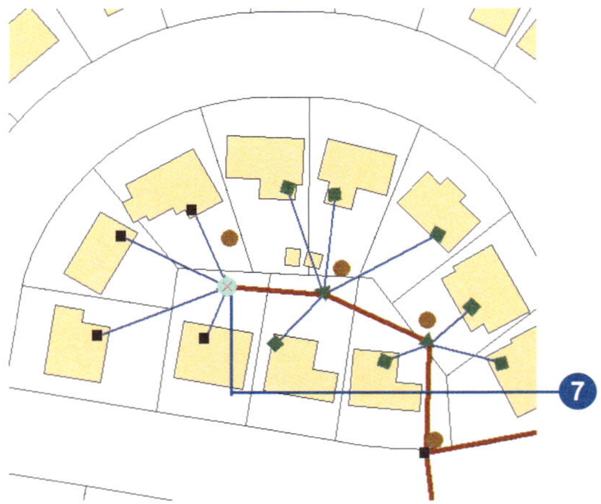

The origin feature doesn't move.

8. Click the origin feature again and click Delete.

   Both the feature and its related feature are deleted.

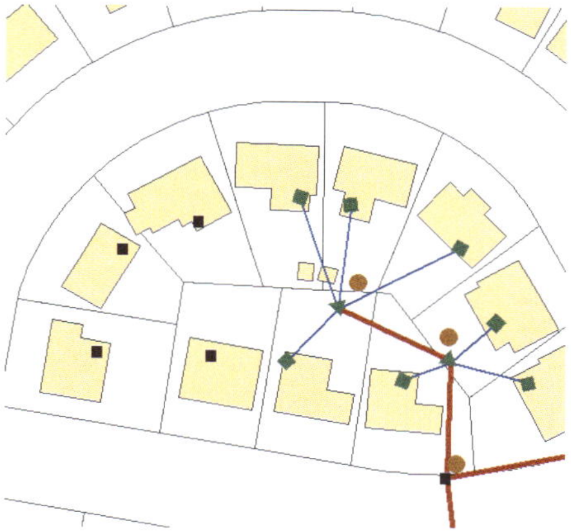

Both the origin feature and its related feature are deleted.

Tip

Relationship rules

Relationship rules can be broken in two ways: when a feature is related to a subtype of the related class for which no valid rule applies or when a cardinality rule is broken.

To learn more about relationship rules, see Building a Geodatabase.

Validating relationships

1. Click the Edit tool and click the feature or features you want to validate.
2. Click Editor and click Validate Features.

   A message box appears telling you how many features are invalid. Only invalid features remain selected. ►

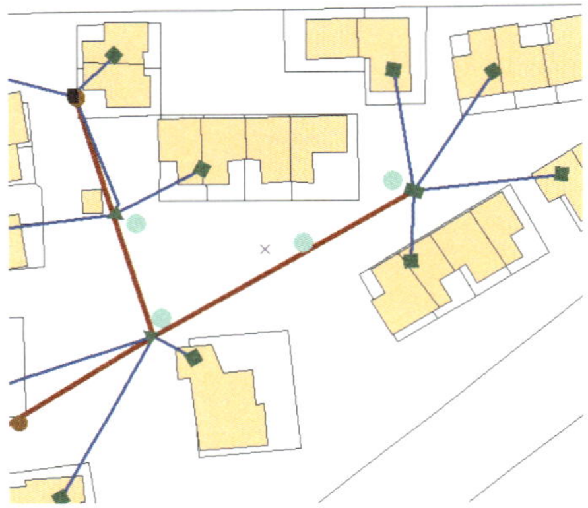

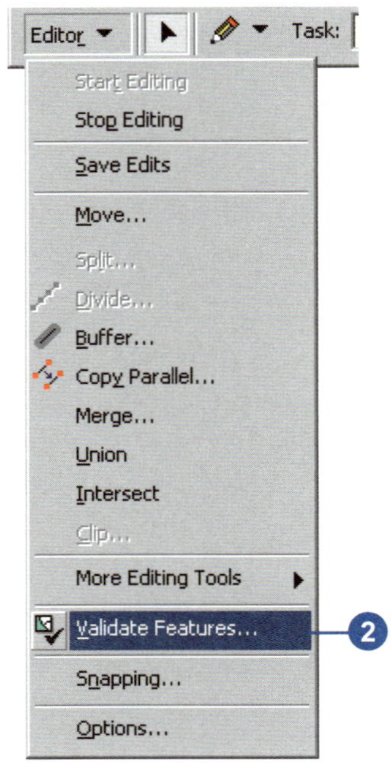

3. Click OK.
4. Click one of the invalid features.
5. Repeat step 2.

   A dialog box appears informing you why the selected feature is invalid.
6. Click OK.
7. Make the necessary edits to the relationships or the related objects to make the feature valid. This may involve adding and deleting relationships or altering the subtype of one or all of the features.
8. Repeat step 2—a message box appears informing you that all the features are valid.
9. Click OK.

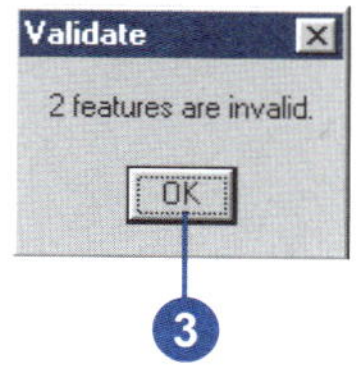

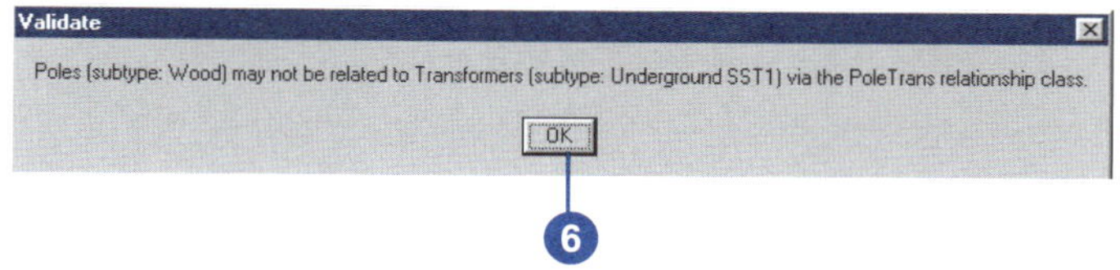

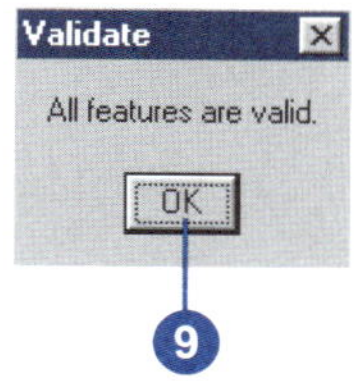

Editing geometric networks 12

IN THIS CHAPTER

- Editing network features
- Creating network edges
- Subsuming network junctions
- The Network Editing toolbar
- Validating network features

In addition to simple features, ArcMap lets you edit collections of features related in a geometric network. In a geodatabase, a geometric network contains special types of network features that enable connectivity tracing, network connection rules, and specialized junction or switch modeling behavior.

Geometric networks are useful for modeling networks of wires, pipes, or natural water flow networks. Networks are built of edge and junction features. Edges model linear features such as pipes, wires, and streams. Junction model nodes in the network—places where edges connect to each other—such as fittings, valves, and hydrants; switches, fuses, and transformers; or confluences, gauging stations, and water quality monitoring devices.

There are two broad categories of network features: simple and complex. Simple edge features are connected to junctions at each end. Snapping another feature along the length of a simple edge feature splits the edge into two simple edges. Complex edge features are connected to junctions at each end but may also have junctions connected to them along their length; they do not split when junctions are added. Simple junction features connect edges. Complex junctions are single custom features that can have internal networks of edges and junctions. A pump station might be modeled as a single complex junction feature in a water network, but it might be composed of a group of pipes, valves, and pumps that have complex internal network connectivity.

Editing network features

Geometric network features store various mechanisms and behaviors that maintain the topological connectivity between them. ArcMap editing capabilities are tightly integrated with the geodatabase when it comes to editing network features.

Creating connectivity

Topological connectivity in a network feature class is based on geometric coincidence. If a junction is added along an edge or one edge is added along another edge, they will become topologically connected to one another.

By using the ArcMap *snapping* environment, you can create new edge and junction features on the fly while maintaining network connectivity. The ArcMap snapping functionality will guarantee geometric coincidence when adding new network features along existing network features.

Simple and complex edges

An edge in a geometric network can be either simple or complex. A simple edge in a geometric network has a 1–1 relationship with edge elements in the logical network. A complex edge has a 1–M relationship with edge elements in the logical network. So one complex edge in the geometric network can represent multiple edges in the logical network.

If you snap a junction or edge along a simple edge, then the edge being snapped to is split both in the logical network and in the geometric network, giving you two edge features. If you snap a junction or an edge along a complex edge, then that edge is split in the logical network but remains a single feature in the geometric network. It will remain a single feature; however, a new vertex is created at the point where the new junction or edge connects to it.

Default junctions

When you snap an edge to another edge where there is no junction, a junction is automatically inserted to establish connectivity. If a default junction type has been specified as part of the connectivity rules for the network, that default junction type is used. If there is no edge–edge rule between these edge types, an orphan junction is inserted, which is stored in the <network>_Junctions feature class.

Similarly, if you create a new edge in the network that is not snapped to an existing junction or edge at both ends, a junction is automatically created and connected to the free end of the new edge. If there is a connectivity rule in place that defines a default junction type for the type of edge that is being added, that default junction type is the junction that is added to the free end of the new feature. If an edge type does not have a default junction type associated with it through a connectivity rule, then an orphan junction is inserted, which is stored in the <network>_Junctions feature class.

Junction subsumption

When you snap a junction to an existing junction, the old junction is subsumed by the new junction. That is, the original junction is deleted from the network and the new junction is inserted in its place. All network connectivity is maintained.

When you create a new edge feature in the network that has an end that does not connect to anything and there is not a connectivity rule stating what type of junction to put at its free end, the network orphan junction type is inserted. This orphan junction can be replaced by snapping another junction to it.

Moving existing network features

When a network edge or junction is moved, the network features to which it is connected respond by rubber-banding and adjusting themselves to maintain connectivity. When you move a network feature and snap it to another network feature, the features may become connected (see below).

Connectivity models

Edit operations that involve adding, deleting, moving, and subsuming network features can all affect the connectivity of a geometric network. Each type of operation may or may not create connectivity, depending on the type of network features involved. The following set of diagrams illustrates various editing scenarios and their resulting connectivity or lack thereof. In these diagrams, use the key below to identify what types of features are illustrated in each scenario:

Orphan Junction

Standard Junction

SEF Simple Edge

CEF Complex Edge

Stretching and *moving:* When stretching or moving junctions, any edges connected to them rubber-band to remain connected. When you snap these junctions to other network features, the following illustration summarizes the network connectivity that results:

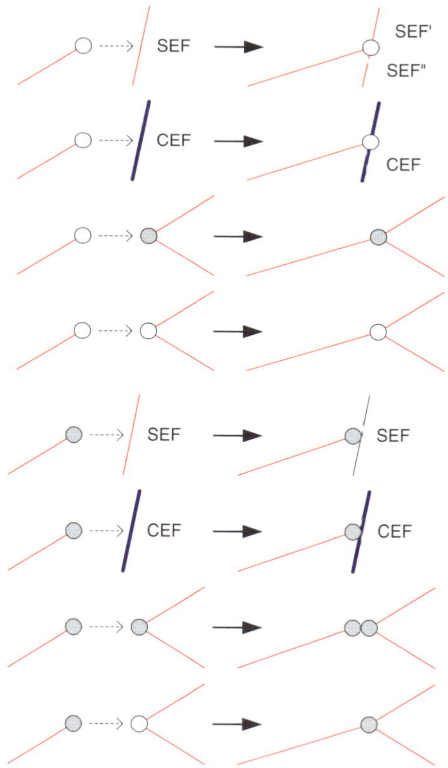

Connectivity behavior when stretching and moving network features

Deleting: Deleting network features can affect those features connected to them. When you delete an edge feature, the edge is physically deleted from the geometric network and logically deleted from the logical network; however, its connected junction features will not be deleted. When deleting junction features, if the junction being deleted is not of an orphan junction type, it will not be physically deleted from the geometric network. Rather than being deleted, the junction will become an orphan junction. When you delete an orphan junction, it is physically deleted from the geometric network. When this happens, depending on what type it is and how many edges are connected to it, some edges may also be deleted. The following illustration summarizes the results of deleting network junctions:

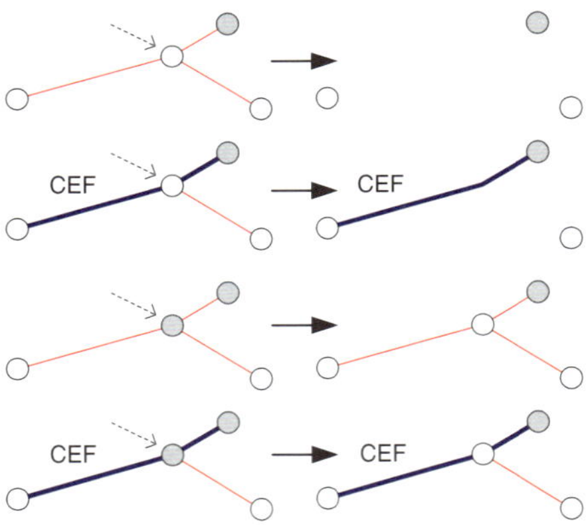

Connectivity behavior when deleting network features

Disconnecting features: The following illustration summarizes how connectivity is affected when disconnecting network edge features and junction features using the Disconnect command in ArcMap:

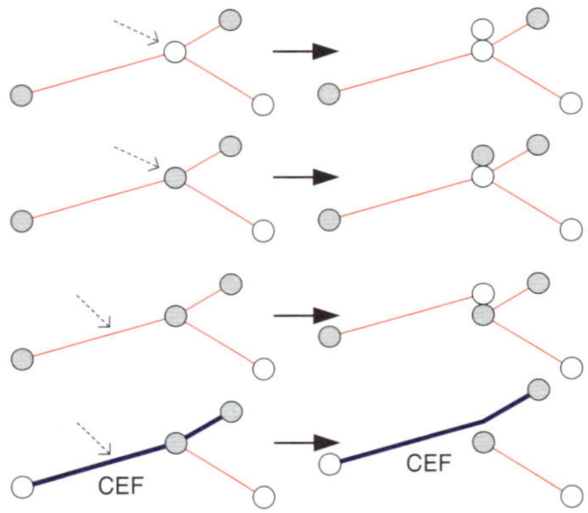

Connectivity behavior when disconnecting network features

Connecting features: The following illustration summarizes how connectivity is affected when connecting network features use the Connect command in ArcMap:

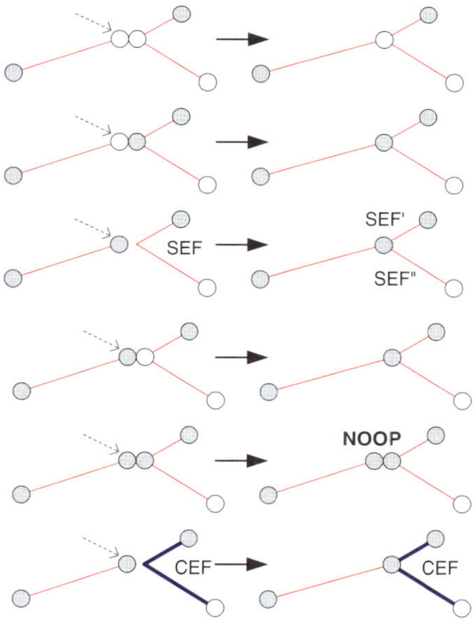

Connectivity behavior when connecting network features. NOOP indicates that the Connect command resulted in no edit operation.

Creating new network features: When creating new network features and snapping them to other network junction and edge features, the resulting connectivity and the effects on the features you connect them to are summarized below:

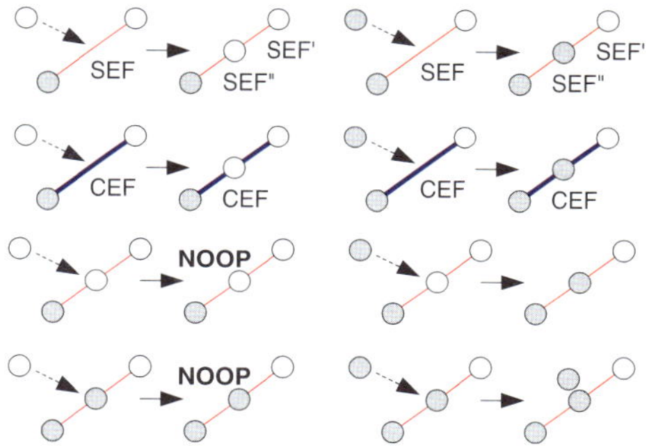

Connectivity behavior when creating new network features. NOOP indicates that the new feature was not created.

Repairing network topology

Connectivity between network features is maintained on the fly as you create, delete, and modify network features. In some circumstances, the association between some network features and their logical elements may become out of sync. This can happen, for example, when using a custom tool that does not correctly handle aborting edit operations.

This kind of network corruption is localized to a collection of features in the network. You will be able to see what features have corrupt topology in two ways: (1) when moving a network feature, if rubber-banding does not occur with other network features it is connected to and the edit operation fails, the topology is corrupt, and (2) reconciling a version with corrupt network features will result in an error (to learn more about reconciling versions, see Chapter 15, 'Working with a versioned geodatabase').

ArcInfo and ArcEditor

The Rebuild Connectivity tool in ArcMap rebuilds connectivity for a set of network features in an extent by re-creating their logical elements. Connectivity is established based on geometric coincidence using the same rules as described in *Building a Geodatabase*.

Performance considerations

Connectivity is established for new network features based on geometric coincidence. When you add or move a feature in a network, each feature class in the network must be analyzed so connectivity can be established. Performing a spatial query against each network class will determine if the new feature or moved feature is coincident with other network features at any point.

If the network is in an ArcSDE geodatabase, then analyzing for connectivity requires a number of spatial queries against the server. By using the edit cache while editing the network, these spatial queries don't go against the server, are much faster, and are not as much of a load on the server. When editing network data in an ArcSDE geodatabase, always use the edit cache. Using the edit cache can make your edits five times faster than without the cache, and it is not as much of a load on the server.

Creating network edges

Through the basic editor tools in ArcMap, you can create network edges in conjunction with your preexisting features. This includes adding network edges at an existing junction, along a complex edge, or along a simple edge.

Tip

Geometric networks and topology

It is important to distinguish geometric networks and their relationships from topology and its associated rules. To learn more about topology, see Chapter 4, 'Editing topology'.

See Also

For more information on the ArcMap snapping environment, see Chapter 3, 'Creating new features'.

Creating a new network edge at an existing junction

1. Add your network feature classes to ArcMap and add the Editor toolbar.
2. Click Editor and click Start Editing.
3. Zoom to the area where you want to add the new feature.
4. Click the Tool Palette dropdown arrow and click the Sketch tool.
5. Click the Task dropdown arrow and click Create New Feature.
6. Click the Target layer dropdown arrow and click the type of edge feature you want to create. ▸

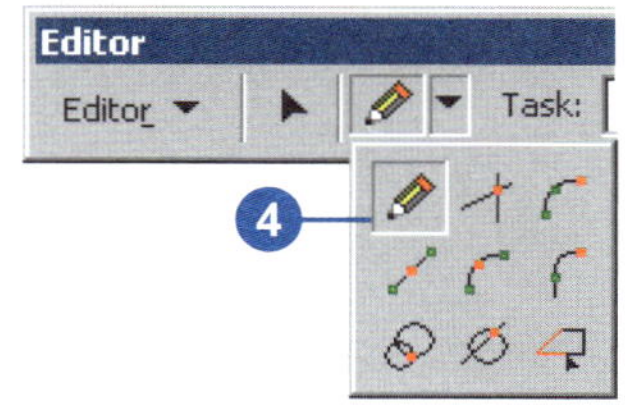

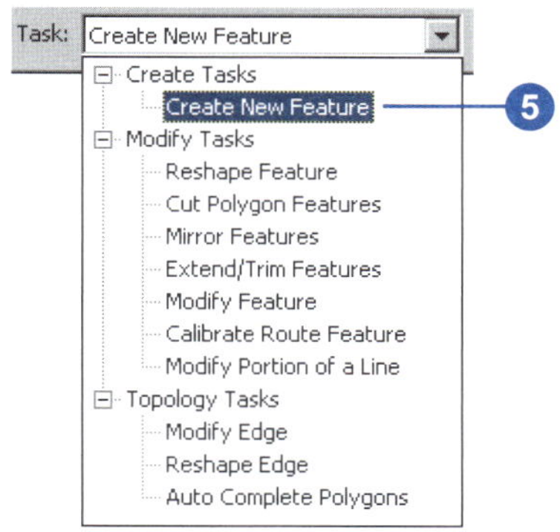

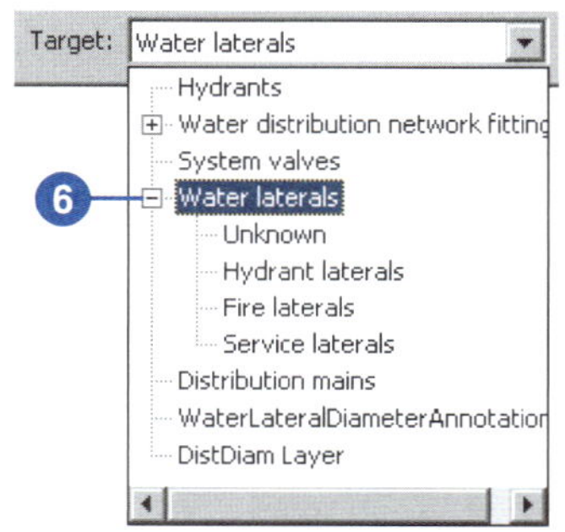

7. Check the appropriate boxes in the Snapping Environment window to set snapping to the vertex of the junction feature class to which you want to snap the new edge.
8. Move the pointer near the junction to which you want to snap this edge until the pointer snaps to it.
9. Click the map to create the new feature's vertices.
10. Double-click the last vertex to finish the feature. ▸

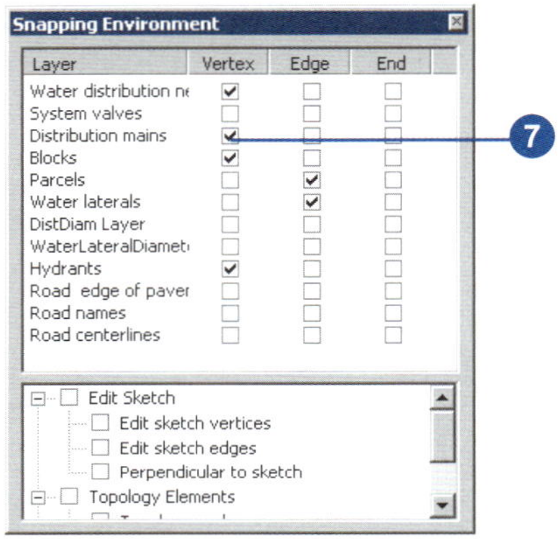

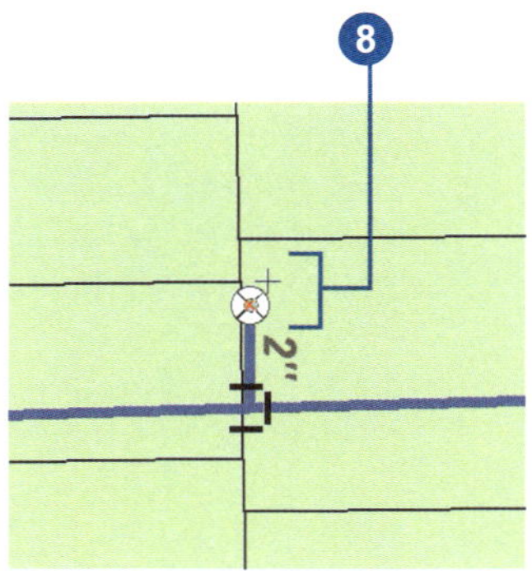

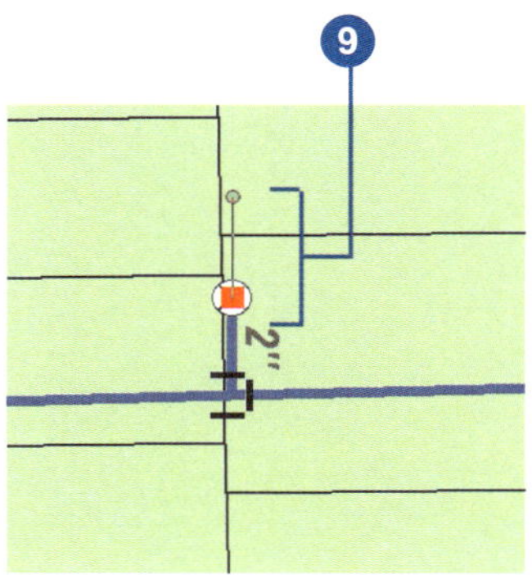

Tip

Default junction

You can specify what type of junction is placed at the free end of new edges by creating an edge–junction rule. For more information on connectivity rules, see Building a Geodatabase.

You have now created a new network edge. Since you snapped it to an existing network junction, it is automatically connected to the network.

If there is an edge–junction rule for the new edge with a default end junction type specified, this junction type will be placed at the free end of the new edge. If there is not an edge–junction rule that specifies a default junction, an orphan junction will be placed at the end of the new edge. For information on how to replace the orphan junction with another junction type, see 'Subsuming network junctions' later in this chapter.

The network junction or orphan junction is added to the end of the edge.

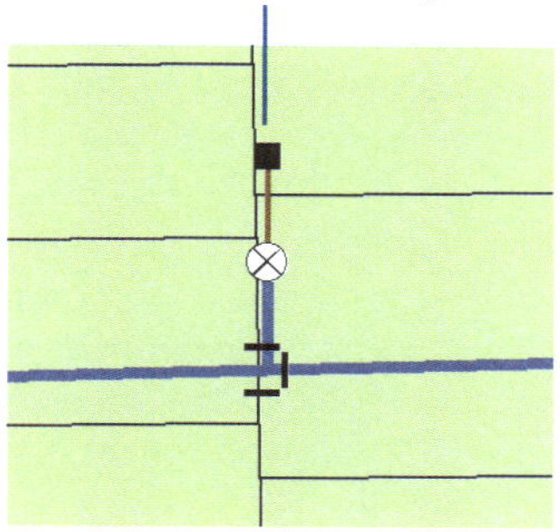

Tip

Adding junctions along complex edges

You can also snap a junction along a complex edge. Similar to snapping an edge, the junction is connected to the network. The complex edge is split in the logical network but remains a single feature.

Creating a new network edge along a complex edge

1. Follow steps 1–6 for 'Creating a new network edge at an existing junction' in this chapter.
2. Check the appropriate boxes in the Snapping Environment window to set snapping to the edge of the complex edge feature class to which you want to snap the new edge.
3. Move the pointer over the complex edge where you want the edge to snap until the pointer snaps to it.
4. Click the map to create the new feature's vertices.
5. Double-click the last vertex to finish the feature. ►

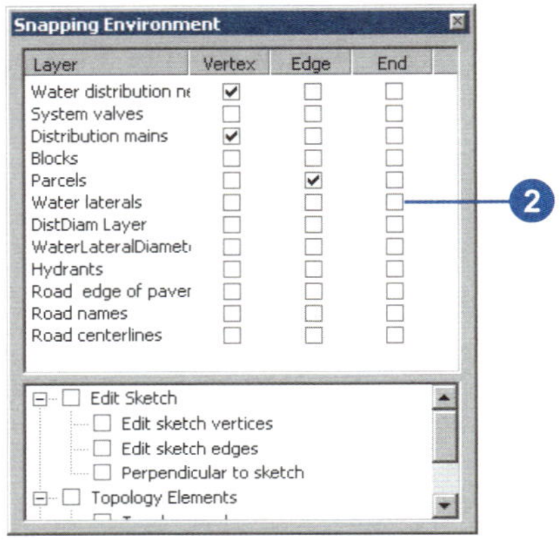

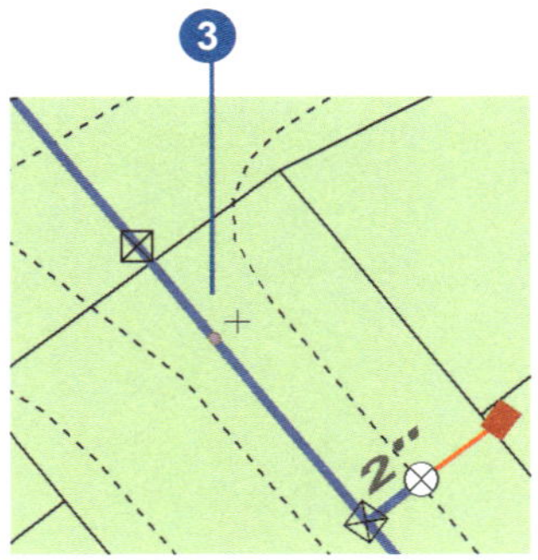

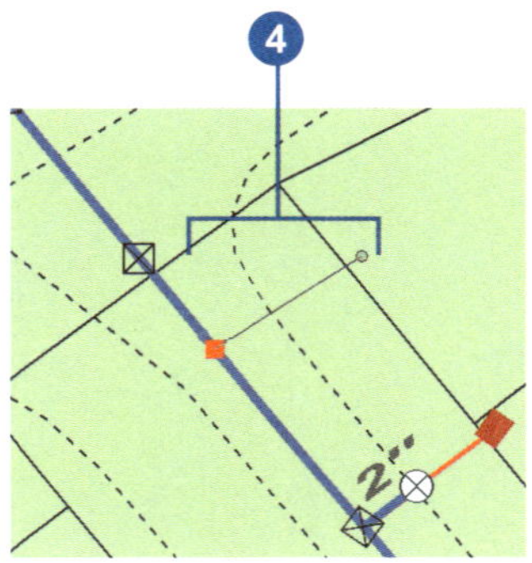

Tip

Default junction

You can specify what type of junction is placed at the free end of new edges by creating an edge–junction rule. For more information on connectivity rules, see Building a Geodatabase.

You have now created a new network edge. Since you snapped it to the edge of an existing edge, if there is an edge–edge connectivity rule between these edges, a new junction is created—the default junction type for that rule. If there is no edge–edge rule, then the new junction is the default network junction.

If there is an edge–junction rule for the new edge that has a default end junction type specified, this junction type will be added. If there is no edge–junction connectivity rule, an orphan junction is added.

Since the edge that was snapped to is a complex edge, it remains as a single feature but is split in the logical network.

6. Click the Edit tool.
7. Click the complex edge to which you snapped your new edge.

The entire edge is selected even though another edge and junction are connected along it. It remains a single feature.

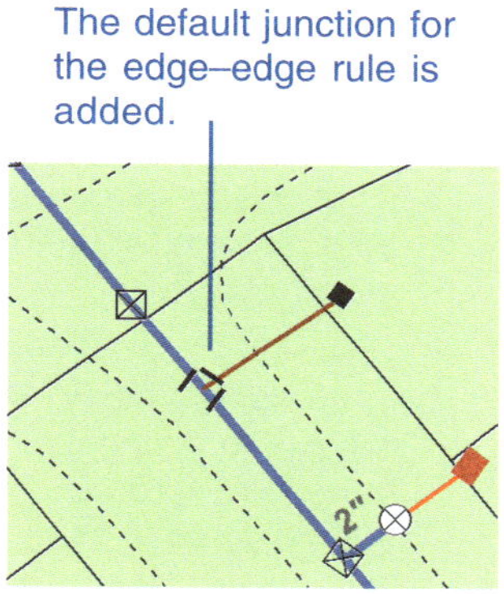

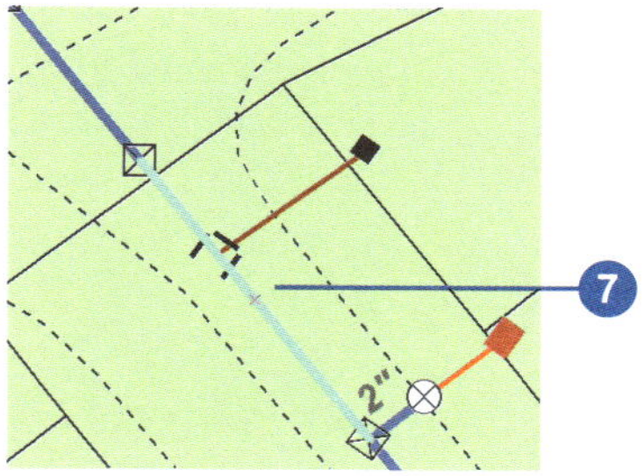

Creating a new network edge along a simple edge

1. Follow steps 1–6 for 'Creating a new network edge at an existing junction' in this chapter.
2. Check the appropriate boxes in the Snapping Environment window to set snapping to the edge of the simple edge feature class to which you want to snap the new edge.
3. Move the pointer near the simple edge to which you want to snap this edge until the pointer snaps to it.
4. Click the map to create the new feature's vertices.
5. Double-click the last vertex to finish the feature. ►

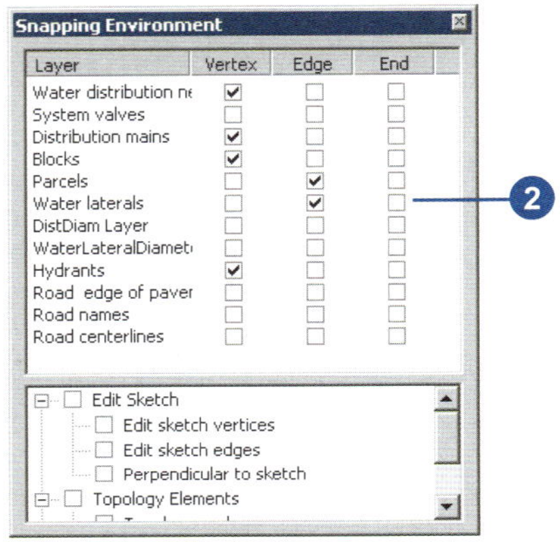

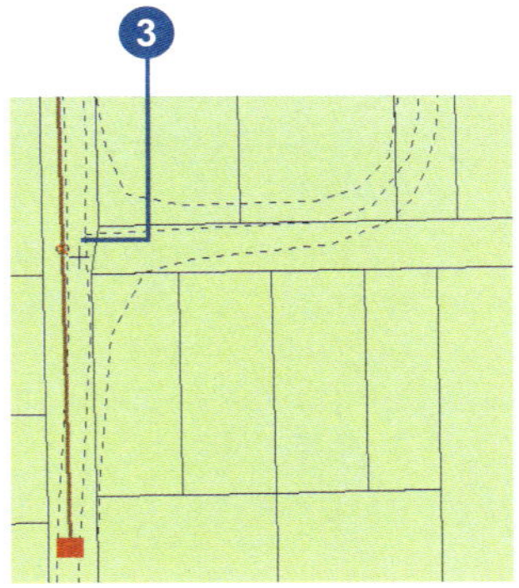

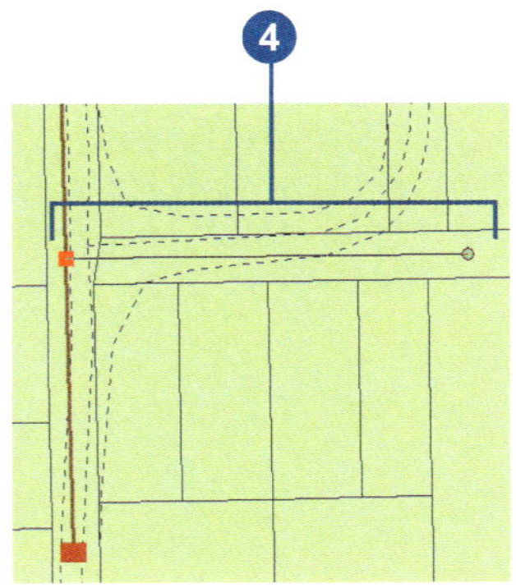

Tip

Adding junctions along simple edges

You can also snap a junction along a simple edge. Similar to snapping an edge, the junction is connected to the network. The simple edge is split into two new features.

Tip

Default junction

You can specify what type of junction is placed at the free end of new edges by creating an edge–junction rule. For more information on connectivity rules, see Building a Geodatabase.

See Also

For more information on split policies and how they affect attribute values, see Building a Geodatabase.

You have now created a new network edge. Since you snapped it to the edge of an existing edge, if there is an edge–edge connectivity rule between these edges, a new junction is created, which is the default junction type for that rule. If there is no edge–edge rule, then the new junction is the default network junction.

If there is an edge–junction rule for the new edge with a default end junction type specified, this junction type will be added. If there is no edge–junction connectivity rule, an orphan junction is added.

Because the edge that was snapped to is a simple edge, it is split into two new edge features. The value of the attributes in the new features is determined by their split policies.

The default junction for the edge–edge rule is added.

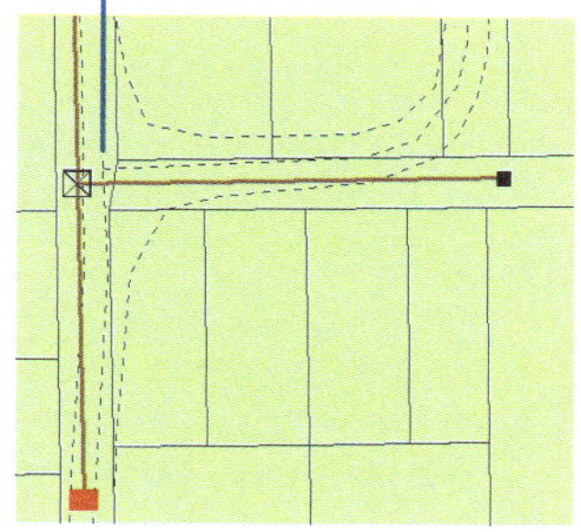

6. Click the Edit tool.

7. Click the simple edge to which you snapped your new edge.

There are now two edges split at the new junction.

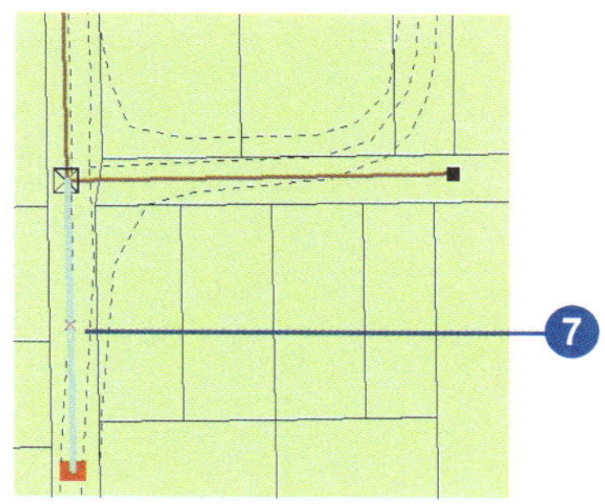

Subsuming network junctions

You may wish to replace network junction attributes with attributes associated with another junction type. For example, in a utilities network, a valve type or pole characteristic might change. Through basic editing functions, the properties of a junction can be subsumed by properties of another type.

1. Follow steps 1–4 for 'Creating a new network edge at an existing junction' in this chapter.
2. Click the Target layer dropdown arrow and click the type of junction feature you want to create.
3. Check the appropriate boxes in the Snapping Environment window to set snapping to the vertex of the junction feature class that you want to subsume.
4. Move the pointer near the junction you want to subsume with a new junction until the pointer snaps to it. ▸

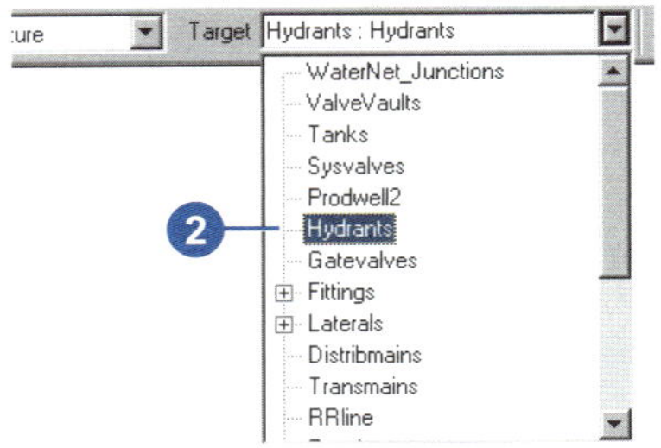

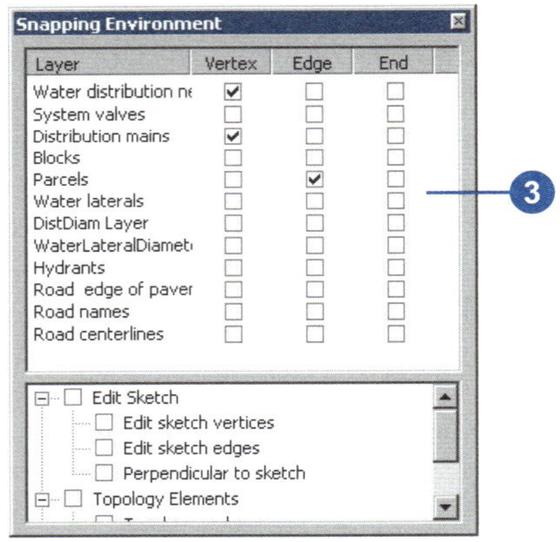

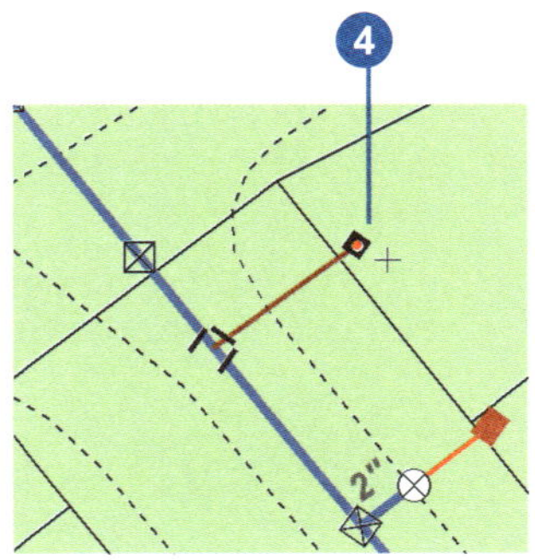

5. Click once to subsume the junction.

   The original junction is deleted and replaced with the new junction; network connectivity is maintained.

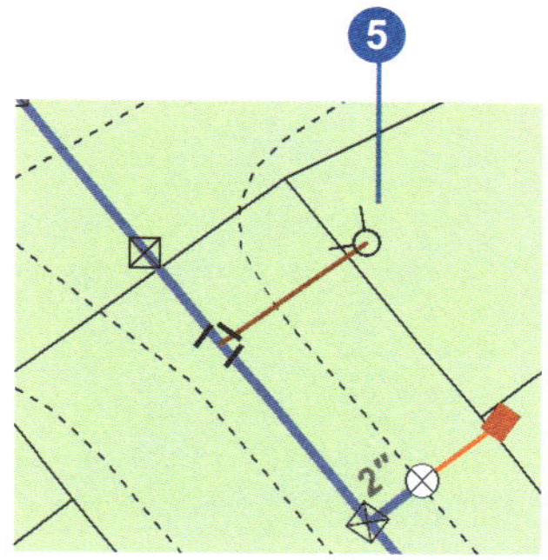

The original junction is deleted and replaced with the new junction.

Tip

Undoing network edits

If you move a network feature, other network features also move. Clicking the Undo button will undo the edits to all the affected features. To learn more about undoing edits, see Chapter 2, 'Editing basics'.

Moving existing network features

1. Follow steps 1 and 2 for 'Creating a new network edge at an existing junction'.
2. Click the Edit tool.
3. Click the network junctions and edges that you want to move.
4. Click and drag the features to the new location.

   Other network elements that are connected to the features rubber-band. This shows how other network elements are affected by moving the selected features.

   All of the features that rubber-banded while you dragged your selected features are automatically updated to maintain network connectivity.

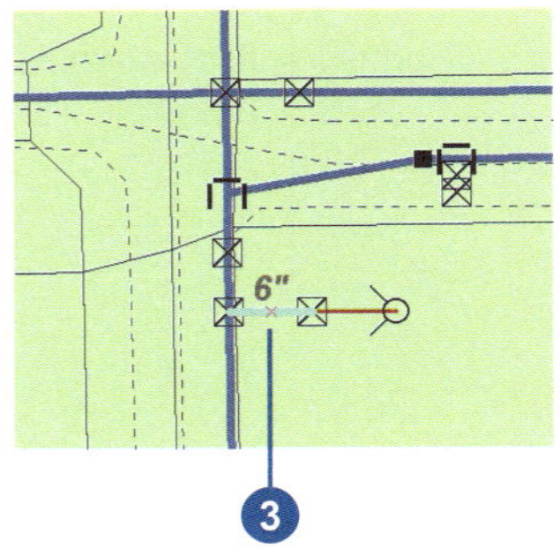

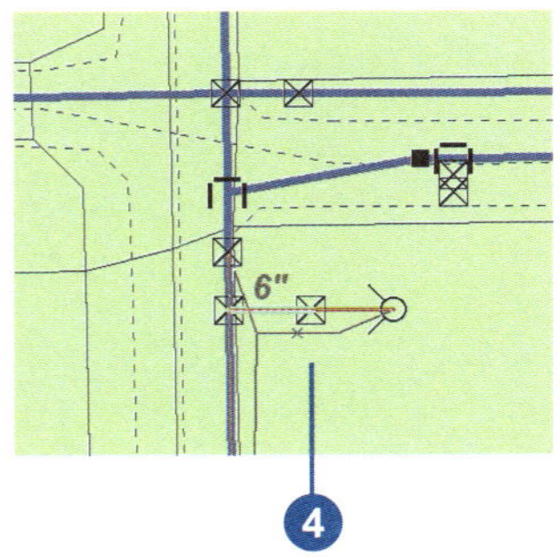

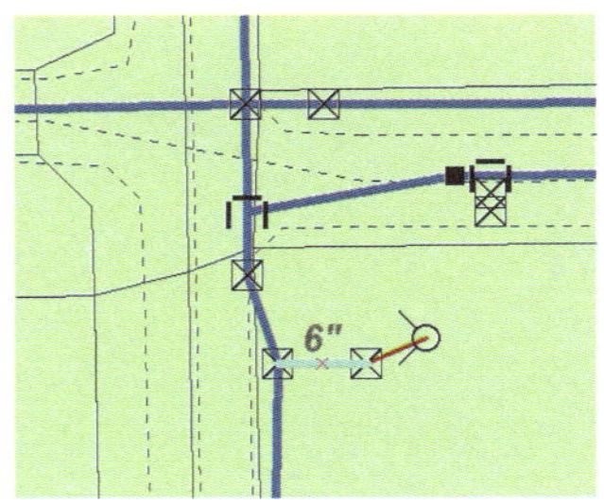

Tip

Ancillary roles

Not all network junction features can have ancillary roles. Those that don't have them can't act as sources or sinks in the network. To learn more about network ancillary roles, see Building a Geodatabase.

See Also

To learn how to set flow direction for a network and other network analysis tools, see Building a Geodatabase.

Altering a junction's ancillary network role

1. Click the Edit tool.
2. Click the network junction whose ancillary role you want to change.
3. Click the Attributes button.
4. Click the value for AncillaryRole.
5. If you want this junction feature to act as a sink in the network, click Sink.

   If you want this junction feature to act as a source in the network, click Source.

   If you don't want this junction feature to be either a source or a sink, click None.
6. Repeat steps 3–5 until all the junctions whose ancillary roles you want to change are updated.
7. Use the tools found in the Network Analysis toolbar to recalculate the flow direction of the network.

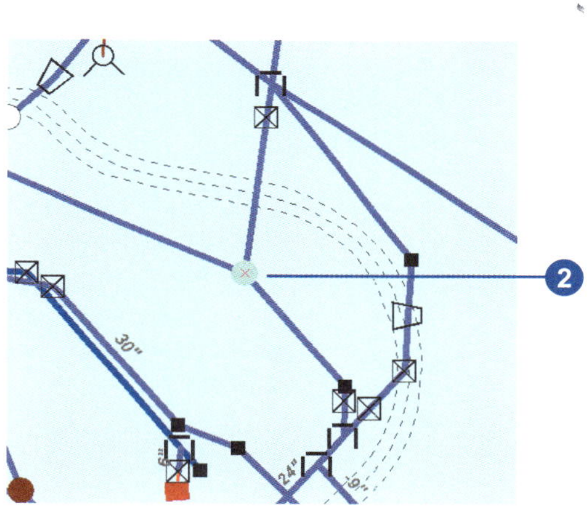

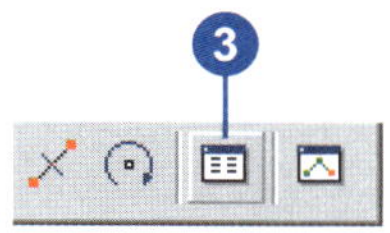

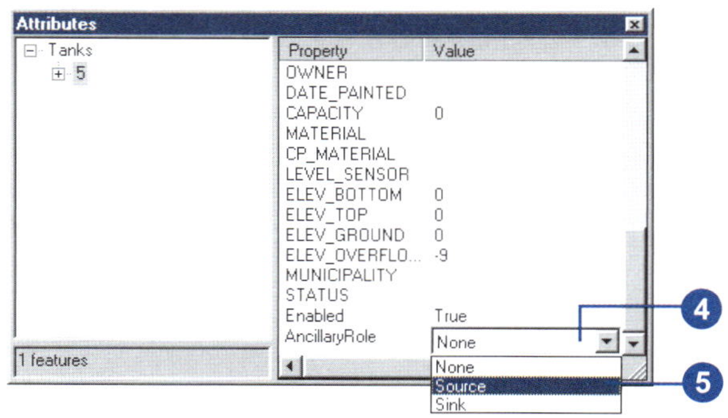

Enabled and disabled network features

If a feature in the network is disabled, it cannot be traced through. To learn more about network tracing, see *Building a Geodatabase*.

Enabling and disabling network features

1. Click the Edit tool.
2. Click the network feature you want to enable or disable.
3. Click the Attributes button.
4. Click the value for Enabled.
5. Click True if you want to enable the feature in the network.

   Click False if you want to disable the feature in the network.

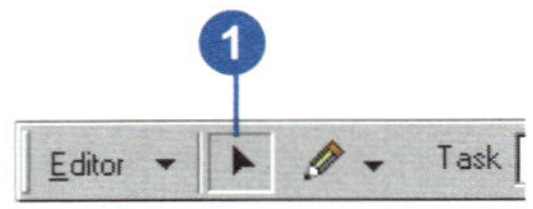

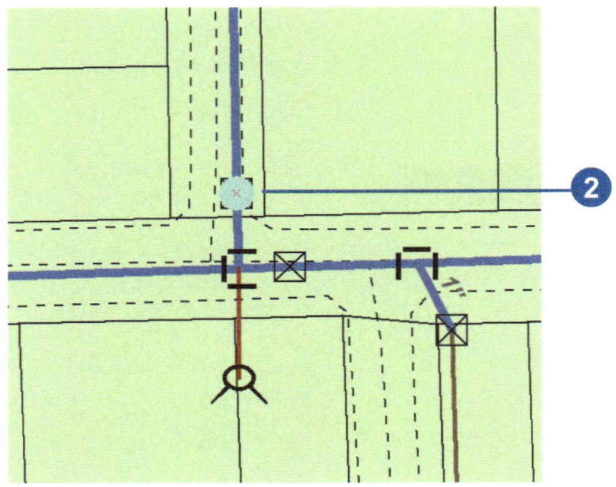

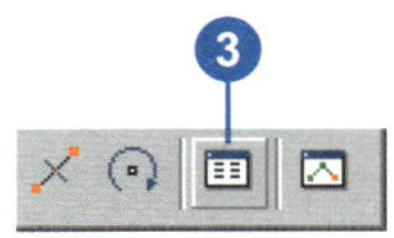

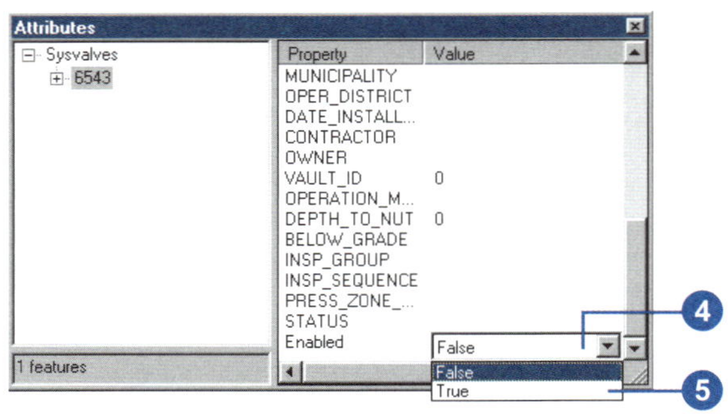

The Network Editing toolbar

The Network Editing toolbar contains tools used for managing geometric network connectivity. The connectivity between coincident network features can be explicitly changed. Instances can arise when parts of a geometric network need to be repaired; a feature having invalid geometry or connectivity between features is incorrect. Features with invalid geometry may exist within feature classes used in building a geometric network. These features can be identified and their geometries repaired. Network connectivity between features may become out of sync or corrupt. These features can be identified and their connectivity repaired and reestablished.

Adding the Network Editing toolbar

1. Click Editor, point to More Editing Tools, and click Network Editing.

   The Network Editing toolbar will appear in the ArcMap window.

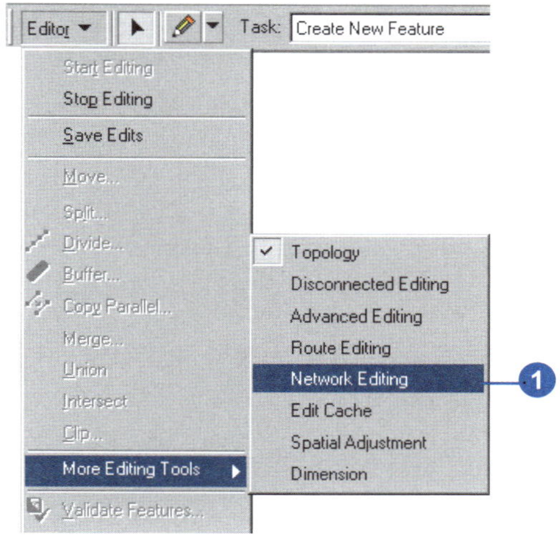

Tip

About connecting and disconnecting network features

In some cases, you may wish to disconnect a feature from the network. Disconnecting a feature does not delete it from the database; it removes the topological associations it has to other features in the network. Similarly, connecting a feature to the network creates topological relationships between the feature and its coincident features.

Disconnecting a feature from the network

1. Select the feature that you want to disconnect from the network.
2. Click the Disconnect tool on the Network Editing toolbar.

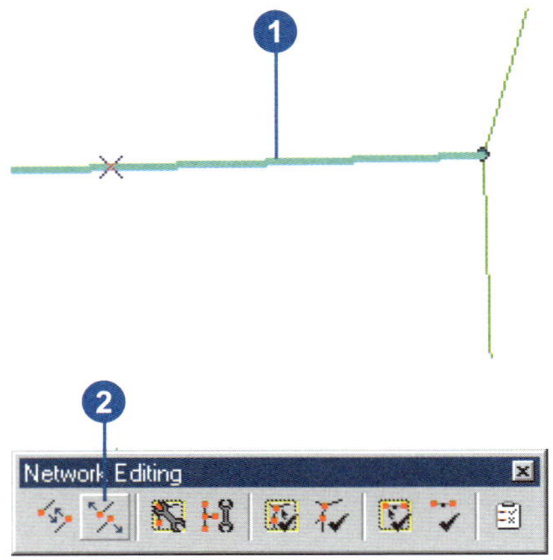

Connecting a feature to the network

1. Select the feature that you want to connect to the network.
2. Click the Connect tool on the Network Editing toolbar.

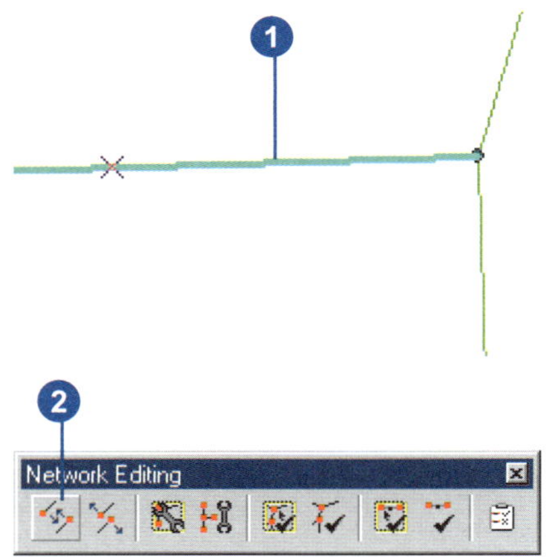

Tip

Assuring network connectivity

The Verify Connectivity tool verifies the connectivity between edges and junctions in the logical network.

The tool searches for features with no corresponding network elements, features with one or more missing network elements, features with duplicate network elements, features associated with corrupt or invalid network elements, and features associated with or connected to a nonexistent network feature.

Tip

Canceling the Verify Connectivity operation

Pressing Esc while defining the area of interest will cancel the operation.

Tip

Verifying the entire geometric network

The Verify Connectivity command will verify the connectivity of the entire network or any subset of selected features.

The Verify Connectivity command

Verifying network connectivity

1. Click one of the feature classes in the geometric network in the ArcMap table of contents.
2. Click the Verify Connectivity tool on the Network Editing toolbar.
3. Click and drag a box around the network features whose connectivity you wish to verify.

   The tool will examine the features and create a selection set of the network features whose connectivity is corrupt. You can then use the Rebuild Connectivity tool to repair the connectivity of the features. If all connectivity is correct, a message box will appear informing you so.

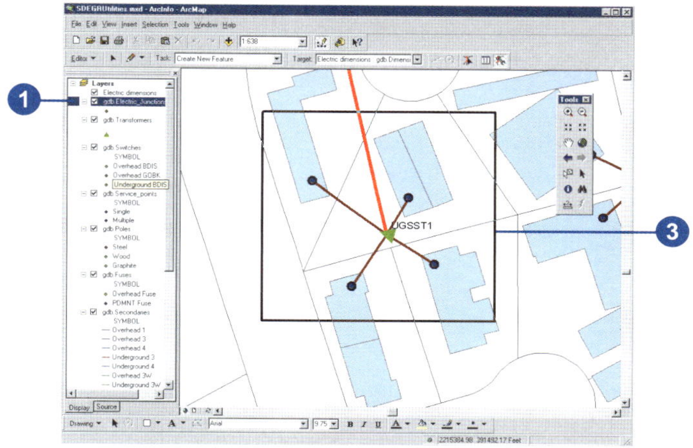

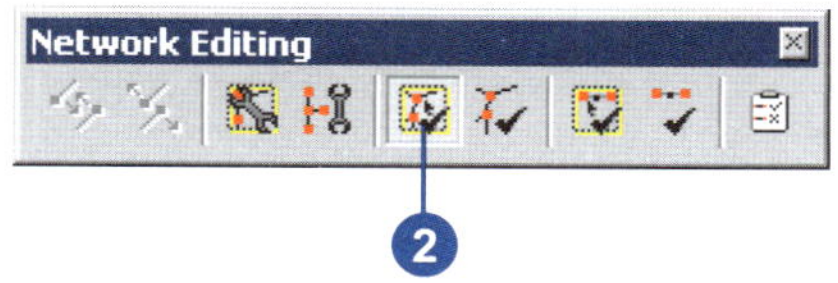

Verify Connectivity Tool

All network connectivity is correct.

OK

Tip

Rebuilding connectivity

During the course of editing, network connectivity is maintained on the fly. You do not need to use the Rebuild Connectivity tool unless the network connectivity has become corrupted for some of your network features.

Rebuilding network connectivity

1. Click one of the feature classes in the geometric network in the ArcMap table of contents.
2. Click the Rebuild Connectivity tool on the Network Editing toolbar.
3. Click and drag a box around the network features whose network connectivity you want to rebuild.

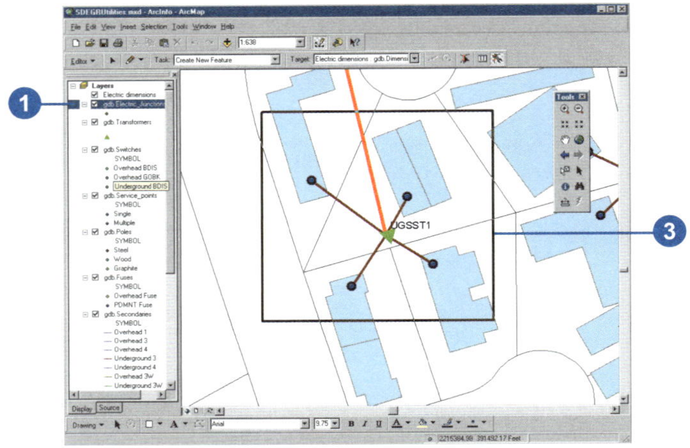

Tip

Ensuring network feature geometry

The Verify Network Feature Geometry tool verifies that the geometry of features that participate in the geometric network are valid shapes. Invalid shapes include features that have empty geometry, edge features that contain multiple parts, edge features that form a closed loop, and edge features that have zero length.

Tip

Canceling the Verify Network Feature Geometry operation

Pressing Esc while defining the area of interest will cancel the operation.

Tip

Verifying the entire geometric network

The Verify Network Feature Geometry command will verify the feature geometry of the entire network or any subset of selected features.

The Verify Network Feature Geometry command

Verifying network feature geometry

1. Click one of the feature classes in the geometric network in the ArcMap table of contents.
2. Click the Verify Network Feature Geometry tool on the Network Editing toolbar.
3. Click and drag a box around the network features whose geometry you wish to verify.

   The tool will examine the features and create a selection set of the network features whose feature geometry is corrupt. You can then use the basic editing tools to correct the geometry of the features. If all feature geometry is correct, a message box will appear informing you so.

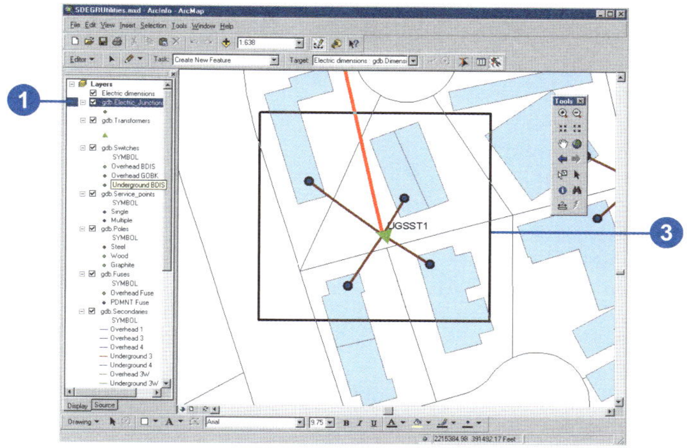

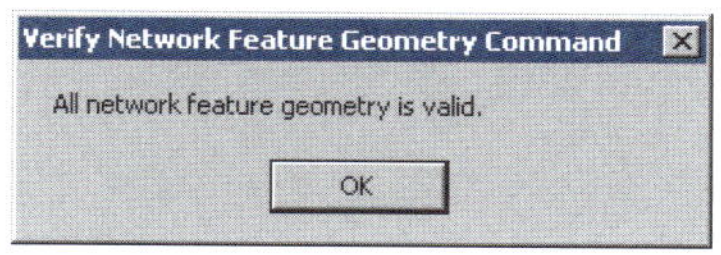

Tip

Repairing connectivity

The Repair Connectivity command will repair connectivity errors in the logical network.

During the process of repairing network connectivity, actions may occur that require review by the user. A message box will appear at the end of the process listing the type of warning, the feature class, and the object ID of the feature.

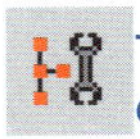
The Repair Connectivity command

Tip

Network Build Errors command

The Network Build Errors command creates a selection of the features with invalid shapes using the error table generated during the network build. Invalid shapes include lines that contain multiple parts, lines that form a closed loop, lines that have zero length and features that do not contain shapes. Once created, the selection set may be used to identify and find the features that cause the particular error. For more information on building geometric networks, see Building a Geodatabase.

The Network Build Errors command

Validating network features

The validation process ensures that the network features follow the connectivity rules. This enables you to find those features that were modified so as to be in conflict with the predetermined network connectivity rules.

For more information on how to create and modify connectivity rules, see *Building a Geodatabase*.

1. Click the Edit tool.
2. Click the network features you want to validate.
3. Click Editor and click Validate Features. ▸

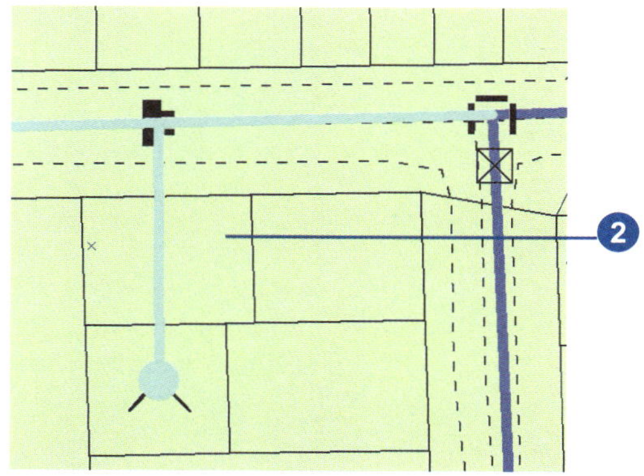

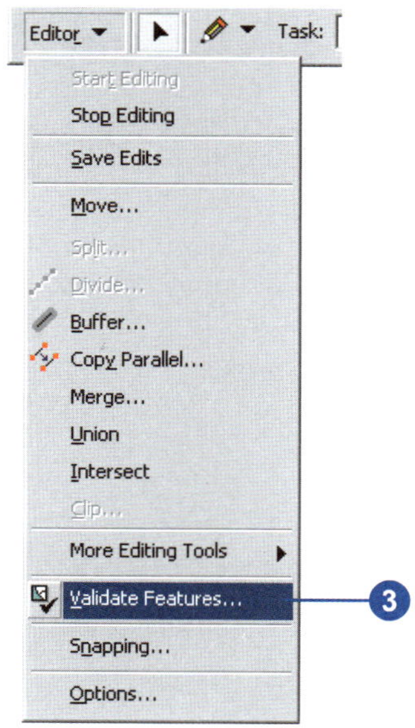

Tip

Validation rules

Network features may have connectivity rules as well as attribute and relationship validation rules associated with them. To learn more about validating attribute and relationship rules, see the tasks outlined earlier in this chapter.

If there are any invalid features, a message box appears telling you how many of the features are invalid. Only those features that are invalid remain selected.

4. Click OK.
5. Click one of the invalid network features.
6. Repeat step 3.

   A dialog box appears informing you why the selected feature is invalid.
7. Click OK.
8. Make the necessary edits to the network to make the feature valid. This may involve performing some of the network editing tasks described earlier in this chapter.
9. Repeat step 3—you should see a message box informing you that all the features are valid.
10. Click OK.

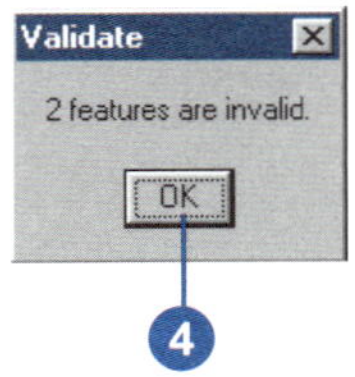

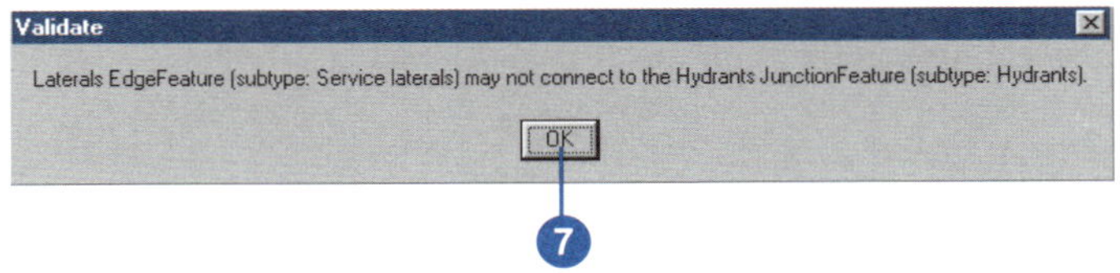

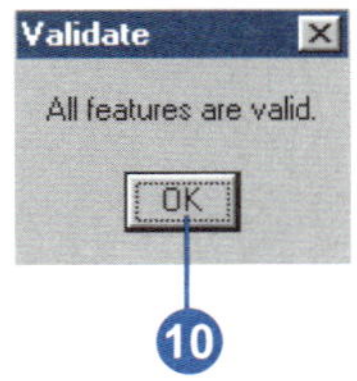

Editing annotation

13

IN THIS CHAPTER

- **Annotation in the geodatabase**
- **Editing annotation**
- **Converting labels to annotation**
- **Converting coverage annotation to geodatabase annotation**

In addition to geometry and location, geographic features may also have some descriptive text associated with them. For example, a feature class that contains streets may have text with the street's name associated with it. Annotation may also be a geographically located piece of text that exists independently of any other feature, such as the name of a mountain range on a physical map.

Annotation refers either to the process of automating text placement or to the text itself. This chapter describes how to create annotation for your feature classes and how to convert annotation that you have in coverages to geodatabase annotation.

Annotation in the geodatabase

Annotation in the *geodatabase* is stored in special feature classes called annotation classes. Unlike points, lines, and polygons, which are stored as ESRI Simple Features, annotation is stored as ESRI Annotation Features.

What is geodatabase annotation?

Like other feature classes in the geodatabase, all features in an annotation class have a geographic location and *attributes* and can either be inside a *feature dataset* or a standalone annotation class. Each annotation feature has its own symbology including font, color, and so on. Annotation need not only be text; it can also include shapes, such as boxes and arrows.

There are two kinds of annotation in the geodatabase: feature-linked annotation and nonfeature-linked annotation. Nonfeature-linked annotation is geographically placed text strings that are not associated with features in the geodatabase. An example of nonfeature-linked annotation is the text on a map for a mountain range. No specific feature represents the mountain range, but it is an area you would want to mark.

Feature-linked annotation is associated with a specific feature in another feature class in the geodatabase. The text in feature-linked annotation reflects the value of a *field* or fields from the feature to which it is linked. The annotation feature class participates in a composite *relationship* with the feature class that it is annotating. The annotation feature class is the destination class in the relationship, while the feature class it is annotating is the origin class. This means the feature controls the location and lifetime of the annotation (see *Building a Geodatabase*).

As an example of feature-linked annotation, a hydrant in a water network may be annotated with its pressure, which is stored in a field in the feature class. In the same network, the water transmission mains may be annotated with their names.

As with other composite relationships, the origin feature controls the destination feature. Therefore, when the origin feature is moved or rotated, the linked annotation moves or rotates with it. When the origin feature is deleted from the *database*, its linked annotation feature is also deleted. If the value of a field from which the annotation text is derived changes in the feature, the annotation feature has special *behaviors* to respond to those changes and to automatically update its text string.

In the water network example, a hydrant may be too close to a busy intersection and may need to be moved by 50 feet. When the hydrant is moved, its linked annotation moves with it. In the same network, the name of a transmission main may change. When the value in its name field is modified, the text stored in its linked annotation feature is automatically changed to reflect this.

Editing annotation

You can use the editing tools in ArcMap to edit both feature-linked annotation and nonfeature-linked annotation. While you can perform the same edit operations on both kinds of annotation, a feature with linked annotation has special behaviors when edited. The annotation feature will respond to changes in its linked feature, such as attribute updates.

You can use the drawing tools in ArcMap to edit the string, the symbology, or the location of any annotation feature. You can also use the drawing tools to create new annotation features.

When you edit features with linked annotation, all of the behaviors that apply to editing features with composite-related objects apply—for example, move and delete. In addition, when you edit the attributes of the feature, the linked annotation is also modified to reflect the changes.

You can use the same tools for editing related objects and breaking and creating new relationships; see Chapter 11, 'Editing related objects'.

Creating new nonfeature-linked annotation

1. Add your annotation feature class to ArcMap and add the Editor and the Draw toolbars.
2. Click Editor and click Start Editing.
3. On the Draw toolbar, click Drawing, point to Active Annotation Target, then check the feature class name of the annotation you want to create.
4. Use the font settings buttons on the Draw toolbar to set the properties of the annotation text you want to create.
5. Click the Create a new text element button.
6. Click the map where you want to add the new annotation feature.
7. Type the annotation's text string and press Enter. ▸

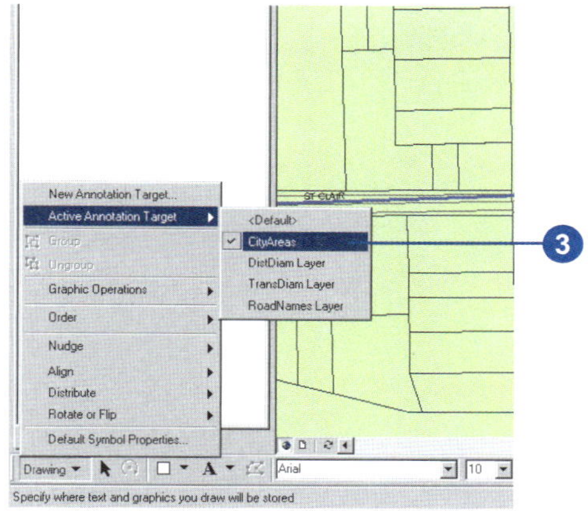

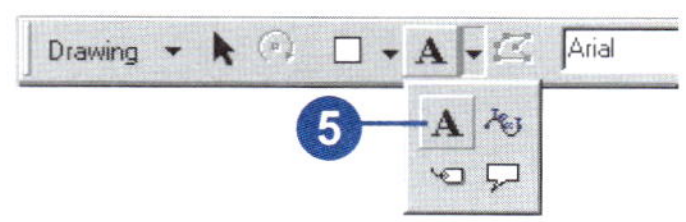

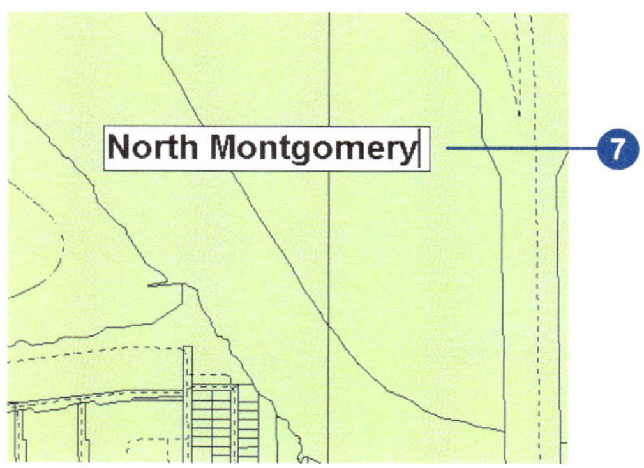

See Also

To learn more about geodatabase annotation, see Building a Geodatabase.

Nonfeature-linked annotation is added to your *map*.

Nonfeature-linked annotation is created.

See Also

The ability to add and edit annotation features with ArcMap is extensive and cannot be fully covered here. If you want to learn more about this tool, see Building a Geodatabase.

Editing an existing annotation feature

1. Click the Select Elements tool.
2. Click the annotation that you want to edit.
3. Right-click the annotation string you want to change and click Properties.
4. Edit the text string as desired.
5. Click OK. ▸

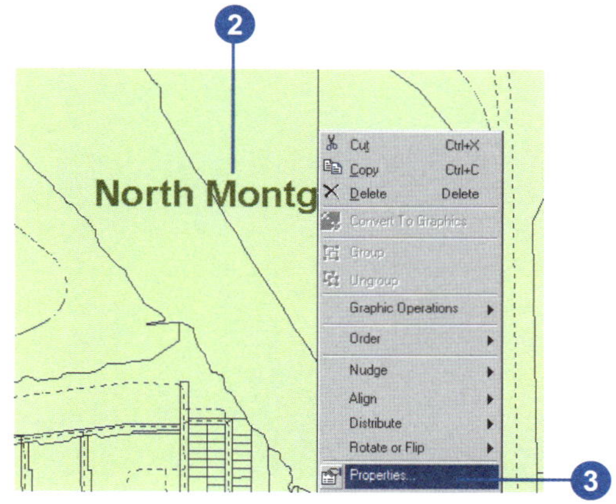

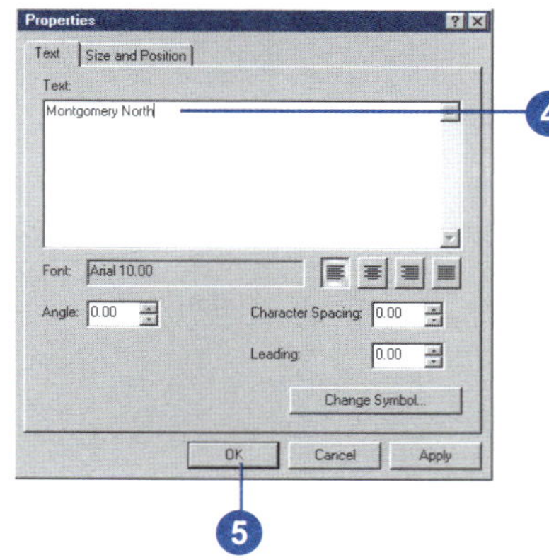

The text is updated on the map.

6. To move the annotation string, click and drag it to a new location.

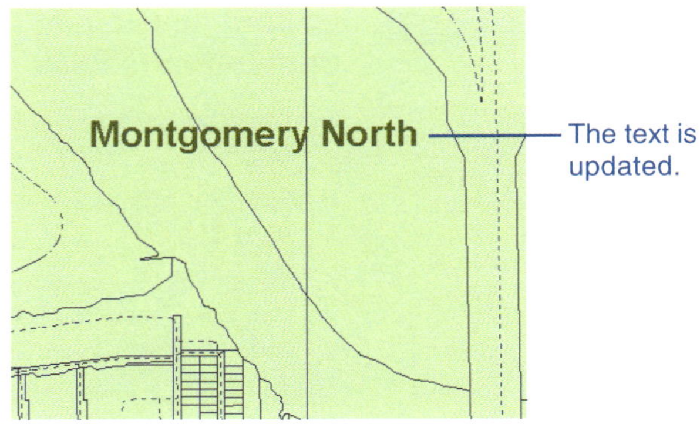

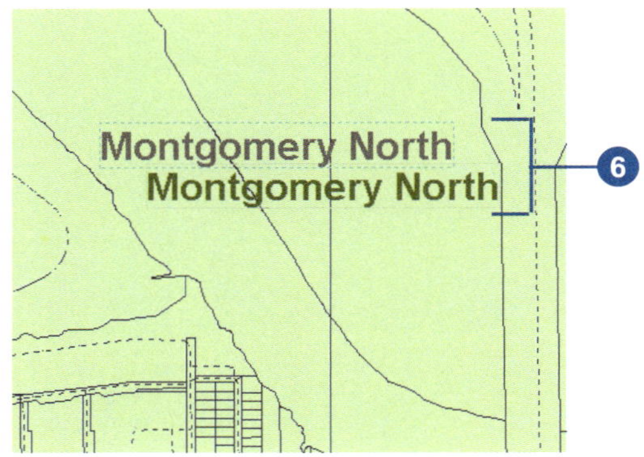

Creating new features with linked annotation

1. Zoom to the area where you want to add the new feature.
2. Click the Tool Palette dropdown arrow and click the Sketch tool.
3. Click the Task dropdown arrow and click Create New Feature.
4. Click the Target layer dropdown arrow and click the type of feature you want to create.
5. Click the map to create the new feature's vertices. ►

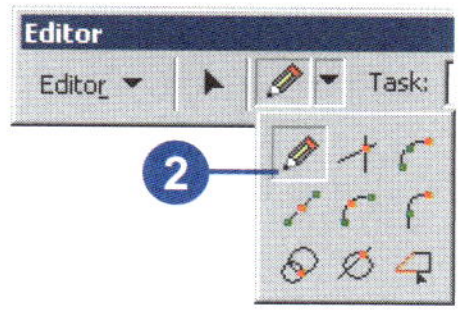

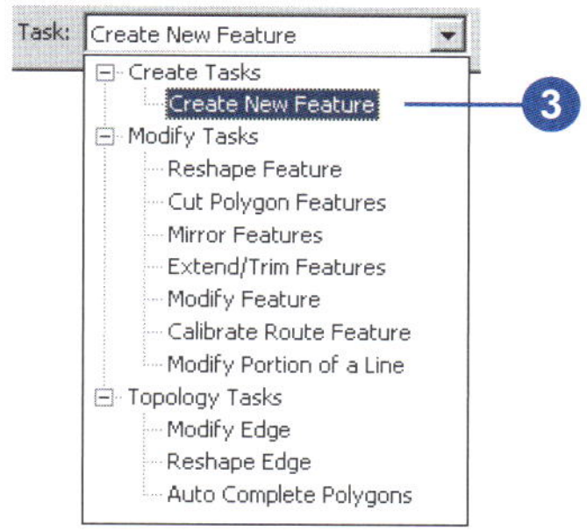

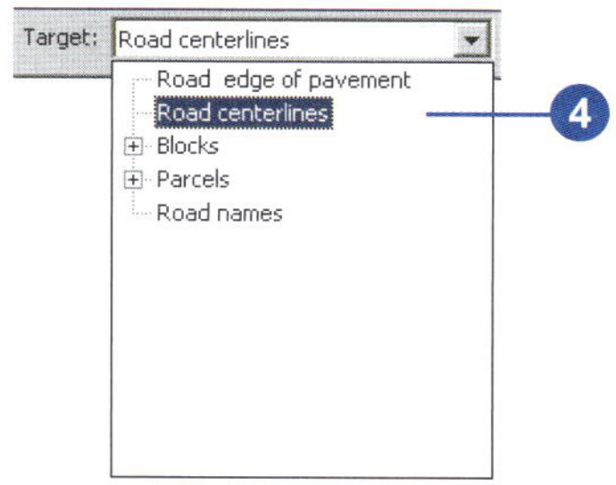

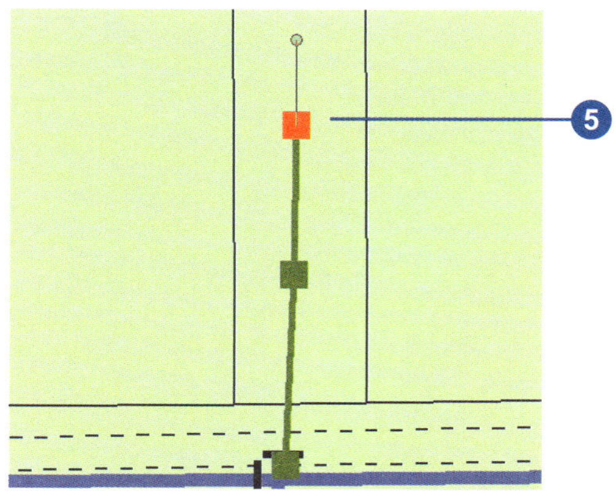

6. Double-click the last vertex to finish the feature.

   An annotation feature is automatically created and linked to the new feature.

   If your feature has default values for the field from which the annotation is derived, the annotation appears and reflects the values from those fields.

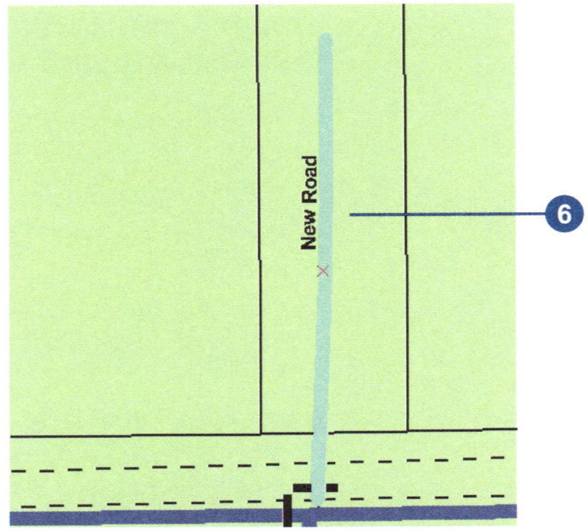

An annotation feature is created along with the new feature.

Tip

Composite relationships

The feature and its annotation participate in a composite relationship with each other. The feature is the origin feature, and the annotation is the destination feature. Editing these features is similar to editing any features that participate in a composite relationship.

Modifying features with linked annotation

1. Click the Edit tool.
2. Click the feature you want to edit.
3. Click the Attributes button.
4. Click the value from which the annotation is derived and type the changes.

   The annotation is automatically updated to reflect these changes.
5. Click the Close button to close the Attributes dialog box. ▸

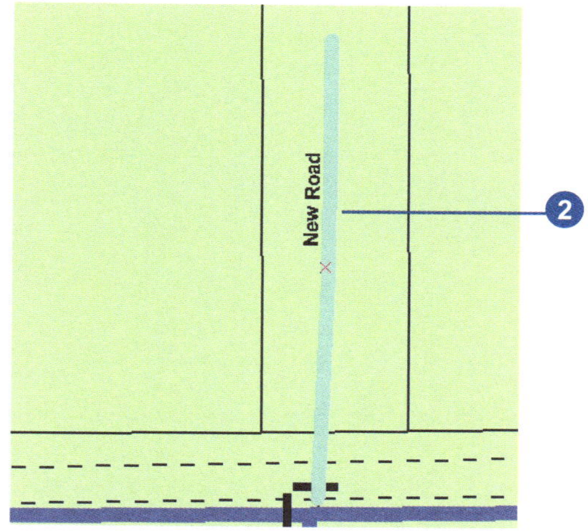

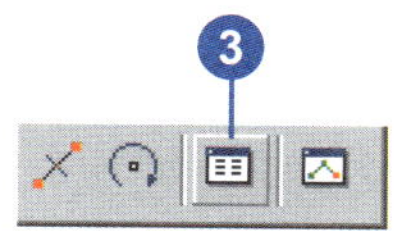

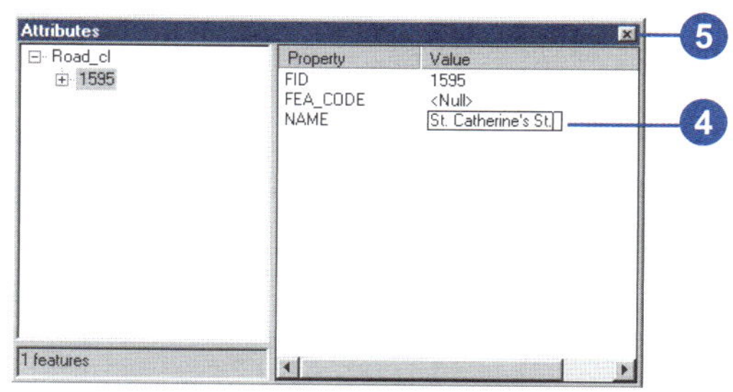

6. Click and drag the feature to a new location.

   The linked annotation feature moves the same x,y distance as the feature you moved. ▸

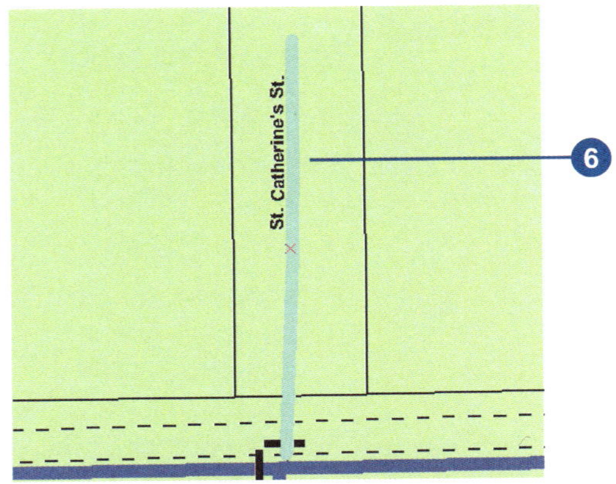

The linked annotation is automatically updated.

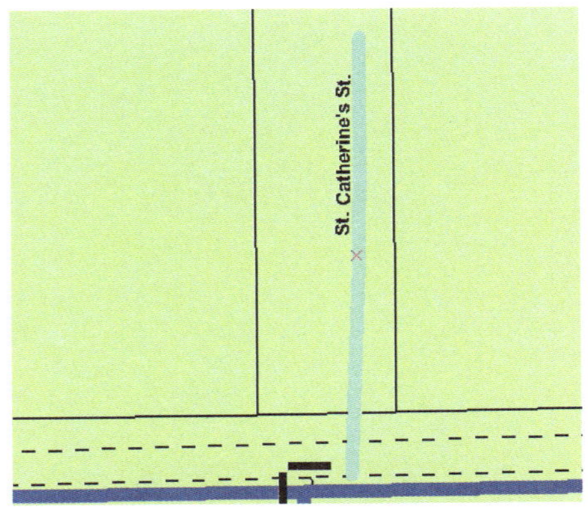

Linked annotation and feature move together to the new location.

7. Click the Rotate tool.
8. Click anywhere on the map and drag the pointer to rotate the feature to the desired location.

   The annotation rotates with the feature.
9. Click the Delete tool.

   The feature you selected, along with its linked annotation, is deleted from the database.

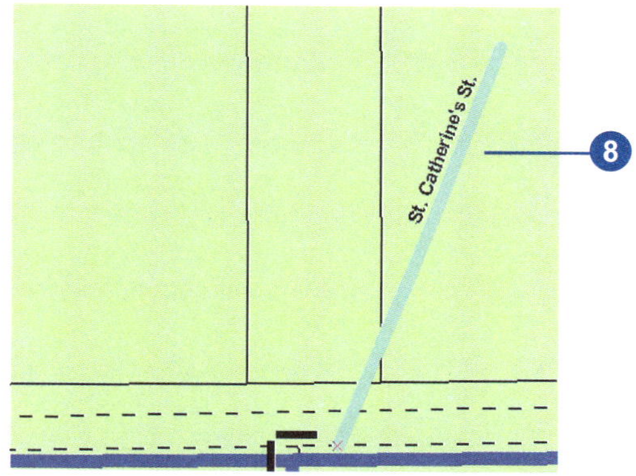

The feature is rotated along with its linked annotation.

Tip

Versioning

When possible, generate your annotation before you version your data.

Generating feature-linked annotation

1. In ArcMap, click the Add Data button to add a feature class and its linked annotation class to your map.
2. Click the Select Features button to select the features for which you want to generate annotation. To create annotation for all of the features, select all of the features.
3. Right-click the feature class in the table of contents.
4. Point to Selection.
5. Click Annotate Selected Features.
6. Check the related annotation classes in which you want to store the annotation.
7. Check the check box to add unplaced labels to the overflow window.
8. Click OK.

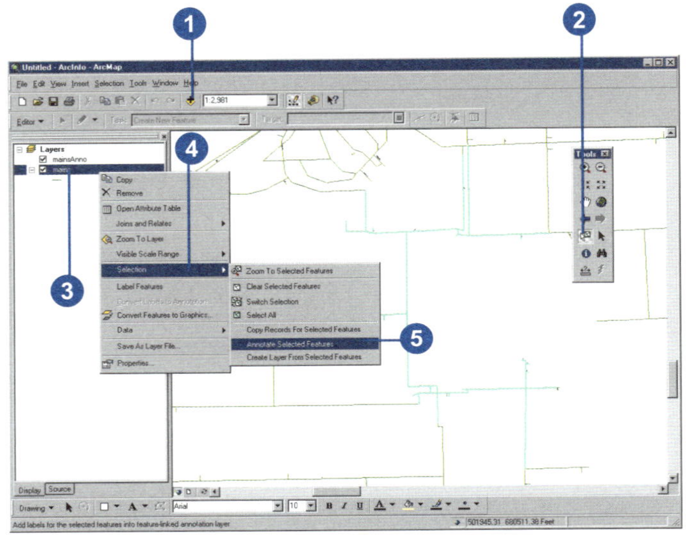

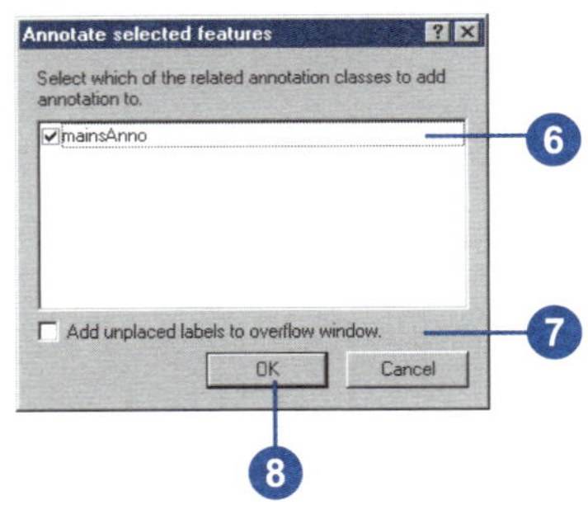

Converting labels to annotation

ArcMap lets you label features stored in a feature class. Once you have created labels, you can store them within the ArcMap document or as graphics in a database or convert them to annotation features.

If you choose to convert them to annotation features, ArcMap will create an annotation class to store the annotation. If you specify that you want them linked to the features, ArcMap will also create the relationship class to maintain the link.

The annotation class and relationship class are created inside the same feature dataset in which the feature class is stored or at the geodatabase level for a standalone feature class.

See Also

For a detailed discussion about labeling maps and the different advanced labeling methods you can use, see Using ArcMap.

1. In ArcMap, click the Add Data button to add the feature class for which you want to create annotation to your map.
2. Label the features in your map.
3. Right-click the feature class in the table of contents.
4. Click Convert Labels to Annotation. ►

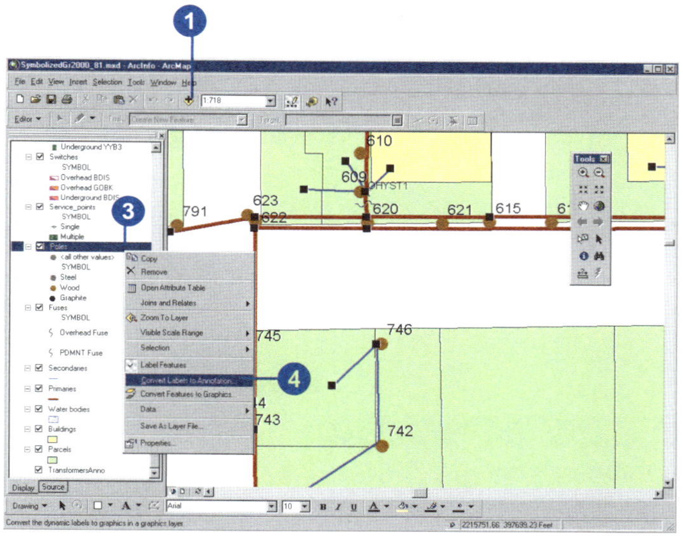

Tip

New annotation class

When all of the labels have been converted to annotation, the new annotation class is automatically added to the map.

Tip

Versioning

When possible, convert your labels to annotation before you version your data.

Tip

Annotation storage

When storing geodatabase annotation in a database management system (DBMS), the row length is between 80 and 100 bytes.

5. Click the feature class for which you want to save labels.
6. Click the third Annotation storage option.
7. Type a name for the new annotation class that will be created to store the annotation.
8. Click All features in the layer to create annotation for all of the features.

   Click Features displayed in the current extent to create annotation for features displayed only in the current extent of the map.

   Click Features currently selected to create annotation for the selected features only.
9. Click Display overlapping labels in the overflow window to display annotation that cannot be created without overlapping others.
10. Click OK.

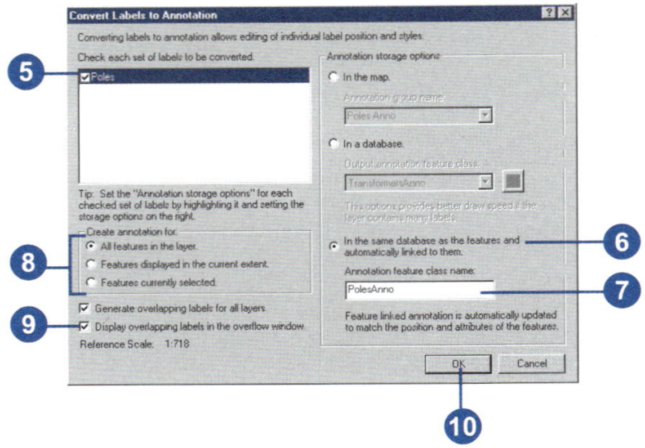

Converting coverage annotation to geodatabase annotation

ArcMap lets you convert annotation stored in a coverage annotation feature class to geodatabase annotation. You must convert the annotation into an existing annotation class in your geodatabase.

If the coverage annotation feature class contains attributes, when you convert it to geodatabase annotation, these attributes will automatically be converted as well.

The command to convert coverage annotation is available through the Customize dialog box in ArcMap.

See Also

For more information on how you can customize ArcMap, see Using ArcMap *and* Exploring ArcObjects.

Adding the Convert Coverage Annotation command to ArcMap

1. In ArcMap, click View, point to Toolbars, and click Customize.
2. Click the Commands tab in the Customize dialog box.
3. Click the Label category.
4. Drag the Convert Coverage Annotation command from the Commands list and drop it on the Standard toolbar.

   The command appears on the toolbar.
5. Click the Close button on the Customize dialog box.

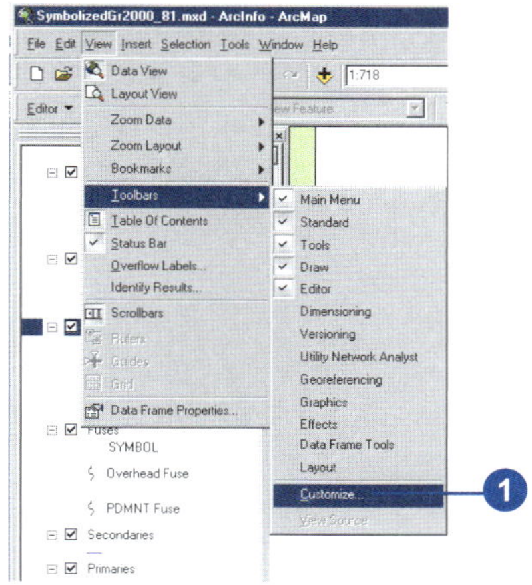

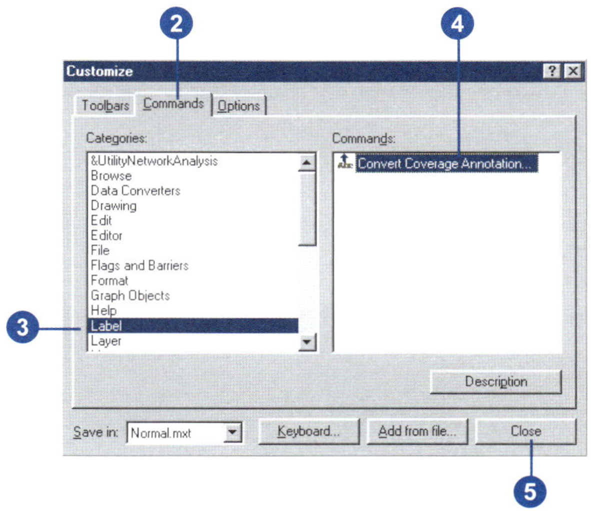

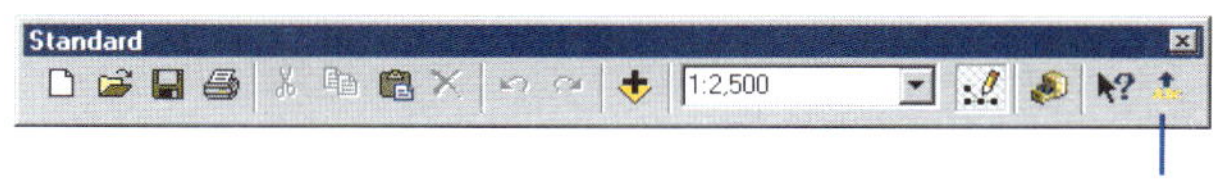

The command appears on the toolbar.

Tip

Annotation class

You can only convert coverage annotation into existing annotation classes. To create an annotation class, see the 'Creating new nonfeature-linked annotation' and the 'Generating feature-linked annotation' tasks in this chapter.

Tip

Annotation storage

When storing geodatabase annotation in a DBMS, the row length is between 80 and 100 bytes.

Tip

Target annotation class

The annotation class into which you convert your coverage annotation must exist in the geodatabase before you perform the conversion. You can create a new annotation class using ArcCatalog.

See Also

For more information on using ArcMap to add feature classes to maps, symbolize annotation, and create plots, see Using ArcMap.

See Also

For more information on symbolizing layers in ArcMap, see Using ArcMap.

Converting coverage annotation

1. Add to your map the coverage annotation feature classes you want to convert.
2. Set the symbology for the coverage annotation feature class. Create test plots of your data to ensure the symbology is correct before continuing to step 3.
3. Click the Convert Coverage Annotation command.
4. Check the annotation classes that you want to convert. You can convert multiple coverage annotation classes into a single geodatabase annotation class.
5. Click the option to convert the annotation into a database.
6. Click the Browse button to browse for an existing annotation class within your geodatabase.
7. Click Convert.
8. Click the Close button to close the Convert Coverage Annotation dialog box when your conversion is complete.

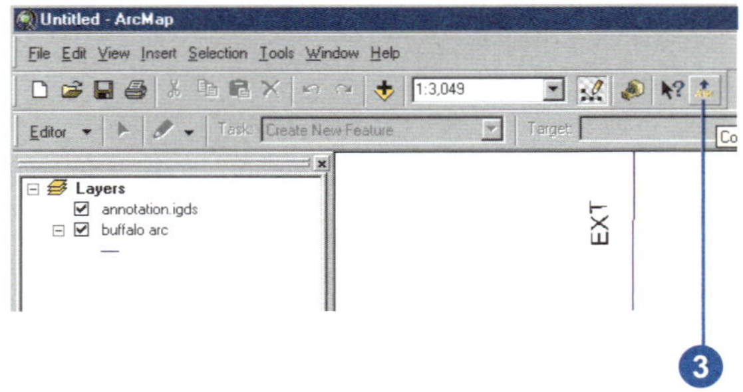

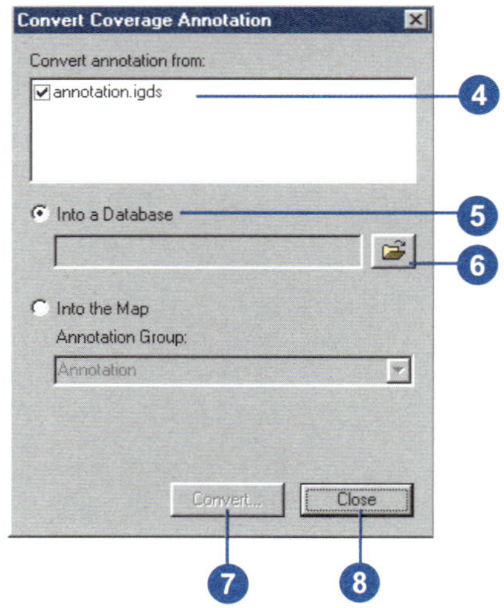

Editing dimension features

14

IN THIS CHAPTER

- **Editing dimension features**
- **Adding the Dimensioning toolbar**
- **Creating dimension features**
- **Modifying dimension features**

Dimension features are used to communicate information about the dimensions of geographic features, or distances between them, on a map. In this respect, they are similar to annotation. However, dimension features exclusively express distance measurements. Dimension features can be used to show the length of a property line, the distance between bridge spans, or the length of a feature along one axis.

Dimension features are stored in dimension feature classes in a geodatabase. Dimension feature classes can have one or more styles to ensure that the dimension features you create are consistent with your mapping standards.

ArcInfo and ArcEditor

Editing dimension features

Dimension features, unlike simple features, know how they are created. A dimension feature requires a specific number of points to be entered into the edit sketch to describe its geometry. The standard edit tools can be used to manually input the points required for these *construction methods*. In addition to the manual construction methods, there are several tools that allow you to create new dimension features from existing dimension features and other features. These tools are collectively called the Autodimension tools.

You can assign a style to a dimension feature when you create it or change an existing dimension feature's style. Dimension features draw and symbolize themselves based on the properties of their assigned style.

Construction methods

The type of dimension feature you are creating will dictate the number of points that are required as input.

The following is a list of dimension types and the number of points required for their construction:

- Simple aligned: two points

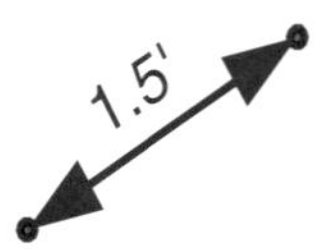

- Aligned: three points

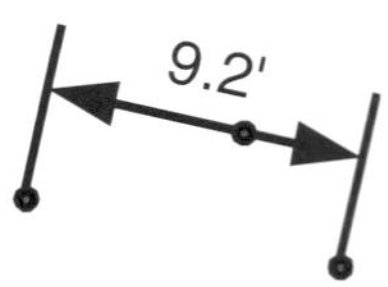

- Linear (horizontal and vertical): three points

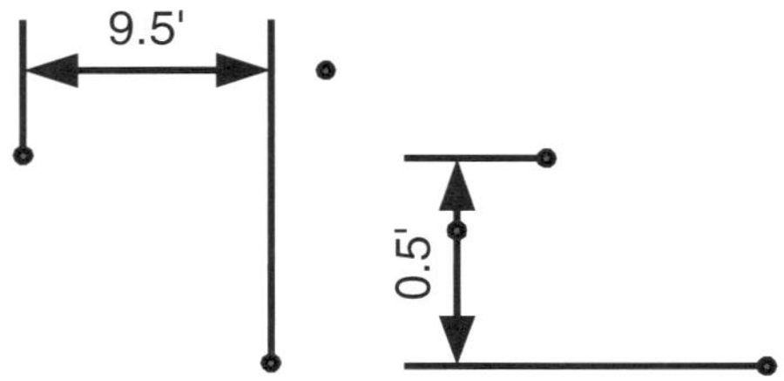

- Rotated linear: four points

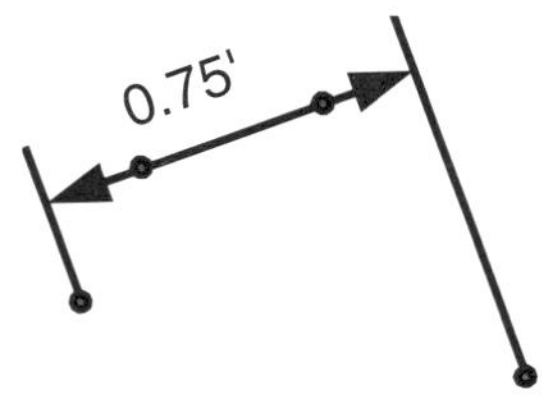

You can specify what type of construction method to use with which to create your dimension feature. The construction method dictates the type of dimension that is created. Each construction method knows how many points are required to create a specific kind of dimension feature. When using these methods, Finish Sketch is automatically called once you have input the correct number of points. The exceptions are the free construction methods.

The free construction methods also know how many points are required for input; however, they do not call Finish Sketch automatically. With the free construction methods, you can add as many points into the edit sketch as you need to construct your dimension feature. When you call Finish Sketch, the type of dimension feature that is created will depend on the number of points in your sketch.

The following summarizes the different construction methods:

- *Simple aligned:* creates simple aligned dimension features. It requires two points as input: the beginning dimension point and the end dimension point. Finish Sketch is automatically called after the second point is input.
- *Aligned:* creates aligned dimension features. It requires three points as input: the beginning dimension point, the end dimension point, and a third point describing the height of the dimension line. Finish Sketch is automatically called after the third point is input.
- *Linear:* creates horizontal and vertical dimension features. It requires three points as input: the beginning dimension point, the end dimension point, and a third point describing the height of the dimension line. The location of the third point relative to the beginning and ending dimension points will dictate whether the dimension feature is horizontal or vertical. Finish Sketch is automatically called after the third point is input.
- *Rotated linear:* creates rotated linear dimension features. It requires four points as input: the beginning dimension point, the end dimension point, a third point describing the height of the dimension line, and a fourth point describing the extension line angle. Finish Sketch is automatically called after the fourth point is input.
- *Free aligned:* creates simple aligned and aligned dimension features. It requires either two or three points as input. If you call Finish Sketch with two points in the edit sketch, a simple aligned dimension feature is created. If you call Finish Sketch with three points in the edit sketch, an aligned dimension feature is created. If you call Finish Sketch with less than two or more than three points in the edit sketch, the edit operation will fail.
- *Free linear:* creates horizontal linear, vertical linear, and rotated linear dimension features. It requires either three or four points as input. If you call Finish Sketch with three points in the edit sketch, a horizontal or vertical linear dimension feature is created. If you call Finish Sketch with four points in the edit sketch, a rotated linear dimension feature is created. If you call Finish Sketch with less than three or more than four points in the edit sketch, the edit operation will fail.

The Autodimension tools

The Autodimension tool palette contains four tools for automatically creating dimension features: Dimension Edge, Perpendicular Dimensions, Baseline Dimension, and Continue Dimension. Using these tools, you can create new dimension features based on existing dimension features or other features on your map.

Dimension Edge works on any type of feature. The Dimension Edge tool will automatically create a dimension whose baseline is defined by a line segment of an existing feature.

Perpendicular Dimensions simultaneously created two dimension features that are perpendicular to one another.

Baseline Dimension and Continue Dimension are both used only on existing dimension features. Baseline Dimension creates a new dimension feature whose beginning dimension point is based on the same beginning point of a previous dimension feature. The Continue Dimension tool creates a new dimension feature whose beginning dimension point is the same as the end dimension point of the existing dimension feature being continued. The Baseline Dimension and Continue Dimension tools create the same type of dimension as the existing dimension they are applied to and assign it the style selected in the Style dropdown list in the Dimensioning toolbar.

Dimension styles

All dimension features are associated with a *dimension style*. When you create a new dimension feature, you must assign it a dimension style. This dimension style must exist in the dimension feature class in which you are creating your new dimension feature. Once a dimension feature is created, it assumes all of the properties of its style. You can use the Attributes dialog box to modify some of those properties; however, some properties, such as the symbology of the dimension feature elements, cannot be modified.

To learn more about dimension styles and how to create them, see *Building a Geodatabase*.

The Dimensioning toolbar

The Autodimension tools and the controls for setting the construction method and assigning a dimension style are located on the Dimensioning toolbar. The controls on the Dimensioning toolbar are only active when you are editing, and the feature class selected in the Editor toolbar's Target dropdown list is a dimension feature class.

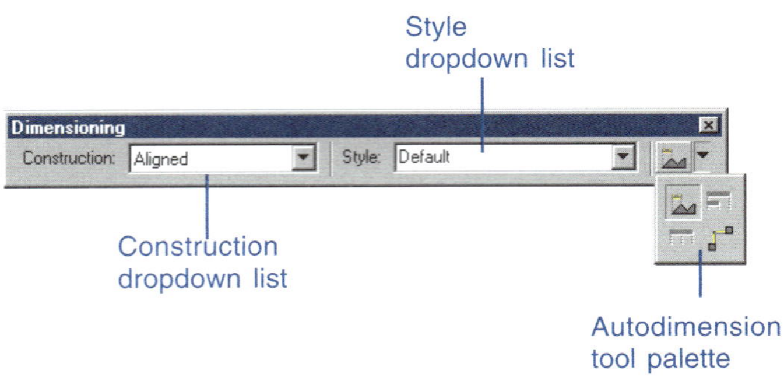

The Construction dropdown list contains all of the methods for constructing dimension features. The construction method dictates the number of points required to construct a dimension feature and the type of dimension feature that is created.

The Styles dropdown list contains all of the styles in the dimension feature class that are selected in the Target dropdown list in the Editor toolbar. New dimension features are created and assigned the style that is selected in the Style dropdown list.

It is important to remember that the Baseline Dimension and Continue Dimension tools will only be active if a dimension feature is selected. The Dimension Edge tool will be active when any feature is selected.

Modifying a dimension feature's geometry

Dimension features not only draw and symbolize themselves based on their assigned style but are also able to regulate the modification of their geometry. By using the editing tools in ArcMap that you use to modify the geometry of other types of features, you can modify a dimension feature's geometry while maintaining the correct configuration of points for a valid dimension feature.

When you are modifying a dimension feature, there are a series of vertices you can pick up and move with the Edit tool and move to alter the dimension feature's geometry. You can't add additional vertices or delete any of the existing vertices. The following diagram illustrates what aspect of a dimension feature is modified when one of these vertices is moved:

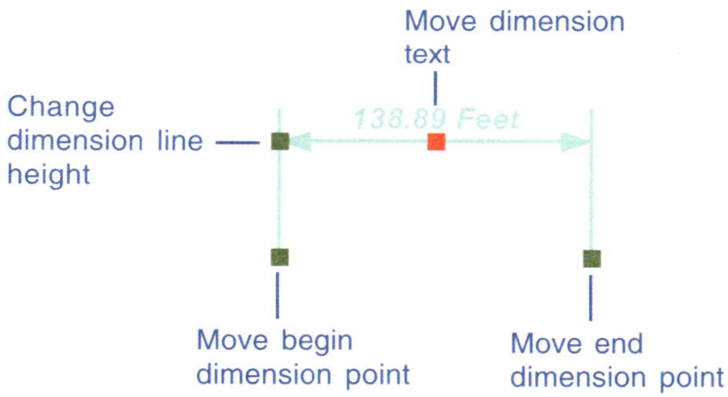

A dimension feature's geometry can be modified by moving a set of vertices while maintaining a valid dimension feature.

You can move a dimension feature's text away from its dimension line. The way the text is shown is dependent on the style chosen for the dimension feature. Some styles have line decoration including a leader line. For these styles, if you move the dimension feature's text far enough from the dimension line that it surpasses the leader line tolerance, then that leader line will automatically be displayed.

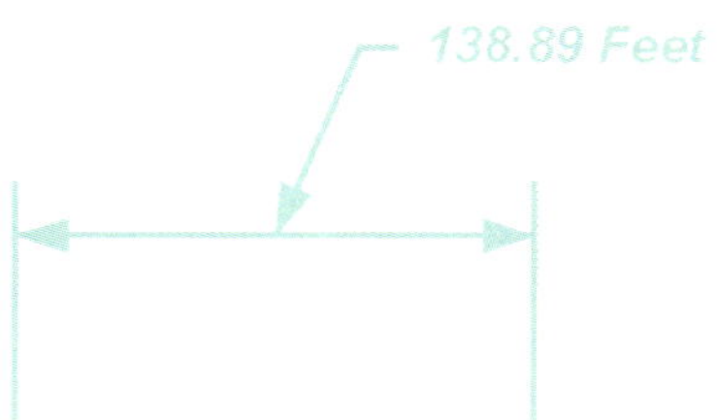

If a dimension feature's style has a text symbol with a leader line, that leader line is drawn when the text is moved farther away from the dimension line than the leader tolerance for the text symbol.

The extension line angle and the other properties of a dimension feature's geometry can be modified by altering the values of some of its fields. The following is a list of the fields you can modify for a dimension feature and how they correspond to its geometry:

Field	Property
BEGINX	x-coordinate of the beginning dimension point
BEGINY	y-coordinate of the beginning dimension point
ENDX	x-coordinate of the end dimension point
ENDY	y-coordinate of the end dimension point
DIMX	x-coordinate of the dimension line height
DIMY	y-coordinate of the dimension line height
TEXTX	x-coordinate of the text point (null if the text hasn't been moved relative to the dimension feature)
TEXTY	y-coordinate of the text point (null if the text hasn't been moved relative to the dimension feature)
EXTANGLE	Extension line angle

For more information on editing a feature's geometry, see Chapter 7, 'Editing existing features'. For more information on text symbols and text decoration, see *Using ArcMap*.

Modifying a dimension feature's properties

A dimension feature gets most of its properties from its style. However, you can override some aspects of a dimension feature's style. The following are the properties that can differ between a dimension feature and its style:

- Dimension line display
- Dimension line arrow symbol display
- Extension line display

For more information about dimension styles, see *Building a Geodatabase*.

In addition to overriding these style properties, you can also change a dimension feature's style, specify a custom value to use for the dimension text instead of the length of the dimension feature and change the extension line angle.

Dimension features can be modified using the Attributes dialog box. Dimension features have a special Attributes dialog box to allow you to easily modify their various properties. However, you can also use the standard Attributes dialog box to modify the properties of a dimension feature or to modify the values of fields that you have added to your dimension feature class.

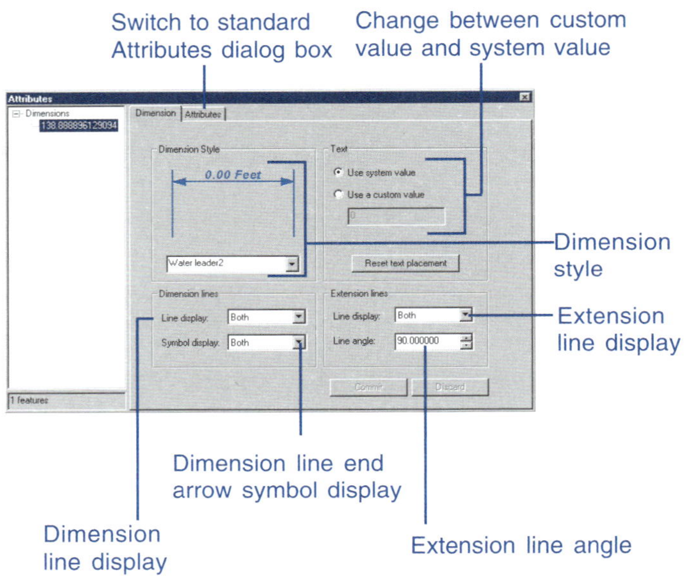

A dimension feature's properties can be modified by a Dimension Attributes dialog box or by using the standard Attributes dialog box.

Each property of a dimension feature that you can change in the dimensioning Attributes dialog box can also be changed by altering the values of some of its fields. A list of the fields that you can modify for a dimension feature and how they correspond to its properties are illustrated on the following page:

Field	Property
STYLEID	ID of the dimension style.
USECUSTOMLENGTH	0 indicates that the feature's length is used for the dimension text; 1 indicates a custom value is used for the dimension text.
CUSTOMLENGTH	Value used for the dimension text if USECUSTOMLENGTH is 1.
DIMDISPLAY	Null indicates both dimension lines are displayed; 1 indicates only the beginning dimension line is displayed; 2 indicates only the end dimension line is displayed; 3 indicates none of the dimension lines are displayed.
EXTDISPLAY	Null indicates both extension lines are displayed; 1 indicates only the beginning extension line is displayed; 2 indicates only the end extension line is displayed; 3 indicates none of the extension lines are displayed.
MARKERDISPLAY	Null indicates both dimension line end arrow markers are displayed; 1 indicates only the beginning dimension line end arrow marker is displayed; 2 indicates only the end dimension line end arrow marker is displayed; 3 indicates none of the dimension line end arrow markers are displayed.

For more information on using the Attributes dialog box in ArcMap, see Chapter 9, 'Editing attributes'.

Adding the Dimensioning toolbar

The Autodimension tools and the controls for setting the construction method and assigning a dimension style are located on the Dimensioning toolbar. The controls on the Dimensioning toolbar are only active when you are editing, and the feature class selected in the Editor toolbar's Target dropdown list is a dimension feature class.

Tip

Adding the toolbar

You can also add the toolbar by clicking the View menu, pointing to Toolbars, then clicking Dimensioning.

1. Right-click the Main menu.
2. Click Dimensioning.
3. Dock the toolbar to the ArcMap window. Now each time you start ArcMap the toolbar will be displayed.

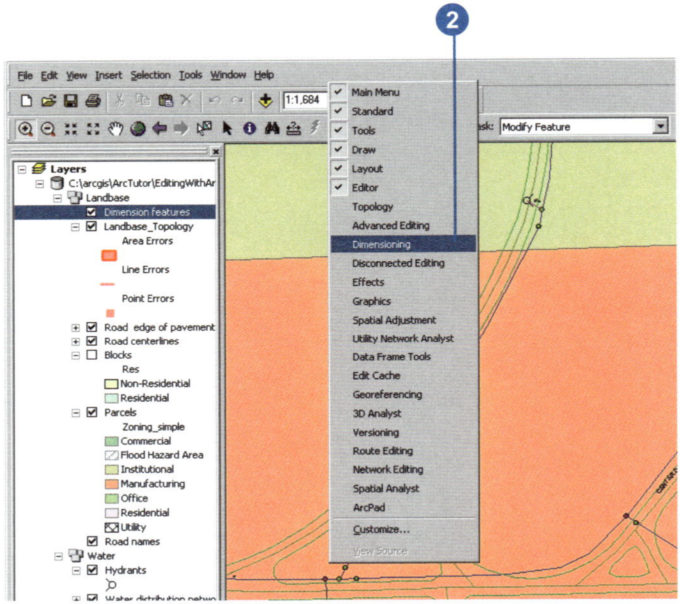

Creating dimension features

By using the tools provided on the basic Editor toolbar and those provided on the Dimensioning toolbar, you can create many types of dimension features.

Tip

The edit sketch

ArcMap contains many tools to help you enter points into your edit sketch. All of the same tools can be used to enter the points required for the various dimension feature construction methods.

Creating a simple aligned dimension feature

1. Add your dimension feature class to ArcMap, then add the Editor toolbar and the Dimensioning toolbar.
2. Click Editor and click Start Editing.
3. Zoom to the area where you want to add the new feature.
4. Click the Tool palette dropdown arrow and click the Sketch tool.
5. Click the Task dropdown arrow and click Create New Feature.
6. Click the Target layer dropdown arrow and click the dimension feature class. ►

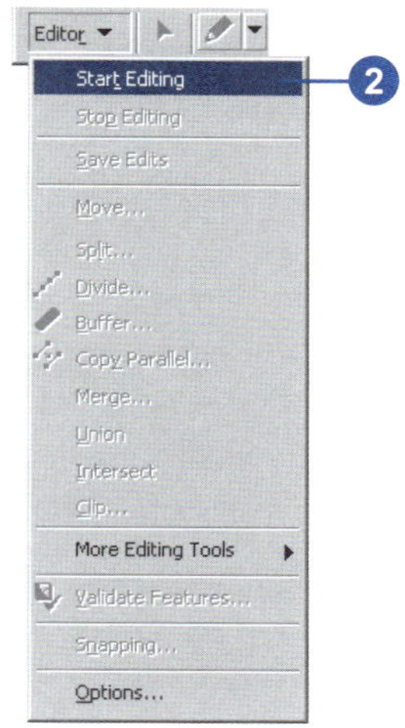

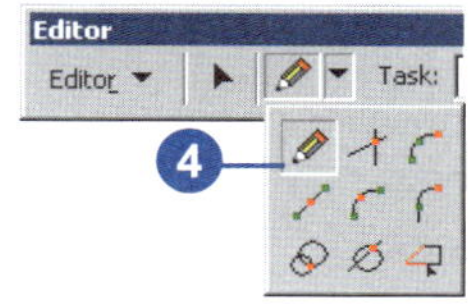

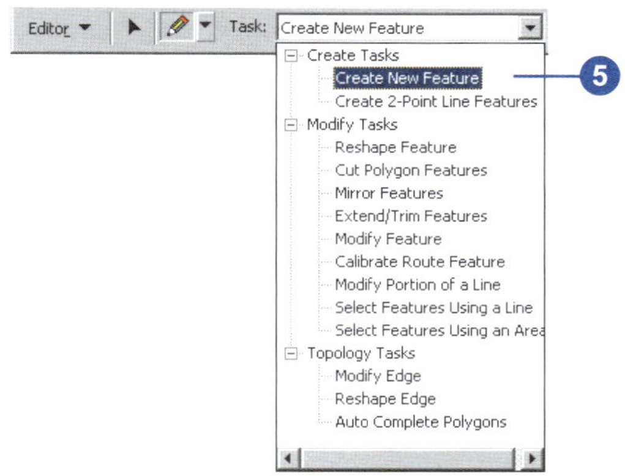

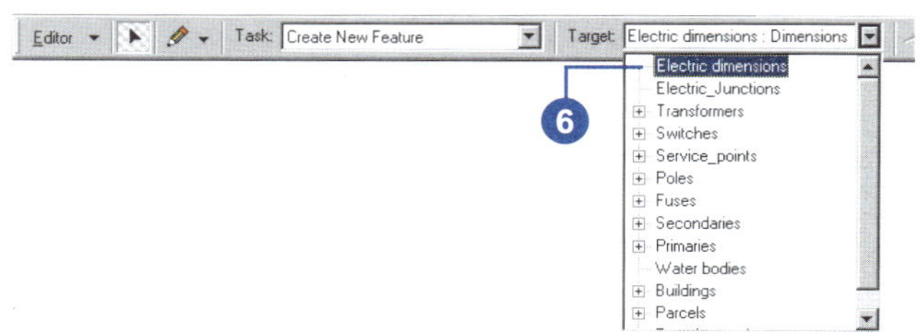

Tip

Edit sketch display

When creating dimension features, the edit sketch will actually show you how the resulting dimension feature will look as you move your mouse.

The exceptions are the free dimension construction methods. With these construction methods, the edit sketch display is the same as that for creating simple features.

Tip

Using the Magnifier Window

To more accurately place your vertices, use the Magnifier Window found under Window on the Main menu. To learn more about the Magnifier Window, see Using ArcMap.

Tip

Adjusting the Snapping Environment

By adjusting the settings in the Snapping Environment dialog box, you can more accurately place your vertices. To learn more about the Snapping Environment, see Chapter 3, 'Creating new features'.

7. Click the Style dropdown arrow and click the style you want your new dimension feature to have.
8. Click the Construction dropdown arrow and click Simple Aligned.
9. Click the map at the beginning dimension point to start the edit sketch.

   As you move the mouse, you will see that the new dimension dynamically draws itself with your mouse location as the end dimension point. ►

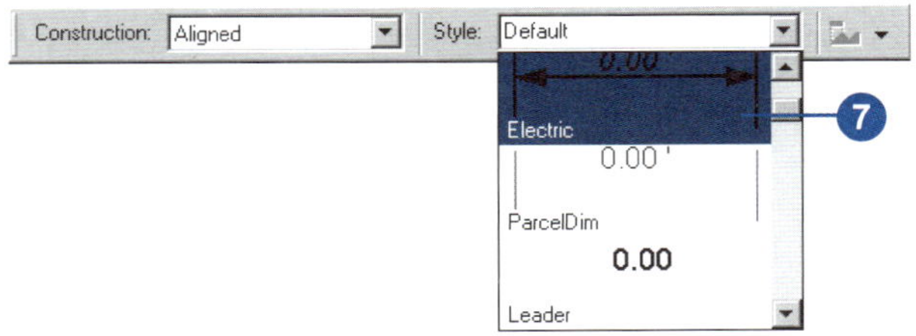

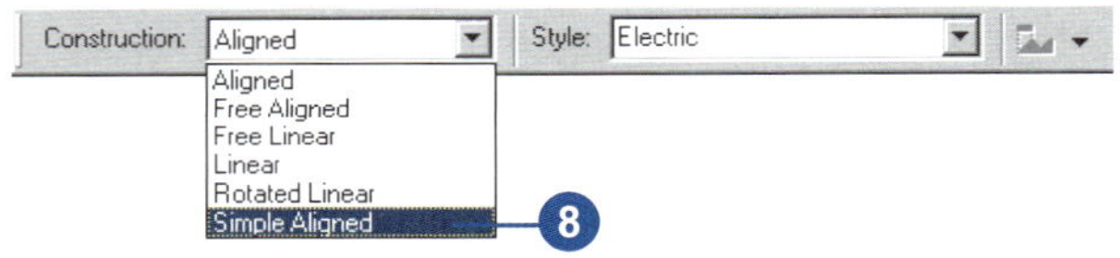

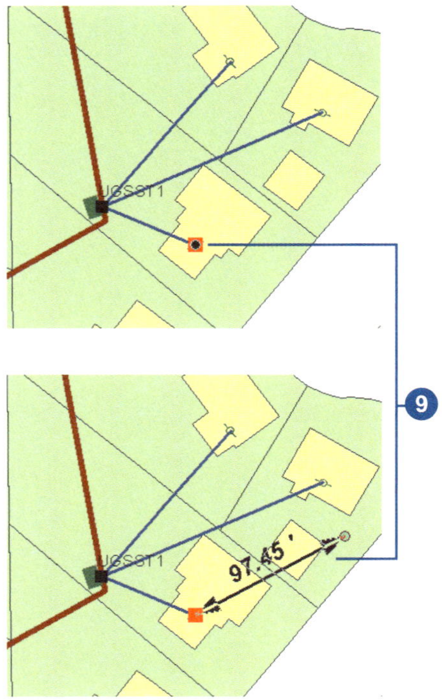

10. Click the map at the end dimension point.

    The sketch is automatically finished and the new simple aligned dimension feature is created with the style you selected.

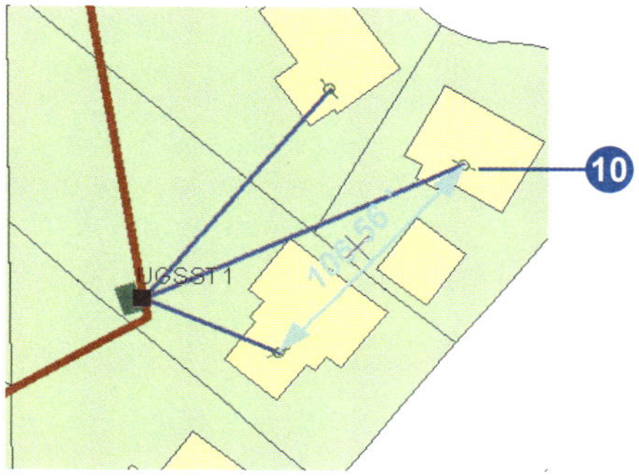

Tip

Aligned dimension features

An aligned dimension feature's dimension line is always parallel to its baseline.

To learn more about the different types of dimension features, see Building a Geodatabase.

Creating an aligned dimension feature

1. Follow steps 1–7 for 'Creating a simple aligned dimension feature' in this chapter.
2. Click the Construction dropdown arrow and click Aligned.
3. Click the map at the beginning dimension point to start the edit sketch.

   As you move the mouse, you will see that the new dimension dynamically draws itself with your mouse location as the end dimension point. ►

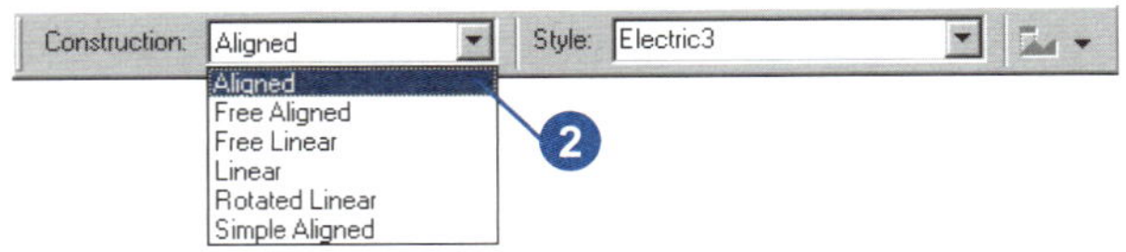

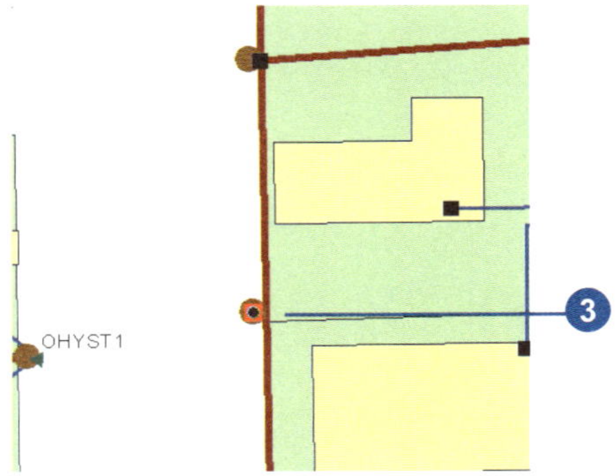

4. Click the map at the end dimension point.

   The new dimension continues to dynamically draw itself; now, however, the beginning and end dimension points are fixed, and the height of the dimension line changes as you move your mouse.

5. Click the map where you want the dimension line to be.

   The sketch is automatically finished, and the new aligned dimension feature is created with the style you selected.

   Since you selected Aligned as your construction method, the dimension line is parallel to the baseline.

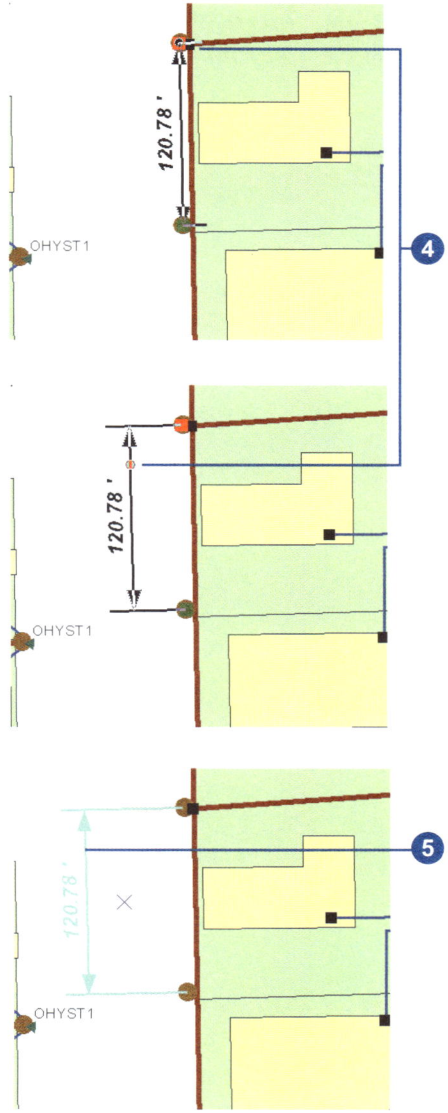

Tip

Linear dimension features

A linear dimension feature's dimension line is generally not parallel to its baseline. Therefore, the distance represented by a linear feature is not the length of the baseline.

To learn more about the different types of dimension features, see Building a Geodatabase.

Creating a linear dimension feature

1. Follow steps 1–7 for 'Creating a simple aligned dimension feature' in this chapter.
2. Click the Construction dropdown arrow and click Linear.
3. Click the map at the beginning dimension point to start the edit sketch.

   As you move the mouse, you will see that the new dimension dynamically draws itself with your mouse location as the end dimension point. ►

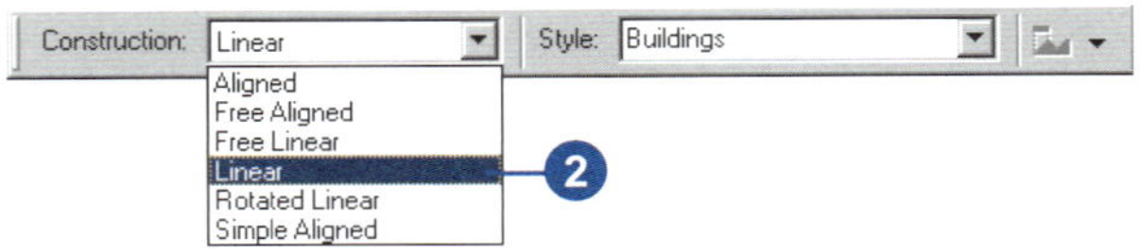

4. Click the map at the end dimension point.

   The new dimension continues to dynamically draw itself; now, however, the beginning and end dimension points are fixed, and the height of the dimension line changes as you move your mouse.

   If you move your mouse to the left or right of the baseline, you will see a vertical linear dimension feature. If you move your mouse above or below the baseline, you will see a horizontal linear dimension feature. ▸

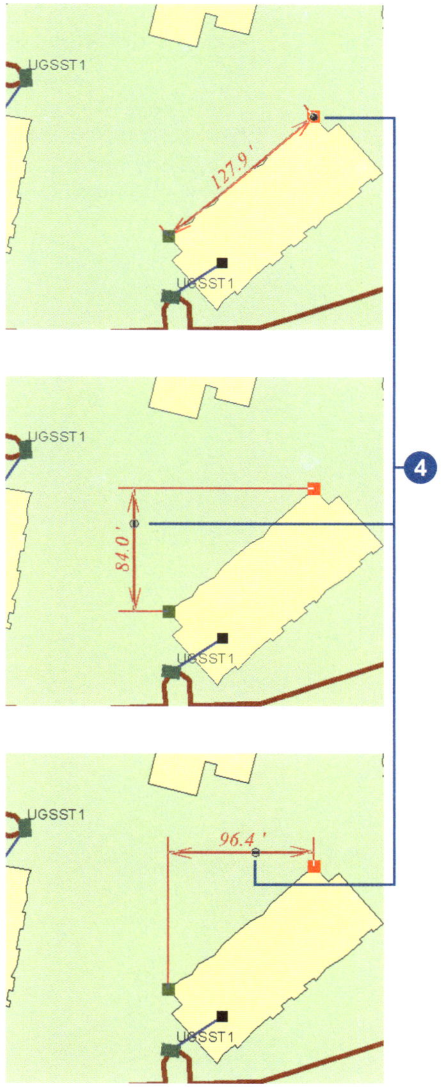

5. Click the map where you want the dimension line to be.

   The sketch is automatically finished, and the new linear dimension feature is created with the style you selected.

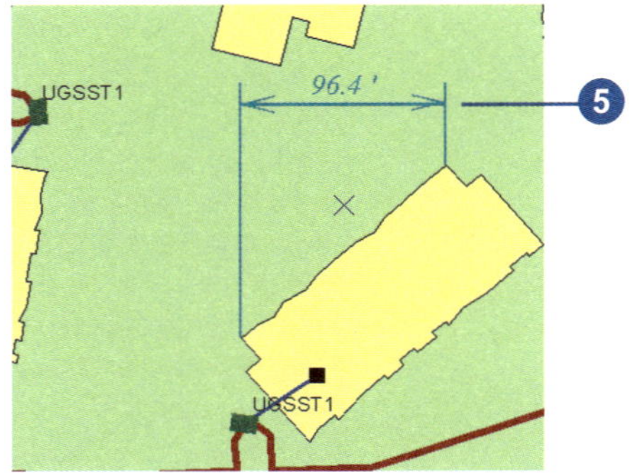

Creating a rotated linear dimension feature

1. Follow steps 1–7 for 'Creating a simple aligned dimension feature' in this chapter.
2. Click the Construction dropdown arrow and click Rotated Linear.
3. Click the map at the beginning dimension point to start the edit sketch.

   As you move the mouse, you will see that the new dimension dynamically draws itself with your mouse location as the end dimension point. ▸

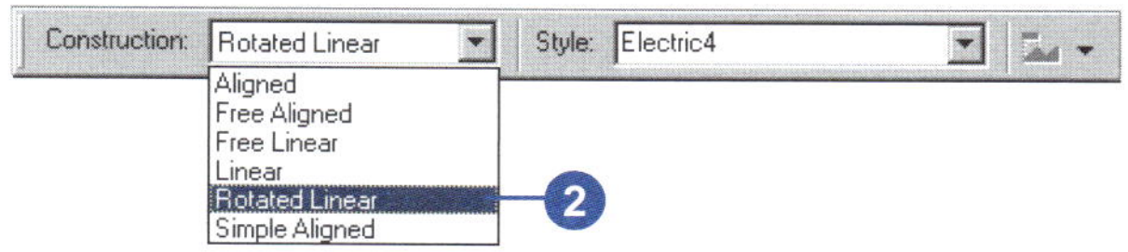

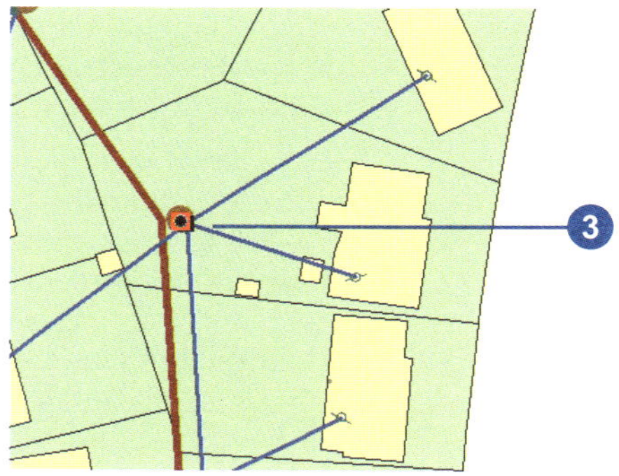

4. Click the map at the end dimension point.

   The new dimension continues to dynamically draw itself; now, however, the beginning and end dimension points are fixed, and the height of the dimension line changes as you move your mouse.

   If you move your mouse above or below the baseline, you will see a horizontal linear dimension feature. If you move your mouse to the left or right of the baseline, you will see a vertical linear dimension feature. ►

5. Click the map where you want the dimension line to be.

   The new dimension continues to dynamically draw itself; now, however, the beginning and end dimension points and dimension line height are fixed, and the angle of the extension lines changes as you move your mouse.

6. Click the map at the angle you want the extension lines to be.

   The sketch is automatically finished, and the new rotated linear dimension feature is created with the style you selected.

Tip

Free construction methods

The free construction methods allow you to enter as many vertices as you like into the edit sketch to help you define your dimension feature points. However, it is your responsibility to delete any vertex from the edit sketch that does not correspond to one of the required dimension feature points before you click Finish Sketch. If you have too few or too many vertices, your edit operation will fail.

Creating a dimension feature with the free aligned construction method

1. Follow steps 1–7 for 'Creating a simple aligned dimension feature' in this chapter.
2. Click the Construction dropdown arrow and click Free Aligned.
3. Click the map at the beginning dimension point to start the edit sketch.

   As you move the cursor, the dimension feature won't dynamically draw itself.
4. Use ArcMap sketch tools and construction methods to enter the end dimension point.

   If you are creating a simple aligned dimension feature, skip to step 6. ►

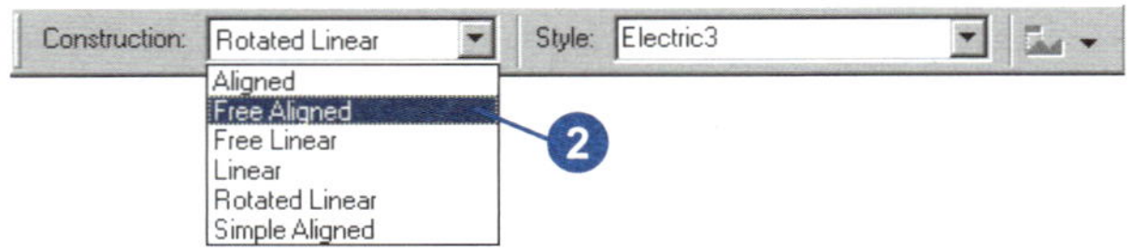

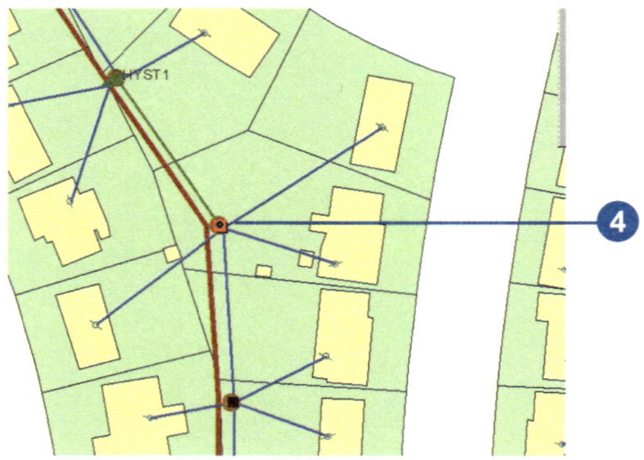

5. Use ArcMap sketch tools and construction methods to enter the point where you want the dimension line to be.

6. If your sketch has more than three vertices or has any vertices that do not represent the beginning or end dimension point or dimension line height, click Delete Vertex to delete them before continuing to step 7. ►

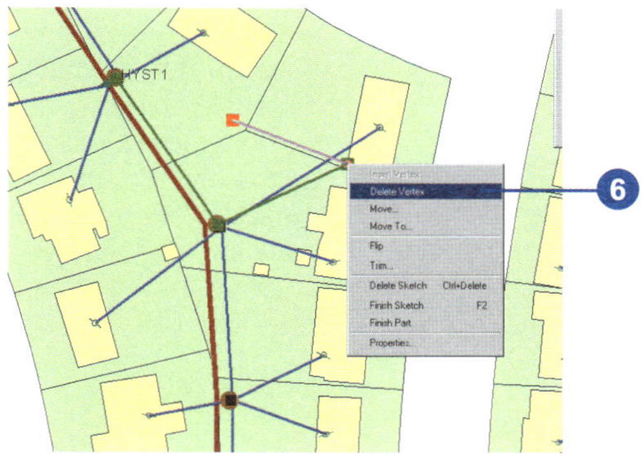

7. Right-click the sketch and click Finish Sketch.

   The new dimension feature is created with the style you selected. If the edit sketch has two points, then a simple aligned feature is created. If the sketch has three points, then an aligned dimension feature is created.

   The dimension feature points that the vertices represent will be determined by the order in which you entered them into the edit sketch. The vertex first entered will be used as the beginning dimension point. The second vertex entered will be used as the end dimension point. If the edit sketch has three vertices, the third vertex will be used as the dimension line height point.

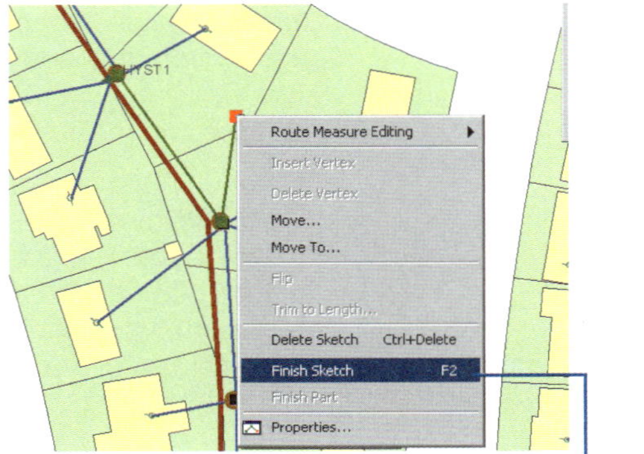

Creating a dimension feature with the free linear construction method

1. Follow steps 1–7 for 'Creating a simple aligned dimension feature' in this chapter.
2. Click the Construction dropdown arrow and click Free Linear.
3. Click the map at the beginning dimension point to start the edit sketch.

   As you move the cursor, the dimension feature won't dynamically draw itself.
4. Use ArcMap sketch tools and construction methods to enter the end dimension point. ▸

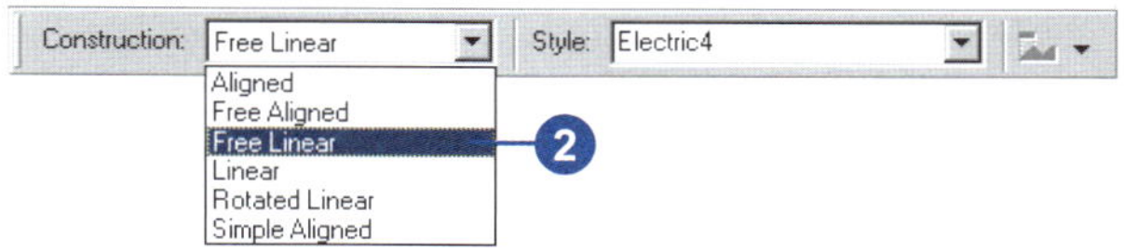

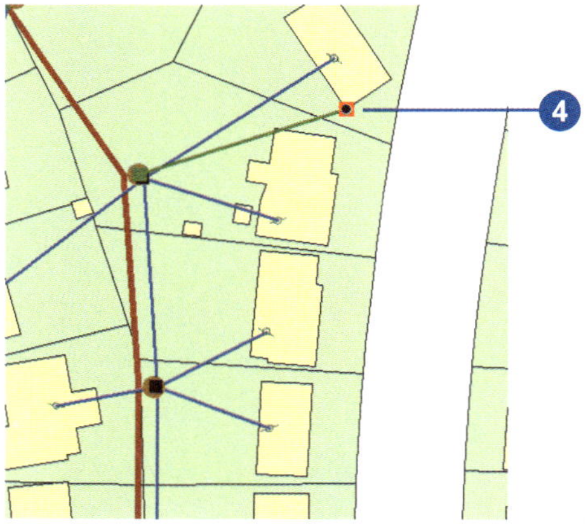

Tip

Extension line angle

When creating rotated linear dimensions, the extension line angle is calculated such that the dimension line is parallel to the line between the third and fourth construction points.

5. Use ArcMap sketch tools and construction methods to enter the point where you want the dimension line to be.

   If you are creating a horizontal or vertical linear dimension feature, skip to step 7.

6. Use ArcMap sketch tools and construction methods to enter the point that describes the extension line angle. ▸

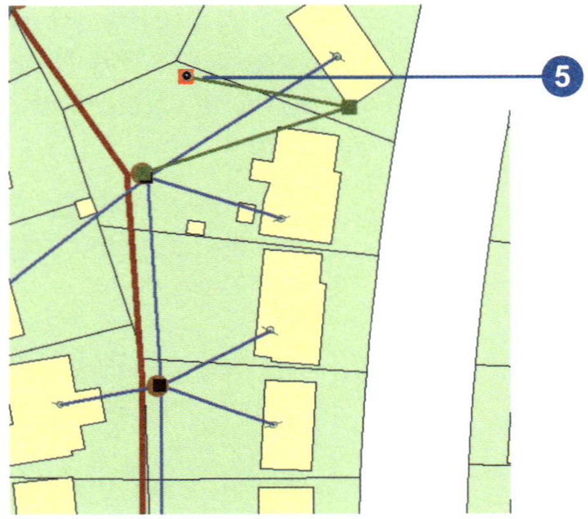

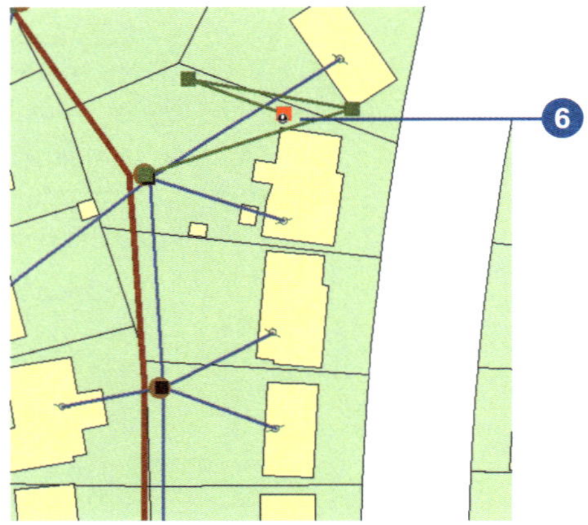

7. If your sketch has more than four vertices or has any vertices that do not represent the beginning or end dimension point, dimension line height, or extension line angle, you must delete them before continuing to step 8.
8. Right-click the sketch and click Finish Sketch.

   The new dimension feature is created with the style you selected. If the edit sketch has three points, then a vertical or horizontal linear dimension feature is created. If the sketch has four points, then a rotated linear dimension feature is created.

   The dimension feature points that the vertices represented will be determined by the order in which you entered them into the edit sketch. The vertex first entered will be used as the beginning dimension point. The second vertex entered will be used as the end dimension point. The third vertex will be used as the dimension line height point. If the edit sketch has four vertices, the fourth vertex will be used to describe the extension line angle.

Tip

Autodimension tools

The Dimension Edge tool only creates linear dimension features by automatically using vertices on existing features for their beginning and end dimension points.

Tip

The Dimension Edge tool and diagonal features

To use the Dimension Edge tool to obtain diagonal measurements, after clicking on the line, position the cursor over the beginning vertex of the line segment. The diagonal measurement will appear.

Creating a dimension feature with the Dimension Edge tool

1. Follow steps 1–7 for 'Creating a simple aligned dimension feature' in this chapter.
2. Select the feature whose edge you want to use as the baseline for your new dimension feature.
3. Click the Tool Palette dropdown arrow and click the Dimension Edge tool.
4. Click the edge you want to use as the baseline for your dimension feature.

   As you move your mouse, the new dimension dynamically draws itself with the beginning and end dimension points fixed at the ends of the edge you clicked; the height of the dimension line changes.

   If you move your mouse to the left or right of the baseline, a vertical linear dimension feature is shown. If you move your mouse above or below the baseline, a horizontal linear dimension feature is shown.
5. Click the map where you want the dimension line to be.

   The sketch is automatically finished and a new linear dimension feature is created with the style you selected.

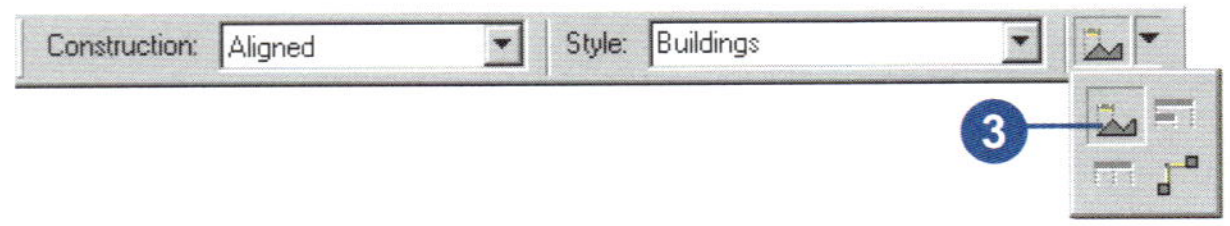

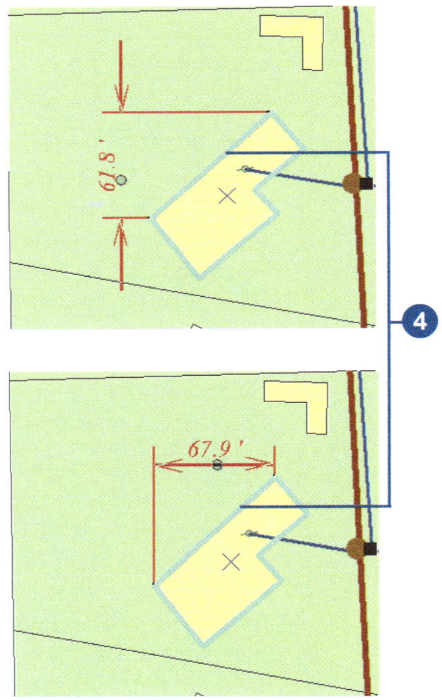

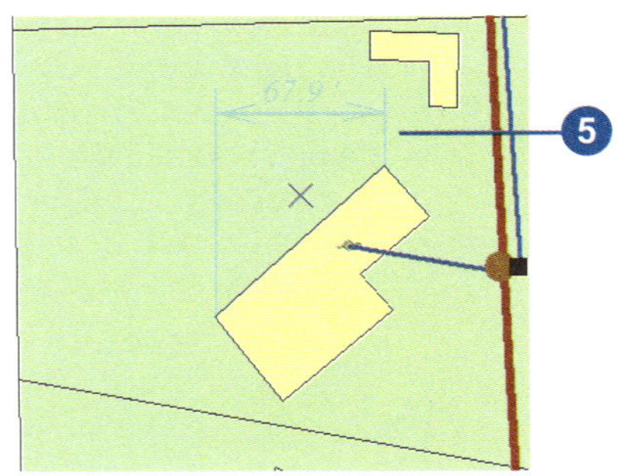

Creating two dimension features with the Perpendicular Dimensions tool

1. Follow steps 1–7 for 'Creating a simple aligned dimension feature' in this chapter.
2. Click the Tool Palette dropdown arrow and click the Perpendicular Dimensions tool.
3. Click the map at the beginning dimension point to start the edit sketch.

   A line will appear on the map that runs through the first vertex placed and the location of the cursor.
4. Click the map again to establish the angle of the first dimension feature.

   As the cursor is moved away from the first line, two dimension features will appear. The first along the established line and the second running perpendicular from the line to the location of the cursor. ▸

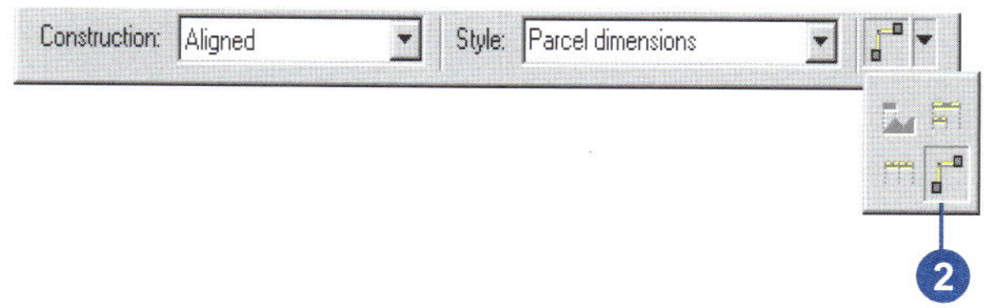

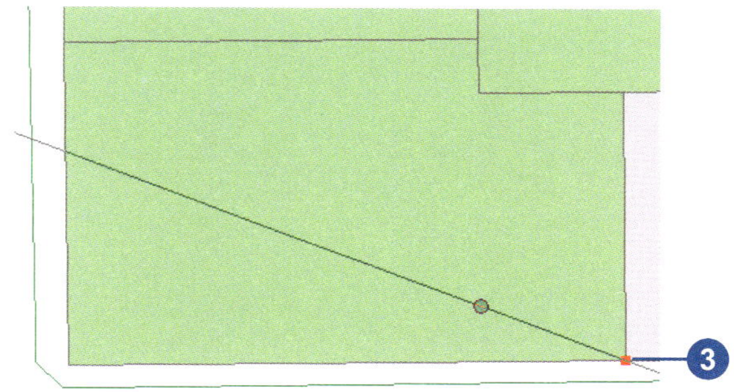

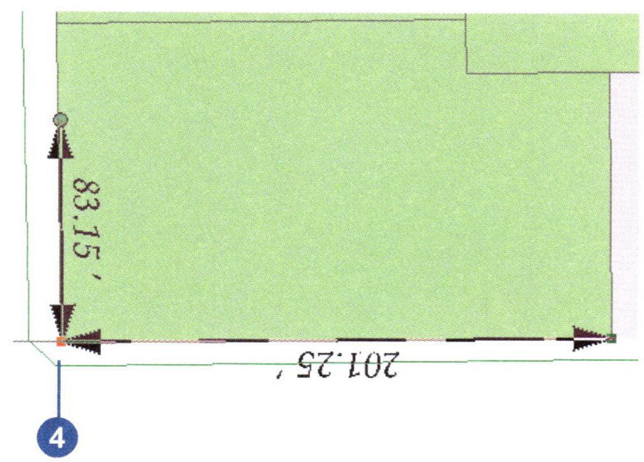

5. Click the map once again to establish the length of the perpendicular measurement.

   This will complete the two dimension features.

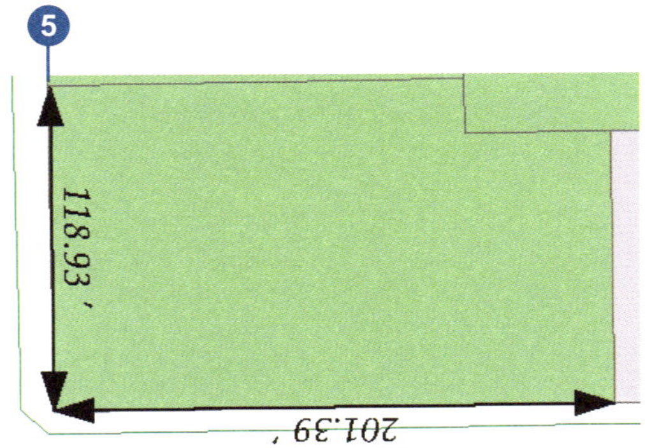

Creating a dimension feature with the Baseline Dimension tool

1. Follow steps 1–7 for 'Creating a simple aligned dimension feature' in this chapter.
2. Select the dimension feature whose beginning dimension point you want to use as the beginning dimension point for your new dimension feature.
3. Click the Autodimension tool palette dropdown arrow and click the Baseline Dimension tool.

   As you move your mouse, the new dimension feature dynamically draws itself with the beginning dimension point fixed at the beginning dimension point of the dimension feature you selected in step 2. The height is fixed at the height of the dimension feature you selected, plus the baseline height for the style you selected in step 2.

   The end dimension point changes as you move your mouse, keeping the baseline for the new dimension feature parallel to the baseline of the dimension feature you selected in step 2. ►

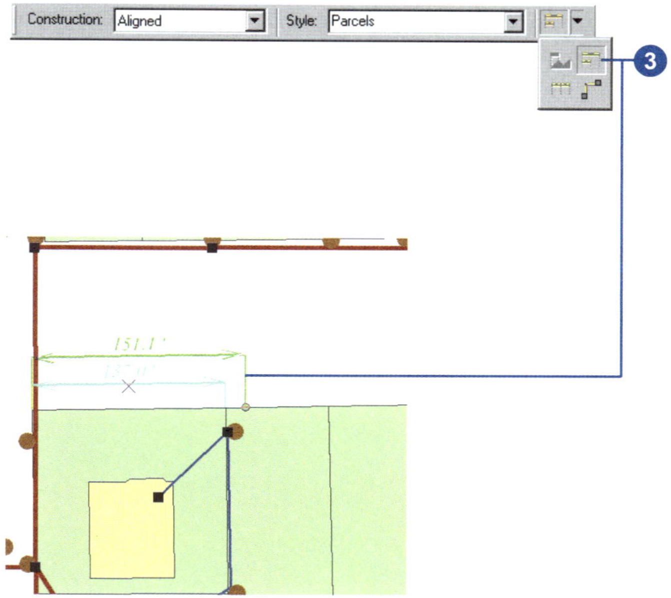

Tip

Baseline height

For a dimension feature created with the Baseline Dimension tool, the height of the dimension line will be controlled by the baseline height property of its style.

The baseline height is only used for creating dimension features. If you change an existing dimension feature's style to a style with a different baseline height, the height of the dimension line will not change.

For more information about styles and how to set the baseline height property, see Building a Geodatabase.

4. Click the map where you want the end dimension point to be.

   The sketch is automatically finished, and a new dimension feature is created with the style you selected. The dimension type will be the same as the dimension feature you selected in step 2.

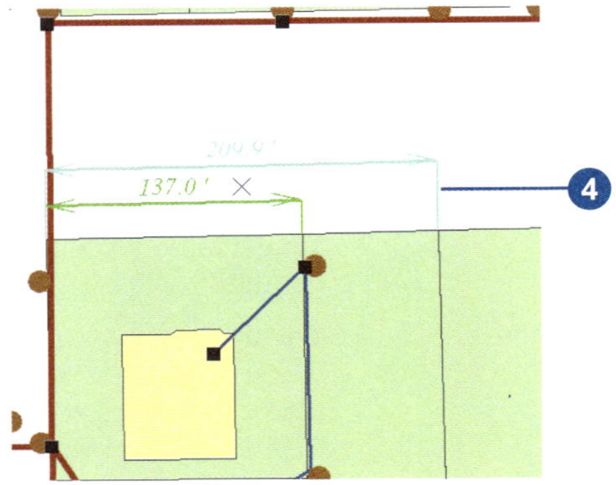

Creating a dimension feature with the Continue Dimension tool

1. Follow steps 1–7 for 'Creating a simple aligned dimension feature' in this chapter.
2. Select the dimension feature whose end dimension point you want to use as the beginning dimension point for your new dimension feature.
3. Click the Tool Palette dropdown arrow and click the Continue Dimension tool.

   As you move your mouse, the new dimension feature dynamically draws itself with the beginning dimension point fixed at the end dimension point of the dimension feature you selected in step 2. The height is also fixed at the height of the dimension feature you selected in step 2.

   The end dimension point changes as you move your mouse, keeping the baseline for the new dimension feature parallel to the baseline of the dimension feature you selected in step 2. ▸

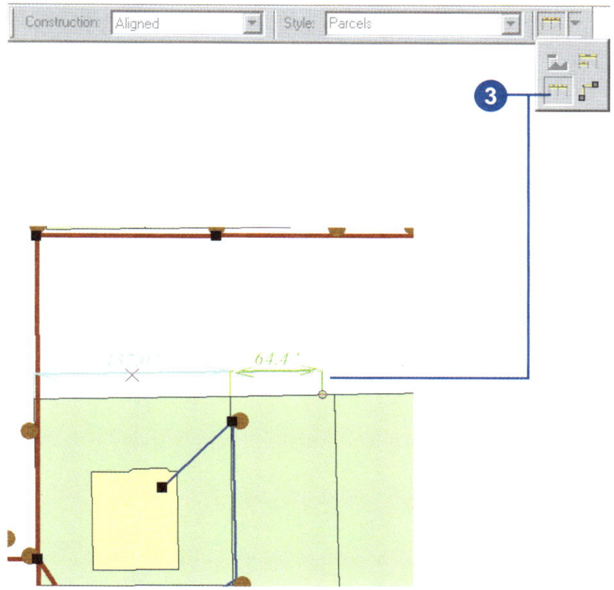

4. Click the map where you want the end dimension point to be.

   The sketch is automatically finished, and a new dimension feature is created with the style you selected. The dimension type will be the same as the dimension feature you selected in step 2.

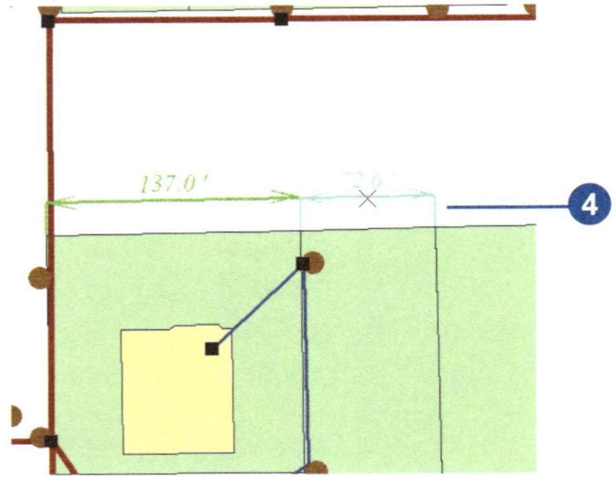

Modifying dimension features

Modifying the dimension feature geometry goes beyond modifying the dimension line height. You can also modify the beginning dimension point, the end dimension point, and the dimension text placement.

In addition to using the Modify tool, you can also use the Attributes dialog box to modify the dimension feature's geometry. You can use the Attributes dialog box to modify:

- Dimension line display
- Dimension line arrow symbol display
- Extension line display
- Extension line angle
- Dimension text value

Tip

Modifying perpendicular dimensions

Once the perpendicular dimensions have been created, they act independently of one another and can be selected and modified as described in this chapter.

Modifying a dimension feature's geometry

1. Click the Task dropdown arrow and click Modify Feature.
2. Click the Edit tool and click the dimension feature whose geometry you want to modify.
3. Position the pointer over the vertex that corresponds to the aspect of the dimension's geometry you want to modify.
4. Click and drag the vertex to the desired location.

   As you move your mouse, the dimension feature dynamically updates itself so you can see how the feature will look after you have modified its geometry. ▸

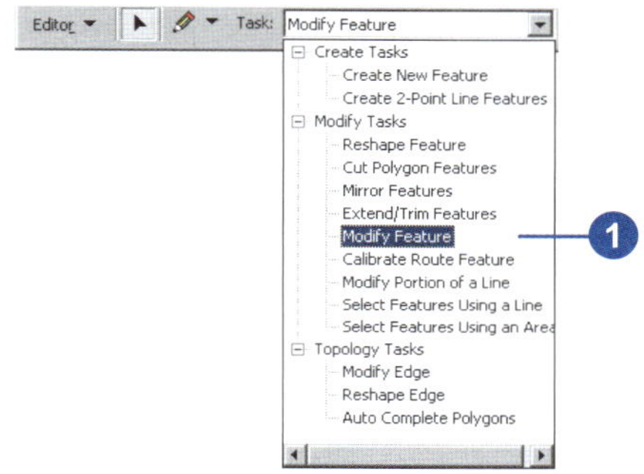

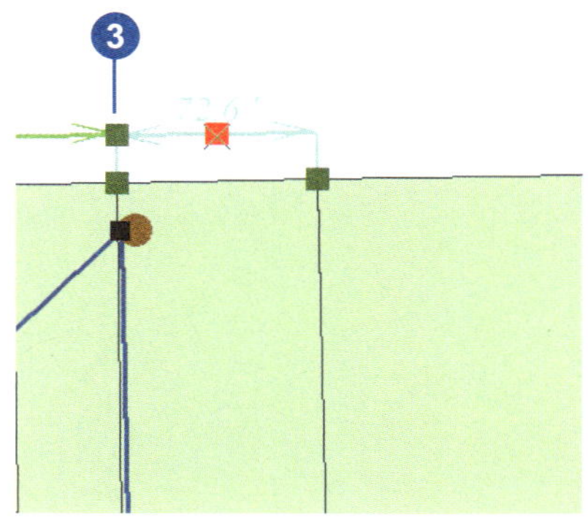

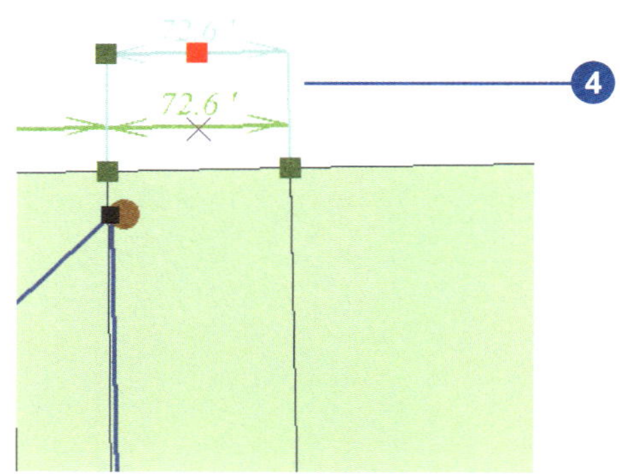

Tip

Modifying a dimension feature's geometry attributes

In addition to using the Modify tool, you can also use the Attributes dialog box to modify the dimension feature's geometry.

5. Right-click over any part of the sketch and click Finish Sketch.

   The dimension feature's geometry is updated.

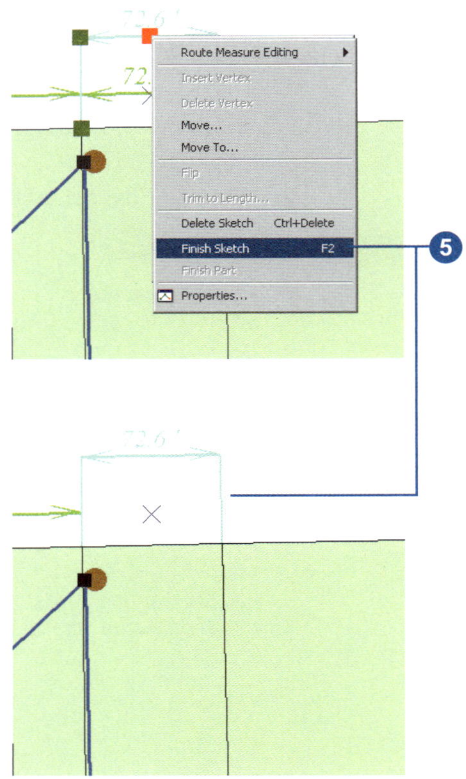

Modifying a dimension feature's style

1. Click the Edit tool and click the dimension feature whose style you want to modify.
2. Click the Attributes button.

   The Attributes dialog box appears. Notice that there is a special Attributes dialog box for modifying the attributes of a dimension feature.
3. Click the Dimension Style dropdown list and click the dimension style you want to assign to this feature. ▸

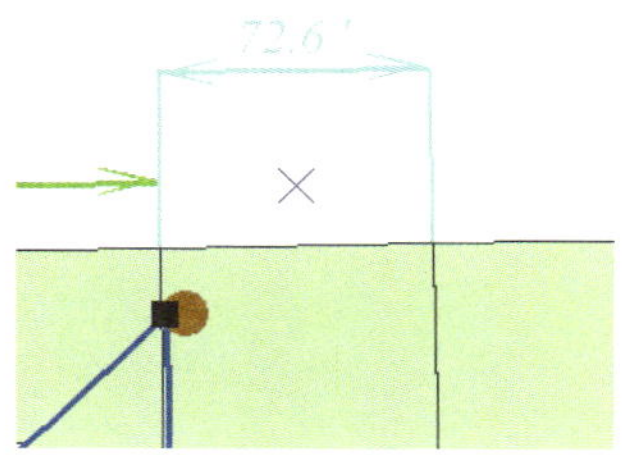

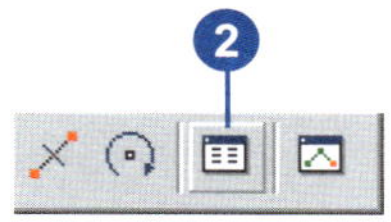

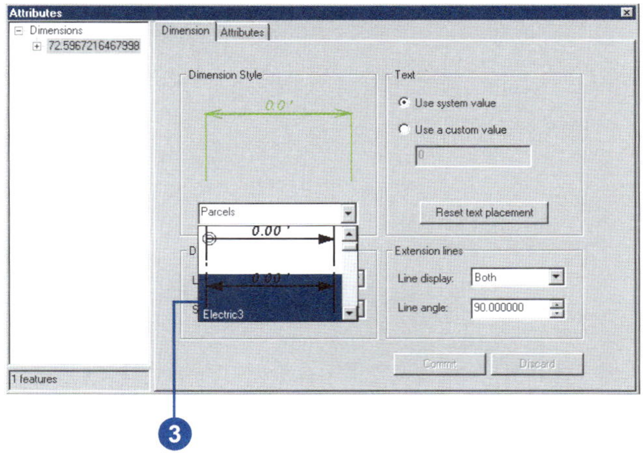

4. Click Commit.

   The dimension feature updates itself to reflect the new style.

5. Click the Close button to close the Attributes dialog box.

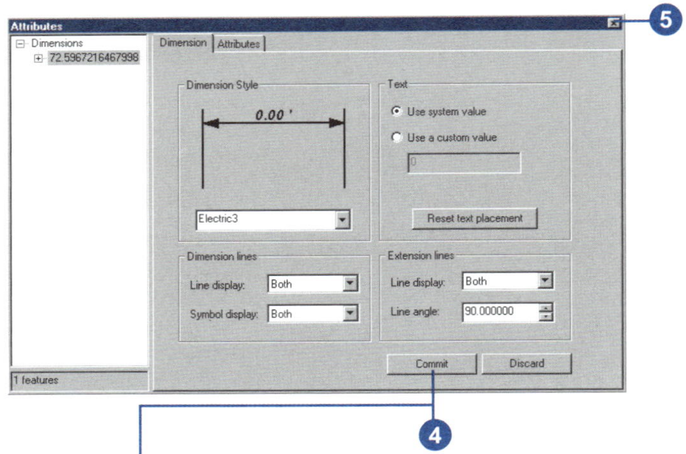

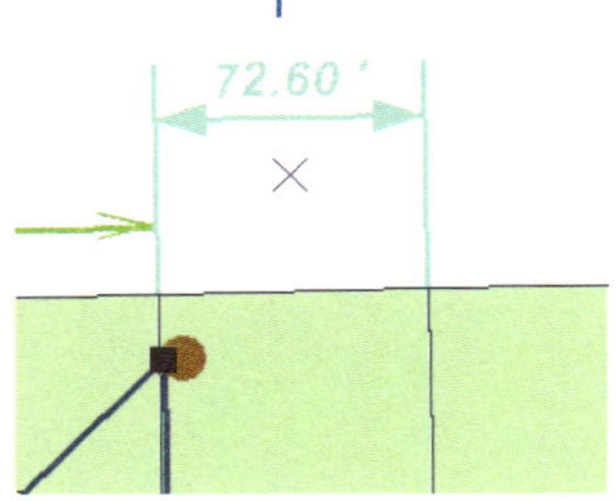

Working with a versioned geodatabase 15

IN THIS CHAPTER

- Integrating versioning with your organization's work flow
- Registering data as versioned
- Creating and administering versions in ArcCatalog
- Working with versions in ArcMap
- Editing and conflict resolution
- Editing a version
- Versioning scenarios

With ArcGIS 8, multiple users can access geographic data in a geodatabase through versioning. Versioning lets users simultaneously create multiple, persistent representations of the database without data replication. Users can edit the same features or rows without explicitly applying locks to prohibit other users from modifying the same data.

An organization can use versioning to manage alternative engineering designs, solve complex what-if scenarios without impacting the corporate database, and create point-in-time representations of the database.

Primarily, versioning simplifies the editing experience. Multiple users can directly modify the database without having to extract data or lock features and rows before editing. If, by chance, the same features are modified, a conflict resolution dialog box guides the user through the process of determining the feature's correct representation and attributes.

Versioned databases may contain topologies. For more information on how versioning affects topologies, see *Building a Geodatabase*.

Versioned databases may also be the checkout databases for disconnected editors. For more on using a versioned database for disconnected editing, see *Building a Geodatabase*.

Integrating versioning with your organization's work flow

The geodatabase and versioning provide organizations with advanced data storage techniques that revolutionize the *work flow* process in many applications where spatial information is used. Engineers can generate design alternatives using the entire database. Spatial analysts can perform complex what-if scenarios without affecting the current representation of the database. Database administrators can create historical snapshots of the database for archiving or database recovery.

In the long run, an organization benefits from implementing a versioned database. The data is centrally located in one corporate database. There is never a need to extract units of the database to update, or lock, map sheets or individual features. These factors simplify the administrative process.

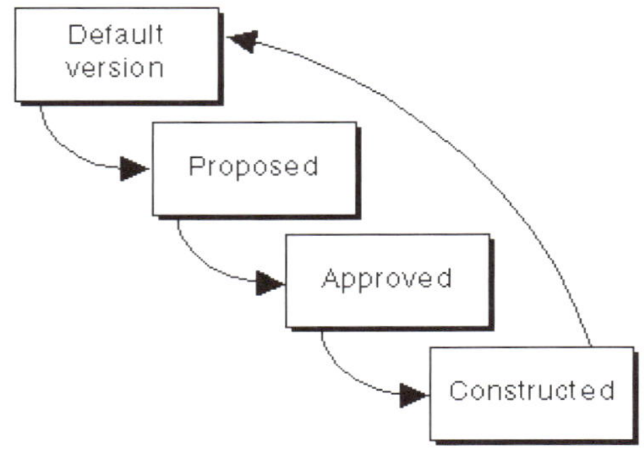

A common work flow process evolving through each stage of a project

The work flow process

The evolution of the work flow process—how projects or *work orders* transpire over time—varies greatly from organization to organization and throughout each sector of the business community. Therefore, the geodatabase's versioning process has been designed to be flexible enough to accommodate the most basic of work flow processes as well as the most complex and to be sufficiently restrictive with or without additional application customization.

Common work flow processes usually progress in discrete stages. At each stage, different requirements or business rules may be enforced. Typically, during each stage of the process, the project or work order is associated by a named stage. For example, within the utility domain, common stages include working, proposed, accepted, under construction, and as built. The process is essentially cyclical. The work order is initially generated and assigned to an engineer and modified over time as it progresses from stage to stage, and finally the changes are posted, or applied, back to the corporate database.

This is one example of how versioning can help simplify the work flow process. Because the work flow process may span days, months, and even years, the corporate database requires continuous availability for daily operations. If a work order applied restrictive locks to the data involved in the process, other database users might not be able to perform their daily work assignments.

To implement your work flow in the geodatabase, versions can be created to correspond with each stage of the work flow process. Alternatively, you may want to create one version for each work order and modify the version's name to represent the current stage as the process proceeds through each step.

The current structure of your organization's work flow significantly influences how the geodatabase's versioning process is implemented to manage your spatial transactions. The flexibility and openness of the system allows you to determine the best solution to meet the requirements of your business processes.

The remaining sections of this chapter will help illustrate how to use ArcCatalog and ArcMap to perform various versioning tasks. In particular, the last section provides examples of how an organization can implement work flow processes using the geodatabase's versioning capabilities. For additional details on managing your organization's work flow with versions, read *Modeling Our World*.

Registering data as versioned

Before editing *feature datasets*, *feature classes*, and *tables*, you must first register the data as versioned in ArcCatalog.

Making a feature class or table multiversioned requires a unique integer *field*. Only the owner of the data may register or unregister the object as versioned.

When unregistering a dataset or feature class as versioned in ArcCatalog, a warning dialog box may appear informing you that outstanding edits still remain in existing versions. Therefore, unregistering the class as versioned will remove all the edits. To preserve the edits, you must compress the database.

Tip

Registering data as versioned

Registering a feature dataset as versioned registers all feature classes within the feature dataset as versioned.

1. In the ArcCatalog tree, right-click the feature dataset, feature class, or table you want to register as versioned.
2. Click Register As Versioned.

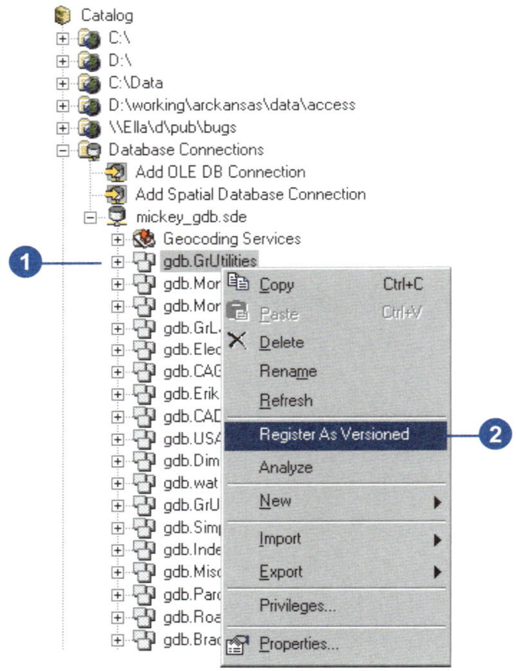

Creating and administering versions in ArcCatalog

ArcCatalog lets you create new versions, rename existing versions, delete versions, and modify version properties. These administrative tasks are accomplished using the Version Manager dialog box.

Initially, the database consists of one version named DEFAULT owned by the ArcSDE administrative user. The new versions that are created are always based on an existing version. When the new version is created, it is identical to the version from which it was derived. Over time, the versions will diverge as changes are made to the parent version and to the new version.

A version consists of several properties: an alphanumeric name, an owner, an optional description, the creation date, the last modified date, the parent version, and the version's permission. ►

Creating a new version

1. Create a new connection to the database in ArcCatalog with the Add SDE Connection dialog box.
2. Right-click your database connection in the Catalog tree and click Versions. ►

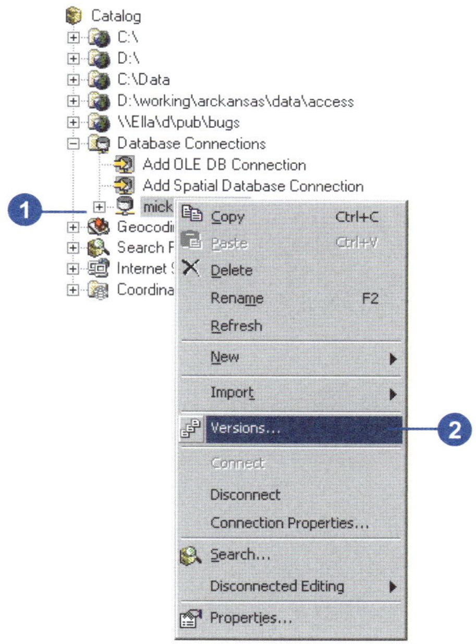

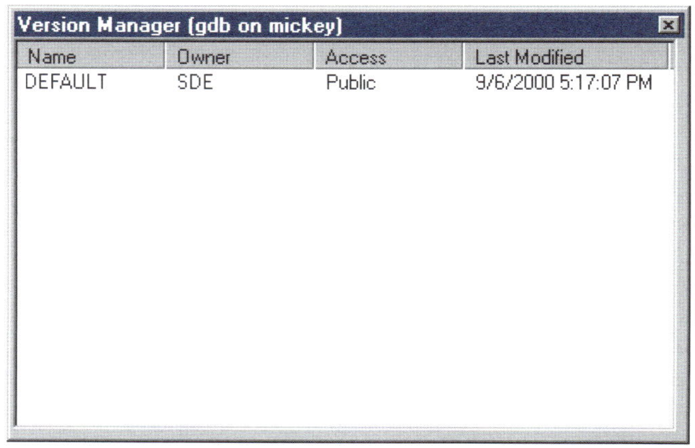

A version's permission can only be changed by its owner. The available permission settings are:

- Private—only the owner may view the version and modify available feature classes.
- Protected—any user may view the version, but only the owner may modify available feature classes.
- Public—any user may view the version and modify available feature classes.

Only the version's owner can rename, delete, or alter the version. A parent version cannot be deleted until all dependent child versions are first deleted.

To improve database performance, the database should be compressed periodically. Compressing the database removes all unreferenced database states and redundant rows. Only the ArcSDE administrator can perform this task. When the *Compress* command is executed, the database is unavailable until compression is completed. For additional details, see the versioning scenarios section at the end of the chapter. ►

3. Right-click a version and click New.
4. Type the new version's name.
5. Type a description.
6. Click the appropriate permission type; the default is Private.
7. Click OK.

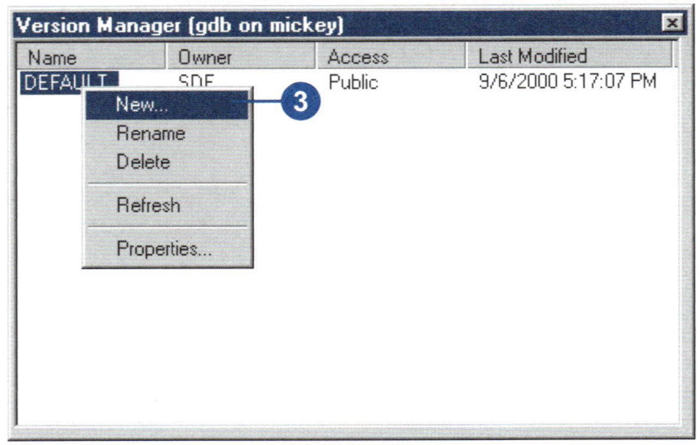

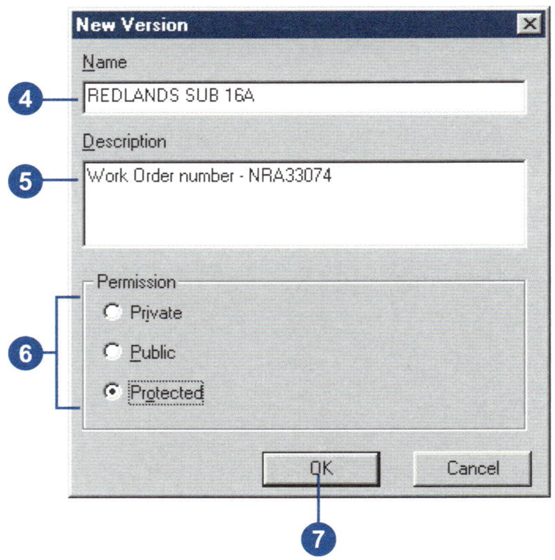

Finally, after compressing the database or editing the data, the Analyze command should be executed to update the database statistics for each dataset or feature class. This will help improve display and query performance.

Tip

Descriptions

Descriptions are useful for providing meaningful information regarding the version's purpose.

Renaming a version

1. Right-click your database connection and click Versions.
2. Right-click the version you want to rename and click Rename.
3. Type a new name and press Enter.

Deleting a version

1. Right-click your database connection and click Versions.
2. Right-click the version you want to delete.
3. Click Delete or press Delete on your keyboard.

Tip

Refresh

Use the Refresh command to update the properties of each version with their current values.

Changing a version's properties

1. Right-click your database connection and click Versions.
2. Right-click a version.
3. Click Properties.
4. Type the new description.
5. Click the new permission type.
6. Click OK.

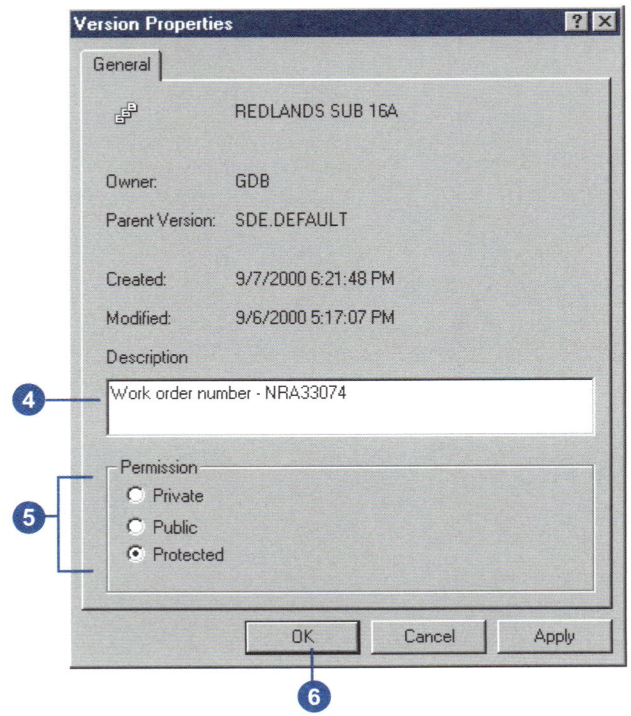

See Also

For more information on how to customize ArcCatalog, see Using ArcCatalog *and* Exploring ArcObjects.

Adding the Compress command to ArcCatalog

1. In ArcCatalog, click View, point to Toolbars, and click Customize.
2. Check Context Menus in the list of toolbars. ►

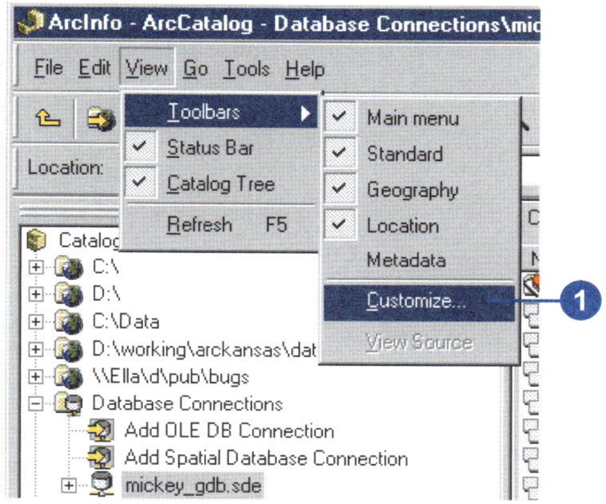

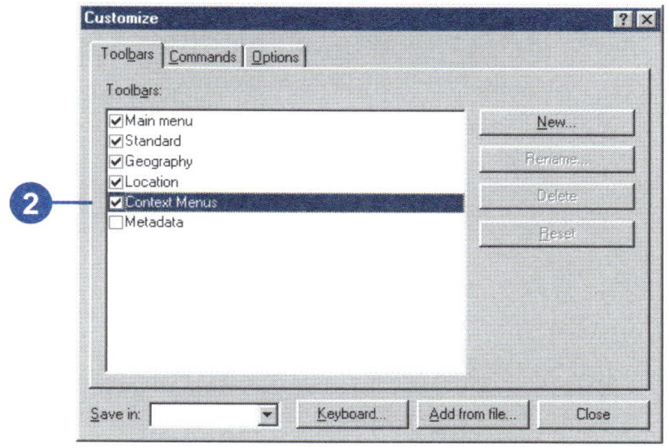

3. Click the Context Menus menu.
4. Click the arrow next to the Remote Database Context Menu.

   The Remote Database Context Menu submenu will remain open.
5. Click the Commands tab in the Customize dialog box.
6. Click Geodatabase tools.
7. Click and drag the Compress Database command from the Commands list and drop it on the Remote Database Context menu submenu.

   The command appears in the context menu.
8. Click Close on the Customize dialog box.

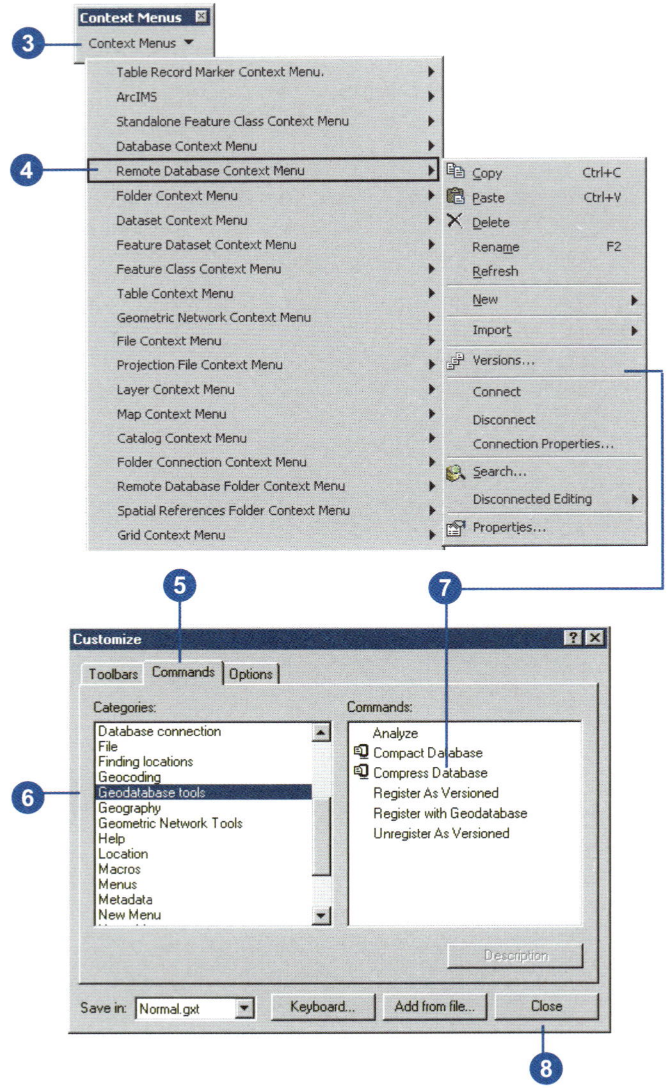

Tip

Analyze

After compressing a database, always analyze your data to update the database statistics.

See Also

For more information on how to administer an ArcSDE database, see Managing ArcSDE Services.

See Also

For more information on how to create connections to databases in ArcCatalog, see Using ArcCatalog.

Compressing the database

1. In ArcCatalog, create a new database connection as the ArcSDE administrative user.
2. Right-click the new database connection and click Compress Database.

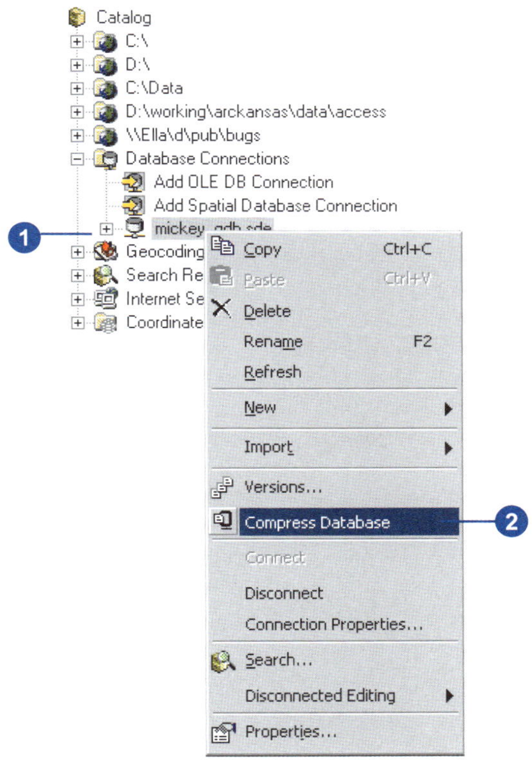

Working with versions in ArcMap

In ArcMap, you can view and work with multiple versions simultaneously, create new versions, and change the feature classes or tables from one version to another version. You can also use the version manager, refresh a version's *workspace* connection, and modify available feature classes in ArcMap.

To create a new version, at least one version must be present in the map. If multiple versions are present, you will need to specify the parent version. The newly created version will then be identical to the parent version.

Changing versions allows you to quickly navigate between two versions by changing the feature classes currently in the map. This simplifies the process of viewing the differences between feature classes or performing an analysis with two versions. ►

Tip

Creating new versions

Create alternative versions as online backups to the original version.

Creating a new version in ArcMap

1. Add the Versioning toolbar to the map.
2. Click the Create New Version button. At least one version is required to be in ArcMap prior to the command becoming enabled.
3. Click the Parent Version dropdown arrow and click the parent version from which you want to create the new version.
4. Type the new version's name.
5. Optionally, type a description.
6. Click the appropriate permission type.
7. Optionally, if you are not currently editing, check the check box to switch the parent version to the new version.
8. Click OK.

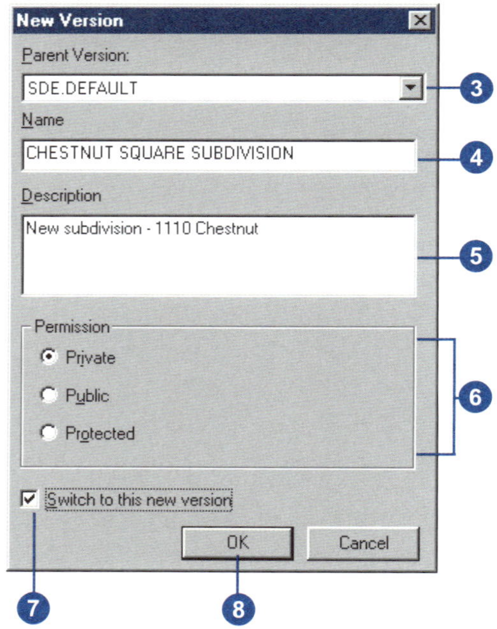

When a version workspace is changed to a different version, all feature classes present in the workspace will represent the target version.

Two methods are available in ArcMap for changing versions. You can change versions from the Versioning toolbar or in the table of contents.

When you work in a multiuser environment, the database may be modified by another user at the same time you're viewing the database. Therefore, the feature classes present in ArcMap may become outdated.

To update the feature classes in ArcMap, you can refresh one or all of the version workspaces present.

To refresh all the versions, click the Refresh button on the Versioning toolbar. To refresh an individual workspace, use the Refresh command on the table of contents context menu.

While you are editing, the Refresh button and the Refresh command for the version's workspace are unavailable. ▸

Tip

The Change Version command

Use the Change Version command instead of adding multiple version workspaces to your map document.

Changing versions

1. Click the Source tab at the bottom of the ArcMap table of contents to list the workspaces in your map.
2. Right-click a version workspace.
3. Click Change Version.
4. Click the version to which you want to change.
5. Click OK.

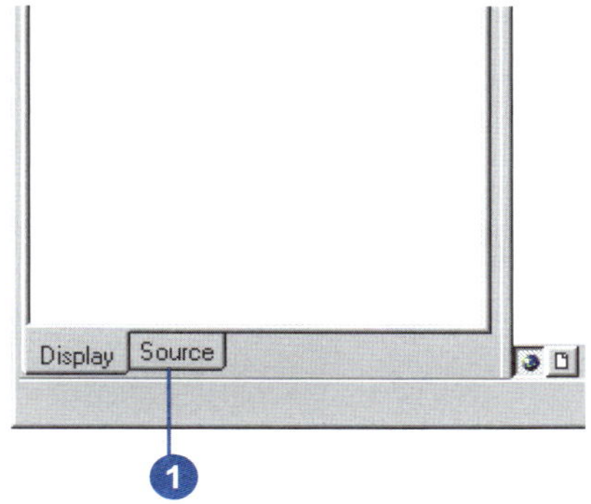

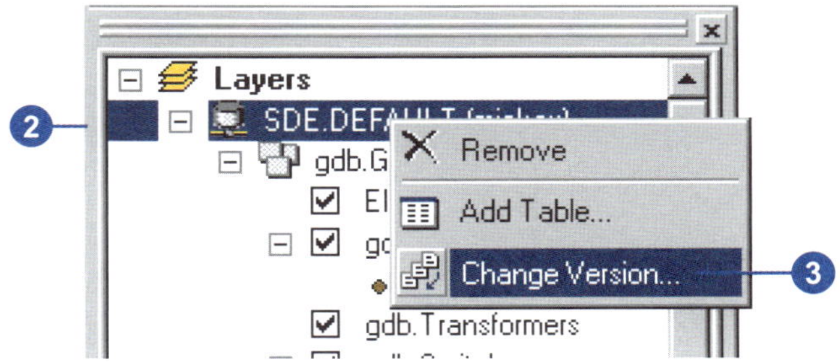

You can have as many versions in the map as needed, but you can only edit one version per *edit session*.

Tip

Preserving a version

If you need to preserve a current representation of the database, create a new version before refreshing.

Refreshing a workspace

1. Click the Refresh button on the Versioning toolbar.

Editing and conflict resolution

The geodatabase is designed to efficiently manage and support long *transactions* using versions. The geodatabase also allows multiple users to edit the same version at the same time. Each edit session in ArcMap is its own representation of the version until you save. Saving the edit session applies your modifications to the version, making these changes immediately accessible in the database.

When multiple users simultaneously edit a version or *reconcile* two versions, *conflicts* can occur. Reconciling is the process of merging two versions. Conflicts occur when the same feature or topologically related features are edited by two or more users and the database is unclear about which representation is valid. Conflicts are rare but can occur when overlapping geographic areas in the database are edited. To ensure database integrity, the geodatabase detects when a feature has been edited in two versions and reports it as a conflict. ArcMap provides the necessary tools for conflict resolution, but your interaction is still required to make the final decision as to the feature's correct representation.

ArcMap provides tools to resolve conflicts as well as the necessary tools to reconcile and *post* versions. The next sections explain these capabilities in more detail.

Geodatabases that contain topologies have additional rules, error detections, and validation techniques. For more information on topologies, see Chapter 4, 'Editing topology' or *Building a Geodatabase*.

Reconcile

The Reconcile button in ArcMap merges all modifications between the current edit session and a target version you select. Any differences between the features in the target version and the features in the edit session are applied to the edit session. Differences can consist of newly inserted, deleted, or updated features. The reconcile process detects these differences and discovers any conflicts. If conflicts exist, a message is displayed followed by the conflict resolution dialog box. Reconciling happens before posting a version to a target version. A target version is any version in the direct ancestry of the version, such as the parent version or the DEFAULT version.

In addition, the reconcile process requires that you are the only user currently editing the version and that you are the only user able to edit the version throughout the reconcile process until you save or post. If another user is simultaneously editing the version or attempts to start editing since you have reconciled, an error message will inform you that the version is currently in use.

The reconcile process requires that you have full permissions to all the feature classes that have been modified in the version being edited. If a feature class is modified in the version for which you do not have update privileges, an error message appears. You will not be able to reconcile the versions; a user with adequate permissions to perform the reconcile must do this for you.

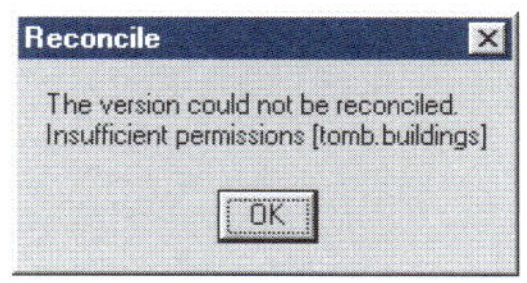

An error message appears when you do not have permissions to a feature class to reconcile versions.

For example, suppose you have completed your changes in a version and need to post the version to the database. You must first reconcile the version with a target version you select, resolve any conflicts if necessary, then post.

Autoreconciliation

When multiple users are simultaneously editing the same version and one user has already saved his or her edit session, ArcMap can notify you when you save that the edit session has been reconciled with the version's current representation. You also have the option to explicitly perform the reconcile and save without notification. You may want the notification to allow further inspection of the results from the reconciliation. It's an opportunity to review the difference introduced from the reconciliation between the edit session and the current version's representation. If conflicts are detected, you will be informed with a warning message, and the save process will fail.

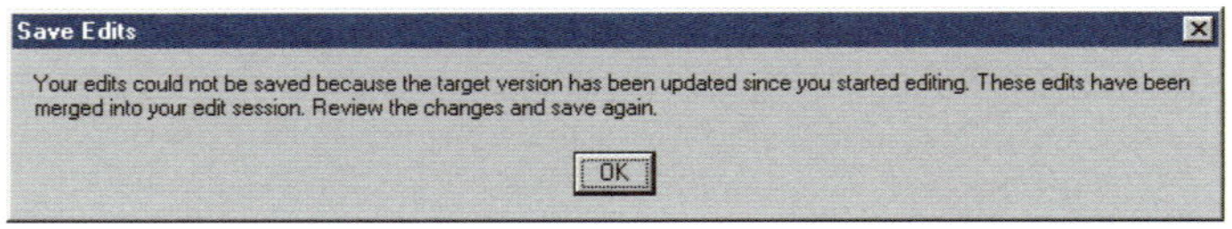

This message indicates when an edit session could not be saved because other users have modified and saved their edit session prior to the current user saving his or her edit session.

Post

You can post a version after you have first performed a reconcile. Once the edit session has reconciled with a target version, clicking the Post button synchronizes the edit session with the reconciled version and performs a save. Posting cannot be undone, as you are applying changes to a version that you are not currently editing. If the reconciled version is modified between reconciling and posting, you will be notified to reconcile again before posting.

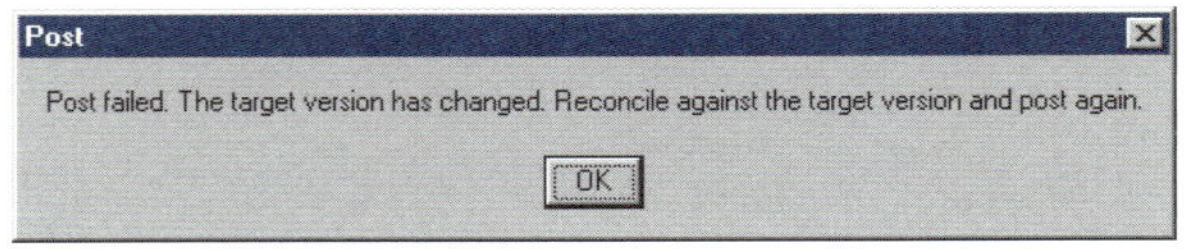

This message indicates that the target version has been modified since the reconciliation; reconcile again before posting.

Conflicts

Conflicts occur when the same feature, or topologically related features or relationship classes, is modified in two versions: the current version being edited and a target version. Conflict detection only occurs during the reconciliation process. If conflicts are detected, a message appears followed by the conflict resolution dialog box.

There are three categories of conflicts: when the same feature has been updated in each version, when the same feature has been updated in one version and deleted in the other or vice versa, and when the same feature has been deleted in one version and updated in the other version.

When conflicts are detected, the parent version's feature representation takes precedence over the edit session's representation. Therefore, all conflicting features in the current edit session are replaced by their representation in the parent version. If multiple users are editing the same version and conflicts are detected, the feature that was first saved, the current version's representation, is preserved by replacing the edit session's feature representation. ArcMap ensures database integrity by forcing you to interactively inspect each conflict and resolve the conflict by replacing the feature in the current version with your edit session's representation.

Conflict resolution

Once conflicts are detected, a conflict resolution dialog box appears, containing all the conflict classes and their features or rows in conflict. The conflict resolution dialog box allows you to interactively resolve conflicts at the level of the feature class or individual feature. Resolving the conflict implies that you will make a decision as to the feature's correct representation; this could mean doing nothing at all if you are satisfied with the current feature's representation.

You can choose from three representations of the conflicting feature or row to resolve the conflict. The pre-edit version is the feature's representation when you initially started editing, before making any changes. The edit session version represents the feature as it existed before you performed the reconcile. The last representation is the conflict version, the feature's representation in the conflicting version.

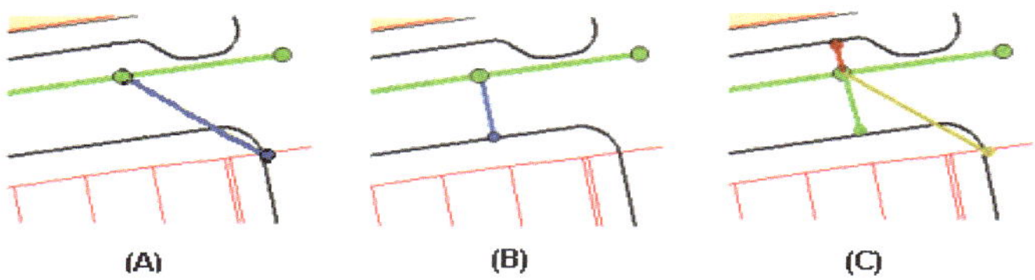

The lateral in blue as it existed prior to editing (A), the lateral after being modified (B), and the three representations during conflict resolution (C).

Selecting a feature class or individual feature displays any of the three representations of the feature in the map. The pre-edit's version is displayed in yellow, the edit session's version is displayed in green, and the conflict's version is displayed in red. You can also optionally enable or disable the display settings for each version—pre-edit, edit session, and conflict—by clicking the Display command on the context menu and checking or unchecking the corresponding version.

When you select a feature in the conflict resolution dialog box, each version's representation of the feature's or row's attributes is listed in the bottom half of the box. A red dot to the left of the field name identifies why the feature is a conflict. For example, if the feature's geometry was edited in each version, a red dot appears next to the shape field. The same principle holds true for attribute conflicts. If a feature has been deleted in either version, <deleted> appears for that version's attribute value. Therefore, a red dot marks each column, signifying that each column is an update/delete or a delete/update conflict.

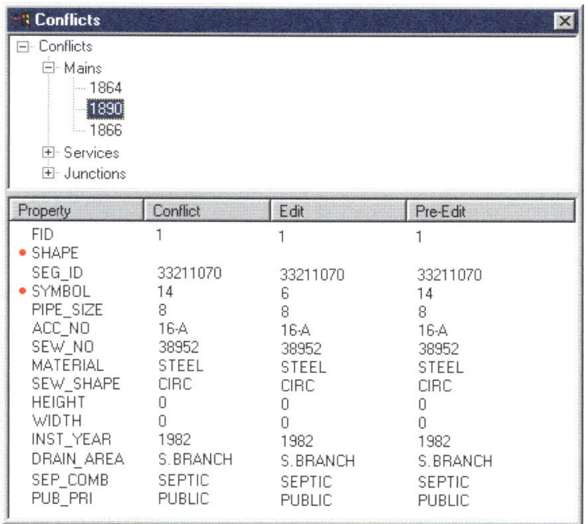

This conflict resolution dialog box shows three feature classes with conflicts and a feature with each of its version's attributes.

Resolving a conflict implies that you made a conscious decision about the feature's correct representation. You can select the feature in the conflict resolution dialog box and replace the current feature in the map with any of the three representations of the feature. This allows you to quickly update and replace conflicting features. If further modifications are required, you can simply use any of the ArcMap editing tools to update the feature.

Conflicts with geometric networks, feature-linked annotation, relationships, and topology

Resolving conflicts with features that are related to other features through *geometric networks*, feature-linked *annotation*, and *relationship classes* is different than resolving conflicts with simple feature classes. Because each of these feature classes has specific geodatabase behaviors that can impact other feature classes, resolving a feature conflict may impact related features.

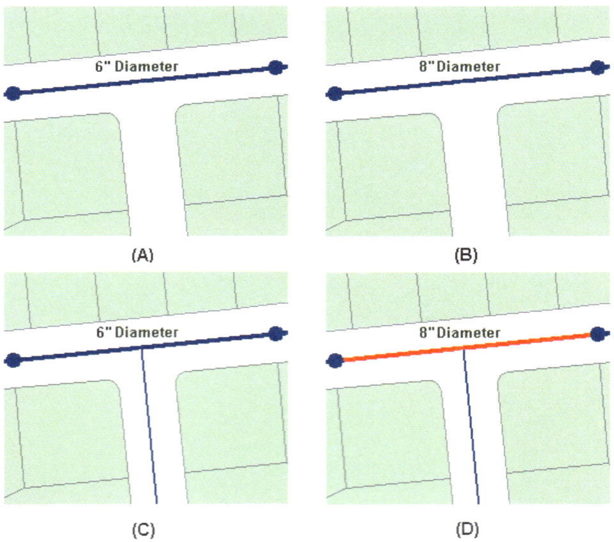

The original water main (A), the water main changed to an 8-inch diameter in the first edit session (B), a new service was inserted in the second edit session (C), and the water main in red is shown as a conflict (D).

When you edit network features, changes to the geometric network and to the logical network may create conflicts. For example, when you add a service to a main, the main will not be physically split in the geometric network but will be split in the *logical network*. Therefore, while you have not directly edited the main's geometry, it has been edited logically. If the target version you are reconciling has also modified the main, then the new service you inserted will create a conflict with the main.

Resolving a conflict involving geometric network feature classes requires understanding how the Replace With command in the conflict resolution dialog box will update the existing network topology present in the edit session.

In the previous example, two users modified the water main—one by changing an attribute and the second by connecting a new service. Resolving the conflict would merely require investigating the differences and seeing that the conflict is valid and no further resolution is required. Since the main contains the correct attribute for the diameter, the new service is correctly connected to the main. But there are cases when resolving conflicts involving a junction feature class will also update the connected network edge.

Working with feature-linked annotation requires remembering one rule: when replacing a feature that has feature-linked annotation, both the feature and the annotation are replaced with the new feature and annotation. You may have to further edit the new annotation. For example, you may encounter a conflict in which you have moved a feature and repositioned its annotation. The conflict version has performed the same edit, moving the feature and rotating the annotation. Your decision is to replace the feature with the conflict version's feature. This action deletes the existing feature-linked annotation, inserts the conflict feature, and creates a new annotation. You will then need to further edit the new annotation by moving and rotating it as necessary.

Relationships have similar dependencies to feature-linked annotation. Deleting a feature from an origin relationship class may trigger a message to delete a feature from the destination relationship class. Therefore, be aware of the ramifications of simply replacing conflicts involving feature classes that participate in relationship classes.

An example of when a conflict can arise between relationship classes is if you were to update the origin class primary field, breaking the relationship in version A. At the same time, in version B, the destination class-related feature is also updated. When you reconcile the versions, since the destination class is dependent on the origin class, a conflict is detected. A similar example is if you were to delete a pole that has a relationship to a transformer, the transformer is also deleted. But in the conflict version, the transformer's attributes are edited. An update/delete conflict would be detected when reconciled.

For more information on topologies, see Chapter 4, 'Editing topology' or *Building a Geodatabase*.

Editing a version

The ArcMap editing toolbar provides the arena for editing versions, reconciling versions, resolving conflicts, and posting versions.

When you start editing, if multiple versions are present in the map, you will have to select one version. Starting an edit session on a version creates a new, unnamed, temporary version that exists until you save or end the edit session. You are the only user who can see your changes until you explicitly save.

When saving an edit session, you have an option to enable or disable autoreconciliation. If enabled, autoreconciliation will automatically reconcile your edit session with the version's current database state and save, making your changes available to others using the database. If autoreconciliation is not enabled, then when you save, your edit session will be reconciled with the version's current database state. A message will inform you that the edit session has been reconciled but has not been saved. This will only occur if a second user has also edited the version and saved since you ►

Enabling and disabling autoreconciliation

1. Click Editor and click Start Editing.
2. Click Editor and click Options.
3. Click the Versioning tab.
4. Check the check box to enable or disable autoreconciliation.
5. Click OK.

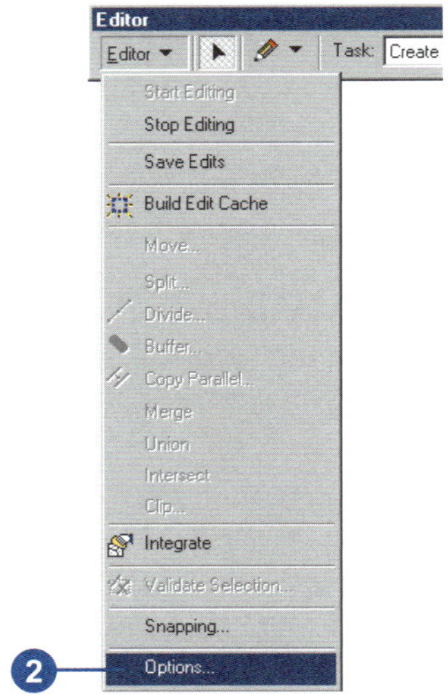

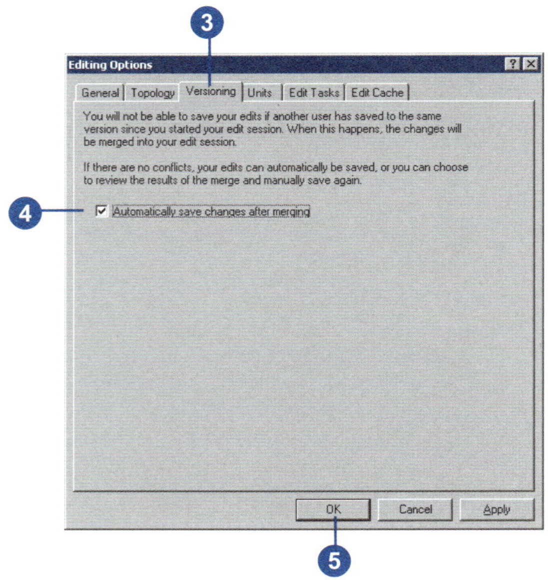

started editing. You will need to save again to make your changes available to others using the database.

Based on your organization's work flow, you may eventually need to reconcile two versions. Reconciliation is the process of merging features from a target version into the current edit session. Reconciliation must be done before posting changes to another version.

During reconciliation, conflicts may be discovered. Conflicts arise when the same feature is updated in each version or updated in one and deleted in the other.

When conflicts arise, an interactive conflict resolution dialog box will provide the tools necessary to resolve the conflicts. For each conflict, you can choose whether to replace the feature in your edit session with the conflict version, the version from your edit session, or the version as it existed at the beginning of your edit session.

Once you have successfully completed the reconciliation, you can post the version. The post operation synchronizes your edit session with the target version. They are then identical.

Reconciling

1. Click the Reconcile button on the Versioning toolbar.
2. Click the target version.
3. Click OK.

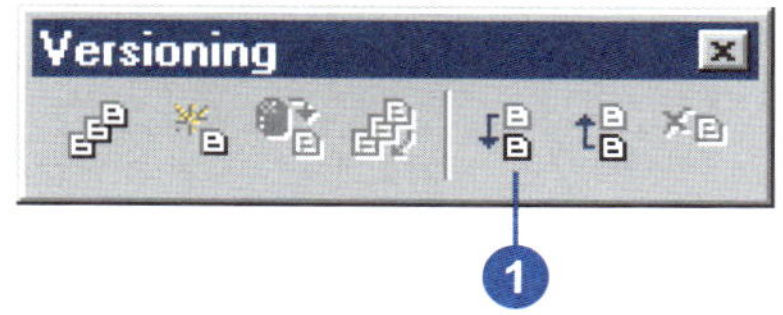

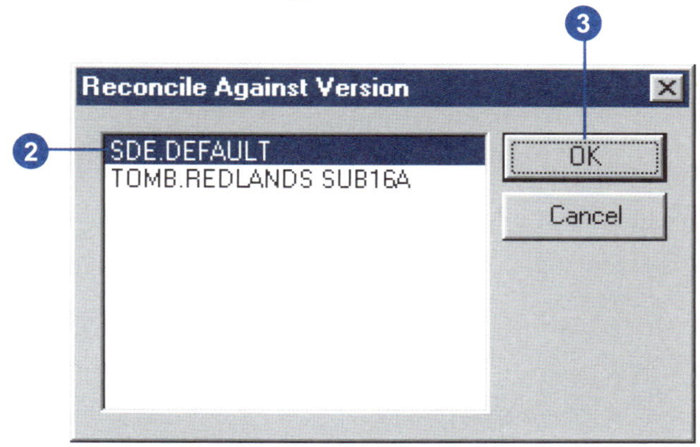

Posting

1. Click the Post button on the Versioning toolbar.

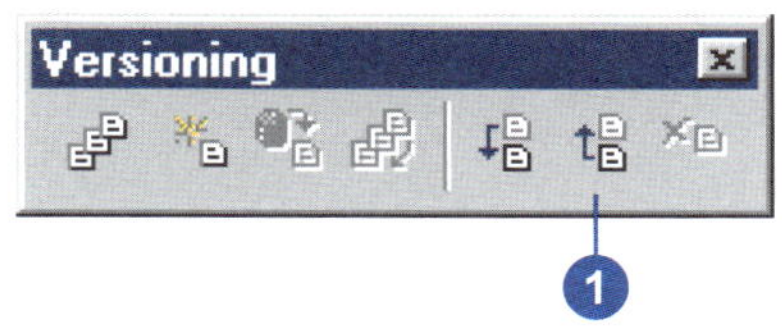

Displaying conflicts

1. Click the Conflicts button on the Versioning toolbar.
2. Right-click Conflicts and click Display.
3. Click the appropriate check box to display each conflict category.
4. Click OK.
5. Click the Close button to close the Conflicts dialog box.

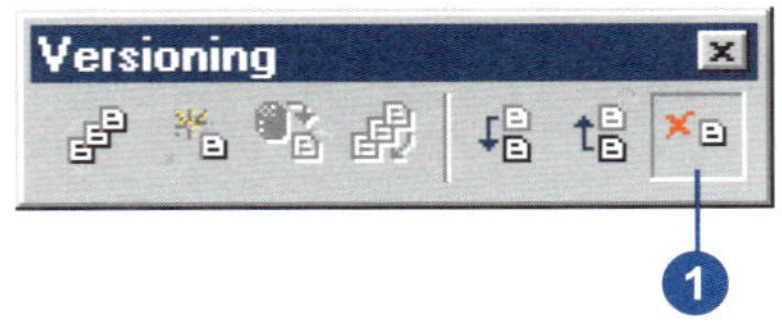

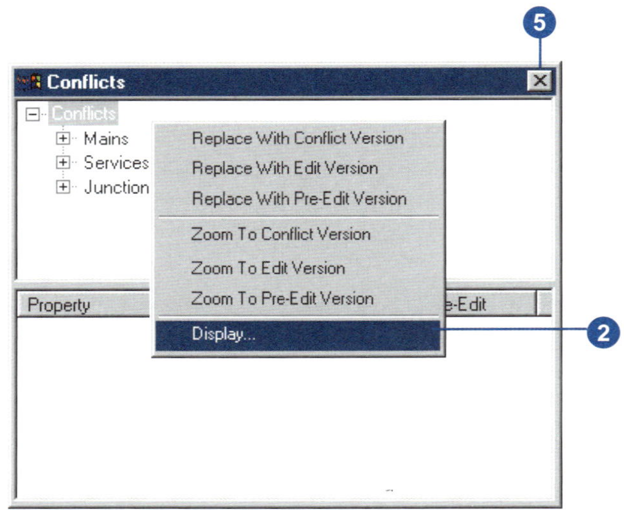

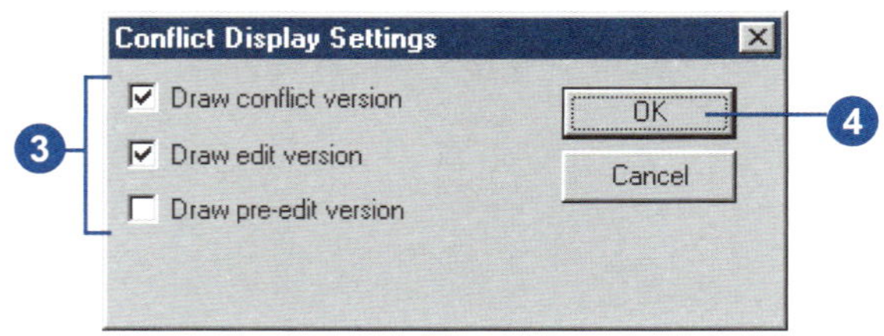

See Also

For information on typology and versioning, see Building a Geodatabase.

Resolving conflicts

1. Click the Conflicts button on the Versioning toolbar.
2. Click a feature class.
3. Click a feature and right-click to display the context menu.
4. Click the appropriate Replace With command to resolve the conflict.
5. Click the Close button to close the Conflicts dialog box.

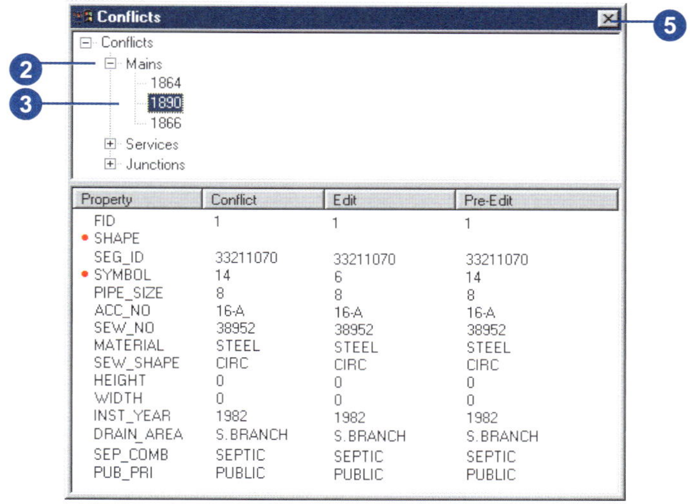

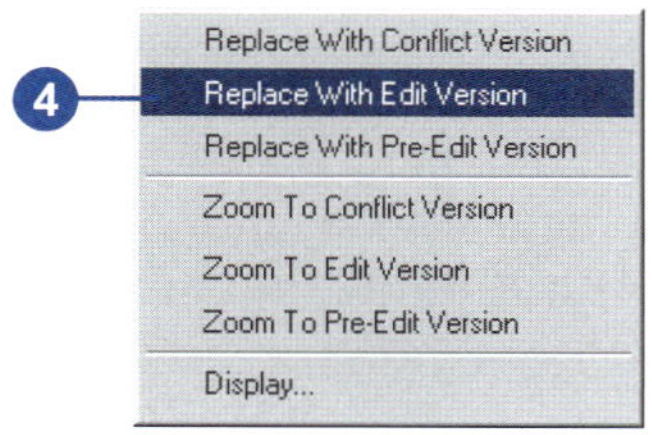

Versioning scenarios

The following scenarios show how an organization can implement its work flow process using a versioned database. These examples demonstrate several techniques available for performing long transactions in a multiuser environment. It is likely that organizations will, in some manner, use each of these techniques, depending on the task.

Scenario 1: Simple database modifications

Task: Multiple users are concurrently editing the database, performing common map sheet changes, such as inserting new features, updating attributes, and removing out-of-date facilities.

Solution: Users can simply connect to the DEFAULT version, simultaneously, start editing, and save their changes when their work is complete. Users do not have to create new versions to modify the database. If another user has edited the DEFAULT version since the current user has started editing, the user saving is notified that the version has been changed, and therefore, the version will need to be saved again. Users may bypass this warning message by enabling autoreconciliation in the ArcMap Options dialog box. Also, if two users modify the same feature during their edit sessions, the second user to save encounters a conflict. The user then has to decide what the feature's correct representation is and save the edit session.

Scenario 2: Transactions spanning multiple days

Task: Update the database to incorporate new and updated facilities in the field, which will likely require multiple edit sessions and a couple of days to complete.

Solution: A user creates and switches to a new version derived from the DEFAULT version. The user starts editing the new version and begins modifying features and saving as required. The user can resume the edit session, as appropriate, the following day or possibly the following week. When the changes are complete and ready to be posted to the DEFAULT version, the user must first click the Reconcile button on the Versioning toolbar. If conflicts are detected, the user can resolve the differences and complete the transaction by clicking the Post button. The posting process applies all the changes in the user's version to the DEFAULT version. The user can then delete the version.

Scenario 3: A work flow process

Task: Create individual versions for each step or stage of the work order and work flow process and post the work order to the database.

Solution: A user or supervisor creates a new version derived from the DEFAULT version. The user starts editing the new version and begins modifying features or creating a new design. When the user has completed the design or proposed modifications, the work order can be submitted to a supervisor for review. At this time, a new version can be created to ensure the preservation of the initial design. The new version can then be further modified or adjusted as required. Once the work order has been approved for construction, another version can be created. The purpose of this version is to reflect any changes that may occur while the work order is being constructed in the field. Finally, as the construction is completed and the new facilities are in service, the work order must be posted to the database. A user can then start editing the work order, perform a reconcile with the DEFAULT version, resolve any conflicts, if necessary, and post.

The solution allows the organization to create new versions of the work order for each step of the project—the initial design or proposed version, a working or accepted version, and a version for the construction phase. Each version is preserved and available to look back on for historical purposes. The final step is to post the constructed version to the database. The project completes a full circle from start to finish, creating individual versions at each step.

Scenario 4: Restricting permissions to the database

Task: The organization's supervisor has restricted write access to the DEFAULT version, requiring managerial review of each user's edits prior to posting the changes to the database.

Solution: To restrict write permissions to the database (the DEFAULT version), the ArcSDE administrative user can set the permission of the DEFAULT version to protected using the version manager. This allows users to continue to view the DEFAULT version but does not allow users to start editing the version. Therefore, users will need to create new versions for editing the database, similar to Scenario 2. When a user has completed and saved the edit session, the ArcSDE administrator can reconcile the version with the DEFAULT version. To accomplish this task, the manager who connects to the database as the ArcSDE administrator starts editing the user's version and clicks the Reconcile button. The process will merge all the changes in the user's version and the DEFAULT version. If conflicts are detected, the manager can resolve the conflicts and save the edit session. Once the edits are acceptable to the manager, the version is ready to be posted to the DEFAULT version. The ArcSDE administrative user can then start editing the version, perform a reconcile, and post the version. The user's version can then be deleted.

Scenario 5: Compressing the database

Task: The geodatabase has been edited for an extended time, and the number of database states and rows in each feature classes' delta tables has significantly increased. How do we improve performance by running the Compress command?

Solution: The Compress command will remove all database states that are no longer referenced by a version and move all the rows in the delta tables, which are common to all versions, to the base table. To achieve the maximum benefit when running the Compress command, you will need to first reconcile, post, and delete each version with the DEFAULT version. Sometimes this may not be a reasonable option based on your organization's work flow. At minimum, to improve performance, simply reconcile each version with the DEFAULT version and save, then perform the compress. This will ensure that all the edits in the DEFAULT version will be compressed from the delta tables to the business table. Remember, the Compress command can still be executed without first reconciling, posting, and deleting each version, but the benefits may not be as noticeable.

Glossary

active data frame

The data frame in a map that is currently being worked on—for example, the data frame to which layers are being added. The active data frame is shown in bold text in the ArcMap table of contents.

alias

Another name for a field in a table.

aligned dimension

A dimension that runs parallel to the baseline and represents the true distance between the beginning and end dimension points.

annotation

1. Descriptive text used to label features. It is used for display, not for analysis.

2. A feature class used to label other features. Information stored for annotation includes a text string, the location at which it is displayed, and a text symbol—color, font, size, and so on—for display.

3. The process of automating text placement or the text associated with a feature or an area on a map.

arc–node topology

Arcs represent linear features and the borders of area features in a coverage. Every arc has a from-node, which is the first vertex in the arc, and a to-node, which is the last vertex. Nodes indicate the endpoints and intersections of arcs. They do not exist as independent features. Together they define the direction of the arc. Arc–node topology defines connectivity in coverages—arcs are connected to each other if they share a common node.

ArcInfo workspace

A file-based collection of coverages, grids, TINs, or shapefiles stored as a directory of folders in the file system.

ArcSDE

A gateway to a multiuser commercial RDBMS—for example, Oracle®, Microsoft® SQL Server™, Informix®, and IBM® DB2®. ArcSDE is an open, high-performance spatial data server that employs client/server architecture to perform efficient spatial operations and manage large, shared geographic data. Was known as a Spatial Database Engine™ (SDE®) before 1999.

asynchronous

Operations or events that do not happen at the same time. In disconnected editing, modifying the properties of a check-out is an asynchronous operation; changes made to the check-out in a master geodatabase do not affect the associated check-out in a check-out geodatabase.

attribute

A characteristic of a map feature. Attributes of a river might include its name, length, average depth, and so on.

attribute domain

A named constraint in the database. An attribute constraint can be applied to a field of a subtype of a feature class or object class to make an attribute rule. Types of attribute domains include range and coded value domains.

attribute table

A database management system (DBMS) or other tabular file containing rows and columns. In ArcInfo, attribute tables are associated with a class of geographic features such as wells or roads. Each row represents a geographic feature. Each column represents one attribute of a feature, with the same column representing the same attribute in each row. See also feature attribute table.

Attributes dialog box

A dialog box that lets you view and edit attributes of features you've selected in ArcMap.

azimuth

An angle measured from north. Often used to define an oblique cylindrical projection or the angle of a geodesic between two points.

behavior

Properties of an object in a geodatabase that describe how it can be edited and drawn. Behavior includes, but is not limited to, validation rules, subtypes, default values, and relationships.

buffer

A zone of a specified distance around features. Both constant- and variable-width buffers can be generated on each feature's attribute values. The resulting buffer zones form polygons that are either inside or outside the specified buffer distance from each feature. Buffers are useful for proximity analysis—for example, to find all stream segments within 300 feet of a proposed logging area.

CAD

See computer-aided design.

CAD feature class

A feature class in a CAD dataset. A CAD feature dataset is composed of feature classes representing all the points, lines, polygons, or annotation in the CAD drawing—for example, a CAD drawing may contain two line layers representing roads and parcel boundaries, respectively. The CAD dataset's line feature class represents all features in both the road and parcel boundary layers.

centroid

The mathematical or geographical center point of a polygon or the midpoint of a line.

check-out

A geodatabase entity that represents data that has been checked out to or from another geodatabase.

check-out geodatabase

A personal or ArcSDE geodatabase that contains data checked out from a master geodatabase.

check-out version

A version created in a check-out ArcSDE geodatabase when data was checked out. This version is created as a child of the synchronization version. Only the edits made to this check-out version can be checked back in to the master geodatabase.

checking in

The process of transferring changes to a master geodatabase.

checking out

The process of copying data to a check-out geodatabase.

circle

A geometric shape for which the distance from the center to any point on the edge is equal.

circular arc

When creating features, it is often necessary to create a circular arc. Instead of being made of numerous vertices, a circular arc has only two vertices as endpoints. ArcMap offers four ways to create a segment that is a circular arc. These include the Arc tool, the Endpoint Arc tool, the Tangent Curve tool, and the Tangent Curve command.

cluster tolerance

The minimum distance between vertices in the topology. Vertices that fall within the cluster tolerance will be snapped together during the validate topology process. The default cluster tolerance is the minimum possible cluster tolerance, based on the precision and extent defined for the spatial reference of the dataset. See also fuzzy tolerance, used analogously in coverage data model.

clustering

A part of the topology validation process in which vertices that fall within the cluster tolerance are snapped together.

coincident

Vertices or boundaries are coincident when they are within the cluster tolerance of one another. See also cluster tolerance.

column

The vertical dimension of a table. A column has a name and a data type applied to all values in the column. See also item, field, and attribute.

compress

The process of shrinking the size of a database or file. Improves database performance by removing redundant rows shared by multiple versions. The process can only be run by the ArcSDE administrator.

computer-aided design

An automated system for the design, drafting, and display of graphically oriented information.

conflict

In the versioning reconciliation process, if the same feature in the edit version and reconciliation version have both been edited, the feature is said to be in conflict. Resolving the conflict requires you to make the decision as to the feature's correct representation using the conflict resolution dialog box.

connectivity

1. In a geodatabase, the state of edges and junctions in a logical network that controls flow, tracing, and pathfinding.

2. The topological identification in a coverage of connected arcs by recording the from- and to-node for each arc. Arcs that share a common node are connected. See also arc–node topology.

connectivity rules

Network rules that constrain the type of network features that may be connected to one another and the number of features of any particular type that can be connected to features of another type. In most networks, not all edge types can logically connect to all junction types. Similarly, not all edge types can logically connect to all other edge types through all junction types. There are two types of connectivity rules: edge–junction and edge–edge.

constraints

In real-world databases, an object's attributes can't have any particular value based solely on what data types and ranges a particular field type in the database allows. In reality, the permissible values are a range or list of values.

construct features

The process of taking selected features from one or more feature classes and creating new features in a target feature class in an edit session. The Construct Features tool uses the input geometries of the selected features to construct polygons or lines following polygon boundaries, depending on the geometry of the target feature class.

contiguity

In coverages, the topological identification of adjacent polygons by recording the left and right polygons of each arc. See also polygon–arc topology.

control points

Points you establish on a paper map to represent known ground points or specific locations. Control points are used to register a paper map before you begin digitizing features on it with a digitizer.

coordinate

A set of numbers that designate location in a given reference system such as x,y in a planar coordinate system or x,y,z in a three-dimensional coordinate system. Coordinates represent locations on the earth's surface relative to other locations.

coordinate system

1. A reference system used to measure horizontal and vertical distances on a planimetric map. A coordinate system is usually defined by a map projection; a spheroid of reference; a datum; one or more standard parallels; a central meridian; and possible shifts in the x- and y-directions to locate x,y positions of point, line, and area features.

2. In ArcInfo, a system with units and characteristics defined by a map projection. A common coordinate system is used to spatially register geographic data for the same area.

coverage

A file-based vector data storage format for storing the location, shape, and attributes of geographic features. A coverage usually represents a single theme, such as soils, streams, roads, or land use. It is one of the primary vector data storage formats for ArcInfo.

A coverage stores geographic features as primary features, such as arcs, nodes, polygons, and label points, and secondary features, such as tics, map extent, links, and annotation. Associated feature attribute tables describe and store attributes of the geographic features.

cracking

A part of the topology validation process in which vertices are created at the intersection of feature edges.

current task

During editing in ArcMap, a setting in the Current Task dropdown list that determines with which task the sketch construction tools—Sketch, Arc, Distance–Distance, and Intersection—will work. The current task is set by clicking a task in the Current Task dropdown list. All tasks in the Current Task dropdown list work with a sketch that you create—for example, the Create New Feature task uses a sketch you create to make a new feature. The Extend/Trim Feature task uses a sketch you create to determine where the selected feature will be extended or trimmed. The Cut Polygon Feature task uses a sketch you create to determine where the polygon will be cut.

custom behavior

Behavior is the implementation of an object class method. ESRI-provided objects have a set of methods associated with them. A developer can choose to override one of these methods or create additional methods. In this instance, the object is said to have custom behavior.

custom feature

A feature with specialized behavior instantiated in a class by a developer.

custom object

Objects that have custom behavior provided by a developer.

dangle tolerance

The minimum length allowed for dangling arcs in coverages in the ArcInfo Clean process. Clean removes dangling arcs that are shorter than the dangle length. Also known as the dangle length.

dangling arc

In coverages, an arc having the same polygon on both its left and right sides and having at least one node that does not connect to any other arc. It often identifies where a polygon does not close properly—for example, undershoot—where arcs don't connect properly, or where an arc was digitized past its intersection with another arc—for example, overshoot. A dangling arc is not always an error—for example, dangling arcs can represent cul-de-sacs in street centerline maps.

data

A collection of related facts usually arranged in a particular format and gathered for a particular purpose.

data frame

In ArcMap, a frame on the map that displays layers occupying the same geographic area. You may have one or more data frames on your map depending upon how you want to organize your data. For instance, one data frame might highlight a study area, and another might provide an overview of where the study area is located.

data integrity

Maintenance of data values according to data model and data type—for example, to maintain integrity, numeric columns will not accept character data.

data source

Any geographic data, such as a coverage, shapefile, raster, or feature class, in a geodatabase.

data type

The characteristic of columns and variables that defines what types of data values they can store. Examples include character, floating point, and integer.

data view

An all-purpose view in ArcMap for exploring, displaying, and querying geographic data. This view hides all map elements such as titles, North arrows, and scalebars. See also layout view.

database

A collection of related files organized for efficient retrieval of information. In the context of an ArcSDE geodatabase, some relational databases group data together in discretely named databases—for example, SQL Server, Sybase®—while others do not—for example, Oracle.

dataset

1. Any feature class, table, or collection of feature classes or tables in the geodatabase.

2. A named collection of logically related data items arranged in a prescribed manner.

decimal degrees

Degrees of latitude and longitude expressed as a decimal rather than in degrees, minutes, and seconds.

default junction type

Two edge types may be connectable through more than one junction type. You can establish which of those junction types is the default for connecting the two edge types. This junction type is the default junction type.

digitizing

1. To encode geographic features in digital form as x,y coordinates.

2. The process of converting the features on a paper map into digital format. When you digitize a map, you use a digitizing tablet, or digitizer, connected to your computer and trace over features with a digitizer puck, which is similar to a mouse. The x,y coordinates of these features are automatically recorded and stored as spatial data.

digitizing mode

Also called absolute mode, digitizing mode is one of the ways in which a digitizing tablet operates. In digitizing mode, the location of the tablet is mapped to a specific location on the screen. Moving the digitizer puck on the tablet surface causes the screen pointer to move to precisely the same position.

dimension contruction methods

Dimension construction methods dictate what type of dimension feature is created and the number of points required to complete the feature's geometry. Construction methods include simple aligned, aligned, linear, rotated linear, free aligned, and free linear.

dimension feature

Dimension features are a special kind of map annotation that show specific lengths or distances on a map. A dimension feature may indicate the length of a side of a building or land parcel or it may indicate the distance between two features such as a fire hydrant and the corner of a building. Dimension features are stored in a dimension feature class.

dimension feature class

In the geodatabase, dimension features are stored in dimension feature classes. Like other feature classes in the geodatabase, all features in a dimension feature class have a geographic location and attributes and can either be inside or outside of a feature dataset.

dimension style

A dimension feature's style describes its symbology, what parts of it are drawn, and how it is drawn. Every time you create a new dimension feature, it is assigned a particular style. A collection of dimension styles is associated with a dimension feature class.

Dimensioning toolbar

A toolbar in ArcMap that facilitates the creation of dimension features.

direct connect

A two-tiered architecture for connecting to spatial databases. Direct connect does not require the ArcSDE application server to connect to a spatial database.

dirty areas

Areas that have been edited after the initial topology validation process. Dirty areas represent regions surrounding features that have been altered and that require an additional topology validation to be performed in order to discover any topology errors that may be present.

disconnected editing

The process of copying data to another geodatabase, editing that data, then merging the changes with the data in the source or master geodatabase.

distance units

The units—for example, feet, miles, meters, or kilometers—ArcMap uses to report measurements, dimensions of shapes, distance tolerances, and offsets.

double precision

Refers to a high level of coordinate accuracy based on the possible number of significant digits that can be stored for each coordinate. ArcInfo datasets can be stored in either single- or double-precision coordinates. Double-precision coverages store up to 15 significant digits per coordinate—typically, 13 to 14 significant digits—retaining the accuracy of much less than one meter at a global extent. See also single precision.

edge (topology)

A line segment in a topology that defines lines or polygon boundaries. Multiple features in one or more feature classes may share topology edges.

edge element

See logical network.

edge–edge rule

A connectivity rule that establishes that an edge of type A may connect to an edge of type B through a junction of type C. Edge–edge rules always involve a junction type.

edge–junction cardinality

A rule may exist that allows an edge of type A to connect to a junction of type B. By default, any number of edges of type A can connect to a single junction of type B. You may want to restrict this. You can specify that between two and five edges of type A can connect to a junction of type B; if there are less than two edges, or more than five edges, the connectivity rule is being violated. Similarly, you can restrict the number of junctions of type C that can connect to any junction of type D. This range of permissible connections is edge–junction cardinality.

edge–junction rule

A connectivity rule that establishes that an edge of type A may connect to a junction of type B.

edit cache

A setting used in spatial data editing in ArcMap that causes the features visible in the current map extent to be held in memory on your local machine. Designed to be used when working with large amounts of data, an edit cache results in faster editing because ArcMap doesn't have to retrieve the data from the server.

edit session

In ArcMap editing takes place within an edit session. An edit session begins when you choose Start Editing from the Editor menu and ends when you choose Stop Editing.

Editor toolbar

A toolbar that lets you create and modify features and their attributes in ArcMap.

ellipse

A geometric shape equivalent to a circle that is viewed obliquely; a flattened circle.

error

Violations of a topology rule detected during the topology validation process.

extent

The coordinates defining the minimum bounding rectangle—that is, xmin, ymin and xmax, ymax—of a data source. All coordinates for the data source fall within this boundary.

feature

1. An object class in a geodatabase that has a field of type geometry. Features are stored in feature classes.

2. A representation of a real-world object.

3. A point, line, or polygon in a coverage or shapefile.

4. A representation of a real-world object in a layer on a map.

feature attribute table

A table used to store attribute information for a specific coverage feature class. ArcInfo maintains the first several items of these tables. Feature attribute tables supported for coverages include the following:

<cover>.PAT	for polygons or points
<cover>.AAT	for arcs
<cover>.NAT	for nodes
<cover>.RAT	for routes
<cover>.SEC	for sections
<cover>.PAT	for regions
<cover>.TAT	for annotation (text)

where <cover> is the coverage name.

feature class

1. The conceptual representation of a category of geographic features. When referring to geographic features, feature classes include point, line, area, and annotation. In a geodatabase, an object class that stores features and has a geometry field type.

2. A classification describing the format of geographic features and supporting data in a coverage. Coverage feature classes for representing geographic features include point, arc, node, route-system, route, section, polygon, and region. One or more coverage features are used to model geographic features—for example, arcs and nodes can be used to model linear features such as street centerlines. The tic, annotation, link, and boundary feature classes provide supporting data for coverage data management and viewing.

3. The collection of all the point, line, or polygon features or annotation in a CAD dataset.

4. In a geodatabase, an object class that stores features and has a field of type geometry.

feature dataset

In geodatabases, a collection of feature classes that share the same spatial reference. Because the feature classes share the same spatial reference, they can participate in topological relationships with each other such as in a geometric network, linear network, or topology. Several feature classes with the same geometry may be stored in the same feature dataset. Object classes and relationship classes can also be stored in a feature dataset.

field

A column in a table. Each field contains the values for a single attribute.

fuzzy tolerance

An extremely small distance used to resolve inexact intersection locations due to the limited arithmetic precision of computers. It defines the resolution of a coverage resulting from the Clean operation or a topological overlay operation such as Union, Intersect, or Clip.

In geodatabase feature classes, this concept is replaced by cluster tolerance.

geocoding

The process of creating geometric representations for locations, such as point features, from descriptions of locations, such as addresses.

geocoding index

An index on geocoding reference data used by geocoding services.

geocoding reference data

Data that a geocoding service uses to determine the geometric representations for locations.

geocoding service

An object that defines a process for creating geometric representations for locations, such as point features, from descriptions of locations, such as addresses.

geodatabase

A geographic database that is hosted inside a relational database management system that provides services for managing geographic data. These services include validation rules, relationships, and topological associations.

geodatabase data model

Geographic data model that represents geographic features as objects in an object-relational database. Features are stored as rows in a table; geometry is stored in a shape field. Supports sophisticated modeling of real-world features. Objects may have custom behavior. Compare to the georelational data model.

geometric network

A geometric network can be thought of as a one-dimensional nonplanar graph, or logical network, that is composed of features. These features are constrained to exist within the network and can, therefore, be considered network features. ArcInfo will automatically maintain the explicit topological relationships between network features in a geometric network.

georelational data model

A geographic data model that represents geographic features as an interrelated set of spatial and descriptive data. The georelational model is the fundamental data model used in coverages. Compare to the geodatabase data model.

index

A data structure used in a database to improve query performance. Feature classes also have spatial indexes that improve spatial query performance.

instance

The name of the process running on the ArcSDE server that allows connections and access to spatial data.

intersect

The topological integration of two spatial datasets that preserves features that fall within the area common to both input datasets. See also union.

IP address

The server's address on the network. The address consists of four numbers, each separated by a ".".

item

1. A column of information in an INFO table.

2. An element in the Catalog tree. The Catalog tree can contain both geographic data sources and nongeographic elements, such as folders, folder connections, and file types.

junction element

See logical network.

label point

A feature class in a coverage used to represent point features and identify polygons.

layer

1. A collection of similar geographic features, such as rivers, lakes, counties, or cities, of a particular area or place for display on a map. A layer references geographic data stored in a data source, such as a coverage, and defines how to display it. You can create and manage layers as you would any other type of data in your database.

2. A feature class in a shared geodatabase managed with SDE 3.

layout view

The view for laying out your map in ArcMap. Layout view shows the virtual page upon which you place and arrange geographic data and map elements, such as titles, legends, and scalebars, for printing. See also data view.

left–right topology

A topological data structure used to represent contiguity between polygons in the coverage data model. Left–right topology supports analysis functions, such as adjacency. See also topology.

linear dimension

A dimension whose length doesn't represent the true distance between the beginning and end dimension points. Linear dimensions can be vertical, horizontal, or rotated. A vertical dimension's line represents the vertical distance between the beginning and end dimension points. A horizontal linear dimension's line represents the horizontal distance between the begin and end dimension points. A rotated linear dimension is a dimension whose line is at some angle to the baseline and whose length represents the length of the dimension line itself, not the baseline.

logical network

A logical network is an abstract representation of a network. A logical network consists of edge, junction, and turn elements and the connectivity between them. You can ask a logical network which elements are connected but you cannot ask it for the geometry of these elements. A logical network does not contain any coordinate data, so you cannot ask it for the location of its elements. For this, you need a geometric network. In a logical network, an edge element is connected to two junction elements—a from-junction and a to-junction—and a junction can have zero or more edges connected to it. A turn has a from-edge, a junction, and a to-edge. Each element can also have many weights associated with it. Weights are typically used to describe the cost to traverse an edge or turn or the cost to pass through a junction.

map

1. A graphical presentation of geographic information. It contains geographic data and other elements, such as a title, North arrow, legend, and scalebar. You can interactively display and query the geographic data on a map and also prepare a printable map by arranging the map elements around the data in a visually pleasing manner.

2. The document used in ArcMap that lets you display and work with geographic data. A map contains one or more layers of geographic data and various supporting map elements, such as scalebars. Layers on a map are contained in data frames. A data frame has properties, such as scale, projection, and extent, and also graphic properties such as where it is located on a map's page. Some maps have one data frame while other more advanced maps may have several data frames.

map document

In ArcMap, the disk-based representation of a map. Map documents can be printed or embedded into other documents. Map documents have an .mxd file extension.

map topology

A temporary set of topological relationships between coincident parts of simple features on a map, used to edit shared parts of multiple features.

map units

The units—for example, feet, miles, meters, or kilometers—in which the coordinates of spatial data are stored.

master check-out version

A version in the master geodatabase, created when data is checked out, that represents the state of the data at the time it was checked out.

master geodatabase

An ArcSDE geodatabase from which data has been checked out.

merge policy

In geodatabases, all attribute domains have a merge policy associated with them. When two features are merged into a single feature in ArcMap, the merge policies dictate what happens to the value of the attribute to which the domain is associated. Standard merge policies are default value, sum, and weighted average.

minimum bounding rectangle

A rectangle, oriented to the x- and y-axes, that bounds a geographic feature or a geographic dataset. It is specified by two coordinates: xmin, ymin and xmax, ymax—for example, the extent defines a minimum bounding rectangle for a coverage.

multipart feature

A feature that is composed of more than one physical part but only references one set of attributes in the database—for example, in a layer of states, the state of Hawaii could be considered a multipart feature. Although composed of many islands, it would be recorded in the database as one feature.

multipoint feature

A feature that consists of more than one point but only references one set of attributes in the database—for example, a system of oil wells might be considered a multipoint feature, as there is a single set of attributes for the main well and multiple well holes.

multiuser geodatabase

A geodatabase in an RDBMS served to client applications—for example, ArcMap—by ArcSDE. Multiuser geodatabases can be very large and support multiple concurrent editors. They are supported on a variety of commercial RDBMSs, including Oracle, Microsoft SQL Server, IBM DB2, and Informix.

network trace

In the most generic sense, a network trace means to navigate through the network following the connectivity of the network for some purpose.

node

An endpoint of a topology edge. Topology nodes may also be introduced along an edge during editing.

null value

The absence of a value. A geographic feature for which there is no associated attribute information.

object

The representation of a real-world entity stored in a geodatabase. An object has properties and behavior.

object class

A collection of objects in the geodatabase that have the same behavior and the same set of attributes. All objects in the geodatabase are stored in object classes.

overshoot

That portion of an arc digitized past its intersection with another arc. See also dangling arc.

pan

To move the viewing window up, down, or sideways to display areas in a geographic dataset that, at the current viewing scale, lie outside the viewing window.

password

The password used for authentication when you log on to an ArcSDE geodatabase.

personal geodatabase

A geodatabase, usually on the same network as the client application—for example, ArcMap—that supports one editor at a time. Personal geodatabases are managed in a Microsoft Jet Engine database. Personal geodatabases can contain geometric networks, linear networks, and topologies and can be used as check-out databases for disconnected editing of geographic data in the field.

planarize

The process of creating multiple line features by splitting longer features at the places where they intersect other line features. This process can be useful when you have nontopological linework that has been spaghetti digitized or imported from a computer-aided design (CAD) drawing.

point

A single x,y coordinate that represents a single geographic feature such as a telephone pole.

point mode digitizing

One of two methods of digitizing features using the ArcMap Editor's Sketch tool or from a paper map using a digitizer. With point mode digitizing, you can create or edit features by digitizing a series of precise points, or vertices. Point mode digitizing is effective when precise digitizing is required—for example, when digitizing a perfectly straight line. See also stream mode digitizing.

polygon

A two-dimensional feature representing an area such as a state or county.

polygon–arc topology

PAT. A coverage polygon is made up of arcs that define the boundary and a label point, which links the polygon feature to an attribute record in the coverage PAT. In the coverage data model, polygons are represented topologically as a list of arcs and a label that make up each polygon. In the geodatabase data model, polygons are stored as simple features, which may participate in topological relationships with other features.

port number

The TCP/IP port number that an ArcSDE geodatabase instance is communicating on.

post

Posting is the process of applying the current edit session to the reconciled target version.

precision (dataset)

The number of system units per one unit of measure in the XY domain of the coordinate system dataset. Precision defines the smallest storable distance between coordinates in the dataset. A spatial reference with a precision of one will store integer values, while a precision of 1,000 will store three decimal places.

preliminary topology

In coverages, refers to incomplete region or polygon topology. Region topology defines region–arc and region–polygon relationships. A topological region has both the region–arc relationship and the region–polygon relationship. A preliminary region has the region–arc relationship but not the region–polygon relationship. In other words, preliminary regions have no polygon topology. Polygon topology defines polygon–arc–label point relationships. A preliminary polygon has the polygon–label point relationship but not the polygon–arc relationship. Coverages with preliminary topology have red in their icons in the Catalog.

projection

A mathematical formula that transforms feature locations from the earth's curved surface to a map's flat surface. A projected coordinate system employs a projection to transform locations expressed as latitude and longitude values to x,y coordinates. Projections cause distortions in one or more of these spatial properties: distance, area, shape, and direction.

property

An attribute of an object defining one of its characteristics or an aspect of its behavior—for example, the Visible property affects whether a control can be seen at run time. You can set a data source's properties using its Properties dialog box.

pseudonode

A node connecting only two edges or a logical split defined in the topology cache while editing. Pseudonodes of the latter sort become a vertex after editing.

pull check-in

A check-in operation initiated from a master geodatabase.

push check-in

A check-in operation initiated from a check-out geodatabase.

pyramids

In raster datasets, reduced resolution layers, or pyramids, record the original data in decreasing levels of resolution. The coarsest level of resolution is used to quickly draw the entire dataset. As you zoom in, layers with finer resolutions are drawn; performance is maintained because you're drawing successively smaller areas.

query

A question or request used for selecting features. A query often appears in the form of a statement or logical expression. In ArcMap, a query contains a field, an operator, and a value.

radius

The distance from the center to the outer edge of a circle or circular curve.

rank

A method of assigning an accuracy value to feature classes to avoid having vertices from a feature class collected with a high level of accuracy being snapped to vertices from a less accurate feature class. Vertices from higher ranking feature classes will not be moved when snapping with vertices with lower ranked feature classes. The highest rank is 1 and you can assign up to 50 different ranks.

raster

Represents any data source that uses a grid structure to store geographic information.

RDBMS

Relational database management system. A database management system with the ability to access data organized in tabular files that can be related to each other by a common field (item). An RDBMS has the capability to recombine the data items from different files, providing powerful tools for data usage. ArcSDE supports several commercial RDBMSs.

reconcile

Reconciling is the process of merging all modified datasets, feature classes, and tables in the current edit session and a second target version. All features and rows that do not conflict are merged into the edit session, replacing the current features or rows. Features that are modified in each version are conflicts and require further resolution via the conflict resolution dialog box.

record

1. In an attribute table, a single row of thematic descriptors. In SQL terms, a record is analogous to a tuple.

2. A logical unit of data in a file—for example, there is one record in the ARC file for each arc in a coverage.

reference data

Tables or feature classes containing address information that geocoding services use to find the locations of addresses.

relate

An operation that establishes a temporary connection between corresponding records in two tables using an item common to both—for example, key attributes. Each record in one table is connected to those records in the other table that share the same value for the common item. See also relational join.

relational join

The operation of relating and physically merging two attribute tables using their common item.

relationship

An association or link between two or more objects in a geodatabase. Relationships can exist between spatial objects (features in feature classes) or nonspatial objects (rows in a table), or between spatial and nonspatial objects.

relationship class

Objects in a real-world system often have particular associations with other objects in the database. These kinds of associations between objects in the geodatabase are called relationships. Relationships can exist between spatial objects (features in feature classes), nonspatial objects (rows in a table), or between spatial and nonspatial objects. While spatial objects are stored in the geodatabase in feature classes, and nonspatial objects are stored in object classes, relationships are stored in relationship classes.

row

1. A record in an attribute table. The horizontal dimension of a table composed of a set of columns containing one data item each.

2. A horizontal group of cells in a raster.

rule (connectivity)

A constraint on the type of network features that may be connected to one another and the number of features of any particular type that can be connected to features of another type. There are two types of connectivity rules: edge–junction and edge–edge rules.

rule (topology)

An instruction to the geodatabase defining the permissible relationships of features within a given feature class or between features in two different feature classes. Topology rules are compared against the features in the feature class during the topology validation process and violations are marked as errors. After the topology validation process, the errors can be corrected by editing the feature class; occasionally, violations of a topology rule may represent acceptable conditions. In these cases, the errors can be marked as exceptions.

scanning

The process of capturing data in raster format with a device called a scanner. Some scanners also use software to convert raster data to vector data.

schema

The structure or design of a database.

schema-only check-out

A type of check-out that creates the schema of the data being checked out in the check-out geodatabase but does not copy any data.

segment

A line that connects vertices—for example, in a sketch of a building, a segment would represent one wall.

select

To choose from a number or group of features or records; to create a separate set or subset.

Selectable layers list

A list on the Selection toolbar that lets you choose from which layers you can select.

For example, suppose you wanted to select a large number of buildings by drawing a box around them but selected a parcel by mistake as you drew the selection box. To avoid this, you might uncheck the Parcels layer in the Selectable layers list so that parcels cannot be selected.

selected set

A subset of the features in a layer or records in a table. ArcMap provides several ways to select features and records graphically or according to their attribute values.

selection anchor

When editing in ArcMap, a small x located in the center of selected features. The selection anchor is used when you move features using snapping. It is the point on the feature or group of features that will be snapped to the snapping location. This is also the point around which your selection will rotate when you use the Rotate tool and around which your feature will scale when you use the Scale tool. You can reposition the selection anchor.

server

The computer where the ArcSDE geodatabase you want to access is located.

shape

The characteristic appearance or visible form of a geographic object. Geographic objects can be represented on a map using one of three basic shapes: points, lines, or polygons.

shapefile

A vector data storage format for storing the location, shape, and attributes of geographic features. A shapefile is stored in a set of related files and contains one feature class.

shared boundary

A segment or boundary common to two features—for example, in a parcel database, adjacent parcels will share a boundary. Another example might be a parcel that shares a boundary on one side with a river. The segment of the river that coincides with the parcel boundary would share the same coordinates as the parcel boundary.

shared vertex

A vertex common to multiple features—for example, in a parcel database, adjacent parcels will share a vertex at the common corner.

simple feature

A feature that implements ESRI Simple Feature. Simple features with shared geometry can be edited with a map topology using ArcView and higher licensed seats of ArcMap

single precision

Refers to a level of coordinate accuracy based on the number of significant digits that can be stored for each coordinate. Single-precision numbers store up to seven significant digits for each coordinate, retaining a precision of ±5 meters in an extent of 1,000,000 meters. ArcInfo datasets can be stored as either single- or double-precision coordinates. See also double precision.

sketch

When editing in ArcMap, a shape that represents a feature's geometry. Every existing feature on a map has an alternate form, a sketch. A sketch lets you see exactly how a feature is composed, with all vertices and segments of the feature visible. To modify a feature, you must modify its sketch. To create a feature, you must first create a sketch. You can only create line and polygon sketches, as points have neither vertices nor segments.

Sketches help complete the current task—for example, the Create New Feature task uses a sketch you create to make a new feature. The Extend/Trim Feature task uses a sketch you create to determine where the selected feature will be extended or trimmed. The Cut Polygon Feature task uses a sketch you create to determine where the polygon will be cut into two features.

sketch constraints

In ArcMap editing, the angle or length limitations you can place on segments you're creating. These commands are available on the Sketch tool context menu—for example, you can set a length constraint that specifies that the length of the segment you're creating will be 50 map units. At whatever angle you create that segment, its length will be constrained to 50 map units.

Angle constraints work in the same way—for example, you can set an angle constraint that specifies that the angle of the segment you're creating will be 45 degrees measured from another feature that already exists. At whatever length you create that segment, its angle will be constrained to 45 degrees.

sketch operations

In ArcMap, editing operations that are performed on an existing sketch. Examples are Insert Vertex, Delete Vertex, Flip, Trim, Delete Sketch, Finish Sketch, and Finish Part. All of these operations are available from the Sketch context menu, which is available when you right-click any part of a sketch using any editing tool.

snapping

The process of moving a feature to coincide exactly with coordinates of another feature within a specified snapping distance or tolerance.

snapping environment

Settings in ArcMap Editor's Snapping Environment window and Editing Options dialog box that help you establish exact locations in relation to other features. You determine the snapping environment by setting a snapping tolerance, snapping properties, and a snapping priority.

snapping priority

During ArcMap editing, the order in which snapping will occur by layer. You can set the snapping priority by dragging the layer names in the Snapping Environment window to new locations.

snapping properties

In ArcMap editing, a combination of a shape to snap to and a method for what part of the shape you will snap to. You can set your snapping properties to have a feature snap to a vertex, edge, or endpoint of features in a specific layer—for example, a layer snapping property might let you snap to the vertices of buildings. A more generic, sketch-specific snapping property might let you snap to the vertices of a sketch you're creating.

snapping tolerance

The distance within which the pointer or a feature will snap to another location during ArcMap editing.

If the location being snapped to (vertex, boundary, midpoint, or connection) is within the distance you set, the pointer will automatically snap. For example, if you want to snap a power line to a utility pole and the snapping tolerance is set to 25 pixels, whenever the power line comes within a 25-pixel range of the pole, it will automatically snap to it. Snapping tolerance can be measured using either map units or pixels.

spatial database

Any DBMS that contains spatial data.

spatial domain

Describes the range and precision of x,y coordinates and z- and m-values that can be stored in a feature dataset or feature class in a geodatabase.

spatial join

A type of spatial analysis in which the attributes of features in two different layers are joined together based on the relative locations of the features.

spatial reference

Describes both the projection and spatial domain extent for a feature dataset or feature class in a geodatabase.

split policy

All attribute domains have a split policy associated with them. When a feature is split into two new features in the ArcMap Editor, the split policies dictate what happens to the value of the attribute to which the domain is associated. Standard split policies are duplicate, default value, and geometry ratio.

SQL

Structured Query Language. A syntax for defining and manipulating data from a relational database. Developed by IBM in the 1970s, it has become an industry standard for query languages in most relational database management systems.

stream mode digitizing

One of the two methods of digitizing features from a paper map. Also known as streaming, stream mode digitizing provides an easy way to capture features when you don't require much precision—for example, to digitize rivers, streams, and contour lines. With stream mode, you create the first vertex of the feature and trace over the rest of the feature with the digitizer puck. You can also use digitize in stream mode with the ArcMap Sketch tool when editing freehand. See also point mode digitizing.

stream tolerance

The minimum distance the pointer must be moved from the last vertex before the next vertex will be created when using the Sketch tool in stream mode.

When streaming, vertices are automatically created at a defined interval as you move the mouse—for example, if the stream tolerance is set to 10 map units, you must move the pointer at least 10 map units before the next vertex will be created. If you move the pointer more than 10 map units, there may be more space between vertices, but there will always be a minimum interval of 10 map units. Stream tolerance is measured in map units.

subtypes

Although all objects in a feature class or object class must have the same behavior and attributes, not all objects have to share the same default values and validation rules. You can group features and objects into subtypes. Subtypes differentiate objects based on their rules.

symbol

A graphic pattern used to represent a feature—for example, line symbols represent arc features; marker symbols, points; shade symbols, polygons; and text symbols, annotation. Many characteristics define symbols including color, size, angle, and pattern.

symbology

The criteria used to determine symbols for the features in a layer. A characteristic of a feature may influence the size, color, and shape of the symbol used.

synchronization version

A verion created in a check-out ArcSDE geodatabase when a check-out is made to that geodatabase. This version is created as a child of the DEFAULT version and represents the state of the data at the time the check-out was created.

table

Information formatted in rows and columns. A set of data elements that has a horizontal dimension (rows) and a vertical dimension (columns) in an RDBMS. A table has a specified number of columns but can have any number of rows. See also attribute table.

table of contents

In ArcMap, lists all the data frames and layers on the map and shows what the features in each layer represent.

tabular data

Descriptive information that is stored in rows and columns and can be linked to map features.

tagged values

Tagged values are used to set additional properties of UML elements—for example, you can set the length, in characters, of a string field by using a tagged value.

target layer

Used in ArcMap editing, a setting in the Target layer dropdown list that determines to which layer new features will be added. The target layer is set by clicking a layer in the Target layer dropdown list. For instance, if you set the target layer to Buildings, any features you create will be part of the Buildings layer. You must set the target layer whenever you're creating new features—whether you're creating them with the Sketch tool, by copying and pasting, or by buffering another feature.

tic

Registration of geographic control points for a coverage representing known locations on the earth's surface. Tics allow all coverage features to be recorded in a common coordinate system such as Universal Transverse Mercator (UTM). Tics are used to register map sheets when they are mounted on a digitizer and to transform the coordinates of a coverage—for example, from digitizer units (inches) to the appropriate values for a coordinate system, which are measured in meters for UTM.

tolerances

A coverage uses many processing tolerances (fuzzy, tic match, dangle length) and editing tolerances (weed, grain, edit distance, snap distance, and nodesnap distance). Stored in a TOL file, ArcInfo uses the values as defaults in many automation, editing, and processing operations. You can edit a coverage's tolerances using its Properties dialog box in ArcCatalog.

topological association

The spatial relationship between features that share geometry, such as boundaries and vertices. When you edit a boundary or vertex shared by two or more features using the topology tools in the ArcMap Editor, the shape of each of those features is updated.

topological feature

A feature that supports network connectivity that is established and maintained based on geometric coincidence.

topology

1. In geodatabases, relationships between connected features in a geometric network or shared borders between features in a topology.

2. In coverages, the spatial relationships between connecting or adjacent features—for example, arcs, nodes, polygons, and points. The topology of an arc includes its from- and to-nodes and its left and right polygons. Topological relationships are built from simple elements into complex elements: points (simplest elements), arcs (sets of connected points), areas (sets of connected arcs), and routes (sets of sections, which are arcs or portions of arcs). Redundant data (coordinates) is eliminated because an arc may represent a linear feature, part of the boundary of an area feature, or both.

topology rules

An instruction to the geodatabase defining the permissible relationships of features within a given feature class or between features in two different feature classes. Topology rules are compared against the features in the feature class during the topology validation process and violations are marked as errors.

After the topology validation process, the errors can be corrected by editing the feature class; occasionally, violations of a topology rule may represent acceptable conditions. In these cases, the errors can be marked as exceptions.

tracing

The building of a set of network elements according to some procedure.

transaction

1. A group of atomic data operations that comprise a complete operational task such as inserting a row into a table.

2. A logical unit of work as defined by a user. Transactions can be data definition (create an object), data manipulation (update an object), or data read (select from an object).

true curve

See circular arc.

undershoot

An arc that does not extend far enough to intersect another arc. See also dangling arc.

union

A topological overlay of two polygonal spatial datasets that preserves features that fall within the spatial extent of either input dataset; that is, all features from both coverages are retained. See also intersect.

username

The identification used for authentication when you log in to an ArcSDE geodatabase.

validate (topology)

The process of comparing the topology rules against the features in the dataset. When you validate a topology, features that violate the rules are marked as error features. Topology validation is typically performed after the initial topology rules have been defined, after the feature classes have been modified, or if additional feature classes or rules have been added to the map topology.

validation rule

Validation rules can be applied to objects in the geodatabase to ensure that their state is consistent with the system that the database is modeling. The geodatabase supports attribute, connectivity, relationship, and custom validation rules.

version

A version is an alternative representation of the database that has an owner, a description, a permission—private, protected, or public—and a parent version. Versions are not affected by changes occurring in other versions of the database.

vertex

1. One of a set of ordered x,y coordinates that defines a line or polygon feature.

2. A point that joins two segments of a feature. For instance, a square building would have four vertices, one at each corner.

virtual page

The map page, as seen in layout view.

wizard

A tool that leads a user step by step through an unusually long, difficult, or complex task.

work flow

An organization's established processes for design, construction, and maintenance of facilities.

work order

One specific task that proceeds through each stage of an organization's work flow processes such as design, acceptance, and construction in the field.

workspace

A container of geographic data. This can be a folder that contains shapefiles, an ArcInfo workspace that contains coverages, a personal geodatabase, or an ArcSDE database connection.

Index

D

E

F

G

I

J

K

L

M

N

O

P

Q

R

S

T

U